Microbial Bioremediation of Non-metals

Current Research

Edited by

Anna-Irini Koukkou

Sector of Organic Chemistry and Biochemistry
Department of Chemistry
University of Ioannina
Ioannina
Greece

Caister Academic Press

Caister Academic Press
Norfolk, UK

www.caister.com

British Library Cataloguing-in-Publication Data
A catalogue record for this book is available from the British Library

ISBN: 978-1-904455-83-7

Printed and bound in Great Britain

Contents

Contributors

Katsumasa Abe
Laboratory of Environmental Biochemistry
Department of Environmental Systems Engineering
Nagaoka University of Technology
Nagaoka, Niigata
Japan

katsuabe@vos.nagaokaut.ac.jp

Amalia S. Afendra
Department of Biological Applications and Technologies
University of Ioannina
Ioannina
Greece

aafendra@cc.uoi.gr

Spiros N. Agathos
Bioengineering Group
Earth & Life Institute
Université Catholique de Louvain
Louvain-la-Neuve
Belgium

spiros.agathos@uclouvain.be

Federico Aulenta
Department of Chemistry
"Sapienza" University of Rome
Rome
Italy

federico.aulenta@uniroma1.it

Asim K. Bej
Department of Biology
University of Alabama at Birmingham
Birmingham, AL
USA

abej@uab.edu

Cristiana Cravo-Laureau
Equipe Environnement et Microbiologie
UMR5254 IPREM CNRS
IBEAS Université de Pau et des Pays de l'Adour
France

cristiana.cravo-laureau@univ-pau.fr

Philippe Cuny
Centre d'Océanologie de Marseille
Université de la Méditerranée
LMGEM, UMR 6117 CNRS
Marseille
France

philippe.cuny@univmed.fr

Constantin Drainas
Sector of Organic Chemistry and Biochemistry
Department of Chemistry
University of Ioannina
Ioannina
Greece

cdrainas@cc.uoi.gr

Beatriz Galán
Environmental Biology Department
Centro de Investigaciones Biológicas (CSIC)
Madrid
Spain

bgalan@cib.csic.es

Esther García
Environmental Biology Department
Centro de Investigaciones Biológicas (CSIC)
Madrid
Spain

egarcia@cib.csic.es

José L. García
Environmental Biology Department
Centro de Investigaciones Biológicas (CSIC)
Madrid
Spain

jlgarcia@cib.csic.es

Franck Gilbert
Ecolab-Laboratoire d'Ecologie
Functionelle et Environnement
UMR 5245 (CNRS-UPS-INPT)
Toulouse, France

franck.gilbert@cict.fr

Vincent Grossi
Laboratoire de Géologie de Lyon
UMR 5276
Université Lyon 1, CNRS, ENS
Lyon, France

vincent.grossi@univ-lyon1.fr

Yves Jouanneau
CEA, DSV, iRTSV, Laboratoire de Chimie et Biologie des Métaux
Grenoble
France

yves.jouanneau@cea.fr

Nicolas Kalogerakis
Department of Environmental Engineering
Technical University of Crete
Chania
Greece

nicolas.kalogerakis@enveng.tuc.gr

Yuki Kasai
MBI Chair "Marine Biosciences" Kamaishi Research Laboratory
Kitasato University
Kamaishi, Iwate
Japan

ykasai@kitasato-u.ac.jp

Yoshio Kera
Laboratory of Environmental Biochemistry
Department of Environmental Systems Engineering
Nagaoka University of Technology
Nagaoka, Niigata
Japan

yoshkera@vos.nagaokaut.ac.jp

Anna-Irini Koukkou
Sector of Organic Chemistry and Biochemistry
Department of Chemistry
University of Ioannina
Ioannina
Greece

akukku@cc.uoi.gr

Serge Krivobok
CEA, DSV, iRTSV, Laboratoire de Chimie et Biologie des Métaux
Grenoble
France

serge.krivobok@ujf-grenoble.fr

Mauro Majone
Department of Chemistry
"Sapienza" University of Rome
Rome
Italy

mauro.majone@uniroma1.it

Florence Martin
CEA, DSV, iRTSV, Laboratoire de Chimie et Biologie des Métaux
Grenoble
France

florence.martin1@cea.fr

Cécile Militon
Centre d'Océanologie de Marseille
Université de la Méditerranée
LMGEM, UMR 6117 CNRS
Marseille
France

cecile.militon@univmed.fr

Nazia Mojib
Department of Biology
University of Alabama at Birmingham
Birmingham, AL
USA

naziamoj@uab.edu

Marco Petrangeli Papini
Department of Chemistry
"Sapienza" University of Rome
Rome
Italy

marco.petrangelipapini@uniroma1.it

Maria Parapouli
Sector of Organic Chemistry and Biochemistry
Department of Chemistry
University of Ioannina
Ioannina
Greece

me00762@cc.uoi.gr

Maria Pouli
Bioengineering Group
Earth & Life Institute
Université Catholique de Louvain
Louvain-la-Neuve
Belgium

poulimaria@yahoo.com

Simona Rossetti
Water Research Institute
National Research Council
Monterotondo, RM
Italy

rossetti@irsa.cnr.it

Michel Sylvestre
Institut National de la Recherche Scientifique
INRS-Institut Armand-Frappier
Laval
Québec
Canada

Michel.Sylvestre@iaf.inrs.ca

Shouji Takahashi
Laboratory of Environmental Biochemistry
Department of Environmental Systems Engineering
Nagaoka University of Technology
Nagaoka, Niigata
Japan

shoutaka@vos.nagaokaut.ac.jp

Valter Tandoi
Water Research Institute
National Research Council
Monterotondo, RM
Italy

tandoi@irsa.cnr.it

Jean-Patrick Toussaint
Institut National de la Recherche Scientifique
INRS-Institut Armand-Frappier
Laval
Québec
Canada

jptoussaint@davidsuzuki.org

Jack T. Trevors
School of Environmental Sciences
University of Guelph
Guelph, ON
Canada

jtrevors@uoguelph.ca

Iria Uhía
Environmental Biology Department
Centro de Investigaciones Biológicas (CSIC)
Madrid
Spain

iriauhia@cib.csic.es

Elpiniki Vandera
Sector of Organic Chemistry and Biochemistry
Department of Chemistry
University of Ioannina
Ioannina
Greece

elpiniki_vandera@yahoo.gr

John C. Willison
CEA, DSV, iRTSV, Laboratoire de Chimie et Biologie des Métaux
Grenoble
France

john.willison@cea.fr

Preface

In recent years many books and review articles on the degradation of organic pollutants have seen the light of publicity. A frequent reader may wonder about the abundance of the printed material on this subject and question the need for further publications. Undoubtedly, for a non-expert in the field this is a reasonable wonder. However, considering that the complexity and metabolic hostility of pollutants provide a selective pressure for the emergence of novel ways to circumvent the adverse effects of xenobiotics in the environment, such publications are always well timed. The present book is a compilation of the most recent progress achieved on the remediation of non-metal pollutants. Owing to the great technological advances and the rapid industrial development, environmental pollution has reached dangerous and, in many cases, catastrophic dimensions for the biosphere. The decontamination of soil, ground-waters and marine environments is a capital demand of societies worldwide, with bioremediation playing the primary role because of its effective and more sustainable approach to pollution abatement. It is well known that microorganisms involved in biodegradation of organic pollutants in the environment are the most prominent solution for decontamination. The best example, recently reported in the ScienceXpress journal, is the biodegradation of oil in the Gulf of Mexico after a leak from a platform of BP causing the worst environmental disaster in US history. Bacteria that developed rapidly in the deep cold water after the leak degraded the oil rapidly without depleting oxygen. Among these bacteria, new and unclassified species were discovered that were closely related to the genus Oceanospirillales. The natural enrichment with a variety of oil-eating microorganisms appears to be one of the major parameters for the rapid decline of the deep-water dispersed oil spill that has been observed (Hazen *et al.* 2010, Science *330*, 204–208, DOI: 10.1126/science.1195979).

This book brings together most of the current knowledge on *in situ* and *ex situ* bioremediation of contaminated sites, presenting the fundamental principles governing biodegradation of environmental organic pollutants and the methodologies for the microbiological and genetic monitoring of such processes. Each chapter is written by experts who have substantially contributed to answering fundamental questions on basic and practical issues on decontamination processes as well as being currently actively engaged in research tackling these problems.

The first chapters of the book focus on current knowledge on microbial biodegradation of various organic pollutants [chloroethenes, steroids, organophosphoric compounds, alkanes, polycyclic aromatic hydrocarbons (PAHs) and polychlorobiphenyls (PCBs)] at the biochemical and genetic level (Chapters 1–8). Numerous topics, including the enzymatic reactions involved in biodegradation, the question of the impact of bioturbation on hydrocarbon dynamics in marine sediments, the structure, function and biodiversity of ring-hydroxylating dioxygenases involved in PAH biodegradation and the strategies that are proposed to engineer PCB-degrading bacterial oxidative enzymes and PCB-degrading plant-microbe systems to degrade PCBs are also covered. Catabolic plasmids and mobile genetic elements involved in the degradation of a number of industrially important xenobiotic pollutants as well as the origin and evolution of these catabolic pathways to different genera are also reviewed (Chapter 9). Next, chapters describing the structure, regulation and diversity of microbial genes encoding biodegradive enzymes, as well as mechanisms of biodegradation of non-metal contaminants including applications of the state-of-the-art molecular technologies to study biodegradative processes are included. Molecular technologies, such as community fingerprinting, molecular detection of degradative genes, metagenomics facilitating the analysis and monitoring of indigenous and inoculated microorganisms in contaminated sites, are examined (Chapters 10 and 11). Finally, appropriate characterization procedures for optimal evaluation and design of the *in situ* remediation strategies are examined (Chapters 1 and 6) and classical *ex situ* technologies (landfarming, composting, biopiling and slurry-phase bioremediation) as well as biopiles and in-vessel composting are also described (Chapters 6 and 12).

This book can serve as a special important reference resource to research scientists, graduate students and other specialists interested on microbial bioremediation of organic pollutants and generally to environmental microbiologists, chemists and engineers.

I am deeply indebted to all the authors who have contributed to this book, a publication that we hope will be a valuable tool for researchers and graduate students who intend to devote their efforts for the protection and preservation of environmental health.

Anna-Irini Koukkou

Reductive Dechlorination of Chloroethenes: From the Laboratory to Field-scale Investigations

1

Federico Aulenta, Mauro Majone, Marco Petrangeli Papini, Simona Rossetti and Valter Tandoi

Abstract

In situ technologies are cost-effective, expanding technologies for the clean-up of soil and groundwater in contaminated sites. On the other hand, these technologies are knowledge-intensive and their application requires thoroughly understanding of the microbiology, ecology, hydrogeology, and geochemistry of contaminated soils and aquifers, under both natural and enhanced conditions.

In this chapter, we summarize current knowledge and future perspectives in the area of microbial anaerobic dechlorination of chlorinated solvents, particularly chloroethenes. Main attention is paid at discussing environmental factors and conditions that influence microbial activity under field conditions. Approaches to stimulate and manipulate the activity of native dechlorinating populations in order to meet target remediation goals are examined. Finally, appropriate characterization procedures for optimal evaluation and design of the *in situ* remediation strategies are examined with main reference to three tools: (a) microbial tools, (b) modelling and (c) microcosms versus field studies.

Introduction

Owing to more than 200 years of industrialization and to the use and presence of dangerous substances in many production processes, Europe is now facing the problem of soil and groundwater contamination. Recently adopted environment policy initiatives on waste, water, air, climate change, chemicals, biodiversity and environmental liability are now enforcing measures for protection of environmental resources against future risks. In particular, the directive on environmental liability (Directive 2004/35/EC) creates a harmonized framework for the liability regime to be applied across the European Union (EU) when contamination creates a significant risk to human health and possible damage of environmental resources. However, it does not apply to historical contamination or to damages that may have occurred prior to its entry into force.

As for previous soil contamination, the Commission adopted the communication 'Towards a thematic strategy on soil protection' in 2002 and a directive proposal is presently under discussion. In this frame, it has been estimated that 3.5 million sites may be potentially contaminated, with 0.5 million sites being really contaminated and needing remediation. Though difficult to estimate, several studies demonstrate significant annual costs of soil contamination in the ranges of €2.4–17.3 billion, not including the damage to the ecological functions of soil as these were not possible to quantify. According to the proposal, member states will perform a preliminary survey and subsequent site investigations to determine which sites are contaminated or not, taken from a pre-established screening list. The costs for preliminary survey, to be carried out within five years after transposition of the directive, are estimated at about €51 million per year for EU25, followed by on-site investigations (up to €240 million yearly for the EU25 over 25 years), to finally conclude if there is indeed a significant risk to the human health or environment. If so, then the site will be classified as contaminated site and introduced in the inventory, to be monitored and remediated according to a National Remediation Programme. Such a tremendous effort will also be combined to other actions dealing with other soil threats: erosion, organic matter decline, compaction, salinization, landslides, sealing, and biodiversity decline.

In the perspective of such an enormous effort, new approaches to site remediation should be developed and implemented in order to increase the 'sustainability' of remediation, both in the environmental and economical sense. Indeed, a recent survey (Spira, 2006) confirmed that the 'dig and dump' (D&D) and 'pump and treat' (P&T) remain the most used approaches to soil and groundwater remediation, respectively. On the contrary, we should move forward from such waste- and energy-intensive approaches, towards sustainable remediation approaches, in order:

- to recover natural functions and potential uses of environmental resources to be remediated (e.g. quantitative and qualitative preservation of groundwater resources);
- to minimize extraction of water and production of wastes to be disposed of; and
- to favour the continued economic use of the site during remediation.

In principle, *in situ* technologies are good candidates to replace D&D and P&T, everywhere possible, because they usually:

- do not require external treatment and water discharge;
- require no or less pumping energy;
- cause minimal disturbance of land use;
- are often based on effective degradation of contaminants, not only on phase transfer;
- can usually be more oriented towards the treatment of the source rather than the plume; and
- can also be effective on separate phases (e.g. source treatment of chlorinated DNAPLs).

On the other hand, some possible drawbacks are:

- difficulties in reaching very low threshold values;
- 'injection' of substances into groundwater is often required, with specific limitations; and
- caution about possible secondary contamination is necessary (e.g. toxic intermediates or side-products).

More generally, the implementation of *in situ* technologies is more site specific and requires very good expertise on underlying processes as well (so called knowledge-intensive approach). Often, no standardized design methodology is available and pilot-scale field tests are required for appropriate scale-up. Thus, appropriate design of *in situ* technologies is more time-consuming and expensive than usual approaches, as well as the overall remediation time is often longer. Thus, *in situ* technologies still suffer of a lack of consensus and regulatory permits are more difficult to obtain.

In order to overcome these possible drawbacks, any contaminated site should be considered and dealt with as a 'case study', where best available scientific and technical knowledge should be used. In general, in order to be achieved this target requires the most accurate 'up-stream' knowledge:

- to perform site-sensitive characterization and to fit it to the processes under evaluation;
- to combine techniques (chemical, geophysical, microbiological, isotopic);
- to look for secondary sources (e.g. DNAPL) and to try to act as closest as possible, in order to minimize volumes under consideration; and
- to calibrate process kinetics to the rate determining step (e.g. dissolution or desorption from separate phases) and to take advantage of down-gradient natural attenuation processes.

As for evaluation of 'down-gradient' impacts, it is necessary:

- to try to preserve natural conditions (organic matter, soil texture, biological activity) as much as possible;
- to evaluate all potential impacts (individuation of possible toxic intermediates, side- or end-products, ecotoxicology tests); and
- to appropriately calibrate monitoring to remediation targets and potential impacts.

Of course, such a general approach needs to be detailed and tuned to different matrixes, contaminants and technologies under evaluation. As an example, some *in situ* technologies (such as soil vapour extraction, air sparging, and in-well stripping) are well established and usually require simpler and less case-sensitive design. However, these technologies are actually based on *in situ* phase transfer of contaminants and on-site treatment of gas phase, i.e. they suffer, to a more limited extent, from some drawbacks of P&T approach. Similarly, other emerging technologies (such as soil flushing and *in situ* thermal treatment) are mainly based on further acceleration of *in situ* phase transfer and on-site treatment. So, they require very good comprehension of physical chemistry, geochemistry and hydrogeology aspects under strongly modified conditions. Finally, other techniques (such as chemical oxidation, electrochemical remediation, bioremediation) are based on induction and/or enhancement of *in situ* degradation of contaminants and also require full comprehension of *in situ* degradation mechanisms, including modifications that are caused on the environmental matrixes to be remediated. In general, all techniques where chemical substances have to be added are considered with most caution because of the possible negative modifications of aquifer and soil natural conditions. On the other hand, it is obvious that the above reported techniques are quite different with respect to induced modifications because thermal or chemical techniques can in principle be expected to cause stronger modifications than biological techniques. As a general rule, techniques that simply accelerate transport and degradation mechanisms that are 'naturally' occurring should be preferred (i.e. enhanced natural attenuation). Moreover, long actions should not necessarily considered as negative, if coherently tuned to down-stream environmental protection and to present or future site uses.

From this short summary, it is evident that the concept of 'sustainability' in the field of site remediation still requires to be detailed and comparative evaluation of techniques mainly remain an empirical exercise. Under FP7, several research actions have been taken in this context. These actions include coordination and demonstration programmes on innovation on remediation technologies, as well as development of certification procedures for advancement of environmental technologies and expert systems.

Here, we will try to apply such general concepts while strictly focusing on one single example, i.e. *in situ* bioremediation of chlorinated aliphatic solvents in groundwater, dealing with its main critical issues and available experimental and modelling tools for appropriate design.

Occurrence, transport and fate of chlorinated solvents

Chlorinated aliphatic solvents are a large family of compounds that are widely used in several industries, e.g. chemical cleaning (e.g. metal, electronic components), dry cleaning, textile dyeing, solvent formulation (rubbers, paint strippers, and industrial paints), and use as chemical reagents (fluorinated hydrocarbons, trichloroacetic acid, polyvinylchloride, etc.). Most common chlorinated aliphatic solvents are chlorinated methanes (such as carbon tetrachloride, chloroform, and methylene dichloride), chlorinated ethanes (such as 1,2-dichloroethane and 1,1,1-trichloroethane), and chlorinated ethenes (such as tetrachloroethene and trichloroethene).

Despite their massive utilization, for most of them a decreasing trend is shown, mainly owing to environmental and health concerns. As an example, in western Europe, the tetrachloroethene (PCE) and trichloroethene (TCE) market decreased during the period 1995–2000 from 86,000 to 70,000 tonnes per year and from 106,000 to 74,000 tonnes per year, respectively. Indeed, owing to improper use and disposal practices, chlorinated aliphatic solvents are common organic contaminants of soil and groundwater in all Europe.

Chlorinated solvents are highly toxic and some are also carcinogenic, so their presence in groundwater is considered to be not acceptable (or at least potentially critical) even when at very low levels, e.g. in Italy maximum concentration limits (MCLs) are in the range of μg/l or less. On the other hand, when chlorinated solvents come into contact with groundwater, they usually form large contaminated plume that can present very high maximum concentration, up to six or seven orders of magnitude higher than MCLs. A survey from several US contaminated sites reports an average plume length of about 600–700 m (up to 2 km) with concentrations ranging from 10 to 106 µg/l (Rice *et al.*, 1999).

This is due to several factors, particularly their presence as a non-aqueous separate phase (NAPL) in the aquifer as well as a certain degree of resistance to biological degradation.

As for the first factor, most chlorinated aliphatic solvents are more dense than water (e.g. neat TCE has a density of 1.46 kg/l) and form a dense non-aqueous liquid phase (DNAPL). DNAPL released to the subsurface can migrate through the soil and reach the groundwater table. Then, an aliquot of DNAPL (referred to as 'free' or 'mobile' DNAPL) moves downward through the aquifer and eventually forms 'pools' of free liquid phase on low permeability surfaces, such as clay layers. On the other hand, as DNAPL flows through porous media, it leaves behind a tail of residual DNAPL that partially fills the pore spaces. Residual DNAPL is held in the pore spaces by capillary forces, depending on several factors, e.g. on interfacial tension between water and the DNAPL, the wettability of the media, capillary pressure, relative permeability of the formation, and the solubility, volatility, density, and viscosity of the DNAPL (Mercer and Cohen, 1990). DNAPL movement is more often controlled by gravity than by hydraulic head in the groundwater. Therefore, a thorough understanding of site geology and hydrogeology is needed before likely locations and pathways of DNAPL movement can be identified.

Both the residual and mobile DNAPL act as long-term sources of groundwater contamination, in that they slowly dissolve into the water path. After that, the hydraulics of the flow system and the retardation characteristics of the contaminants will determine the movement of dissolved phase contaminants in the groundwater, as function of typical mechanisms, i.e. advection, mechanical dispersion, diffusion, and adsorption. Full comprehension of these factors and the role they play at the local scale are necessary to describe the transport of chlorinated solvents through the aquifer and eventually implement appropriate remediation actions. A detailed description of transport mechanisms of chlorinated solvents and DNAPLs and can be found elsewhere (Ahmed and Chen, 2006; Lee and Chrysikopoulos, 2006).

Biodegradation mechanisms of chlorinated solvents

In the aquifer, chlorinated aliphatic solvents can be degraded by several reactions either abiotic (chemical reactions) or biotic (biological reactions), the latter being usually predominant. Biological degradation reactions can happen under different environmental conditions (aerobic or anaerobic), different mechanisms (oxidation or reduction) and different role of involved microorganisms (metabolic or cometabolic reactions).

Cometabolic processes are those where the contaminant is degraded (as a co-substrate) by microorganisms growing on another substrate (primary substrate). In cometabolic processes degradation is fortuitously mediated by enzymes or cofactors produced during growth on the primary substrate. Contaminant degradation is not directly linked to growth or energy production (i.e. microorganisms take no advantage from their cometabolism). On the contrary, in metabolic processes the contaminant degradation is coupled to the growth of microorganisms, in that the contaminant acts as a primary carbon source or as terminal electron acceptor. Owing to their fortuitous nature, cometabolic processes are usually characterized by far lower pollutant transformation rate and yield than metabolic processes. Moreover, being growth-linked processes, metabolic degradations are autocatalytic whereas cometabolic degradations are not.

Both cometabolic and metabolic processes act by either oxidation or reduction mechanisms. Oxidation requires an external electron acceptor that can be either oxygen (aerobic) or other oxidized compounds like nitrate (anaerobic) and typically involves the complete mineralization of the organic backbone, resulting in inorganic compounds like CO_2, H_2O and HCl. On the other hand, the reduction requires an external electron donor and involves a step-wise dechlorination, down to non-chlorinated end products (e.g. ethene and/or ethane). Besides environmental conditions, the type of degradation reaction depends on the initial chlorination degree of the molecule. PCE appears to be resistant to metabolism of aerobic microorganisms owing to its highly oxidized nature (full substitution of hydrogen atoms by chlorine atoms). Indeed, cometabolic oxidation of PCE is only reported by toluene-*o*-xylene mono-oxygenase of *Pseudomonas stutzeri* OX1 (Ryoo *et al.*, 2000) whereas evidences of metabolic oxidation have not been reported yet.

On the contrary, PCE can be transformed through sequential reductive dechlorination (RD), via TCE, DCEs (dichloroethenes), VC (vinyl chloride), ETH (ethene), and even ethane formation (Middeldorp *et al.*, 1999). TCE has similar behaviour to PCE whereas DCEs and VC are less available to the reduction mechanisms. Thus, although complete RD of PCE or TCE to ETH or ethane has been observed in the field and in laboratory experiments (Carter and Jewell, 1993; de Bruin *et al.*, 1992; DiStefano *et al.*, 1991; Freedman and Gossett, 1989), DCEs or VC reduction is often the rate limiting step in the overall RD sequence from PCE to ETH (Tandoi *et al.*, 1994; Vogel *et al.*, 1987). Indeed, RD is strongly inhibited by the presence of higher chlorinated ethenes (Aulenta *et al.*, 2005a; Maymo-Gatell *et al.*, 1997) and has a half-saturation constant (K_S) higher than PCE RD (Haston and McCarty, 1999; He *et al.*, 2002). Possible VC accumulation and persistence during RD in a aquifer is of special concern since VC is more toxic and mobile than its parent compounds. On the other hand, DCEs and VC are less oxidized than their parent compounds, so they are more available to oxidation, according to the metabolic or cometabolic pathways. Thus, a sequence of anaerobic reductive and aerobic oxidative conditions can be the optimal combination for a fast and complete degradation of chlorinated ethenes (Fig. 1.1).

RD requires that an appropriate electron donor. Several different electron donors, including methanol, butyrate, lactate, benzoate (Aulenta *et al.*, 2005b; DiStefano *et al.*, 1992; Fennell *et al.*, 1997; Yang and McCarty, 2002) have been shown to support enhanced RD of PCE and TCE, both in field and in laboratory studies. Recently, also complex organic compounds have been used in laboratory experiments as slowly fermenting low-cost electron donors for the RD of PCE (Aulenta *et al.*, 2008c; Brennan *et al.*, 2006; Sun *et al.*, 2009; Yu and Semprini, 2009). Nevertheless, in most cases the actual electron donor used in the RD is the H_2 produced during fermentation of these organic compounds (Ballapragada *et al.*, 1997; He *et al.*, 2002). In each step, one chlorine atom will be substituted by a hydrogen atom along with two electrons (indeed, this type of RD is also called reductive hydrogenolysis). There is also evidence that some dechlorinating bacteria can use acetate as electron donor (He *et al.*, 2002) and in this case H_2 may be used for acetate formation through homoacetogenesis.

Thus, the choice of fermentable organic substrates is mainly based on rate and yield of their conversion to H_2 as well as on the consequent competition for H_2 that can establish between dechlorinators and other H_2-utilizing microorganisms, like hydrogenophilic methanogens or homoacetogens (Ballapragada *et al.*, 1997; Aulenta *et al.*, 2008a; Fennell *et al.*, 1997; Mazur and Jones, 2001; Smatlak *et al.*, 1996; Yang and McCarty, 1998) (Fig. 1.2). As an example, in a methanol-fed culture (Aulenta *et al.*, 2002), electrons ultimately transferred to ethene through PCE RD were only 4% of the total fed whereas the remaining

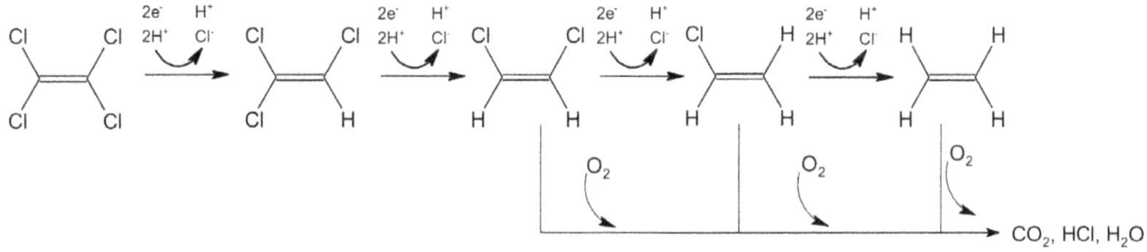

Figure 1.1 Schematic diagram of anaerobic and/or aerobic sequence for biodegradation of tetrachloroethene (PCE). TCE, trichloroethene.

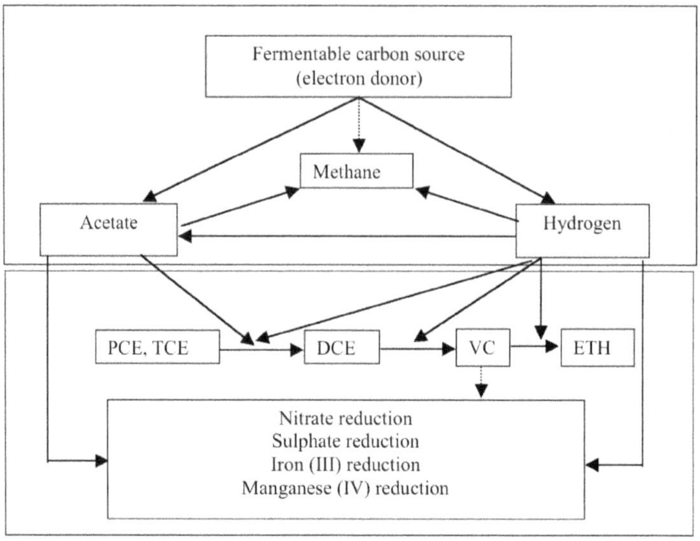

Figure 1.2 Electron flow from electron donors to electron acceptors in RD systems.

were used for direct methane formation from methanol (15%), acetate formation due to both acetogenenesis and homoacetogenesis (72%, where homoacetogenesis contribution was less than 19%), and growth (9%). Other studies demonstrated that dechlorinators have the potential to out-compete other H_2-utilizers when H_2 is present at low concentrations, owing to their higher affinity for this electron donor (Ballapragada *et al.*, 1997; Smatlak *et al.*, 1996; Yang and McCarty, 1998). Consequently, substrates which are slowly fermented, and only under low hydrogen partial pressures would result in a competitive advantage for dechlorinators. If present, nitrate and sulphate can also act as sinks for the reducing equivalents, both coming from the fermentable organic carbon source and the produced molecular hydrogen, thus increasing the range of possible metabolisms directly or indirectly in competition with RD (Fig. 1.2). In particular, RD is usually reported not to occur when nitrate is present whereas it is possible under sulphate reducing conditions (Aulenta *et al.*, 2007b, 2008a).

A possible drawback associated to the use of fermentable organic substrates to stimulate *in situ* the RD of chlorinated solvents is the production and accumulation in the subsurface of large amounts of fermentation products such as acetate or propionate and resulting deterioration of groundwater quality. In principle, direct hydrogen addition may offer some advantages over the use of slowly fermentable organics. Indeed, hydrogen does not leave any environmentally harmful residues in the subsurface, is generally less expensive than most of the organic compounds listed above, and generates less biomass in the subsurface thus poorly affecting groundwater flow patterns. Therefore, technologies based on direct hydrogen subsurface addition have been recently developed, based on passive dissolution through membranes (Fang *et al.*, 2002; Ma *et al.*, 2003), low-pressure sparging, or hydrogen-generating electrodes (Aulenta *et al.*, 2008d; Newell *et al.*, 1997). In principle, the direct use of hydrogen can speed up the dechlorination rate because hydrogen formation from fermentation can be the rate determining step of RD when carbon sources are used (Aulenta *et al.*, 2004). However, several aspects may still limit the use of direct hydrogen addition such as its poor solubility in water and its tendency, once injected in the subsurface, to rapidly escape from the contamination plume. Moreover the presence of high levels of H_2 could provide a selective advantage to methanogens and eventually result in the marginalization of dechlorinators.

A ground-breaking alternative to the above described stimulatory approaches is the use of solid-state electrodes to directly and selectively transfer electrons (instead of chemicals) to dechlorinating communities growing as biofilms at the electrode surface. This would probably represent the 'cleanest' and most efficient way to stimulate and control microbial reductive dechlorination reactions in a contaminated aquifer. This possibility was demonstrated in a recent studies conducted in our laboratory (Aulenta *et al.*, 2007a, 2008b, 2009a,b, 2010; Aulenta and Majone, 2010).

RD of chlorinated ethanes proceeds similarly to that of chlorinated ethenes, e.g. RD of 1,1,2,2-tetrachloroethane (TeCA) proceeds via formation of trichloroethane (TCA), dichloroethane (DCA), chloroethane (CA), and eventually ethane. On the other hand, chlorinated ethanes can also undergo a different type of RD, dichloroelimination, where two vicinal chlorine atoms are lost with concurrent formation of a C=C double bond. This requires the input of two electrons for each reductive step, thus the ratio between chlorine released and reducing electrons is double than for hydrogenolysis. Some chlorinated ethanes can be also transformed via dehydrochlorination, where vicinal hydrogen and chlorine atoms are lost to form a C=C double bond and hydrochloric acid. Dehydrochlorination does not require the input of electrons and it is considered to occur under abiotic conditions. Both dehydrochlorination and dichloroelimination cause that a chlorinated ethene is formed from the chlorinated ethane precursor with one or two chlorine atoms less, respectively. Anyway, the behaviour of chlorinated ethanes is far less known than that of chlorinated ethenes under both lab- and field-scale conditions.

In situ bioremediation of chlorinated solvents

Presently, RD is the most promising mechanism for enhanced *in situ* remediation of chlorinated ethenes in contaminated groundwater (Morse *et al.*, 1997). A survey of 93 sites worldwide (mostly USA, but also including UK, The Netherlands, Japan) has been published by the Environmental Security Technology Certification Program, ESTCP (www.estcp.org), where enhanced in site anaerobic bioremediation has been applied or is presently going on (21 sites at full scale, 9 sites at pilot and full scale, 59 sites at pilot scale). *In situ* enhanced RD was accomplished by stimulating the activity of native dechlorinating populations through the addition of electron donors to provide the electrons required for RD. Several substrates were used as the electron donors, including single soluble compounds (14 lactate, 3 butyrate, 3 acetate, 6 others), soluble or emulsified mixtures of substrates (15 molasses, 10 vegetable oils), slow-release soluble polymers (35 HRC®, a polylactate-based commercialized by Regenesis Inc.), organic solids (three mulch, one chitin), and gas (three molecular hydrogen). Of course, each type of substrate required the appropriate and different delivery systems.

Several protocols have been proposed to drive the evaluation processes and preliminary design in the perspective of *in situ* enhanced bioremediation through RD. As an example, the RABITT protocol (Morse *et al.*, 1997) is based on preliminary site assessment by using a ranking system, that include contaminant, geochemical and hydrogeological data. Hydraulic conductivity is considered as the most important parameter to understand whether substrate addition can be appropriately managed. Then, a treatability study has to be performed by using microcosms and/or field studies. A similar approach is also advised in 'Principles and practice of enhanced anaerobic bioremediation of chlorinated solvents' (available online at www.estcp.org), where a preliminary screening system is based on 'red flags' (occurrence of negative conditions that exclude effective application of RD).

More in general, a typical flow-sheet of evaluation procedure that should be adopted for application of RD is presented in Fig. 1.3. In the following, some key points of this flow-sheet are outlined with reference to treatability studies (microbial tools and microcosms), modelling and field tests.

Microorganisms responsible for metabolic chloroethenes dechlorination and their individuation/quantification through biomolecular tools

Several bacteria have been isolated that can couple the RD of chlorinated aliphatic compounds to energy conservation and growth; therefore they

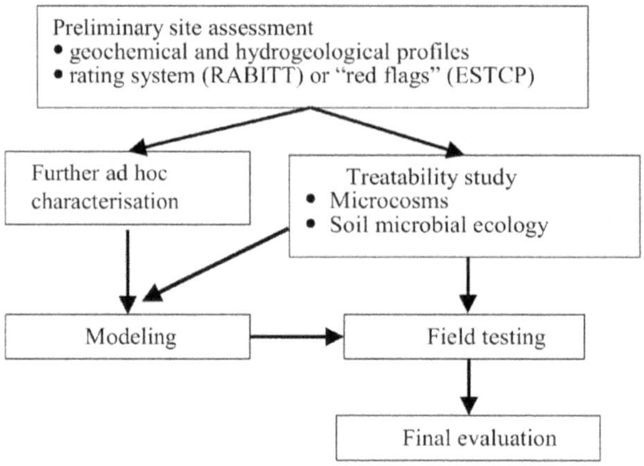

Figure 1.3 *In situ* RD evaluation flow chart.

are of greater interest for *in situ* bioremediation (El Fantroussi *et al.*, 1998). These microorganisms differ in their electron donor requirements, kinetics, end-points of dechlorination, and maximum concentration of chlorinated solvent tolerated (Table 1.1). Several strains are quite restrictive in terms of electron donor requirements, such as *Dehalobacter* and *Dehalococcoides* that can only utilize H_2. On the other hand, other strains (*Dehalospirillum, Desulfitobacterium*) are quite versatile utilizing a broad spectrum of electron donors. *Desulforomonas* spp. are unique since they are the only strains that can utilize acetate as electron donor for PCE dechlorination. Only members of the genus *Dehalococcoides* seem to be able to drive the dechlorination of chloroethenes to harmless ethene. Hendrickson and colleagues (Hendrickson *et al.*, 2002) tested samples from 24 chloroethene-dechlorinating sites and a positive correlation between ethene formation and the presence of members of the *Dehalococcoides* group was observed. However, as reported in Table 1.1, different *Dehalococcoides* species exhibit distinct dechlorinating abilities, and none of the isolated strains can gain energy from dechlorination of all chloroethenes (i.e. PCE, TCE, DCEs, VC).

Moreover, *Dehalococcoides* FL2 share 99.9% similar 16S rRNA gene sequence with strain BAV1, and share an identical 16S rRNA gene sequence with strain CBDB1 (He *et al.*, 2005; Kube *et al.*, 2005) an isolate that dechlorinates chloroaromatic compounds but cannot dechlorinate chloroethenes. The authors concluded that 16S rRNA gene-based analyses are insufficient to predict dechlorination activity and distinguish between some members of the *Dehalococcoides* group. This information is relevant for bioremediation practitioners because the detection of *Dehalococcoides* 16S rRNA gene sequences is often interpreted as evidence that native populations are capable of full dechlorination of PCE to ethene.

Hopefully, in the next future, additional information about the dechlorination potential of a contaminated site could be provided by detection of genes directly involved in the dechlorination process of interest. In addition, more studies, aimed at understanding the distribution and relative importance of dechlorinating microorganisms other than *Dehalococcoides*, are needed. Indeed, in another approach, such as bio-enhanced DNAPL dissolution, the partial conversion of PCE to cDCE can be highly beneficial without the need to drive the dechlorination to completion, if coupled to an aerobic polishing step (Adamson *et al.*, 2003, 2004).

Considering both the high specificity of microorganisms involved responsible for RD and high complexity of anaerobic microbial consortia, many molecular techniques and combinations of these techniques are applied in scientific research for the identification and quantification of

Table 1.1 Some properties of tetrachloroethene (PCE)- and trichloroethene (TCE)-dehalorespiring bacteria

Microorganism (reference)	Ultimate electron donors[1]	End product of dechlorination	PCE tolerated (mM)
Desulfitobacterium Viet1	H_2	TCE	nr
Desulfitobacterium PCE1	H_2	TCE	nr
Desulfitobacterium frappieri TCE1	H_2	cDCE	nr
Desulfitobacterium PCE-S	H_2	cDCE	nr
Desulfitobacterium Y51	H_2	cDCE	Saturation
Clostridium bifermentans DPH-1	H_2	cDCE	Saturation
Dehalobacter restrictus PER-K23 and TEA	H_2	cDCE	<0.2
Enterobacter agglomerans MS-1	Acetate[8]	cDCE	Saturation
Desulfuromonas chloroethenica TT4B	Acetate	cDCE	<0.1
Desulfuromonas michiganensis BB1 and BRS1	Acetate	cDCE	Saturation
Geobacter lovelyi SZ	Acetate	cDCE	nr
Sulfurospirillum multivorans[2]	H_2	cDCE	<0.16
Sulfurospirillum halorespirans PCE-M2[T]	H_2	cDCE	nr
Dehalococcoides ethenogenes 195[3]	H_2	Ethene	<0.7
Dehalococcoides FL2[4]	H_2	Ethene	nr
Dehalococcoides BAV[5]	H_2	Ethene	nr
Dehalococcoides VS[6]	H_2	Ethene	nr
Dehalococcoides GT[7]	H_2	Ethene	nr

[1]Indicated is the ability to use H_2 or acetate as an electron donor for reductive dechlorination. However, *Dehalococcoides* and *Dehalobacter* are the only species that are restricted to the use of H_2. Other species can also use a variety of simple organic compounds.

[2]Formerly known as *Dehalospirillum*.

[3]VC dechlorination to ethene is a cometabolic process.

[4]PCE dechlorination to TCE and VC dechlorination to ETH are cometabolic processes.

[5]PCE and TCE are dechlorinated only cometabolically.

[6]PCE is not dechlorinated and TCE is dechlorinated only cometabolically.

[7]PCE was not dechlorinated.

[8]H_2 was not tested.

nr, not reported.

members of dechlorinating microbial communities. However, to date, PCR (polymerase chain reaction), qPCR (quantitative PCR) and PLFA (phospholipid fatty acids) analysis, are the only molecular tools on practice use at field level on a somewhat regular basis. Certainly there have been field applications of other techniques but mostly in the context of field research, not as a routinely adopted monitoring or assessment tool. Recently, *in situ* methodologies, such as FISH (fluorescence *in situ* hybridization) and CARD-FISH (catalysed reporter deposition-FISH), that give additional information on activity and on the actual biodiversity and structure of the microbial communities, have been shown to be reliable and easily applicable tools for the molecular monitoring of known dechlorinators present in both laboratory bacterial enrichments and contaminated sites.

While use of these tools is still quite limited, it is evident that their application will grow, especially as evidence for the value added increases and as protocols and methodology become more standardized and automated. However, because the lack of an elective molecular tool, the use of multiple molecular approaches is recommended to get multiple lines of evidence especially for the analysis of site samples. A summary of the main available techniques utilized for the molecular identification of dechlorinating microbial communities, the related obtainable information and their current applications is reported in Table 1.2.

Table 1.2 Main molecular tools for microbiological characterization of dechlorinating microbial communities

Technique	Obtainable information	Current application
Direct and nested PCR (16S rRNA gene)	Qualitative: presence/absence of 16S rRNA genes of interest	F
qPCR (16S rRNA gene; mRNA; functional genes)	Quantitative: presence of specific organisms of interest	F
Clone library (16S rRNA and functional genes)	It provides data on the gene biodiversity. Very useful for FISH probe design	R
DGGE	Qualitative: presence/absence of 16S rRNA and/or functional genes of interest	F, its application is declining
FISH	Quantitative; information on activity, spatial distribution and the whole biodiversity of the sample	R
PLFA	Monitoring of individual groups of organisms and for total biomass determination; it can be quantitative	F

PCR, polymerase chain reaction; F, field application; R, research application; DGGE, denaturing gradient gel electrophoresis; FISH, fluorescence *in situ* hybridization; PFLA, phospholipid fatty acid analysis.

PCR-based molecular methods

Specific PCR primers have been developed for qPCR for *Dehalococcoides* and other known dechlorinators such as *Dehalobacter* and *Desulfitobacterium* (Smits *et al.*, 2004). To date, qPCR is mainly utilized to detect DNA but there is the potential also for mRNA and expressed proteins and consequently to detect the activity of specific functional microbial groups. This technique can be applied also to functional genes, such as reductive dehalogenase genes (Lee *et al.*, 2006). However, qPCR–based studies have been shown to be subject to significant experimental variability which leads both to inaccurate estimates of gene abundance and to inappropriate comparisons between data sets from different studies (Smith *et al.*, 2006). Denaturing gradient gel electrophoresis (DGGE), being able to distinguish closely related strains, has been applied in research studies and at field sites for the identification of specific micro-organism but its application is declining, because DGGE does not have the advantages of qPCR being not quantitative and because of the limited information due to the short amplicon length.

An important research molecular technique is represented by 16S rRNA gene clone library. Even if it will remain a research tool because of being a time-consuming and expensive technique, it provides useful information about the phylogenetic placement of the microorganisms that are associated with bioremediation processes. Microbial community structure studies have been conducted on several dechlorinating enrichments (Bowman *et al.*, 1993; Richardson *et al.*, 2002; Dennis *et al.*, 2003; Rossetti *et al.*, 2003, 2008; Duhamel *et al.*, 2004; Gu *et al.*, 2004; Freeborn *et al.*, 2005). 16S rRNA gene sequences produced by the clonal approach can be also utilized for the design of specific FISH probes. However, only few reports of clone library application coupled with FISH analysis are available (Rossetti *et al.*, 2003, 2008). In these studies, novel oligonucleotide probes of key-organisms relevant in bioremediation processes were obtained and the application of the so called 'full cycle 16S rRNA approach' permitted to gain more information into the abundance and distribution of species involved in dechlorination processes. The availability of oligonucleotide probes for a wider range of organisms or functional genes would help to notably increase the field use of FISH analysis.

In situ identification methods

Fluorescence *in situ* hybridization

Fluorescence *in situ* hybridization has proven to be a powerful molecular method for identification, visualization, and quantification of organisms of interest in microbial communities. Specific oligonucleotide probes have been recently designed and applied for the detection and the enumeration by FISH of *Dehalococcoides* (Yang and Zeyer, 2003; Aulenta *et al.*, 2004). It

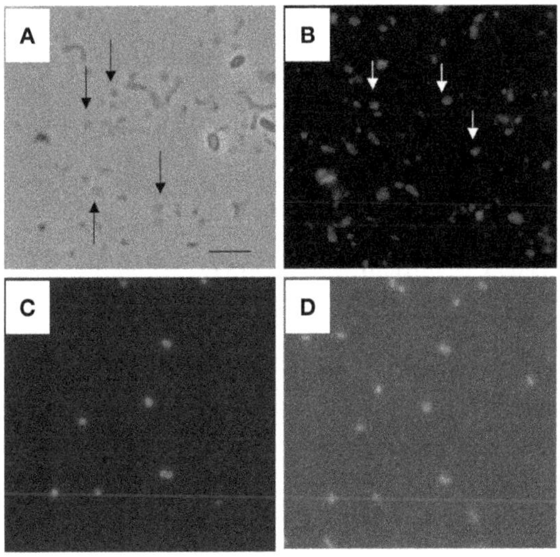

Figure 1.4 Appearance of *Dehalococcoides* spp. in (A) phase-contrast microscopy and (B) after conventional fluorescence *in situ* hybridization (FISH) with probe DHE1259t (modified from Aulenta *et al*., 2004). Application of catalysed reporter deposition (CARD)-FISH on a trichloroethene (TCE) dechlorinating microbial consortium: (C) *Dehalococcoides* cells after hybridization with probe DHE1259t; (D) same microscopic field after DAPI staining (Fazi *et al*., 2007).

has been shown that FISH can be easily applied to actively degrading dehalogenating enrichments (Fig. 1.4) and it was found that FISH quantification data were in agreement with the quantitative estimation of *Dehalococcoides* spp. performed by kinetic batch tests under non-limiting hydrogen and PCE concentrations (Aulenta *et al.*, 2004). FISH was successfully applied for the quantification of *Dehalococcoides* species on other actively dechlorinating enrichments (Rossetti *et al.*, 2008; Aulenta *et al.*, 2009b). Oligonucleotide probes have been also designed and applied for *Desulfitobacterium* subsp. in enrichment cultures (Yang *et al.*, 2005; Rossetti *et al.*, 2008).

Differently from the molecular characterization of active dechlorinating microbial consortia, FISH application on environmental samples can be strongly affected by many limitations owing to the presence of high particulate loads, that highly interfere with visualization by epifluorescence microscopy, and to the usual low activity state of bacteria (low ribosome content and consequently low fluorescence signals). However these limitations have been shown to be adequately circumvented by introducing samples pretreatments (cell extraction method; Barra Caracciolo *et al.*, 2005) and by utilizing methodological improvements of FISH (i.e. CARD-FISH). Even if conventional FISH was already successfully applied during a field test where the dechlorination activity was stimulated by the addition of lactate (data not shown), the improvement of detection efficiency obtainable by CARD-FISH can strengthen the application potentiality under more stringent conditions. In particular, CARD-FISH allowed to visualize and quantify *Dehalococcoides* cells in chlorinated ethenes and ethanes contaminated groundwater where its presence ranged between 3% and 6% of the total DAPI-stained cells (Fazi *et al.*, 2008).

Alternative molecular approaches

PLFA analysis

PLFA (phospholipid fatty acid analysis) analysis was the first molecular technique commercialized for environmental applications. PLFA analysis is based on the extraction of 'signature' lipid biomarkers from the cell membranes and walls of microorganisms. Because certain bacterial groups make 'signature' fatty acids, a profile of the PLFA can be used to characterize the microbial community according to fatty acids quality and concentration found in the sample. This technique can produce information about the overall community structure and viability; however, with the increasing availability of alternative more specific methodologies (i.e. qPCR and clone libraries), the utility of this tool is minimal and might be useful only for particular applications (i.e. identification of environmental limitation such as nutrient deficiency or other stress).

Modelling the kinetics of microbial reductive dechlorination

Many models for contaminant degradation in groundwater were originally developed for petroleum hydrocarbons. These compounds serve as electron donors and are mineralized by heterotrophic bacteria that use oxygen or nitrate as electron acceptors. In contrast chlorinated solvents generally serve as electron acceptors in anaerobic environments and may be dechlorinated only by a few highly specialized organisms that use a limited number of electron donors, in competition with other bacteria (Fennell and Gossett, 2003). In spite of these differences, most widely used modelling packages have been adapted typically to model the reductive dechlorination of chlorinated solvents [e.g. BIOCHLOR (Aziz *et al.*, 2000), BIOPLUMEIII (Rifai *et al.*, 1995) and UTCHEM (De Blanc *et al.*, 1996)] also including the possible presence of a continuous contamination source (e.g. DNAPL). On the other hand, they use very simple mathematical expressions (e.g. first order decay, Monod) to describe step-by-step RD, while neglecting several other factors (e.g. inhibition, fermentation and competition).

On the other hand, several models have been specifically proposed in the last years to describe the kinetics of anaerobic reductive dechlorination (RD) of chlorinated solvents such as PCE or TCE in a range of environmental conditions. The remarkable progress on the knowledge on physiology of bacteria performing such reactions (e.g. involved enzymes, electron donor and nutrient requirements, energy conservation mechanisms) and the role of syntrophic or competitive relationships in anaerobic mixed cultures (e.g. fermentation, methanogenesis, sulfate reduction, nitrate reduction) has typically resulted in an increased degree of complexity of such mathematical models.

Depending on the number of processes which are taken into account, models for RD can be classified into two major groups. To the first group belong those model which only describe the transformation of chlorinated compounds, assuming unlimited electron donor availability; to the second group belong those models which describe dechlorination under limited electron donor conditions and that generally account for the fermentation of the added electron donor and for the presence of metabolisms in competition for both the added electron donor and the fermentation products. Obviously, the choice of the most appropriate model greatly depends, among the others, on site conditions and remediation strategy adopted.

Reductive dechlorination reactions under unlimited electron donor concentration

The most straightforward approach to model the RD of PCE to ethene is to use a Monod term for each step of this series reaction (Skeen *et al.*, 1995). According to this model (Table 1.3, line 1), the same population (X_{RD}) (Table 1.3, line 1) maybe responsible for the complete dechlorination of PCE to ethene or two different populations (one dechlorinating PCE to cis-DCE and one dechlorinating cis-DCE to ethene) (Table 1.3, lines 2–3) (Bagley, 1998; Fennell and Gossett, 1998).

This assumption was also based on the evidence (Rosner *et al.*, 1997) that PCE degradation occurs at different enzymatic sites than cis-DCE and VC degradation. Bagley also recognized that, if *Dehalococcoides*-type microorganisms were present in the system, the two groups could be collapsed to one group. More recent studies (Cupples *et al.*, 2004a,b; Yu and Semprini, 2004; Yu *et al.*, 2005), involving PCE-to-ethene dechlorinating cultures containing *Dehalococcoides* spp., have demonstrated that models including competitive inhibition between chloroethenes are able to better predict experimental data than simple Monod-type models. In general, higher chlorinated ethenes inhibit the dechlorination of lower chlorinated ethenes; whereas lower chloroethenes have only a weak effect on dechlorination of higher chlorinated

Table 1.3 Typical rate equations used to model RD under unlimited electron donor concentrations

No.	Reaction	Monod-type rate expression
1	PCE→TCE→…to ETH	$k_{RD}^{PCE} \cdot X_{RD} \cdot \left(\frac{PCE}{PCE + K_{PCE}}\right)$
2	PCE→TCE→…to cDCE	$k_{RD}^{PCE} \cdot X_{RD}^{PCE} \cdot \left(\frac{PCE}{PCE + K_{PCE}}\right)$
3	cDCE→VC→…to ETH	$k_{RD}^{cDCE} \cdot X_{RD}^{cDCE} \cdot \left(\frac{cDCE}{cDCE + K_{cDCE}}\right)$
4	VC→ETH	$k_{RD}^{VC} \cdot X_{RD} \cdot \left(\frac{VC}{VC + K_{VC} \cdot \left(1 + PCE/K_i^{PCE} + TCE/K_i^{TCE} + cDCE/K_i^{cDCE}\right)}\right)$
5	PCE→TCE→…cDCE	$k_{RD}^{PCE} \cdot X_{RD} \cdot \left(\frac{PCE}{PCE + PCE^2/K_i + K_{PCE}}\right)$

ethenes. As an example, PCE dechlorination rate was not influenced by the presence of TCE, cDCE, or VC, whereas VC dechlorination rate was competitively inhibited by all these compounds (Table 1.3, line 4). In addition, Yu and Semprini (2004) also observed that at high TCE or PCE concentrations, as those typically found in the proximity of a DNAPL-source zone, substrate inhibition effects also occurs and, accordingly, Haldane inhibition kinetics was proposed (Table 1.3, line 5).

Reductive dechlorination under limiting electron donor and electron acceptor concentrations and influence of competitive metabolisms

The majority of isolated dechlorinating organisms use H_2 as the ultimate electron donor for RD. In anaerobic systems fed with an organic electron donor, H_2 is typically present at very low concentrations, hence the assumption of unlimited electron donor concentration may no longer be valid, and the influence of H_2 on the RD rate needs to be accounted for. On the contrary, in anaerobic systems fed with an organic electron donor, acetate is typically present at non-limiting concentrations and therefore its influence on RD rate can be often neglected, even when acetotrophic dechlorination is being considered.

In order to adequately describe the effect of H_2 on RD rate, some authors (Aulenta *et al.*, 2006a; Bagley, 1998) have used double-Monod terms, with dependence on both chloroethene and H_2 concentrations (Table 1.4, line 1). Other researchers have also introduced the concept of H_2 threshold level for dechlorination, i.e. the lowest H_2 concentration which support RD (Table 1.4, line 2).

Under H_2 limiting concentrations, the production of H_2 form fermentation of organic electron donors must be also described. H_2 is typically produced from fatty acids and alcohols under conditions that involve highly complex interactions of syntrophic bacteria. Both kinetics and thermodynamics govern the degradation of these electron donors. Fennell and Gossett (1998) developed a model which could well describe the production of hydrogen from fermentation of volatile fatty acids (VFA) and alcohols and its subsequent use for reductive dechlorination of PCE to ETH. The model described the H_2-releasing degradation of organic electron donors using

Table 1.4 Rate equations used to model fermentation reactions and RD under H_2-limited concentrations

No.	Reaction	Rate expression
1	PCE→TCE→... to ethene	$k_{RD}^{PCE}\cdot X_{RD}\cdot\left(\frac{PCE}{PCE+K_{PCE}}\right)$
2	PCE→TCE→...to ethene	$k_{RD}^{PCE}\cdot X_{RD}\cdot\left(\frac{PCE}{PCE+K_{PCE}}\right)\left(\frac{H_2-H_2^*}{H_2-H_2^*+K_{H2}}\right)$
3	VFA or alcohols→H_2 and acetate	$k_{donor}\cdot X_{donor}\cdot\frac{S\cdot\Phi}{S+K_{donor}}$ with $\Phi=1-\exp\left(\frac{\Delta G_{rxn}-\Delta G_{critical}}{RT}\right)$

Monod kinetics, but with the inclusion of the thermodynamic influence of product formation (e.g. H_2 and acetate) on the rate of fermentation. The term Φ (Table 1.4, line 3), which appears in the model, is a measure of the distance of the reaction from thermodynamic equilibrium. If the electron donor concentration is high relative to the concentration of the products of the reaction, H_2 and acetate, the driving force is high, then Φ approaches one and the fermentation reaction is limited primarily by its intrinsic kinetics. After the electron donor concentration has decreased and H_2 and acetate have increased, the driving force is lessened, the value of Φ approaches zero and the fermentation is limited primarily by thermodynamics.

Microbial competition for H_2, either produced from the fermentation of organic donors or directly provided, is a major issue in the application of RD. Under field conditions, the main metabolisms in competition with RD for H_2 use are methanogenesis, sulfate reduction, acetogenesis, and Fe and Mn reduction. Fennell and Gossett (2003) have developed a model which accounts for the competition for H_2 between dechlorinators, sulfate reducers, acetogenesis, and methanogenesis. Rate expressions for these processes are quite similar in the form to those used for RD (Table 1.4 line 2), but present salient differences in kinetic (e.g. half-velocity coefficient, maximum utilization rate) and thermodynamic parameters (e.g. threshold concentration) for hydrogen use, between dechlorinators, methanogens, and acetogens. Such differences make it possible to develop a strategy for addressing the competition for H_2 in favour of dechlorinators.

In contrast, much more controversial is the problem of competition between RD and sulfate reduction, which present similar threshold values. At present, the question of competition between dechlorinating and iron-reducing bacteria has not yet been addressed to a great extent. Modelling this type of competition maybe complicated by the fact that the competitive electron acceptor (Fe oxides) is not soluble but associated with the aquifer matrix.

As a final remark, it should be noticed that most of the knowledge regarding the kinetics of chlorinated solvent dechlorination, the physiology of dechlorinating microorganisms, and the microbial competitive aspects of the process has been obtained from suspended-growth systems, whereas only limited information is available for dechlorinating biofilms, even though these are present in most systems, either natural (e.g. contaminated soils and aquifers), or engineered (e.g. permeable barriers). In biofilm systems, the transport of substrates contributes to determine overall process performance along with intrinsic reaction kinetics. In particular, for dechlorinating biofilms, the presence of mass-transport resistances for the primary substrate (e.g. fermentable carbon sources), the electron donor (i.e. H_2) or the electron acceptors (i.e. chloroethenes) may control both the kinetics of the RD steps and the competition for the primary electron donor as well as for H_2 among the different trophic groups (Aulenta *et al.*, 2006a).

Microcosms versus field tests

The reported knowledge on environmental factors controlling RD and on involved microorganisms makes it possible to have a preliminary evaluation of possible application of RD at a specific site, based on the knowledge of hydrogeological and geochemical conditions. However, even such a deep knowledge is not usually enough to appropriately foresee the contaminant fate under natural or modified conditions at the specific site without confirmation by a microcosm study and/or a field test.

With reference to microcosm studies (soil and groundwater samples are incubated in serum bottles, in the presence or not of various amendments and monitored for some months), they still represent the most common way to test the natural response of native microbial population and/or possible stimulation of RD under modified conditions, that are potentially achievable in groundwater by addition of substrates. Note that this is only true to the extent that the microcosm soils are representative of the site (this can be particularly problematic when the microbial populations are heterogeneously distributed). Microcosms are particularly useful:

- to verify the presence of native microorganisms able to dechlorinate and their possible end-products (with main reference to DCE conversion to VC and ethene);
- to choose the best reducing substrate and/or growth factors;
- to evaluate the influence of competing metabolisms and their effects on substrate dosage;
- to evaluate possible effects of co-contaminants and other substances in the groundwater (e.g. inhibition), including possible accumulation of toxic intermediates or side-products; and
- to verify the need and the effectiveness of further enhancement through addition of a specialized inoculum (bioaugmentation).

Usually, several carbon sources have to be tested, e.g. those that can be fermented under low or high H_2 partial pressures (e.g. butyric acid or lactic acid, respectively), along with specific growth factors (vitamin B_{12}, yeast extract). Modifications of the RABITT protocol have been recently tested to include molecular hydrogen as a direct electron donor for RD as well as bioaugmentation with specialized inocula (Aulenta *et al.*, 2005a). In general, if chlorinated solvents are not completely degraded after six to nine months of incubation in microcosms amended with any substrate, even when appropriate redox conditions and electron donor availability are maintained, then appropriate native dechlorinating organisms may not be present. For chlorinated ethenes, a lack of degradation past *cis*-DCE in the microcosms indicates that it is unlikely that dechlorination in the field will proceed past *cis*-DCE to VC and ethene, and bioaugmentation may be required.

On the other hand, even when degradation occurs, the choice of optimal electron donor is not straightforward. In a recent microcosm study, not only different rates and end-products were obtained from different substrates but also accumulation of fermentation products, Fe and Mn release, and residual groundwater ecotoxicity were highly affected (Aulenta *et al.*, 2007b). As an example, compared with other electron donors, lactate initially supported RD at the highest rate but it was also consumed very fast. As a consequence of its fast consumption, high levels of VFA and dissolved metals accumulated in lactate-amended microcosms. The accumulation of these side products not only represented a wastage of reducing equivalents but also determined the deterioration of groundwater quality and the increase of groundwater ecotoxicity (tested through the germination of *Lepidum sativum*). Among the electron donors tested, the H_2+ acetate mixture, yielded the most promising results in terms of extent of dechlorination, negligible accumulation of byproducts such as VFA, Fe(II) and Mn(II), and residual groundwater toxicity. The average amount of donor that was required for the removal of 1 μmol of chloride was 2.13 meq/μmol Cl^- for lactate, 1.01 meq/μmol Cl^- for butyrate, and 0.39 meq/μmol Cl^- for H_2+ acetate mixture. FISH analysis also confirmed that H_2+ acetate-amended microcosms were mostly dominated by *Dehalococcoides* spp., while a higher biodiversity was observed in the cultures fed with lactate.

Moreover, comparison between microcosms performed in the presence of groundwater and defined mineral medium, could provide additional information whether other groundwater constituents may have any further effects on reductive dechlorination reactions. Indeed, the presence in the aquifer of multiple, and in many cases unknown, co-contaminants may drastically reduce our ability to predict the outcome of a bio-remedial action. As an example, in a PCE-enriched culture containing *Dehalococcoides* spp. as the putative dechlorinating organism (Aulenta *et al.*, 2005a), we observed that a 1,1,2,2-TeCA spike had a negative effect on the last step of dechlorination (i.e. from VC to ETH). Adamson and Parkin (2001) observed that a PCE–degrading, methanogenic, lactate-enriched culture was able to degrade cometabolically both carbon tetrachloride (CT) and 1,1,1-TCA despite no previous exposure to these compounds. However, while the presence of <20 μM of 1,1,1-TCA had little effect on PCE removal, the addition of 10–15 μM of CT negatively impacted both the PCE and VC dechlorination steps.

In some other cases, the presence of co-contaminants may even be beneficial to dechlorinating microorganism. For instance, organic co-contaminants may act as electron donors for dechlorination reactions or may lead to the establishment of complex mixed cultures in which organisms growing on co-contaminants provide dechlorinators with nutritional factors through lysis or excretion. To address the problem of complex contaminated aquifers, bioremediation studies will have to be preferably carried out using real groundwater.

The primary disadvantage of microcosms is that the tests may not always accurately reflect subsurface conditions in the field. Microcosms are usually performed in batch for very long residence time (that can be quite different from field conditions) and there is evidence that accumulation of metabolic intermediates and/or daughter products can reduce the rate and extent of RD (Aulenta *et al.*, 2005b). Also, aquifers are extremely heterogeneous systems, even at a very small scale, in terms of both geochemical composition, microbial and contaminant distribution. Such a complexity may not be captured by a single microcosm study, especially if DNAPL is present. This suggests that microcosm studies for candidate sites should be as expansive as possible, including collection of microcosm samples from a number of locations and/or compositing samples from multiple locations. However, performing an expansive microcosm study will be relatively expensive owing to the increased number of sample cores needed and related laboratory analyses.

In any case, caution is advised when interpreting microcosm results as being indicative of what can be achieved in the field. According to the technical document from a ESTCP survey (2004), information for field application, that microcosms *cannot* usually provide, include the following:

- rates of dechlorination that will occur under field conditions;
- efficiency of substrate utilization that may occur under field conditions;
- acclimation periods in the field; and
- any field scale phenomenon such as rates of increased DNAPL dissolution.

As an example, one study (Aulenta *et al.*, 2007c) reports a comparison between microcosms and field test results in a small and contained area of a shallow aquifer. Both studies provided indications of the presence at the site of native dechlorinating populations that were able to degrade 1,1,2,2-TeCA and other chlorinated pollutants (Aulenta *et al.*, 2006b; 2007b). On the contrary, microcosms did not provide any evidence of DNAPL presence whereas groundwater recirculation and substrate addition caused a strong increase in 1,1,2,2-TeCA concentration during field test operation. Most probably, the aquifer material used for microcosms (one single core sample) did not contain chlorinated solvents as free phase, indicating that DNAPL dissolution effects occurring in the field can randomly be predicted by microcosm studies.

Moreover, complete reduction of sulfate was obtained in microcosms in around 70 days whereas negligible reduction of sulfate was observed in the field. Possibly, the higher concentration of 1,1,2,2-TeCA caused by DNAPL enhanced dissolution in the field proved to be toxic to native sulfate-reducing microorganisms. It is noteworthy that sulfate-reduction was a major sink of substrate (lactate) in microcosms, so showing that substrate dosage in field could be overestimated. Moreover, inhibition of sulfate-reduction also changed both the lactate fermentation and the 1,1,2,2-TeCA RD pathways.

Overall, bioremediation of DNAPL sources through reductive dechlorination presents significant challenges, often closely related to the highly complex architecture of the sources which makes their localization and treatment very difficult; moreover, as already discussed, DNAPL sources typically result in extremely high contaminant concentrations which, in some cases may prove toxic to bacteria. An extensive review of factors and conditions affecting DNAPL bioremediation can be found in a document produced by the American Environmental Protection Agency (EPA, 2003).

Thus, although more expensive and time-consuming, small-scale field tests should be also conducted along with microcosm studies to assess the potential for *in situ* bioremediation of chlorinated solvents at specific sites, particularly when the presence of a DNAPL is anticipated. As a pilot test evaluates a much larger, more representative volume of the aquifer, it makes it possible that microorganisms grow and become more active and widely distributed in the treatment zone, even if they were initially present in only a relatively small portion of the aquifer. In this respect, indicative information about the costs of both field tests and microcosms studies obtained from the application of the RABITT protocol at different sites has been recently published (available online at ESTCP website). This document indicated a total cost in the range of \$84,000–111,000 or \$124,000–154,000 for a field test involving a 10 m or 60 m drilling, respectively. As for the microcosm studies, the document indicated a total cost in the range of \$77,000–94,000 (10 m depth of drilling) or \$94,000–111,000 (60 m depth of drilling).

In order to be successful, these field tests need a research-quality approach including careful design, operation, and monitoring. Moreover, the design of field test is quite more variable than microcosm procedure, depending on remedial target, aquifer characteristics, and chosen substrate (either soluble or solid substrate). Dealing with the addition of soluble electron donors, a small-scale field test is usually based on a pilot-scale hydraulic system which is used to extract groundwater and re-inject it, after amendment in a mixing chamber. The design of the field test should be aimed at creating an hydraulically controlled reaction volume in the aquifer; controlling residence time in the reaction volume; obtaining good mixing of contaminated groundwater with substrates and good distribution of amended groundwater in the reaction volume, and carefully monitoring inside and outside the test area. The choice of the system arrangement should be strictly dependent on aquifer characteristics. Appropriate extraction flow rates (usually in the range of l/min) and field scale (usually 10–20 m) have to be chosen in order to create the appropriate hydraulically controlled reaction volume in the aquifer and to control residence time in the reaction volume (based on microcosm results and at least of 30 days). A conservative tracer test should be usually performed before adding substrates along with hydrodynamic modelling in order to confirm good control of the fluid dynamics of the system as well as good substrate distribution. Where the choice of an extraction-re-injection system is hindered by local constraints such as regulatory permits, internal recirculation in a single well can be also performed.

In addition to small-scale pilot tests, singe well 'push-pull' tests may also be used (Fisher *et al.*, 1999; Newell *et al.*, 2000). This test consists of injecting ('pushing') a unit volume of contaminated groundwater with known contaminant and conservative tracer concentrations into the aquifer via a single well. This unit volume is then sampled ('pulled') at discrete time periods and analysed for the ratio of parent and dechlorination products. The change in parent and daughter product concentrations over time can be used to calculate field degradation rates and the extent of dechlorination. Conservative tracers are used to account for the effects of dilution, and non-conservative tracers are used to account for sorption. The volume of groundwater injected, and the timing and spacing of the sampling periods, should be designed such as that the injected groundwater remains within the immediate vicinity of the test well over the duration of the test, generally from 48 hours up to two or three months (Hageman *et al.*, 2001). Therefore, these tests may not be practical in high flow aquifers whereas are quite suitable for testing insoluble or slow-release substrates in low flow aquifers. Moreover, applying this method for chlorinated solvents is complicated by the lag time typically needed to develop microbial populations capable of degrading them. Therefore, 'push-pull' substrate utilization tests should be applied only after the aquifer has been conditioned to appropriate anaerobic conditions, usually on the order of two to four months after injection of a substrate (if necessary).

In general, field tests provide a greater level of confidence in estimating the *in situ* extent and rate of dechlorination, and provide more information for design purposes (e.g. injection well spacing, injection pressures and frequency, substrate loading requirements). In particular situations, field tests may preclude the need for laboratory studies.

On the other hand, they are usually done with a single pre-chosen substrate that could had better optimized by a microcosm study. Moreover, field tests require a large array of caution procedures to avoid that malfunctioning will increase any environmental and health risks (sometimes, including physical containment of test area); thus specific permits are required and are often the rate determining step of the overall design procedure. These permits could be more easily obtained and some cautions avoided in the presence of clear and good microcosms results.

References

Adamson, D.T., and Parkin, G.F. (2001). Product distribution during transformation of multiple contaminants by a high-rate, tetrachlorethene-dechlorinating enrichment culture. Biodegradation *12*, 337–348.

Adamson, D.T., McDade, J.M., and Hughes, J.B. (2003). Inoculation of a DNAPL source zone to initiate reductive dechlorination of PCE. Environ. Sci. Technol. *37*, 2525–2533.

Adamson, D.T., Lyon, D.Y., and Hughes, J.B. (2004). Flux and product distribution during biological treatment of tetrachloroethene dense non-aqueous-phase liquid. Environ. Sci. Technol. *38*, 2021–2028.

Ahmed, A.A., and Chen, D. (2006). Field scale DNAPLs transport under nonequilibrium sorption conditions. J. Environ. Sci. Health A Tox. Hazard. Subst. Environ. Eng. *41*, 273–290.

Aulenta, F., and Majone, M. (2010). Bioelectrochemical systems (BES) for subsurface remediation. In: Bioelectrochemical Systems, From Extracellular Electron Transfer to Biotechnological Application, Rabaey, K., Angenent, L.T., Schroder, U., and Keller, J., eds. (IWA Publishing: London), pp. 305–322.

Aulenta, F., Majone, M., Verbo, P., and Tandoi, V. (2002). Complete dechlorination of tetrachloroethene to ethene in presence of methanogenesis and acetogenesis by an anaerobic sediment microcosm. Biodegradation *13*, 411–424.

Aulenta, F., Rossetti, S., Majone, M., and Tandoi, V. (2004). Detection and quantitative estimation of *Dehalococcoides* spp. in a dechlorinating bioreactor by a combination of fluorescent in situ hybridisation (FISH) and kinetic analysis. Appl. Microbiol. Biotechnol. *64*, 206–212.

Aulenta, F., Fina, A., Potalivo, M., Petrangeli Papini, M., Rossetti, S., and Majone, M. (2005a). Anaerobic transformation of tetrachloroethane, perchloroethylene, and their mixtures by mixed-cultures enriched from contaminated soils and sediments. Wat. Sci. Technol. *52*, 357–362.

Aulenta, F., Gossett, J.M., Papini, M.P., Rossetti, S., and Majone, M. (2005b). Comparative study of methanol, butyrate, and hydrogen as electron donors for long-term dechlorination of tetrachloroethene in mixed anerobic cultures. Biotechnol. Bioeng. *91*, 743–753.

Aulenta, F., Di Tomassi, C., Cupo, C., Papini, M.P., and Majone, M. (2006a). Influence of hydrogen on the reductive dechlorination of tetrachloroethene (PCE) to ethene in a methanogenic biofilm reactor: Role of mass transport phenomena. J. Chem. Technol. Biotechnol. *81*, 1520–1529.

Aulenta, F., Potalivo, M., Majone, M., Papini, M. P., and Tandoi, V. (2006b). Anaerobic bioremediation of groundwater containing a mixture of 1,1,2,2-tetrachloroethane and chloroethenes. Biodegradation *17*, 193–206.

Aulenta, F., Catervi, A., Majone, M., Panero, S., Reale, P., and Rossetti, S. (2007a). Electron transfer from a solid-state electrode assisted by methyl viologen sustains efficient microbial reductive dechlorination of TCE. Environ. Sci. Technol. *41*, 2554–2559.

Aulenta, F., Pera, A., Rossetti, S., Petrangeli Papini, M., and Majone, M. (2007b). Relevance of side reactions in anaerobic reductive dechlorination microcosms amended with different electron donors. Wat. Res. *41*, 27–38.

Aulenta, F., Canosa, A., Leccese, M., Petrangeli Papini, M., Majone, M., and Viotti, P. (2007c). Field study of *in situ* anaerobic bioremediation of a chlorinated solvent source zone. Ind. Eng. Chem. Res. *46*, 6812–6819.

Aulenta, F., Beccari, M., Majone, M., Papini, M. P., and Tandoi, V. (2008a). Competition for H2 between sulfate reduction and dechlorination in butyrate-fed anaerobic cultures. Process Biochem. *43*, 161–168.

Aulenta, F., Canosa, A., Majone, M., Panero, S., Reale, P., and Rossetti, S. (2008b). Trichloroethene dechlorination and H2 evolution are alternative biological pathways of electric charge utilization by a dechlorinating culture in a bioelectrochemical system. Environ. Sci. Technol. *42*, 6185–6190.

Aulenta, F., Fuoco, M., Canosa, A., Petrangeli Papini, M., and Majone, M. (2008c). Use of poly-beta-hydroxy-butyrate as a slow-release electron donor for the microbial reductive dechlorination of TCE. Wat. Sci. Technol. *57*, 921–925.

Aulenta, F., Reale, P., Catervi, A., Panero, S., and Majone, M. (2008d). Kinetics of trichloroethene dechlorination and methane formation by a mixed anaerobic culture in a bio-electrochemical system. Electrochim. Acta *53*, 5300–5305.

Aulenta, F., Canosa, A., De Roma, L., Reale, P., Panero, S., Rossetti, S., and Majone, M. (2009a). Influence of mediator immobilization on the electrochemically assisted microbial dechlorination of trichloroethene (TCE) and cis-dichloroethene (cis-DCE). J. Chem. Technol. Biotechnol. *84*, 864–870.

Aulenta, F., Canosa, A., Reale, P., Rossetti, S., Panero, S., and Majone, M. (2009b). Microbial reductive dechlorination of trichloroethene to ethene with electrodes serving as electron donors without the external addition of redox mediators. Biotechnol. Bioeng. *103*, 85–91.

Aulenta, F., Reale, P., Canosa, A., Rossetti, S., Panero, S., and Majone, M. (2010). Characterization of an electro-active biocathode capable of dechlorinating trichloroethene and cis-dichloroethene to ethene. Biosens. Bioelectron. *25*, 1796–1802.

Aziz, C.E., Newell, C.J., Gonzales, J.R., Haas, P.E., Clement, T.P., and Sun, Y. (2000). Natural attenuation decision support system, User's Manual Version, 1.0. BIOCHLORY – US EPA, EPA/600R-00/008.

Bagley, D.M. (1998). Systematic approach for modeling tetrachloroethene biodegradation. J. Environ. Eng. *124*, 1076–1086.

Ballapragada, B.S., Stensel, H.D., Puhakka, J.A., and Ferguson, J.F. (1997). Effect of hydrogen on reductive dechlorination of chlorinated ethenes. Environ. Sci. Technol. *31*, 1728–1734.

Barra Caracciolo, A., Grenni, P., Cupo, C., and Rossetti, S. (2005). *In situ* analysis of native microbial communities in complex samples with high particulate loads. FEMS Microbiol. Lett. *253*, 55–58.

Bowman, J.P., Jimenez, L., Rosario, I., Hazen, T.C., and Sayler, G.S. (1993). Characterization of the methanotrophic bacterial community present in a trichloroethylene-contaminated subsurface groundwater site. Appl. Environ. Microbiol. *59*, 2380–2387.

Brennan, R.A., Sanford, R.A., and Werth, C.J. (2006). Chitin and corncobs as electron donor sources for the reductive dechlorination of tetrachloroethene. Wat. Res. *40*, 2125–2134.

de Bruin, W.P., Kotterman, M.J., Posthumus, M.A., Schraa, G., and Zehnder, A.J. (1992). Complete biological reductive transformation of tetrachloroethene to ethane. Appl. Environ. Microbiol. *58*, 1996–2000.

Carter, S.R., and Jewell, W.J. (1993). Biotransformation of Tetrachloroethylene by anaerobic attached-films at low-temperatures. Wat. Res. *27*, 607–615.

Cupples, A.M., Spormann, A.M., and McCarty, P.L. (2004a). Comparative evaluation of chloroethene dechlorination to ethene by *Dehalococcoides*-like microorganisms. Environ. Sci. Technol. *38*, 4768–4774.

Cupples, A.M., Spormann, A.M., and McCarty, P.L. (2004b). Vinyl chloride and cis-dichloroethene dechlorination kinetics and microorganism growth under substrate limiting conditions. Environ. Sci. Technol. *38*, 1102–1107.

De Blanc, P.C., McKinney, D.C., Speitel, G.E., Jr., Sepehrnoori, K., and Delshad, M. (1996). A 3-D NAPL flow and biodegradation model. In Non-acqueous Phase Liquids (NAPLs) in Subsurface Enviornments: Assessment and Remediation, Reddi, N.L., ed. (ASCE: Washington), pp. 468–489.

Dennis, P.C., Sleep, B.E., Fulthorpe, R.R., and Liss, S.N. (2003). Phylogenetic analysis of bacterial populations in an anaerobic microbial consortium capable of degrading saturation concentrations of tetrachloroethylene. Can. J. Microbiol. *49*, 15–27.

DiStefano, T.D., Gossett, J.M., and Zinder, S.H. (1991). Reductive dechlorination of high concentrations of tetrachloroethene to ethene by an anaerobic enrichment culture in the absence of methanogenesis. Appl. Environ. Microbiol. *57*, 2287–2292.

DiStefano, T.D., Gossett, J.M., and Zinder, S.H. (1992). Hydrogen as an electron donor for dechlorination of tetrachloroethene by an anaerobic mixed culture. Appl. Environ. Microbiol. *58*, 3622–3629.

Duhamel, M., Mo, K., and Edwards, E.A. (2004). Characterization of a highly enriched *dehalococcoides*-containing culture that grows on vinyl chloride and trichloroethene. Appl. Environ. Microbiol. *70*, 5538–5545.

El Fantroussi, S., Naveau, H., and Agathos, S.N. (1998). Anaerobic dechlorinating bacteria. Biotechnol. Prog. *14*, 167–188.

EPA (2003). The DNAPL remediation challenge: is there a case for source depletion? In Report EPA/600/R-03/143, National Risk Management Research Laboratory Expert Panel, ed. (U.S. Environmental Protection Agency, Cincinnati, OH).

Fang, Y., Hozalski, R.M., Clapp, L.W., Novak, P.J., and Semmens, M.J. (2002). Passive dissolution of hydrogen gas into groundwater using hollow-fiber membranes. Wat. Res. *36*, 3533–3542.

Fazi, S., Aulenta, F., Majone, M., and Rossetti, S. (2008). Improved quantification of *Dehalococcoides* species by fluorescent *in situ* hybridization and catalysed reporter deposition (CARD-FISH). Sys. Appl. Microbiol. *31*, 62–67.

Fennell, D.E., and Gossett, J.M. (1998). Modeling the production of and competition for hydrogen in a dechlorinating culture. Environ. Sci. Technol. 32, 2450–2460.

Fennell, D.E., and Gossett, J.M. (2003). Microcosms for site-specific evaluation of enhanced biological reductive dehalogenation, In Dehalogenation,

Microbial Processes and Environmental Applications, Haggblom, M., and Bossert, I.D., eds. (Kluwer Academic Publishers: Boston), pp. 385–420.

Fennell, D.E., Gossett, J.M., and Zinder, S.H. (1997). Comparison of butyric acid, ethanol, lactic acid, and propionic acid as hydrogen donors for the reductive dechlorination of tetrachloroethene. Environ. Sci. Technol. *31*, 918–926.

Fisher, R.T., Newell, C.J., Haas, P.E., and Hughes, J.B. (1999). Treatability studies of hydrogen-enhanced bioremediation of chlorinated solvent-impacted media. In Engineered Approaches for *in situ* Bioremediation of Chlorinated Solvent Contamination, Leeson, A., and Alleman, B., eds. Fifth International *in situ* and Site Bioremediation Conference, 19–22 April 1999, Battelle Press: San Diego, California, pp. 185–190.

Freeborn, R.A., West, K.A., Bhupathiraju, V.K., Chauhan, S., Rahm, B.G., Richardson, R.E., and Alvarez-Cohen, L. (2005). Phylogenetic analysis of TCE-dechlorinating consortia enriched on a variety of electron donors. Environ. Sci. Technol. *39*, 8358–8368.

Freedman, D.L., and Gossett, J.M. (1989). Biological reductive dechlorination of tetrachloroethylene and trichloroethylene to ethylene under methanogenic conditions. Appl. Environ. Microbiol. *55*, 2144–2151.

Gu, A.Z., Hedlund, B.P., Staley, J.T., Strand, S.E., and Stensel, H.D. (2004). Analysis and comparison of the microbial community structures of two enrichment cultures capable of reductively dechlorinating TCE and cis-DCE. Environ. Microbiol. *6*, 45–54.

Hageman, K.J., Istok, J.D., Field, J.A., Buscheck, T.E., and Semprini, L. (2001). *In situ* anaerobic transformation of trichlorofluoroethene in trichloroethene-contaminated groundwater. Environ. Sci. Technol. *35*, 1729–1735.

Haston, Z.C., and McCarty, P.L. (1999). Chlorinated ethene half-velocity coefficients (KS) for reductive dehalogenation. Environ. Sci. Technol. 33, 223–226.

He, J., Sung, Y., Dollhopf, M.E., Fathepure, B.Z., Tiedje, J.M., and Loffler, F. E. (2002). Acetate versus hydrogen as direct electron donors to stimulate the microbial reductive dechlorination process at chloroethene-contaminated sites. Environ. Sci. Technol. *36*, 3945–3952.

He, J., Sung, Y., Krajmalnik-Brown, R., Ritalahti, K.M., and Loffler, F.E. (2005). Isolation and characterization of *Dehalococcoides* sp. strain FL2, a trichloroethene (TCE)- and 1,2-dichloroethene-respiring anaerobe. Environ. Microbiol. *7*, 1442–1450.

Hendrickson, E.R., Payne, J.A., Young, R.M., Starr, M.G., Perry, M.P., Fahnestock, S., Ellis, D.E., and Ebersole, R.C. (2002). Molecular analysis of *Dehalococcoides* 16S ribosomal DNA from chloroethene-contaminated sites throughout North America and Europe. Appl. Environ. Microbiol. *68*, 485–495.

Kube, M., Beck, A., Zinder, S.H., Kuhl, H., Reinhardt, R., and Adrian, L. (2005). Genome sequence of the chlorinated compound-respiring bacterium *Dehalococcoides* species strain CBDB1. Nat. Biotechnol. 23, 1269–1273.

Lee, K.Y., and Chrysikopoulos, C.V. (2006). Dissolution of a multicomponent DNAPL pool in an experimental aquifer. J. Hazard. Mater. *128*, 218–226.

Lee, P.K., Johnson, D.R., Holmes, V.F., He, J., and Alvarez-Cohen, L. (2006). Reductive dehalogenase gene expression as a biomarker for physiological activity of *Dehalococcoides* spp. Appl. Environ. Microbiol. *72*, 6161–6168.

Ma, X., Novak, P.J., Clapp, L.W., Semmens, M.J., and Hozalski, R.M. (2003). Evaluation of polyethylene hollow-fiber membranes for hydrogen delivery to support reductive dechlorination in a soil column. Wat. Res. *37*, 2905–2918.

Maymo-Gatell, X., Chien, Y., Gossett, J.M., and Zinder, S.H. (1997). Isolation of a bacterium that reductively dechlorinates tetrachloroethene to ethene. Science *276*, 1568–1571.

Mazur, C.S., and Jones, W.J. (2001). Hydrogen concentrations in sulfate-reducing estuarine sediments during PCE dehalogenation. Environ. Sci. Technol. *35*, 4783–4788.

Mercer, J.W., and Cohen, R.M. (1990). A review of immiscible fluids in the subsurface: Properties, models, characterization and remediation. J. Cont. Hydrol. *6*, 107–163.

Middeldorp, P.J., Luijten, M.L., van de Pas, B.A., Van Eekert, M.H., Kengen, S.W., Schraa, G., and Stams, A. (1999). Anaerobic Microbial Reductive Dehalogenation of Chlorinated Ethenes Biorem. Journal 3, 151–169.

Morse, J.J., Alleman, B.C., Gossett, J.M., Zinder, S.H., Fennell, D.E., Sewell, G.W., and Vogel, C.M. (1997). A treatability test for evaluating the potential applicability of the reductive anaerobic biological *in situ* treatment technology (RABITT) to remediate chloroethenes. A draft technical protocol developed for the Environmental Security Technology Certification Program, Department of Defence (US Government Printing Office: Washington DC).

Newell, C.J., Fisher, R.T., and Hughes, J.B. (1997). Direct hydrogen addition for the *in situ* biodegradation of chlorinated solvents. Paper presented at: NGWA petroleum hydrocarbons conference (Houston, TX).

Newell, C.J., Haas, P.E., Hughes, J.B., and Khan, T.A. (2000). Results from two direct hydrogen delivery field tests for enhanced dechlorination. In Bioremediation and Phytoremediation of Chlorinated and Recalcitrant Compounds, Wickramanayake, G.B., Gavaskar, A.R., Alleman, B.C., and Magar, V.S., eds. The Second International Conference on Remediation of Chlorinated and Recalcitrant Compounds Conference, Monterey, California, 22–25 May 2000 (Battelle Press: Columbus, OH) pp. 31–38.

Rice, D., Bartow, G., and Mcnab, W. (1999). Historical case analysis of chlorinated volatile organic compound plumes. Internet training UCRL-AR-133361.

Richardson, R.E., Bhupathiraju, V.K., Song, D.L., Goulet, T.A., and Alvarez-Cohen, L. (2002). Phylogenetic characterization of microbial communities that reductively dechlorinate TCE based upon a combination of molecular techniques. Environ. Sci. Technol. *36*, 2652–2662.

Rifai, H.S., Newell, C.J., Miller, R.N., Taffinder, S., and Rounsaville, M. (1995). Simulation of natural attenuation with multiple electron acceptors (Battelle Press: Colombus, OH).

Rosner, B.M., McCarty, P.L., and Spormann, A. M. (1997). *In vitro* studies on reductive vinyl chloride dehalogenation by an anaerobic mixed culture. Appl. Environ. Microbiol. *63*, 4139–4144.

Rossetti, S., Blackall, L. L., Majone, M., Hugenholtz, P., Plumb, J. J., and Tandoi, V. (2003). Kinetic and phylogenetic characterization of an anaerobic dechlorinating microbial community. Microbiology *149*, 459–469.

Rossetti, S., Aulenta, F., Majone, M., Crocetti, G., and Tandoi, V. (2008). Structure analysis and performance of a microbial community from a contaminated aquifer involved in the complete reductive dechlorination of 1,1,2,2-tetrachloroethane to ethene. Biotechnol. Bioeng. *100*, 240–249.

Ryoo, D., Shim, H., Canada, K., Barbieri, P., and Wood, T.K. (2000). Aerobic degradation of tetrachloroethylene by toluene-o-xylene monooxygenase of *Pseudomonas stutzeri* OX1. Nat. Biotechnol. *18*, 775–778.

Skeen, R.S., Gao, J., and Hooker, B.S. (1995). Kinetics of chlorinated ethylene dehalogenation under methanogenic conditions. Biotechnol. Bioeng. *48*, 659–666.

Smatlak, C.R., Gossett, J.M., and Zinder, S.H. (1996). Comparative kinetics of hydrogen utilization for reductive dechlorination of tetrachloroethene and methanogenesis in an anaerobic enrichment culture. Environ. Sci. Technol. *30*, 2850–2858.

Smith, C.J., Nedwell, D.B., Dong, L.F., and Osborn, A.M. (2006). Evaluation of quantitative polymerase chain reaction-based approaches for determining gene copy and gene transcript numbers in environmental samples. Environ. Microbiol. *8*, 804–815.

Smits, T.H., Devenoges, C., Szynalski, K., Maillard, J., and Holliger, C. (2004). Development of a real-time PCR method for quantification of the three genera *Dehalobacter, Dehalococcoides,* and *Desulfitobacterium* in microbial communities. J. Microbiol. Methods *57*, 369–378.

Spira, Y. (2006). Eurodemo aims, strategy, and results in the European Context, Paper presented at: EURODEMO Workshop – Paving the way to innovation in remediation technology (Rimini, Italy).

Sun, M., Sheng, G.P., Mu, Z.X., Liu, X.W., Chen, Y.Z., Wang, H.L., and Yu, H.Q. (2009). Manipulating the hydrogen production from acetate in a microbial electrolysis cell-microbial fuel cell-coupled system. J. Power Sources *191*, 338–343.

Tandoi, V., DiStefano, T.D., Bowser, P.A., Gossett, J.M., and Zinder, S.H. (1994). Reductive dehalogenation of chlorinated ethenes and halogenated ethanes by a high-rate anaerobic enrichment culture. Environ. Sci. Technol. *28*, 973–979.

Vogel, T.M., Criddle, C.S., and McCarty, P.L. (1987). Transformations of halogenated aliphatic compounds. Environ. Sci. Technol. *21*, 722–736.

Yang, Y., and McCarty, P.L. (1998). Competition for hydrogen within a chlorinated solvent dehalogenating anaerobic mixed culture. Environ. Sci. Technol. *32*, 3591–3597.

Yang, Y., and McCarty, P.L. (2002). Comparison between donor substrates for biologically enhanced tetrachloroethene DNAPL dissolution. Environ. Sci. Technol. *36*, 3400–3404.

Yang, Y., and Zeyer, J. (2003). Specific detection of *Dehalococcoides* species by fluorescence *in situ* hybridization with 16S rRNA-targeted oligonucleotide probes. Appl. Environ. Microbiol. *69*, 2879–2883.

Yang, Y., Pesaro, M., Sigler, W., and Zeyer, J. (2005). Identification of microorganisms involved in reductive dehalogenation of chlorinated ethenes in an anaerobic microbial community. Wat. Res. *39*, 3954–3966.

Yu, S., and Semprini, L. (2004). Kinetics and modeling of reductive dechlorination at high PCE and TCE concentrations. Biotechnol. Bioeng. *88*, 451–464.

Yu, S., and Semprini, L. (2009). Enhanced reductive dechlorination of PCE DNAPL with TBOS as a slow-release electron donor. J. Haz. Mat. *167*, 97–104.

Yu, S., Dolan, M.E., and Semprini, L. (2005). Kinetics and inhibition of reductive dechlorination of chlorinated ethylenes by two different mixed cultures. Environ. Sci. Technol. *39*, 195–205.

Bacterial Degradation of Cholesterol and Other Contaminant Steroids

2

José L. García, Iria Uhía, Esther García and Beatriz Galán

Abstract

Cholesterol is a steroid highly abundant in the environment that plays a major role in the global carbon cycle. Many synthetic steroidic compounds, such as some sex hormones, frequently appear in municipal and industrial wastewaters, acting as environmental pollutants with strong metabolic activities negatively affecting the ecosystems. Since these compounds are common carbon sources for many different microorganisms their aerobic and anaerobic mineralization has been extensively studied. The interest of these studies lies on the biotechnological applications of sterol-transforming enzymes for the industrial synthesis of sex hormones and corticoids. Very recently the catabolism of cholesterol has acquired a high relevance because it is involved in the infectivity of *Mycobacterium tuberculosis*. This review describes the current knowledge on the catabolism of cholesterol and related steroids in bacteria at both the biochemical and genetic level.

Introduction

Steroids are a class of terpenoid lipids characterized by a carbon skeleton of four fused rings, labelled A to D, with great importance in biology, medicine and chemistry (Fig. 2.1). The steroid nucleus is virtually flat and relatively rigid so that the fused ring does not allow rotation in CC bonds. Methyl groups usually occur in positions C-10 and C-13, and in the C-17 position may appear an alkyl side-chain (Moss, 1989). The naturally occurring sterols include phytosterols in plants, ergosterol in fungi and yeast and cholesterol in animal cells (Fig. 2.1). All play an important role as stabilizing agents of the cell membranes and also as precursors to a variety of products with specific biological activities. Cholesterol is one of the most common molecules in nature, as a component of animal cell membranes, but it has been also found in some prokaryotes (Hayami *et al.*, 1979). This polycyclic alcohol has been extensively studied as an essential component of all animal tissues (Staytor and Bloch, 1965). The metabolism of cholesterol is crucial, being the immediate precursor of a large number of substances such as vitamins, steroid hormones and bile acids (Miller, 1988; Björkhem and Eggertsen, 2001). The ubiquity of cholesterol and related sterols in the environment has made them a common carbon source for many different microorganisms. Since 1913 it has been well known that there are soil microorganisms capable of using cholesterol as the sole carbon and energy source (Söhngen, 1913). However, a detailed knowledge of the microbial degradation of cholesterol is still challenging because of its low solubility in water and its complex chemical structure, with a low number of functional groups and the presence of four alicyclic rings and two quaternary carbon atoms.

Among the thousands of new pollutants discharged into the environment as a result of human activity, the appearance of many molecules with a steroid structure is very frequently observed in different ecosystems. The potent metabolic activities of many of these compounds affect a large number of cellular processes (some of them as essential as cell proliferation, cell cycle control, basic mechanisms of intercellular recognition,

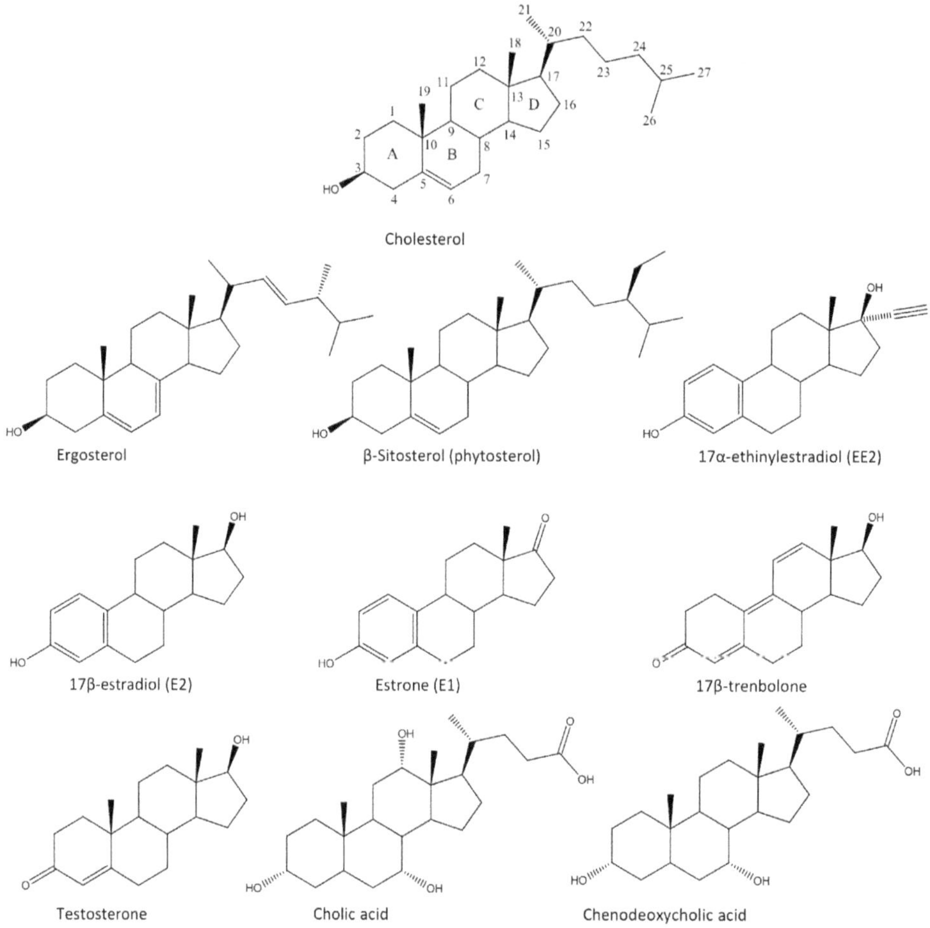

Figure 2.1 Molecular structures of some steroid compounds.

etc.) so that their presence and accumulation in certain ecological niches can disrupt the endocrine system of animals and humans (Hotchkiss *et al.*, 2008). This pollution comes partly as a result of several industrial activities, but also from municipal effluents. In this regard, the pharmaceutical and personal care products (PPCPs) have been classified as an emerging class of contaminants of particular relevance in the aquatic environment (Gagné *et al.*, 2006). Among these products there are many compounds of pharmaceutical origin with steroid nature. For example, the synthetic oral contraceptive 17α-ethinylestradiol (EE2) (Fig. 2.1) is held responsible, along with estrogen 17β-estradiol (E2) and estrone (E1) (Fig. 2.1), of causing vitellogenin production (feminization) in male fish (Daughton and Ternes, 1999) and the development of intersex gonads (ovotestis) in fish and amphibians (Hutchinson *et al.*, 2006). Levels of EE2 sufficient to cause vitellogenin induction in fish have also been implied in causing a substantial decrease in the sustainability of the wild fish populations (Kidd *et al.*, 2007). This phenomenon of feminization was first observed in pools of sewage in the mid-1980s (Routledge *et al.*, 1998). The steroids 17α-trenbolone and 17β-trenbolone (Fig. 2.1) were detected in effluents from cattle concentrated animal feeding operations where trenbolone acetate implants were used to stimulate weight gain (Durhan *et al.*, 2006).

It is important to point out that the most frequent situation is the exposure not to single but mixtures of different endocrine-disrupting chemicals in the environment (Kolpin *et al.*, 2002). *In vitro* studies found that mixtures of xenoestrogens cause additive activation of the estrogen receptor (Payne *et al.*, 2000; Payne *et al.*, 2001; Silva *et al.*, 2002). It was also demonstrated that chemicals present at levels below their individual no-observed-effect concentrations contribute to a cumulative effect of the mixture (Silva *et al.*, 2002). *In vivo* studies showed that a mixture of five estrogenic compounds was capable of inducing vitellogenin in male fish corresponding to predictions based on the concept of concentration addition (Brian *et al.*, 2005).

Furthermore, it should be noted that cholesterol is the main component of lanolin, and this and other related sterols are natural contaminants fairly resistant to the anaerobic treatment that is carried out on the effluents from wool-processing industry (Poole and Cord-Ruwisch, 2004). The amount of estrogen from animal sources is less important than that from sewage, but manure waste contribute significantly to its entry into the environment (Hanselman *et al.*, 2003, Raman *et al.*, 2004). Manure from livestock and poultry is also considered a source of environmental testosterone (Lee and Liu, 2002).

Finally, since natural steroids can be the starting material for other synthetic compounds, the bacterial catabolism of steroids has attracted considerable attention as a potential means of producing bioactive steroids from natural low-cost sterols. Moreover, the characterization of these metabolic pathways could be useful to develop new organisms by metabolic engineering with a potential use as bioremediation agents. Therefore, this chapter will revise the accumulated biochemical and genetic information of the bacterial pathways that degrade or transform cholesterol and other related steroids.

Catabolism of cholesterol

As mentioned above, cholesterol and other related compounds are frequently found in the biosphere, not only because of their natural abundance, but also because of their high resistance to microbial degradation. The fact that cholesterol is a recalcitrant molecule is attributed to the low number of functional groups (one CC double bond and a single hydroxyl group), its low solubility in water (3×10^{-8} M) and the complexity of its spatial conformation. The high hydrophobicity and low volatility of steroid compounds leads to their high absorption to solid phases. Because of their high rate of persistence, cholesterol and some related compounds, such as coprostanol, are ubiquitous and have been used as reference biomarkers for environmental pollution analysis (Veiga *et al.*, 2005).

The search for microorganisms capable of degrading cholesterol was started over 70 years ago. In the early 20th century it was observed that several species of *Mycobacterium* could use cholesterol as the sole source of carbon and energy (Söhngen, 1913). Later on, it was recognized that microorganisms belonging to the genus *Proactinomyces,* like *Rhodococcus erythropolis* (formerly *Nocardia erytrhopolis*), have the ability to degrade cholesterol and a potential capacity for the synthesis of new steroid derivatives (Turfitt, 1944, 1948). Moreover, some species of *Azotobacter* transformed cholesterol in cholest-4-en-3-one or 7-dehydrocholesterol, and have the ability to hydrolyse the side chain of cholesterol generating methylheptanone that was accumulated in the culture medium (Horvath and Kramli, 1947). Whitmarsh (1964) showed that certain species of *Nocardia* isolated from soil were capable of hydrolysing the cholesterol side chain transforming this molecule in different steroid derivatives, with similar characteristics to sex hormones. Many different bacteria belonging to the genera *Nocardia, Arthrobacter, Bacillus, Brevibacterium, Corynebacterium, Streptomyces, Microbacterium, Serratia, Achromobacter, Pseudomonas,* or *Protaminobacter,* among others, have been reported to accomplish partial or complete cholesterol degradation (Tak, 1942; Brown and Peterson, 1966; Arima *et al.*, 1969; Nagasawa *et al.*, 1969; Chipley *et al.*, 1975; Martin, 1977; Ferreira and Tracey, 1984). Unfortunately, one of the best studied steroid degraders, *Pseudomonas* sp. NCIB 10590 (Owen *et al.*, 1983), identified by 16S analysis as *Stenotrophomonas maltophilia* (B. Galán, personal communication) has apparently lost these capacities when deposited in the culture collection. This suggests that the classification of steroid degrader strains would need to be confirmed by using novel molecular genetic approaches.

Aerobic degradation of cholesterol

Although the bacterial catabolism of cholesterol has not been fully elucidated in any of the bacterial strains degrading sterols, the metabolic pathway can be postulated by combining the biochemical and genetic studies carried out in several organisms (Fig. 2.2) (Martin, 1977; Schoeman and Martin, 1980; Owen *et al.*, 1983; Kieslich, 1985; Sedlaczek and Smith, 1988; Szentirmai, 1990). To facilitate its comprehension the process can be divided into three major steps as presented below.

Transformation of cholesterol in cholest-4-en-3-one

It is generally assumed that the first reaction in the aerobic metabolism of cholesterol is its oxidation to cholest-4-en-3-one through two sequential reactions, i.e. the oxidation of cholesterol to cholest-5-en-3-one (Fig. 2.2) followed by its isomerization to cholest-4-en-3-one (Fig. 2.2). The enzymes responsible of this activity are cholesterol oxidases or 3β-hydroxysteroid dehydrogenase/isomerases.

Cholesterol oxidase (EC 1.1.3.6) (hereinafter abbreviated as CO) is a monomeric flavoenzyme that catalyses the oxidation and isomerization of cholesterol to cholest-4-en-3-one and has been well characterized structurally and chemically. The oxidation of cholesterol requires FAD that is concomitantly reduced. Regeneration of the oxidized cofactor is achieved by the reduction of O_2 to hydrogen peroxide. The enzyme is extracellular and occurs as secreted and/or cell-surface-associated forms, depending on the microorganism and growth conditions. It has been shown that CO is an interfacial enzyme that binds transiently to the membrane surface during catalysis (Kreit and Sampson, 2009).

Three-dimensional structures of some COs have been solved showing that the enzymes from *Streptomyces* subsp. SA-COO (SCO) (Yue *et al.*, 1999; Lario *et al.*, 2003) and from *Brevibacterium sterolicum* (BCO) (Vrielink *et al.*, 1991; Li *et al.*, 1993) are almost identical in sequence and structure and are classified as type I oxidases. Members of this class belong to the glucose-methanol-choline (GMC) oxidoreductase family of flavoenzymes where the FAD cofactor is non-covalently bound and have two conserved residues (His447 and Asn485) involved in substrate oxidation. However, a third CO crystallized from a different *Brevibacterium* strain (Coulombe *et al.*, 2001) has been classified as type II oxidase belonging to the family of vanillyl-alcohol oxidase (VAO) enzymes and does not share sequence similarity with the other two. This CO contains a FAD covalently bound to a His residue that is a key factor in the modulation of the redox properties (Motteran *et al.*, 2001).

Both classes of CO share the same catalytic activity but show significant differences in their redox and kinetic properties. In general, CO exhibits a broad range of steroid specificities and can oxidize a number of hydroxysterols, including sterols, steroid hormones and bile acids, but the presence of a 3β-hydroxyl group is an important requirement for activity. Also for the optimum activity, a C2 chain at C-17 is preferred. The oxidation of cholesterol in BCO works via a ping-pong mechanism (binary complex mechanism), whereas the catalytic pathway of SCO is sequential and follows a ternary complex mechanism (Pollegioni *et al.*, 1999).

Remarkably, some COs, like those from *Burkholderia cepacia* ST-200, *Pseudomonas* subsp., and *Chromobacterium* subsp. DS-1, oxidize cholesterol mainly to 6β-hydroperoxy-cholest-4-en-3-one and hydrogen peroxide instead of cholest-4-en-3-one (Doukyu and Aono, 1999; Doukyu *et al.*, 2008, 2009). This intermediate has not been associated so far to any cholesterol degradation pathway; in fact, *B. cepacia* ST-200 (formerly *Pseudomonas* sp. ST-200s) is unable to grow in 6β-hydroperoxy-cholest-4-en-3-one, suggesting that cholesterol is not the natural substrate of this enzyme or that its transformation has other purposes than energy production (Doukyu and Aono, 1999; Doukyu, 2009).

Besides the nutritional role, other biological roles have been ascribed to COs. COs are produced in pathogenic microorganisms that may require this enzyme for infection of macrophages probably because of its ability to alter the physical structure of the membrane owing to the conversion of cholesterol to cholest-4-en-3-one (Ghoshroy *et al.*, 1997). COs have been also suggested to act as fungal sensors although the

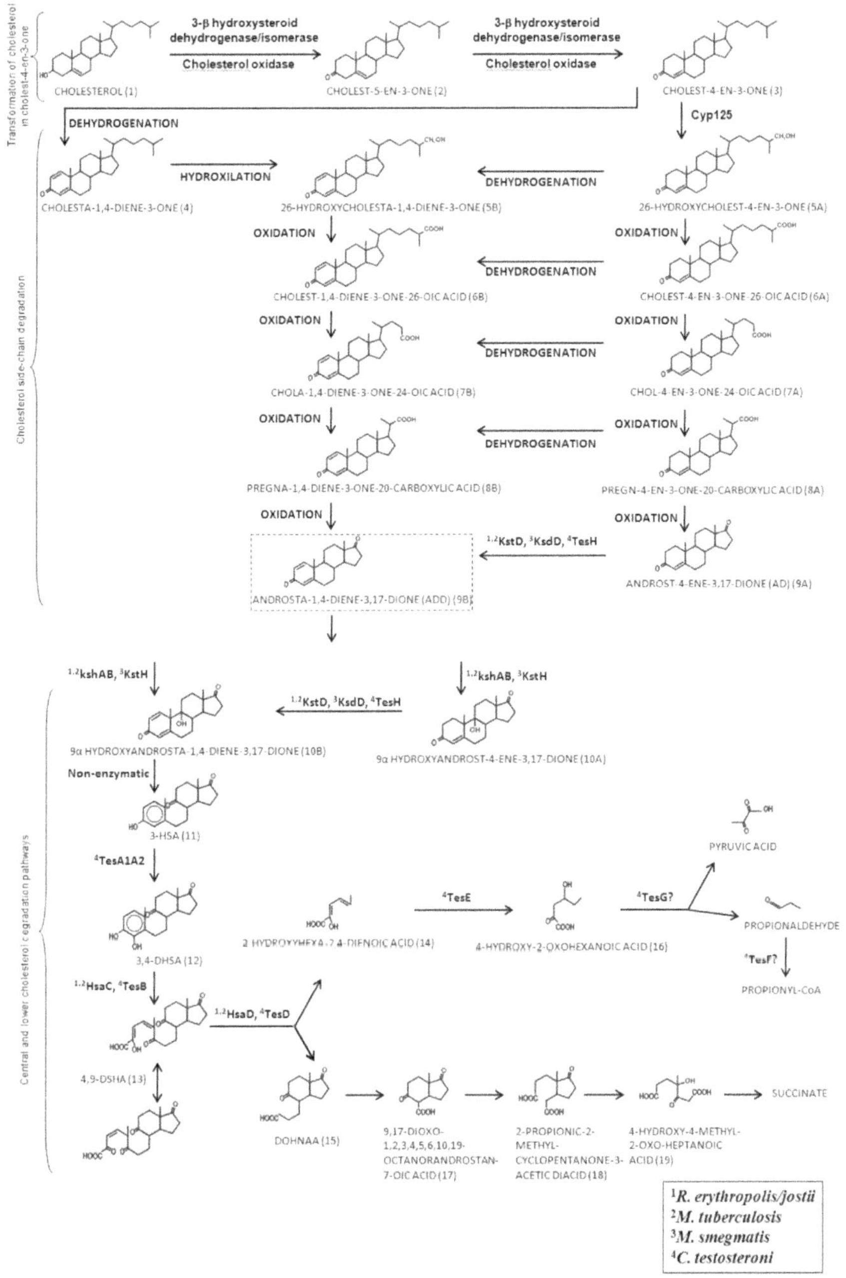

Figure 2.2 Proposed pathway for cholesterol degradation under aerobic conditions. The three major steps in which the process can be divided are indicated by brackets. Cholest-4-en-3-one (3) or any of the subsequent metabolites from degradation of the side chain up to (and including) AD (9A) may undergo a dehydrogenation reaction to introduce a double bond in the position 1, leading to compound cholesta-1,4-diene-3-one (4) in the case of cholest-4-en-3-one, or to the corresponding 1,2-dehydro derivatives for other molecules (B). The side-chain degradation of this compounds will be identical to that of the cholest-4-en-3-one to the common intermediate 9α-hydroxyandrosta-1,4-diene-3,17-dione (10B). The compound androsta-1,4-diene-3,17-dione (ADD) (9B), that can be considered as the central metabolite in the degradation of many steroids, is surrounded by a dashed line. The microorganisms from which the enzymes implicated in different steps were identified are indicated by numbers. 3-HSA (11), 3-hydroxy-9,10-secoandrosta-1,3,5(10)-trien-9,17-dione; 3,4-DHSA (12), 3,4-dihydroxy-9,10-secoandrosta-1,3,5(10)-trien-9,17-dione; 4,9-DSHA (13), 4,5,9,10-*diseco*-3-hydroxy-5,9,17-trioxoandrosta-1(10),2-diene-4-oic acid; DOHNAA (15), 9,17-dioxo-1,2,3,4,10,19-hexanorandrostan-5-oic acid.

signalling mechanism is not clear (Martín and Aparicio, 2009).

The 3β-hydroxy-Δ5-steroid dehydrogenase/isomerase belongs to a family of enzymes that catalyse the conversion of 3β-hydroxy-5-ene-steroids to 3-oxo-4-ene-steroids through an oxidation reaction and an isomerization, using NAD or NADP as cofactors. Gene database analysis revealed the existence of a protein superfamily that in addition to the bacterial cholesterol dehydrogenases includes mammalian 3β-hydroxysteroid dehydrogenases, plant dihydroflavonol reductases, bacterial UDP-galactose-4-epimerases and viral 3β-hydroxysteroid dehydrogenase (Baker and Blasco, 1992). All of them present a glycine-rich sequence Gly-X-X-Gly-X-X-Gly, similar to the Rossman fold sequence Gly-X-Gly-X-X-Gly for AMP binding and an atypical active site motif Tyr-X-X-X-Lys. It is worth to mention that whereas the viral 3β-hydroxysteroid dehydrogenases are genetically close to the mammalian enzymes, the bacterial dehydrogenases are more distantly related to the mammalian enzymes (Yang *et al.*, 2007). The NAD(P)-dependent dehydrogenase from *Nocardia* subsp. (NAD(P)-CDH), which belongs to this family of proteins, has been well characterized and appears to be responsible for the transformation of cholesterol to cholest-4-en-3-one in this cholesterol degrader (Horinouchi *et al.*, 1991). Recent genetic and biochemical studies have demonstrated the direct involvement of the 3β-hydroxysteroid dehydrogenase encoded by the *Rv1106c* gene from *Mycobacterium tuberculosis*, in cholesterol catabolism (Yang *et al.*, 2007).

Cholesterol side-chain degradation

Most probably, simultaneously but independently of the first oxidative process, the long aliphatic chain of cholesterol is removed via a process similar to the β-oxidation of fatty acids, leading to the formation of a steroid C22-oic acid intermediate (Fig. 2.2) with the concomitant release of several molecules of propionyl-CoA and acetyl-CoA. Two recent studies have pointed out that the cytochrome Cyp125 from *Rhodococcus jostii* RHA1 (*ro04679* coding gene) or from *M. tuberculosis* (*Rv3545c* coding gene) initiates the side-chain degradation (Capyk *et al.*, 2009; Rosłoniec *et al.*, 2009). This cytochrome belongs to the P450 family and catalyses the hydroxylation at carbon C-26 yielding 26-hydroxycholest-4-en-3-one (Fig. 2.2). In a more recent study, cytochrome Cyp125 from *M. tuberculosis* CDC1551 was proved to be involved in the hydroxylation at C-27 and in the further oxidation to cholest-4-en-3-one-27-oic acid (Ouellet *et al.*, 2010).

Central and lower cholesterol degradation pathways

Once the aliphatic chain has been oxidized, cholesterol catabolism follows a common pathway described for C19 steroids. Briefly, 3-ketosteroids-Δ1-dehydrogenases, as KsdD in *Mycobacterium smegmatis* (Brzostek *et al.*, 2005), KstD in *R. erythropolis* (van der Geize *et al.*, 2000, 2001, 2002a) and *M. tuberculosis* (Knol *et al.*, 2008) or TesH in *Comamonas testosteroni* (Horinouchi *et al.*, 2003b) transform the androst-4-ene-3,17-dione (AD) (Fig. 2.2) in androsta-1,4-diene-3,17-dione (ADD) (Fig. 2.2), that could be considered as the central metabolite in the degradation of many steroids. Later on, a 9α-hydroxylation takes place catalysed by KstH in *M. smegmatis* (Andor *et al.*, 2006) and KshAB in *R. erythropolis* (van der Geize *et al.*, 2001, 2002b, 2008) and *M. tuberculosis* (Capyk *et al.*, 2009), followed by a non-enzymatic transformation of the 9α-hydroxyandrosta-1,4-diene-3,17-dione (Fig. 2.2) to 3-hydroxy-9,10-secoandrosta-1,3,5(10)-trien-9,17-dione (3-HSA) (Fig. 2.2). The subsequent hydroxylation of 3-HSA by enzymes such as the two-component oxygenase TesA1A2 in *C. testosteroni* (Horinouchi *et al.*, 2004a) leads to the production of 3,4-dihydroxy-9,10-secoandrosta-1,3,5(10)-trien-9,17-dione (3,4-DHSA) (Fig. 2.2), a catechol derivative that can be attacked by a *meta* extradiol dioxygenase [e.g. TesB in *C. testosteroni* (Horinouchi *et al.*, 2001), HsaC in *R. jostii* RHA1 (van der Geize *et al.*, 2007) and HsaC in *M. tuberculosis* (Yam *et al.*, 2009)] yielding 4,5,9,10-*diseco*-3-hydroxy-5,9,17-trioxoandrosta-1(10),2-diene-4-oic acid (4,9-DSHA) (Fig. 2.2). This compound is then hydrolysed by TesD in *C. testosteroni* (Horinouchi *et al.*, 2003a) and HsaD in *M. tuberculosis* (Lack *et al.*, 2008) or *R. jostii* RHA1 (van der Geize *et al.*, 2007) yielding acids 2-hydroxyhexa-2,4-dienoic (Fig. 2.2) and 9,17-dioxo-1,2,3,4,10,19-hexanorandrostan-5-oic

(DOHNAA) (Fig. 2.2). The following steps, which most probably involve the *tesE*, *tesF*, and *tesG* catabolic genes in *C. testosteroni* (Horinouchi *et al.*, 2005), lead to metabolites that enter in the central metabolic pathways (Kieslich, 1985). However, the enzymes involved in this downwards pathway have not been yet investigated. Probably, the 2-hydroxyhexa-2,4-dienoic acid is metabolized to 4-hydroxy-2-oxo-hexanoic acid (this step would be catalysed by TesE) (Fig. 2.2) that eventually will break into pyruvic acid and propionaldehyde (probably by the action of TesG), that is transformed into propionyl-CoA by the action of TesF. Moreover, it has been proposed that the DOHNAA would be further transformed to succinic acid and it is assumed that the first step in the degradation of this compound could be the elimination of a propionyl moiety, to produce 9,17-dioxo-1,2,3,4,5,6,10,19-octanorandrostan-7-oic acid (Fig. 2.2) through a typical β-oxidation (Kieslich, 1985).

It is worth to mention that cholesterol metabolism might differ in the various actinobacteria tested since they do not share an obligatory order for side chain and ring structure modifications. In this sense, while the Δ*cyp125* mutants of BCG (*Bacillus* Calmette-Guerin, an inactivated form of *M. bovis*) and *R. rhodochrous* RG32 transformed cholesterol to cholest-4-en-3-one, the corresponding mutant of *R. jostii* RHA1 did not (Capyk *et al.*, 2009). Furthermore, steroid ring degradation cannot proceed with substrates having a full-length aliphatic side chain in the BCG mutant since it is not able to grow or further transform cholest-4-en-3-one, while *R. jostii* RHA1 mutant grows in this compound. It was suggested that growth is supported by catabolism of the steroid rings but still remains unclear whether this effect could be due by the different specificities of the ring degrading enzymes in the two genera or the presence of multiple steroid degradation clusters in rhodococci (Capyk *et al.*, 2009).

Anoxic metabolism of cholesterol

The best known transformation of steroids in anoxic habitats occurs during the enterohepatic circulation in mammals by intestinal anaerobic bacteria. Cleavage of alkylaryl ether linkages, dehydroxylations, and oxidation or reduction at C-17 were found, but a breakdown of the steroid nucleus by these bacteria to gain energy was not described (Groh, 1993). Taylor *et al.* (1981) showed for the first time that bacteria were capable of degrading cholesterol under denitrifying conditions. The most frequent alteration of cholesterol in anoxic habitats is a biogenic reduction of its double bond yielding coprostanol (Li *et al.*, 1995; Harder and Probian, 1997). Many intestinal fermenting bacteria are able to catalyse this reaction (Groh *et al.*, 1993; Freier *et al.*, 1994), and the reductase responsible for this reaction has been assayed in resting cells of *Eubacterium coprostanoligenes* (Li et al., 1995).

A denitrifying β-protobacterium, strain 72Chol, was the first bacterium isolated and cultured on cholesterol under strictly anoxic conditions able to complete degrade cholesterol to carbon dioxide (Harder and Probian, 1997). At the biochemical level the anoxic metabolism of cholesterol has been mainly studied in another denitrifying β-protobacterium, *Sterolibacterium denitrificans*, now used as a model organism (Tarlera and Denner, 2003). The first steps of the anoxic metabolism of cholesterol are similar to those described for the aerobic catabolism (Chiang *et al.*, 2008). Firstly the hydroxyl group at C-3 is oxidized into the keto group leading cholest-5-en-3-one followed by a subsequent isomerization to yield cholest-4-en-3-one. These two reactions are catalysed by the bifunctional enzyme AcmA that belongs to the short-chain dehydrogenase reductase (SDR) superfamily (Chiang *et al.*, 2008). The enzyme contains a highly conserved NAD(P)-binding motif (G-X-G-(X)2-G(X)10-G), where X is any amino acid) at its N-terminus. The second step, oxidation of cholest-4-en-3-one to cholesta-1,4-diene-3-one, is catalysed by AcmB, a cholest-4-en-3-one-Δ1-dehydrogenase. The enzyme is very similar to the flavin adenine dinucleotide-dependent 3-ketosteroid-Δ1-dehydrogenase from *Nocardia coralline* (Itagaki *et al.*, 1990). In contrast with the aerobic pathway where cholest-4-en-3-one is hydroxylated by a monooxygenase at C-26 methyl group to produce the corresponding primary alcohol, the anaerobic hydroxylation of the cholesterol side chain is oxygen-independent. This step is catalysed by an enzyme belonging to the molybdenum-containing hydroxylase family and involves the hydroxylation of C-25 of cholest-4-en-3-one or cholesta-1,4-diene-3-one (Hille, 2005).

Bacterial catabolism of other steroid compounds

As commented above, natural steroid compounds derived from plants, animals or humans have been delivered into the environment over thousands of years. Nevertheless, the release to the ecosystem of non-natural steroid compounds produced by chemical industry is increasing, especially because of the growing population and intensive farming. These compounds together with natural steroid hormones are secreted in urine and mainly enter the environment by waste water treatment plant (WWTP) effluents. The most important steroids are E2 and E1 estrogens, testosterone, and the synthetic EE2 compound. These substances consist of four carbon rings what makes them stable in the environment. Nevertheless, several bacteria have acquired pathways to make use of these steroid compounds as the sole source of carbon and energy both under aerobic and anaerobic circumstances. Under aerobic conditions, steroids can be degraded by different bacterial species (Kieslich, 1985; Fujii *et al.*, 2002, 2003; Yoshimoto *et al.*, 2004). In such conditions the breakup of C–C bonds and the oxidation of quaternary carbon atoms are dependent on oxygen (Fahrbach, 2006).

Testosterone catabolism

The catabolic pathway of testosterone in *C. testosteroni*, formerly *Pseudomonas testosteroni* (Talalay *et al.*, 1952; Tamaoka *et al.*, 1987) was studied in great detail since the 1960s (Sih, *et al.*, 1962, 1965; Sih and Whitlock, 1968), but the first whole degradation pathway of testosterone was presented many years later by Kieslich (1985). This degradative pathway, which has been named as *9,10-seco* pathway by Fahrbach (2006), has been mainly studied in the *C. testosteroni* TA441 strain by Horinouchi *et al.* (2001, 2003a,b, 2004a,b, 2008).

Testosterone is initially oxidized to AD (Fig. 2.2) by the 3β,17β-hydroxysteroid dehydrogenase (3β17β HSDH) that is an inducible enzyme (Cabrera *et al.*, 2000). This enzyme catalyses the reversible oxidation of both 3- and 17-hydroxy groups of some hydroxysteroids (Benach *et al.*, 2002). *C. testosteroni* also has the enzyme HsdA (3α-hydroxysteroid dehydrogenase/carbonyl reductase, 3α-HSD/CR) that is functional towards testosterone and a variety of steroid substrates, including the steroid antibiotic fusidic acid (Maser *et al.*, 2001). This enzyme also catalyses the carbonyl reduction of non-steroidal aldehydes and ketones, such as the novel insecticide NKI42255, and it has been suggested to contribute to important defence strategies of *C. testosteroni* against natural and synthetic toxic compounds (Maser *et al.*, 2001).

Upon induction with testosterone, *C. testosteroni* synthesizes three different dehydrogenases able to promote the desaturation of ring A of 3-ketosteroids: one Δ(1)-dehydrogenase and two Δ(4)-dehydrogenases acting, respectively, on compounds with A:B ring fusion in *trans* (Δ4(5α)-dehydrogenase) or in *cis* (Δ4(5β)-dehydrogenase) (Florin *et al.*, 1996). The Δ(1)-dehydrogenase catalyses the transdiaxal removal of hydrogens from positions 1 and 2, while Δ(4)-dehydrogenases promote the elimination of hydrogens located at positions 4 and 5.

The Δ(1)-dehydrogenase is used to transform AD into ADD (Fig. 2.2). After the 9α-hydroxylation of the C-9 atom, the degradation proceeds by the breakup of the C–C bond between C-9 and C-10, which leads to ring B cleavage rendering 3-HSA (Fig. 2.2), the metabolite that gave the name to the pathway. The phenolic ring A is then opened by *meta*-cleavage yielding 4,9-DSHA (Fig. 2.2). This compound is hydrolysed to 2-hydroxyhexa-2,4-dienoic acid (Fig. 2.2) and DOHNAA (Fig. 2.2). Only the keto form of 2-hydroxyhexa-2,4-dienoic acid is further metabolized to 4-hydroxy-2-oxo-hexanoic acid (Fig. 2.2), which is finally cleaved to propionaldehyde and pyruvic acid. The compound DOHNAA is thought to be further degraded to succinic acid (Schubert *et al.*, 1969). Nevertheless these last steps have to be further investigated (Horinouchi *et al.*, 2005; Fahrbach, 2006). Particularly, the quaternary carbons atom in the product of step 11 as well as the following tertiary alcohol pose a problem for their further degradation (Hylemon and Harder, 1999).

Finally, it is also worth to mention that C. testsoteroni has a steroid isomerase (isomerase; 3-oxosteroid Δ5-Δ4-isomerase) that is able to catalyse the isomerization of the double bond of androst-5-ene-3,17-dione into conjugation with the carbonyl group (Kuliopulos *et al.*, 1987).

Horinouchi *et al.* (2001, 2003a,b, 2004a,b, 2006, 2010) have demonstrated that steroid degradation genes of *C. testosteroni* TA441 are encoded in several gene clusters. One cluster contains the *meta*-cleavage enzyme encoding gene (*tesB*) and *orf1-2-3*, whereas a second cluster consists of *orf18*, *orf17*, *orf11*, *orf12*, *tesIHA2* and *tesA1DEFG* (*tesA2* to *orf18* and *tesA1* to *tesG* are placed in opposite directions). In addition, an analysis of transposon mutants that show low steroid degradation ability revealed that the *orf4-5-21-22-23-25-26-27-28-30-31-32-33* and the *tesR* gene, which are located downstream of *orf3*, are involved in DOHNAA degradation. TesB is a *meta*-cleavage enzyme which is induced in strain TA441 during growth on testosterone. TesA1 and A2 are necessary for hydroxylation of 3-HSA to 3,4-DHSA. TesH and TesI are, respectively, the Δ(1)- and Δ(4)(5α)-dehydrogenases of the 3-ketosteroid. *orf17* probably encodes the reductase component of the 9α-hydroxylase (TesH) for ADD. ORF18 was shown to be necessary to transform DOHNAA. TesD is the hydrolase that acts the product of *meta*-cleavage reaction, and TesEFG are probably the hydratase, aldolase, and dehydrogenase required for the degradation of 2-hydroxyhexa-2,4-dienoic acid, one of the products generated by TesD.

The *hsdA* gene encoding the 3α-HSD/CR (258 aa residues) (Maser *et al.*, 2001), the gene encoding the 3β17β HSDH (254 aa residues) (Abalain *et al.*, 1993), and the gene encoding the Δ5–3-ketosteroid isomerase (125 aa residues) (Kuliopolus *et al.*, 1987) have been cloned and expressed. However, the gene encoding the Δ(4)(5β)-dehydrogenase of *C. testosteroni* remains to be identified. As the complete genome of *C. testosteroni* strain CNB-2 has been recently described (Ma *et al.*, 2009), the new information derived from these studies will pave the way to better understand the metabolism of testosterone in this microorganism.

Catabolism of bile salts

In the intestinal tract, bacteria survive in the presence of bile (Begley *et al.*, 2005) and transform the primary bile acids cholate and chenodeoxycholate into the secondary bile acids deoxycholate and lithocholate, respectively, via reductive dehydroxylation (Hylemon and Harder, 1999; Ridlon *et al.*, 2006). When bile acids are released into the environment, they can be degraded completely by numerous Gram-positive and Gram-negative bacteria (Hayakawa, 1982), but very few degradation pathways have been studied at genetic level (Leppik, 1989).

C. testosteroni TA441 is able to degrade cholic acid and other bile salts (Horinuchi *et al.*, 2008). In the DNA region upstream of the *tesB* gene, two adjacent genes (*steAB*) likely to be co-transcribed and required for cholic acid degradation were identified. The SteA dehydrogenase converts 7α,12α-dihydroxyandrosta-1,4-diene-3,17-dione to 7α-hydroxyandrosta-1,4-diene-3,12,17-trione, and the SteB hydrogenase converts the latter to 7α,12β-dihydroxyandrosta-1,4-diene-3,17-dione. Inversion of the hydroxyl group at C-12 is indispensable for the subsequent effective B–ring cleavage of the androstane compound. In addition to the compounds already mentioned, 12α-hydroxyandrosta-1,4,6-triene-3,17-dione and 12β-hydroxyandrosta-1,4,6-triene-3,17-dione were identified as minor intermediate compounds in cholic acid degradation.

Recently, a *Pseudomonas* sp. strain Chol1, which is able to grow aerobically and anaerobically (denitrifying) with cholate and other bile salts as the sole source of carbon and energy, was isolated (Philipp *et al.*, 2006). Under aerobic conditions, strain Chol1 grows by transforming cholate to 7,12-dihydroxy-1,4-androstadiene-3,17-dione (DHADD). The conversion of cholate into DHADD proceeds via A-ring oxidation and degradation of the acyl side chain (Birkenmaier *et al.*, 2007). DHADD is further transformed to 3,7,12-trihydroxy-9,10-*seco*-1,3,5(10)-androstatriene-9,17-dione (THSATD) by an oxygenase-dependent reaction. The *seco*steroid THSATD is then degraded to CO_2. The degradation pathway in strain Chol1 thus follows the unifying scheme proposed by Hayakawa (1982).

Catabolism of 17β-estradiol (E2) and estrone (E1)

Bacteria that use the 9,10-*seco* pathway are unable to metabolize the E2 or E1 estrogens, since the presence of the C-19 carbon atom is a prerequisite for the ring B cleavage. Thus, when 19-nor-AD is incubated with *C. testosteroni*, estrone accumulates in the medium (Levy and Talalay, 1959).

The degradation pathways of estrogens are not thoroughly described yet. While a lot of studies were made to assess the degradation rate of estrogens, hardly any experiments to define the catabolic pathway and intermediate products have been reported. It is, hence, not yet clear how E2 is degraded. It is accepted that the first step of the degradation of E2 is the oxidation of the C-17 to a keto-group with the formation of E1 (2.1). This step is suggested to be a rather fast step as it consists of a simple oxidation which can be done by ubiquitous enzymes. The further mineralization and the factors favouring the degradation are still controversial. The degradation of E1 must involve a ring cleavage reaction that is supposed to be a slow enzymatic step.

Many years ago, Coombe *et al.* (1966) proposed a degradative pathway by incubating *Nocardia* (isolated from soil) in a mineral salt medium with E1 as sole carbon source. The most important step of this pathway was the cleavage of ring A by a dioxygenase with the formation of product I (Fig. 2.3). Although product I was not detected, this compound was proposed as intermediate because the pyridine carboxylic acid (product II) which was detected can only be synthetized via formation of product I. Further degradation to product IV via product III (these intermediates were also detected) is also consistent with the formation of product I. However, degradation from product I to IV, and further catabolism from IV to CO_2 have not been determined precisely. It was supposed that ring B is the second one to be cleaved. The fact that ring A is the first one to be cleaved is supported by the results of Jürgens *et al.* (2002) and Layton *et al.* (2000). Jürgens *et al.* (2002) found that $^{14}CO_2$ was formed but at slower rate than the oxidation from ^{14}C-labelled E2 to E1. In contrast, Lee and Liu (2002) have proposed a totally different degradation pathway for E2 by investigating

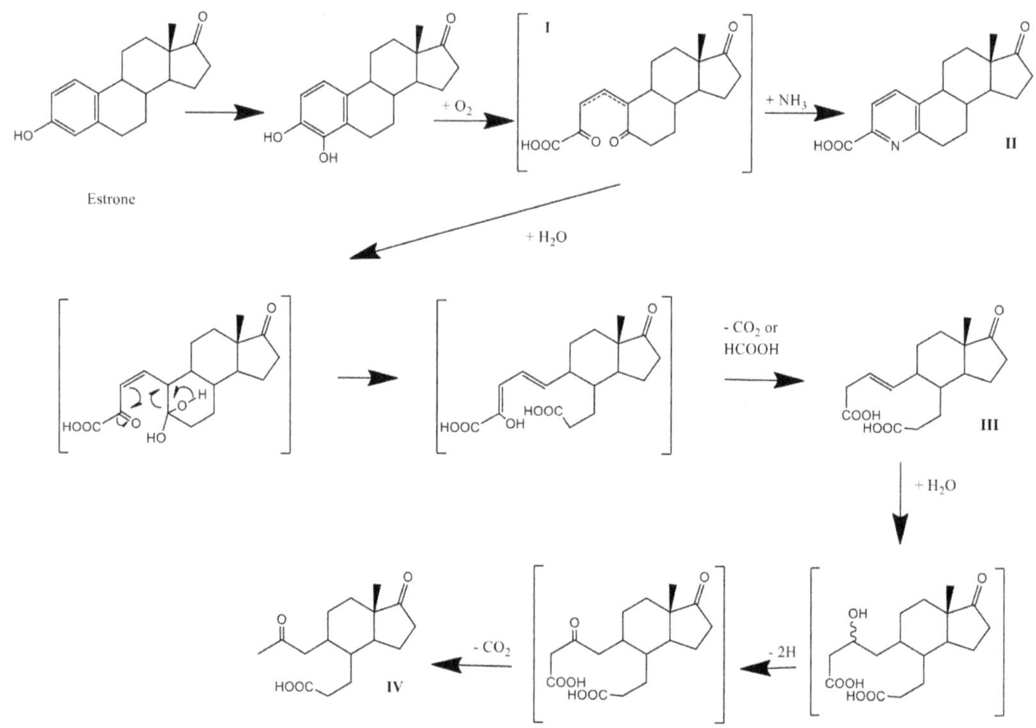

Figure 2.3 Proposed degradation pathway of estrone (Coombe *et al.*, 1966).

the supernatant of activated sludge growing on mineral salts and on high E2 concentrations. Thus, the E2 degradation pathway is not yet clear and both mechanisms may work, perhaps depending on the environmental conditions and the bacterial strains.

Catabolism of 17α-ethinylestradiol (EE2)

A very peculiar degradation pathway for EE2 was proposed by Ren *et al.* (2007). Firstly, it was suggested that EE2 was converted to E1. However, this is very unlikely because the ethinyl group at C-17 makes this position very recalcitrant to microbial attack. Secondly, a pathway similar to the one of testosterone with the initial cleavage of ring B was suggested for the further degradation of E1. This cleavage is, however, not possible if the C-10 is not methylated as it occurs in testosterone (Fahrbach, 2006). In addition, Coombe *et al.* (1966) already quoted a paper from Sih and Wang (1963) that found out that the organisms endowed with the *seco*-pathway were not able to metabolize E1. This means that bacteria that degrade EE2 to E1 in a first step (if they really exist) should not be able to degrade E1 in this way.

Yi and Harper (2007) were interested in the pathway of the co-metabolic transformation of EE2 of an enriched nitrifying community from a WWTP. They suggested that the molecule is probably attacked at the aromatic ring A where the electron density is highest. It was found by NMR that the formation of a dihydroxy-EE2 (Arcand-Hoy *et al.*, 1998) or the cleavage of ring A could be the first step of the EE2 degradation. In contrast to the ring A cleavage of E1 (Coombe *et al.*, 1966), ring A of EE2 was assumed to be cleaved between C-1 and C-2 as well as between C-3 and C-4 to form 3-ethynyl-3a,6,7-trimethyl-2,3,3a,4,5,5a,8,9,9a,9b-decahydro-1*H*-cyclopenta[a]naphthalene-3-ol (ETDC). However, these findings are not certain as former studies showed that the ammonium monooxygenase (i.e. the enzyme which is expected to be responsible for this cleavage) is inhibited by acetylene which also contains a C-C triple bond like EE2. This could also indicate that the cleavage starts at ring A because this is the opposite site of ring D (which contains the ethinyl group) which is transformed. In addition, the relatively small acetylene molecule cannot really be compared with a large steroid hormone and it was not clear at which concentration acetylene inhibited the enzyme.

On the other hand, Vader *et al.* (2000) claimed that the degradation of EE2 leads to hydrophilic substances which can be assumed to be hydroxylated derivatives. It was shown that such substances have a much lower estrogenic activity than the starting product. In the degradation pathway of E2 it was never argued that ring A could be most vulnerable owing to its highest electron density. This could be a new approach for an unknown degradation pathway or a further argument that the ring A cleavage postulated by Coombe *et al.* (1966) could be the most likely pathway.

Anaerobic catabolism

Anoxic environments arise where organic matter is degraded through microbial activity, and oxygen has only limited access, for example, during the denitrification step in wastewater treatment or in sediments of lakes and rivers. Steroids that occur in such environments have in common with aliphatic hydrocarbons and monoaromatic compounds a chemical inertness that makes them recalcitrant substrates for bacteria (Fahrbach, 2006).

Very few data are available about the biochemical reaction principles in steroid metabolism carried out in the absence of molecular oxygen. Quaternary carbon atoms are hardly degradable without the co-substrate oxygen and detailed knowledge about the involved mechanisms is still lacking (Hylemon and Harder, 1999; Kniemeyer *et al.*, 1999; Probian *et al.*, 2003). Besides energy gaining, co-metabolic pathways could also play a role in the anaerobic degradation of steroids (Fahrbach, 2006).

The mineralization of E2 has been reported in the denitrifying step of WWTP and in anoxic river sediments, but the responsible bacteria and further oxidation products of E1 are still unknown (Andersen *et al.*, 2003; Joss *et al.*, 2004). In fact, the loss of E2 was only accompanied by a corresponding accumulation of E1 (Jürgens *et al.*, 2002). Czajka and Londry (2006) also reported an oxidation of E2 to E1 but no further degradation under different redox conditions (even with nitrate as the electron acceptor), and suggested an accumulation of estrogens in anoxic environments. An inhibitory or toxic effect of the reductant sodium sulfide on denitrifying bacteria in the mineral medium used by Czajka and Londry (2006) could not be ruled out.

Accumulation of androgens such as AD most probably caused by microbial conversion of phytosteroids, was observed in anoxic river sediments that receive effluent from paper mills (Jenkins *et al.*, 2003, 2004). Unfortunately, no information about the mineralization of estrogens and androgens under anoxic conditions is available in the literature.

A Gram-negative denitrifying β-proteobacterium, *Denitratisoma oestradiolicum*, was isolated from activated sludge of a municipal wastewater treatment plant using E2 as sole source of carbon and energy (Fahrbach *et al.*, 2006). More recently, a steroidal hormone-degrading γ-proteobacterium, *S. denitrificans*, was isolated from anoxic digested sludge on E2 or testosterone as the sole source of carbon and energy with nitrate as the electron acceptor that represents the first known bacterium to grow anaerobically on both E2 (C18) and testosterone (C19) (Fahrbach *et al.*, 2008). Under anoxic conditions, *S. denitrificans* oxidizes testosterone to 1-dehydrotestosterone, which is then transformed to ADD. In general, the initial steps of anoxic testosterone degradation by *S. denitrificans* are similar to those of the oxic pathway demonstrated in *C. testosteroni* (Chiang *et al.*, 2010).

The steroid uptake system

Many nutrient compounds are not able to diffuse to the bacterial cytoplasm and need to be transported by specific uptake systems before been metabolized. This appears to be also the case for steroids where a few specific transport systems have been characterized. A pioneering work was carried by Watanabe and Po (1974), who analysed an inducible transport system for testosterone in membrane vesicles prepared from cultures of *P. testosteroni*. More recently, genome sequence analyses revealed that diverse actinobacteria, including members of the *Mycobacterium*, *Nocardia*, *Rhodococcus*, and *Streptomyces* genera, harbour *mce* transporter operons involved in steroid uptake. The *mce* operon contains 9–13 genes that encode two transmembrane proteins homologous to the permease subunits of ABC transporters. *M. tuberculosis* contains four *mce* homologous operons (Casali and Riley, 2007) that encode ATP-driven transport systems. A common ATPase, MceG, belonging to the Mlk family, provides the energy required for substrate translocation (Joshi *et al.*, 2006). The *mceG* gene is located far away from the *mce* operons (Mohn *et al.*, 2008). The Mce4 transporter represents the major cholesterol import system of *M. tuberculosis* because mutants lacking either the *mce4* operon or the *mceG* gene suffer a severe restriction for cholesterol dependent-growth (Pandey and Sassetti, 2008). The *mce4 locus* also encodes a system for steroids uptake in *R. jostii* RHA1 (Mohn *et al.*, 2008). Previous findings showed that *mce4 locus* of *R. jostii* RHA1 is up-regulated when growing in cholesterol, and mutants with deletions in some of the genes belonging to this operon lost their ability to grow in this compound (van der Geize *et al.*, 2007). Up to now, the *mce4 locus* is associated with steroid metabolism genes in all bacteria whose genome sequence is available. On the other hand, this uptake system is also essential for growth of *R. jostii* RHA1 in cholesterol related compounds such as 5α-cholestanol, 5α-cholestanone, and β-sitosterol (Mohn *et al.*, 2008). Nevertheless, mutants of the *mce4* uptake system were unaffected for growth on steroid compounds having shorter polar side chains (Mohn *et al.*, 2008). The available data indicate that only side chains of at least eight carbons are recognized by Mce4 steroid uptake system. Finally, the reason why the Mce system requires so many more proteins than do classical ABC transporters remains unclear. In this sense, it has been suggested that Mce4A-F proteins might form a complex that would mediate movement of the substrate across cell walls (Mohn *et al.*, 2008).

A gene involved in the transport of bile acids has been identified and characterized in Eubacterium sp. strain VPI 12708 (Mallonee and Hylemon, 1996). This gene, named *baiG*, encodes a polypeptide showing a significant homology to several membrane transport proteins and forms part of an operon involved in the bile acid 7α-dehydroxylation pathway. Interestingly, when this gene was cloned into *Escherichia coli* it conferred the host an energy-dependent bile acid uptake activity.

Regulation of the steroid catabolism

From the very beginning it was recognized that the expression of several genes/enzymes involved in steroid degradation is induced by their respective steroid substrates (Marcus and Talalay, 1956; Möbus *et al.*, 1997; Skowasch *et al.*, 2002). Studies on the mechanisms regulating the testosterone-inducible gene expression revealed that regulatory proteins were probably involved in microbial steroid catabolism (Thomas *et al.*, 1989; Cabrera *et al.*, 1997; Xiong and Maser, 2001; Horinouchi *et al.*, 2001, 2004b; Xiong *et al.*, 2003a,b; Pruneda-Paz *et al.*, 2004).

The first molecular determinant identified in prokaryotic steroid signalling was the transcriptional repressor RepA of *C. testosteroni* which acts as a negative regulator by binding to the upstream region of the *hsdA* gene (Xiong and Maser, 2001). In fact, two genes involved in HsdA regulation were identified in the vicinity of *hsdA* and reported to act as a two repressor system that controls *hsdA* gene expression. RepA was found to bind to two operator sequences upstream of *hsdA*, thereby preventing *hsdA* transcription (Xiong and Maser, 2001). The expression of *hsdA* is closely connected to the distance between two 10 bp operator elements (OP1 and OP2) to which the RepA protein binds (Xiong *et al.*, 2003a). Repressor B (RepB) turned out to bind to the mRNA of 3α-HSD/CR, thereby interfering with *hsdA* translation (Xiong *et al.*, 2003b). Testosterone was shown to specifically bind to RepB, thereby causing the release of RepB from the *hsdA* mRNA and allowing *hsdA* translation.

On the other hand, the regulator TesR and its homologue TeiR were shown to be necessary for the induction of steroid degradation in *C. testosteroni* TA441 (Horinuchi *et al.*, 2004b), and *C. testosteroni* ATCC11996 (Pruneda-Paz *et al.*, 2004), respectively. TeiR has 391 amino acids and shows homology between residues 327 and 380 to the helix–turn–helix DNA binding domain of LuxR and between residues 192 and 227 to the PAS sensor domain (Pruneda-Paz *et al.*, 2004). This domain structure resembles that of TraR, a specific regulator involved in quorum sensing in *Agrobacterium tumefaciens* (Vannini *et al.*, 2002).

TeiR is tightly controlled at the transcriptional level by the presence of testosterone in the culture medium. A *teiR*-disrupted mutant strain was completely unable to use testosterone as the sole carbon and energy source (Pruneda-Paz *et al.*, 2004). In addition, the expression of several steroid-inducible genes was abolished in this mutant suggesting that TeiR positively regulates the transcription of genes involved in steroid degradation in *C. testosteroni* (Pruneda-Paz *et al.*, 2004). More recently, it has been demonstrated that TeiR is a membrane protein with asymmetrical localization at one of the cell poles of *C. testosteroni* that consists of three different functional domains, an N-terminal domain important for membrane association, a central steroid binding domain, and a C-terminal domain mediating TeiR function (Göhler *et al.*, 2008). TeiR works as a kinase in the steroid signalling chain and responds to a variety of different steroids other than testosterone. Knock-out mutants of the *teiR* gene revealed that TeiR provides swimming and twitching motility of *C. testosteroni* to the steroid substrate source (Göhler *et al.*, 2008). TeiR regulates steroid catabolic genes interacting with their promoters and this interaction requires the presence of a testosterone-derived metabolite to induce the system (Linares *et al.*, 2008).

Very few data are available upon the regulation of genes involved in bile acid degradation. The gene responsible for bile acid 7-dehydroxylation in *Eubacterium* sp. strain VPI 12708 appears to be inducible by this compound and contains at its 3′ terminus a large element with dyad symmetry of unknown function (Coleman *et al.*, 1988).

Recently, Kendall *et al.* (2007) identified a repressor, named KstR, belonging to the TetR family of transcriptional regulators and involved in the regulation of 83 genes in *M. smegmatis* related with cholesterol and lipid metabolism. The operator motif of the KstR dimer is a 14 bp conserved sequence (TnnAACnnGTTnnA, where n is any nucleotide) located in the upstream region of the KstR-regulated genes. In fact the binding of purified KstR protein to its operator was demonstrated using its own promoter in *M. tuberculosis* as DNA probe. The repression mediated by KstR is alleviated by the binding of an inducer molecule that still remains unknown. Using the microarray data from *M. smegmatis*, a KstR regulon was also deduced for the human pathogen *M. tuberculosis* due the high degree of sequence similarity. In this bacterium, KstR is involved in the transcriptional repression of 74 genes. The KstR regulon in *M. tuberculosis* contains a large number of genes that have been found to be essential for the survival of this pathogen in both macrophages and mice (Schnappinger *et al.*, 2003).

More recently, another transcriptional repressor has been described, KstR2, which controls the expression of other additional 15 genes related to cholesterol metabolism in *M. tuberculosis* and *M. smegmatis*, thus indicating that this catabolic pathway appears to be controlled by at least two regulatory proteins (Kendall *et al.*, 2010).

Finally, it is worth noting that the inducible expression of the gene encoding the 3/17β-hydroxysteroid dehydrogenase enzyme of *C. testosteroni* was shown to be dependent on the presence of other carbon sources in the growth medium, revealing the implication of a catabolite repression control (Cabrera *et al.*, 2000).

Pathogenic implications of cholesterol metabolism

Cholesterol has recently been identified as an important lipid for mycobacterial infection by various mycobacteria (de Chastellier and Thilo, 2006). The cholesterol present in the host cell, which is relatively abundant, is an important growth substrate for the bacterium in the different stages during the infection process like intracellular growth or intracellular persistence. In this sense, the human pathogen *M. tuberculosis* can degrade cholesterol as a sole carbon and energy source (Pandey and Sassetti, 2008). Experiments performed with ^{14}C-radiolabeled cholesterol showed that carbon from the sterol ring system was preferentially converted to CO_2 to be used for energy generation via tricarboxilic acid (TCA) cycle. In contrast, the carbon coming from the side-chain is not converted into CO_2 but is assimilated into the mycobacterial lipids, most probably because C-26 is removed as a propionyl-CoA which is efficiently assimilated into branched chain lipids (Cox *et al.*, 1999).

It is stated that *M. tuberculosis* growing in the host obtains its energy from host lipids rather than carbohydrates. In this sense, the *M. tuberculosis* genome contains more than 250 genes involved in fatty acid metabolism, compared with an estimated number of 50 genes in *Escherichia coli* (Dubnau *et al.*, 2005). Current data indicate that numerous genes involved in lipid metabolism are up-regulated during growth in the mammalian macrophage and the mouse lung (Dubnau *et al.*, 2005).

Some genes that are critical for the survival of *M. tuberculosis* in the macrophages are involved in cholesterol degradation (Rengarajan *et al.*, 2005; van der Geize *et al.*, 2007). The Mce4 transport system, that is required for cholesterol import into bacterial cells, plays an important role in pathogenesis (Pandey and Sassetti, 2008). A mutant lacking the Mce4 transporter fails to persist in the lungs of chronically infected mice and cannot grow in γ-interferon (IFN-γ)-activated macrophages (Pandey and Sassetti, 2008). Furthermore, the ability of Mce4 proteins to bind cholesterol could act as a signal allowing the pathogen to recognize the host by modifying the host membrane (Mohn *et al.*, 2008).

The *igr* locus, which has been identified because of its importance for intracellular growth of *M. tuberculosis* in macrophages and mice, is absolutely required for cholesterol metabolism, since *igr*-deficient bacteria cannot grow using cholesterol as a primary carbon source (Chang *et al.*, 2007). This locus consists of a single operon containing genes for a putative cytochrome P450 (*igrA*), two acyl coenzyme A dehydrogenases (*igrBC*), a conserved hypothetical protein (*igrD*), a putative enoyl coenzyme A hydratase (*igrE*) and a lipid carrier protein (*igrF*) (Chang *et al.*, 2007). A homologous operon is present in *R. jostii* RHA1 and its expression is shown to be induced during growth in cholesterol (van der Geize *et al.*, 2007). The *igr* cluster is encompassing the *mce4* genes, which encode the uptake system for cholesterol and related steroids (Mohn *et al.*, 2008).

The *igrA* gene is homologous to the *Rv3545* from *M. tuberculosis* and *ro04679* from *R. jostii* RHA1. These two genes encode the Cyp125 P450-like cytochrome Cyp125, a steroid 26-hydroxylase involved in the first reaction for steroid side-chain degradation (Capyk *et al.*, 2009; Rosloniec *et al.*, 2009).

It has been demonstrated that a *M. tuberculosis* strain lacking the *kshA* and *kshB* genes, that encode a two-component system catalysing the NADH dependent 9α-hydroxylation of 3-ketosteroids (Petrusma *et al.*, 2009), failed to use cholesterol and 4-adrostene-3,17-dione and its growth and survival is attenuated in macrophages and mice (Hu *et al.*, 2009).

ChoE cholesterol oxidase from *Rhodococcus equi*, a primary pathogen of horses and an opportunistic pathogen of humans, has been described as the major membrane-damaging factor (Navas *et al.*, 2001), although other authors do not consider it crucial for pathogenesis (Pei *et al.*, 2006). ChoD cholesterol oxidase from *M. tuberculosis* is important for the growth of the pathogen in peritoneal macrophages and lungs of mice (Brzostek *et al.*, 2007). The complete genome sequence of *M. leprae*, an obligate intracellular pathogen, shows a dramatic reduction of functional genes with a coding capacity less than 50%. Despite this massive reduction of genes, the *Bacillus* of leprosy has managed to preserve a minimum set of genes, most of them shared with *M. tuberculosis*, which allow its survival in the host resulting in the subsequent pathological manifestations. One of such conserved genes whose expression has been shown *in vivo* in *M. leprae* is the putative orthologue of *choD* (Marques *et al.*, 2008). Although *M. leprae* does not appear to contain the metabolic machinery necessary to fully catabolize cholesterol, it is proposed that cholesterol is a necessary factor for growth in macrophages (Kato, 1978).

Future prospects

Despite of the large number of works carried out during the last century, the bacterial catabolism of cholesterol and other steroids is still far for being completely understood. For instance, the enzymes involved in the degradation of the steroid side chain as well as those responsible for the last steps of the catabolic pathway remain to be precisely established. Moreover, the physiological role played by bacterial cholesterol oxidases is still unclear. Very few data are available about the proteins responsible for the bacterial uptake of steroids and this field will require supplementary attention since the possibility to create by metabolic engineering heterologous recombinant bacteria able to transform steroids for biotechnological purposes will depend of the possibility of conferring these bacteria-specific steroid transport capacities. As stated above, only a limited number of regulatory proteins involved in the metabolism of steroids have been identified and partially characterized. In fact the real inducers of these regulators remain to be elucidated since it has not been determined yet whether the steroids or their degradative intermediates are their effectors. Taking into account that the availability of pathway intermediates is restricted the identification of such putative inducers appears to be a difficult task. The knowledge on the bacterial degradation of steroids in anoxic environments is even more limited and considering that some steroid pollutants can be accumulated in such niches for long periods of time to opening this area will require further efforts. While powerful technological approaches based on modern *omic* technologies (genomics, metagenomics, transcriptomics, etc.) will be useful to solve many of the questions raised here, the complexity of the pathways and their intermediates still demand to carry out huge efforts by using conventional biochemical and genetic tools to be able to decipher all the steps and processes involved in steroid catabolism.

Acknowledgements

We thank Dr E. Díaz and Dr M.A. Prieto for the critical reading of the manuscript and helpful discussions. This work was supported by grants from the Ministry of Science and Innovation (BFU2006-15214-C03–01, BFU2009-11545-C03–03), an I3P pre-doctoral fellowship from the Consejo Superior de Investigaciones Científicas and a FPU predoctoral fellowship from the Spanish Ministry of Science and Innovation.

References

Abalain, J.H., Di Stefano, S., Amet, Y., Quemener, E., Abalain-Colloc, M.L., and Floch, H.H. (1993). Cloning, DNA sequencing and expression of (3–17) β hydroxysteroid dehydrogenase from *Pseudomonas testosteroni*. J. Steroid. Biochem. Mol. Biol. *44*, 133–139.

Andersen, H.R., Siegrist, H., Halling-Sørensen, B., and Ternes, T. (2003). Fate of estrogens in a municipal sewage treatment plant. Environ. Sci. Technol. *37*, 4021–4026.

Andor, A., Jekkel, A., Hopwood, D.A., Jeanplong, F., Ilkoy, E., Konya, A., Kurucz, I., and Ambrus, G. (2006). Generation of useful insertionally blocked sterol degradation pathway mutants of fast-growing Mycobacteria and cloning, characterization, and expression of the terminal oxygenase of the 3-ketosteroid 9α-hydroxylase in *Mycobacterium smegmatis* mc2155. Appl. Environ. Microbiol. *72*, 6554–6559.

Arima, K., Nagasawa, M., Bae, M., and Tamura, G. (1969). Microbial transformation of sterols. Part I. Decomposition of cholesterol by microorganisms. Agric. Biol. Chem. *33*, 1636–1643.

Arcand-Hoy, L.D., Nimrod A.C., and Benson, W.H. (1998). Endocrine-modulating substances in the environment estrogenic effects of pharmaceutical products. Int. J. Toxicol. *17*, 139–158.

Baker, M.E., and Blasco, R. (1992). Expansion of the mammalian 3β-hydroxysteroid dehydrogenase/plant dihydroflavonol reductase superfamily to include a bacterial cholesterol dehydrogenase, a bacterial UDP-galactose-4-epimerase, and open reading frames in vaccinia virus and fish lymphocystis disease virus. FEBS Lett. *301*, 89–93.

Begley, M., Gahan, C.G., and Hill, C. (2005). The interaction between bacteria and bile. FEMS Microbiol. Rev. *29*, 625–651.

Benach, J., Filling, C., Oppermann, U.C., Roversi, P., Bricogne, G., Berndt, K.D., Jörnvall, H., and Ladenstein, R. (2002). Structure of bacterial 3β/17β-hydroxysteroid dehydrogenase at 1.2 A resolution: a model for multiple steroid recognition. Biochemistry *41*, 14659–14668.

Birkenmaier, A., Holert, J., Erdbrink, H., Moeller, H.M., Friemel, A., Schoenenberger, R., Suter, M.J., Klebensberger, J., and Philipp, B. (2007). Biochemical and genetic investigation of initial reactions in aerobic degradation of the bile acid cholate in *Pseudomonas* sp. strain Chol1. J. Bacteriol. *189*, 7165–7173.

Björkhem, I., and Eggertsen, G. (2001). Genes involved in initial steps of bile acid synthesis. Curr. Opin. Lipidol. *12*, 97–103.

Brian, J.V., Harris, C.A., Scholze, M., Backhaus, T., Booy, P., Lamoree, M., Pojana, G., Jonkers, N., Runnalls, T., Bonfa, A., *et al.* (2005). Accurate prediction of the response of freshwater fish to a mixture of estrogenic chemicals. Environ. Health Perspec. *113*, 721–728.

Brown, R.L., and Peterson, G.E. (1966). Cholesterol oxidation by soil Actinomycetes. J. Gen. Microbiol. *45*, 441–450.

Brzostek, A., Sliwiński, T., Rumijowska-Galewicz, A., Korycka-Machała, M., and Dziadek, J. (2005). Identification and targeted disruption of the gene encoding the main 3-ketosteroid dehydrogenase in *Mycobacterium smegmatis*. Microbiology *151*, 2393–402.

Brzostek, A., Dziadek, B., Rumijowska-Galewicz, A., Pawelczyk, J., and Dziadek, J. (2007). Cholesterol oxidase is required for virulence of *Mycobacterium tuberculosis*. FEMS Microbiol. Lett. *275*, 106–112.

Cabrera, J.E, Panzetta-Dutari, G., Pruneda Paz, J.L., and Genti-Raimondi, S. (1997). A new *Comamonas testosteroni* steroid-inducible gene: cloning and sequence analysis. J. Steroid. Biochem. Mol. Biol. *63*, 91–98.

Cabrera, J.E., Pruneda Paz, J.L., Genti-Raimondi, S. (2000). Steroid-inducible transcription of the 3β/17β-hydroxysteroid dehydrogenase gene (3β/17β-hsd) in *Comamonas testosteroni*. J. Steroid. Biochem. Mol. Biol. *73*, 147–52.

Capyk, J.K., Kalscheuer, R., Stewart, G.R., Liu, J., Kwon, H., Zhao, R., Okamoto, S., Jacobs, W.R., Jr., Eltis L.D., and Mohn, W.W. (2009). Mycobacterial cytochrome P450 125 (Cyp125) catalyzes the terminal hydroxylation of C27-steroids. J. Biol. Chem. *284*, 35534–35542.

Casali, N., and Riley, L.W. (2007). A phylogenomic analysis of the Actinomycetales *mce* operons. BCM Genomics *8*, 60.

Chang, J.C., Harik, N.S., Liao, R.P., and Sherman, D.R. (2007). Identification of mycobacterial genes that alter growth and pathology in macrophages and in mice. J. Infect. Dis. *196*, 788–795.

de Chastellier, C., and Thilo, L. (2006), Cholesterol depletion in *Mycobacterium avium*-infected macrophages overcomes the block in phagosome maturation and leads to the reversible sequestration of viable mycobacteria in phagolysosome-derived autophagic vacuoles. Cell. Microbiol. *8*, 242–256.

Chiang, Y., Ismail, W., Heintz, D., Schaeffer, C., van Dorsselaer, A., and Fuchs, G. (2008). Study of anoxic and oxic cholesterol metabolis by *Steloribacterium denitrificans*. J. Bacteriol. *190*, 905–914.

Chiang, Y.R, Fang, J.Y., Ismail, W., and Wang, P.H. (2010). Initial steps in anoxic testosterone degradation by *Steroidobacter denitrificans*. Microbiology (DOI 10.1099/mic.0.037788–0).

Chipley, J.R., Dreyfuss, M.S., and Smucker, R.A. (1975). Cholesterol metabolism by *Mycobacterium*. Microbios *12*, 199–207.

Coleman, J. P., White, W.B., Lijewski, M., and Hylemon, P. B. (1988). Nucleotide sequence and regulation of a gene involved in bile acid 7-dehydroxylation by *Eubacterium* sp. strain VPI 12708. J. Bacteriol. *175*, 2070–2077.

Coombe, R.G., Tsong, Y.Y., Hamilton, P.B., and Sih, C.J. (1966). Mechanisms of steroid oxidation by microorganisms. Oxidative cleavage of estrone. J. Biol. Chem. *241*, 1587–1595.

Coulombe, R., Yue, K.Q., Ghisla, S., and Vrielink, A. (2001). Oxygen access to the active site of cholesterol oxidase through a narrow channel is gated by an Arg-Glu pair. J. Biol. Chem. *276*, 30435–30441.

Cox, J.S., Chen, B., McNeil, M., and Jacobs, W.R. Jr. (1999). Complex lipid determines tissue-specific replication of *Mycobacterium tuberculosis* in mice. Nature. *402*, 79–83.

Czajka, C.P., and Londry, K.L. (2006). Anaerobic biotransformation of estrogens. Sci. Total. Environ. *367*, 932–941.

Daughton, C.G., and Ternes, T.A. (1999). Pharmaceuticals and personal care products in the environment: Agents of subtle change? Environ. Health. Perspect. *107*, 907–938.

Doukyu, N. (2009). Characteristics and biotechnological applications of microbial cholesterol oxidases. Appl. Microbiol. Biotechnol. *83*, 825–837.

Doukyu, N., and Aono, R. (1999). Two moles of O2 consumption and one mole of H2O2 formation during cholesterol peroxidation with cholesterol oxidase from *Pseudomonas* sp. strain ST-200. Biochem. J. *341*, 621–627.

Doukyu, N., Shibata, K., Ogino, H., and Sagermann, M. (2008). Purification and characterization of *Chromobacterium* sp. DS-1 cholesterol oxidase with thermal, organic solvent, and detergent tolerance. Appl. Microbiol. Biotechnol. *80*, 59–70.

Dubnau, E., Chan, J., Mohan, V.P., and Smith, I. (2005). Responses of *Mycobacterium tuberculosis* to growth in the mouse lung. Infect. Immun. *73*, 3754–3757.

Durhan, E.J., Lambright, C.S., Makynen, E.A., Lazorchak, J., Hartig, P.C., Wilson, V.S., Gray, L. E., and Ankley, G.T. (2006). Identification of metabolites of trenbolone acetate in androgenic runoff from a beef feedlot. Environ. Health Perspect. *114*, 65–68.

Fahrbach, M. (2006). Anaerobic degradation of steroid hormones by novel denitrifying bacteria. PhD thesis, RWTH Aachen University, Germany.

Fahrbach, M., Kuever, J., Meinke, R., Kämpfer, P., and Hollender, J. (2006). *Denitratisoma oestradiolicum* gen. nov., sp. nov., a 17β-oestradiol-degrading, denitrifying betaproteobacterium. Int. J. Syst. Evol. Microbiol. *56*, 1547–1552.

Fahrbach, M., Kuever, J., Remesch, M., Huber, B.E., Kämpfer, P., Dott, W., and Hollender, J. (2008). *Steroidobacter denitrificans* gen. nov., sp. nov., a steroidal hormone-degrading gammaproteobacterium. Int. J. Syst. Evol. Microbiol. *58*, 2215–2223.

Ferreira, N.P., and Tracey, R.P. (1984). Numerical taxonomy of cholesterol-degrading soil bacteria. J. Appl. Microbiol. *57*, 429–446.

Florin, C., Kholer, T., Grandguillot, M., and Plesiat, P. (1996). *Comamonas testosteroni* 3-ketosteroid-Δ4(5α)-dehydrogenase: gene and protein characterization. J. Bacteriol. *178*, 3322–3330

Freier, T.A., Beitz, D.C., Li, L., and Hartman, P.A. (1994). Characterization of *Eubacterium coprostanoligenes* sp. nov., a cholesterol-reducing anaerobe. Int. J. Syst. Bacteriol. *44*, 137–142.

Fujii, K., Kikuchi, S., Satomi, M., Ushio-Sata, N., and Morita, N. (2002). Degradation of 17β-estradiol by a Gram-negative bacterium isolated from activated sludge in a sewage treatment plant in Tokyo, Japan. Appl. Environ. Microbiol. *68*, 2057–2060.

Fujii, K., Satomi, M., Morita, N., Motomura, T., Tanaka, T., and Kikuchi, S. (2003). *Novosphingobium tardaugens* sp. nov., an oestradiol-degrading bacterium isolated from activated sludge of a sewage treatment plant in Tokyo. Int. J. Syst. Environ. Microbiol. *53*, 47–52.

Gagné, F., Blaise, C., and André, C. (2006). Occurrence of pharmaceutical products in a municipal effluent and toxicity to rainbow trout (*Oncorhynchus mykiss*) hepatocytes. Ecotox. Environ. Saf. *64*, 329–336.

van der Geize, R., Hessels, G.I., van Gerwen, R., Vrijbloed, J.W., van der Meijden, P., and Dijkhuizen, L. (2000). Targeted disruption of the kstD gene encoding a 3-ketosteroid Δ1-dehydrogenase isoenzyme of *Rhodococcus erythropolis* strain SQ1. Appl. Environ. Microbiol. *66*, 2029–2036.

van der Geize, R., Hessels, G.I., van Gerwen, R., van der Meijden, P., and Dijkhuizen, L. (2001). Unmarked gene deletion mutagenesis of kstD, encoding 3-ketosteroid Δ1-dehydrogenase, in *Rhodococcus erythropolis* SQ1 using *sacB* as counter-selectable marker. FEMS. Microbiol. Lett. *205*, 197–202.

van der Geize, R., Hessels, G.I., and Dijkhuizen, L. (2002a). Molecular and functional characterization of the *kstD2* gene of *Rhodococcus erythropolis* SQ1 encoding a second 3-ketosteroid Δ1-dehydrogenase isoenzyme. Microbiology *148*, 3285–3292.

van der Geize, R., Hessels, G.I., van Gerwen, R., van der Meijden, P., and Dijkhuizen, L. (2002b). Molecular and functional characterization of *kshA* and *kshB*, encoding two components of 3-ketosteroid 9α-hydroxylase, a class IA monooxygenase, in *Rhodococcus erythropolis* strain SQ1. Mol. Microbiol. *45*, 1007–1018.

van der Geize, R., Yam, K., Heuser, T., Wilbrink, M.H., Hara, H., Anderton, M.C., Sim, E., Dijkhuizen, L., Davies, J.E., Mohn, W.W., *et al.* (2007). A gene cluster encoding cholesterol catabolism in a soil actinomycete provides insight into *Mycobacterium tuberculosis* survival in macrophages. Proc. Natl. Acad. Sci. U.S.A. *104*, 1947–1952.

van der Geize, R., Hessels, G.I., Nienhuis-Kuiper, M., and Dijkhuizen, L. (2008). Characterization of a second *Rhodococcus erythropolis* SQ1 3-ketosteroid 9α-hydroxylase activity comprising a terminal oxygenase homologue, KshA2, active with oxygenase-reductase component KshB. Appl. Environ. Microbiol. *74*, 7197–7203.

Ghoshroy, K.B., Zhu, W., and Sampson, N.S. (1997). Investigation of membrane disruption in the reaction catalyzed by cholesterol oxidase. Biochemistry *36*, 6133–6140.

Göhler, A., Xiong, G., Paulsen, S., Trentmann, G., and Maser, E. (2008). Testosterone-inducible regulator is a kinase that drives steroid sensing and metabolism in *Comamonas testosteroni*. J. Biol. Chem. *283*, 17380–17390.

Groh, H., Schade, K., and Hörhold-Schubert, C. (1993). Steroid metabolism with intestinal microorganisms. J. Basic Microbiol. *33*, 59–72.

Hanselman, T.A., Graetz, D.A., Wilkie, A.C. (2003). Manure-borne estrogens as potential environmental contaminants: a review. Environ. Sci. Technol. *37*, 5471–5478.

Harder, J., and Probian, C. (1997). Anaerobic mineralization of cholesterol by a novel type of denitrifying bacterium. Arch. Microbiol. *167*, 269–274.

Hayakawa, S. (1982). Microbial transformation of bile acids. A unified scheme for bile acid degradation, and hydroxylation of bile acids. Z. Allg. Mikrobiol. *22*, 309–326.

Hayami, M., Okabe, A., Sacia, K., Hayashi, H., and Kanemasa, Y. (1979) Presence and synthesis of cholesterol in stable staphylococcal L-forms. J. Bacteriol. *140*, 859–863.

Hille, R. (2005). Molybdenum-containing hydroxylases. Arch. Biochem. Biophys, *433*, 107–116.

Horinouchi, S., Ishizuka, H., and Beppu, T. (1991) Cloning, nucleotide sequence, and transcriptional analysis of the NAD(P)-dependent cholesterol dehydrogenase gene from a *Nocardia* sp. and its hyperexpression in *Streptomyces* spp. Appl Environ Microbiol 57, 1386–1393.

Horinouchi, M., Yamamoto, T., Taguchi, H., Arai, H., and Kudo, T. (2001). *Meta*-cleavage enzyme gene tesB is neccessary for testosterone degradation in *Comamonas testosteroni* TA441. Microbiology *147*, 3367–3375.

Horinouchi, M., Hayashi, T., Koshino, H., Yamamoto, T., and Kudo, T. (2003a). Gene encoding the hydrolase for the product of the *meta*-cleavage reaction in testosterone degradation by *Comamonas testosteroni*. Appl. Environ. Microbiol. *69*, 2139–2152.

Horinouchi, M., Hayashi, T., Yamamoto, T., and Kudo, T. (2003b). A new bacterial steroid degradation gene cluster in *Comamonas testosteroni* TA441 which consists of aromatic-compound degradation genes for seco-steroids and 3-ketosteroid dehydrogenase genes. Appl. Environ. Microbiol. *69*, 4421–4430.

Horinouchi, M., Hayashi, T., and Kudo, T. (2004a). The genes encoding the hydroxylase of 3-hydroxy-9,10-secoandrosta-1,3,5 -triene-9,17-dione in steroid degradation in *Comamonas testosteroni* TA441. J. Steroid. Biochem. Mol. Biol. *92*, 143–154.

Horinouchi, M., Kurita, T., Yamamoto, T., Hatori, E., Hayashi, T., and Kudo, T. (2004b). Steroid degradation gene cluster of *Comamonas testosteroni* consisting of 18 putative genes from *meta*-cleavage enzyme gene tesB to regulator gene tesR. Biochem. Biophys. Res. Commun. *324*, 597–604.

Horinouchi, M., Hayashi, T., Koshino, H., Kurita, T., and Kudo, T. (2005). Identification of 9,17-dioxo-1,2,3,4,10,19-hexanorandrostan-5-oic acid, 4-hydroxy-2-oxohexanoic acid, and 2-hydroxyhexa-2,4-dienoic acid and related enzymes involved in testosterone degradation in *Comamonas testosteroni* TA441. Appl. Environ. Microbiol. *71*, 5275–5281.

Horinouchi, M., Hayashi, T., Koshino, H., and Kudo, T. (2006). ORF18-disrupted mutant of *Comamonas testosteroni* TA441 accumulates significant amounts of 9,17-dioxo-1,2,3,4,10,19-hexanorandrostan-5-oic acid and its derivatives after incubation with steroids. J. Steroid Biochem. Mol. Biol. *101*, 78–84.

Horinouchi, M., Hayashi, T., Koshino, H., Malon, M., Yamamoto, T., and Kudo, T. (2008). Identification of genes involved in inversion of stereochemistry of a C-12 hydroxyl group in the catabolism of cholic acid by *Comamonas testosteroni* TA441. J. Bacteriol. *190*, 5545–5554.

Horinouchi, M., Kurita, T., Hayashi, T., and Kudo, T. (2010). Steroid degradation genes in *Comamonas testosteroni* TA441: Isolation of genes encoding a Δ4 -isomerase and 3α- and 3β-dehydrogenases and evidence for a 100kb steroid degradation gene hot spot. J. Steroid. Biochem. Mol. Biol. [Epub ahead of print.]

Horvath, J., and Kramli, A. (1947). Microbiological oxidation of cholesterol with *Azotobacter*. Nature *160*, 639.

Hotchkiss, A.K., Rider, C.V., Blystone, C.R., Wilson, V.S., Hartig, P.C., Ankley, G.T., Foster, P.M., Gray, C.L., and Gray, L.E. (2008). Fifteen years after 'wingspread' environmental endocrine disrupters and human and wildlife health: where we are today and where we need to go. Toxicol. Sci. *105*, 235–259.

Hu, Y., van der Geize, R., Besra, G.S., Gurcha, S.S., Liu, A., Rohde, M., Singh, M., and Coates, A. (2009). 3-ketosteroid 9α-hydroxylase is an essential factor in the pathogenesis of *Mycobacterium tuberculosis*. Mol. Microbiol. 75, 107–121.

Hutchinson, T.H., Ankley, G.T., Segner, H., and Tyler, C.R. (2006). Screening and testing for endocrine disruption in fish-biomarkers as "signposts," not "traffic lights," in risk assessment. Environ. Health Perspect. *114*, 106–114.

Hylemon, P.B., and Harder, J. (1999). Biotransformation of monoterpenes, bile acids, and other isoprenoids in anaerobic ecosystems. FEMS Microbiol. Rev. *22*, 475–488.

Itagaki, E., Wakabayashi, T., and Hatta, H. (1990). Purification and characterization of 3-ketosteroid-Δ1-dehydrogenase from *Nocardia coralline*. Biochim. Biophys. Acta *1038*, 60–67.

Jenkins, R.L., Wilson, E.M., Angus, R.A., Howell, W.M., and Kirk, M. (2003). Androstenedione and progesterone in the sediment of a river receiving paper mill effluent. Toxicol. Sci. *73*, 53–59.

Jenkins, R.L., Wilson, E.M., Angus, R.A., Howell, W.M., Kirk, M., Moore, R., Nance, M., and Brown, A. (2004). Production of androgens by microbial transformation of progesterone *in vitro*: A model for androgen production in rivers receiving paper mill effluent. Environ. Health Perspect. *112*, 1508–1511.

Joshi, S.M., Pandey, A.K., Capite, N., Fortune, S.M., Rubin, E.J., and Sassetti, C.M. (2006). Characterization of mycobacterial virulence genes through genetic interaction mapping. Proc. Natl. Acad. Sci. U.S.A. *103*, 11760–11765.

Joss, A., Andersen, H., Ternes T., Richle, P.R., and Siegrist, H. (2004). Removal of estrogens in municipal wastewater treatment under aerobic and anaerobic conditions: consequences for plant optimization. Environ. Sci. Technol. *38,* 3047–3055.

Jürgens, M.D., Holthaus, K.I.E., Johnson, A.C., Smith, J.J.L., Hetheridge, M., and Williams, R.J. (2002). The potential for estradiol and ethinylestradiol degradation in English rivers. Environ. Toxicol. Chem. *21,* 480–488.

Kato, L. (1978). Cholesterol, a factor which is required for growth of mycobacteria from leprous tissues. Int. J. Lepr. Other. Mycobact. Dis. *46,* 133–143.

Kendall, S. L., Withers, M., Soffair, C.N., Moreland, N.J., Gurcha, S., Sidders, B., Frita, R., Ten Bokum, A., Besra, G.S., Lott, J.S., *et al.* (2007) A highly conserved transcriptional repressor controls a large regulon involved in lipid degradation in *Mycobacterium smegmatis* and *Mycobacterium tuberculosis.* Mol. Microbiol. *65,* 684–99.

Kendall, S.L, Burgess, P., Balhana, R., Withers, M., Ten Bokum, A., Lott, J.S., Gao, C., Uhia, C.I., and Stoker, N.G. (2010). Cholesterol utilisation in mycobacteria is controlled by two TetR-type transcriptional regulators; *kstR* and *kstR2.* Microbiology *156,* 1362–1371.

Kidd, K.A., Blanchfield, P.J., Mills, K.H., Palace, V.P., Evans, R.E., Lazorchak, J.M., and Flick, R.W. (2007). Collapse of a fish population after exposure to a synthetic estrogen. Proc. Natl. Acad. Sci. U.S.A. *104,* 8897–8901.

Kieslich, K. (1985). Microbial side-chain degradation of sterols. J. Basic Microbiol. *7,* 461–474.

Kniemeyer, O., Probian, C., Rosselló-Mora, A., and Harder, J. (1999). Anaerobic mineralization of quaternary carbon atoms: Isolation of denitrifying bacteria on dimethylmalonate. Appl. Environ. Microbiol. *65,* 3319–3324.

Knol, J., Bodewits, K., Hessels, G., Dijkhuizen, L., and van der Geize, R. (2008). 3-Keto-5α-steroid 1-dehydrogenase from *Rhodococcus* erythropolys SQ1 and its orthologue in *Mycobacterium tuberculosis* H37Rv are highly specific enzymes that function in cholesterol catabolism. Biochem. J. *410,* 339–346.

Kolpin, D.W., Furlong, E.T., Meyer, M.T., Thurman, E.M., Zaugg, S.D., Barber, L.B., and Buxton, H.T. (2002). Pharmaceuticals, hormones, and other organic wastewater contaminants in U.S. streams, 1999–2000: A national reconnaissance. Environ. Sci. Technol. *36,* 1202–1211.

Kreit, J., and Sampson, N.S. (2009). Cholesterol oxidase: physiological functions. FEBS J. *276,* 6844–6856.

Kuliopulos, A., Shortle, D., and Talalay, P. (1987). Isolation and sequencing of the gene encoding delta 5-3-ketosteroid isomerase of *Pseudomonas testosteroni*: overexpression of the protein. Proc. Natl. Acad. Sci. U.S.A. *84,* 8893–8897.

Lack, N., Lowe, E.D., Liu, J., Eltis, L.D., Noble, M E. M., Sima, E., and Westwooda, I.M. (2008). Structure of HsaD, a steroid-degrading hydrolase, from *Mycobacterium tuberculosis.* Acta Cryst. *64,* 2–7.

Lario, P.I., Sampson, N., and Vrielink, A. (2003). Sub-atomic resolution crystal structure of cholesterol oxidase: What atomic resolution crystallography reveals about enzyme mechanism and the role of the FAD cofactor in redox activity. J. Mol. Biol. *326,* 1635–1650.

Layton, A.C., Gregory, B.W., Seward, J.R., Schultz, T.W., and Sayler, G.S. (2000). Mineralization of steroidal hormones by biosolids in wastewater treatment systems in Tennessee U.S.A. Environ. Sci. Technol. *34,* 3925–3931.

Lee, H.B., and Liu, D. (2002). Degradation of 17 β-estradiol and its metabolites by sewage bacteria. Water Air Soil Pollution *134,* 353–368.

Leppik, R.A. (1989). The genetics of bile acid degradation in *Pseudomonas* spp.: location and cloning of catabolic genes. J. Gen. Microbiol. *135,* 1989–1996.

Levy, R.H., and Talalay, P. (1959). Bacterial oxidation of steroids. I. Ring A dehydrogenations by intact cells. J. Biol. Chem. *234,* 2009–2013.

Li, J., Vrielink, A., Brick, P., and Blow, D.M. (1993). Crystal structure of cholesterol oxidase complexed with a steroid substrate: Implications for flavin adenine dinucleotide dependent alcohol oxidases. Biochemistry *32,* 11507–11515.

Li, L., Freier, A.T., Hartman, P.A., Young, J.W., and Beitz, D.C. (1995). A resting-cell assay for cholesterol reductase activity in *Eubacterium coprostanoligenes* ATCC 51222. Appl. Microbiol. Biotechnol. *43,* 887–892.

Linares, M., Pruneda-Paz, J.L., Reyna, L., and Genti-Raimondi, S. (2008). Regulation of testosterone degradation in *Comamonas testosteroni.* J. Steroid. Biochem. Mol. Biol. *112,* 145–150.

Ma, Y.F., Zhang, Y., Zhang, J.Y., Chen, D.W., Zhu, Y., Zheng, H., Wang, S.Y., Jiang, C.Y., Zhao, G.P., and Liu, S.J. (2009). The complete genome of *Comamonas testosteroni* reveals its genetic adaptations to changing environments. Appl. Environ. Microbiol. *75,* 6812–6819.

Mallonee, D.H., and Hylemon, P. B. (1996). Sequencing and expression of a gene encoding a bile acid transporter from *Eubacterium* sp. strain VPI 12708. J. Bacteriol. *178,* 7053–7058.

Marcus, P.I., and Talalay, P. (1956). Induction and purification of α- and β-hydroxysteroid dehydrogenases. J. Biol. Chem. *218,* 661–674.

Marques, M.A., Neves-Ferreira, A.G., da Silveira, E.K., Valente, R.H., Chapeaurouge, A., Perales, J., da Silva Bernardes, R., Dobos, K.M., Spencer, J.S., Brennan, P.J, *et al.* (2008). Deciphering the proteomic profile of *Mycobacterium leprae* cell envelope. Proteomics *8,* 2477–2491.

Martin, C.K.A. (1977). Microbial cleavage of sterol side chains. Adv. Appl. Microbiol. *22,* 29–58.

Martín, J.F., and Aparicio, J.F. (2009). Enzymology of the polyenes pimaricin and candicidin biosynthesis. Methods Enzymol. *459,* 215–242

Maser, E., Xiong, G., Grimm, C., Ficner, R., and Reuter, K. (2001). 3α-hydroxysteroid dehydrogenase/carbonyl reductase from *Comamonas testosteroni*: biological significance, three-dimensional structure and gene regulation. Chem. Biol. Interact. 130–132, 707–722.

Miller, W.L. (1988). Molecular biology of steroid hormone synthesis. Endocr. Rev. *9,* 295–318.

Möbus, E., Jahn, M., Schmid, R., Jahn, D., and Maser, E. (1997). Testosterone regulated expression of enzymes involved in steroid and hydrocarbon catabolism in *Comamonas testosteroni*. J. Bacteriol. *179*, 5951–5955.

Mohn, W.W., van der Geize, R., Stewart, G.R., Okamoto, S., Liu, J., Dijkhuizen, L., and Eltis, L.D. (2008). The actinobacterial *mce4 locus* encodes a steroid transporter. J. Biol. Chem. *283*, 35368–35371.

Moss, G.P. (1989) Nomenclature of steroids. Pure. Appl. Chem. *61*, 1783–1822.

Motteran, L., Pilone, M.P., Molla, G., Ghisla, S., and Pollegioni, L. (2001). Cholesterol oxidase from *Brevibacterium sterolicum*. J. Biol. Chem. *276*, 18024–18030.

Nagasawa, M., Bae, M., Tamura, G., and Arima, K. (1969). Microbial transformation of steroids. Part II. Cleavage of sterols side chains by microorganisms. Agric. Biol. Chem. *33*, 1644–1650.

Navas, J., Gonzalez-Zorn, B., Ladron, N., Garrido, P., and Vazquez-Boland, J.A. (2001). Identification and mutagenesis by allelic exchange of choE, encoding a cholesterol oxidase from the intracellular pathogen *Rhodococcus equi*. J. Bacteriol. *183*, 4796–4805.

Ouellet, H., Guan, S., Johnston, J.B., Chow, E.D., Kells, P.M., Burlingame, A.L., Cox, J.S., Podust, L.M., and de Montellano, P.R. (2010). *Mycobacterium tuberculosis* CYP125A1, a steroid C27 monooxygenase that detoxifies intracellularly generated cholest-4-en-3-one. Mol. Microbiol. 77, 730–742.

Owen, R.W., Mason, A.N., and Bilton, R.F. (1983). The degradation of cholesterol by *Pseudomonas* sp. NCIB 10590 under aerobic conditions. J. Lipid. Res. *24*, 1500–1511.

Pandey, A.K., and Sassetti, C.M. (2008). Mycobacterial persistence requires the utilization of host cholesterol. Proc. Natl. Acad. Sci. U.S.A. 105, 4376–4380.

Payne, J., Rajapakse, N., Wilkins, M., and Kortenkamp, A. (2000). Prediction and assessment of the effects of mixtures of four xenoestrogens. Environ. Health Perspect. *108*, 983–987.

Payne, J., Scholze, M., and Kortenkamp, A. (2001). Mixtures of four organochlorines enhance human breast cancer cell proliferation. Environ. Health Perspect. *109*, 391–397.

Pei, Y., Dupont, C., Sydor, T., Haas, A., and Prescott, J.F. (2006). Cholesterol oxidase (ChoE) is not important in the virulence of *Rhodococcus equi*. Vet. Microbiol. *118*, 240–246.

Petrusma, M., Dijkhuizen, L., and van der Geize, R. (2009) *Rhodococcus rhodochrous* DSM 43269 3-ketosteroid 9α-hydroxylase, a two component iron–sulfur-containning monoxigenase with a subtle steroid substrate specificity. Appl. Environ. Microbiol. 75, 5300–5307.

Philipp, B., H., Erdbrink, M.J., Suter, F., and Schink, B. (2006). Degradation of and sensitivity to cholate in *Pseudomonas* sp. strain Chol1. Arch. Microbiol. *185*, 192–201.

Pollegioni, L., Wels, G., Pilone, M.S., and Ghisla, S. (1999). Kinetic mechanisms of cholesterol oxidase from *Streptomyces hygroscopicus* and *Brevibacterium sterolicum*. Eur. J. Biochem. *264*, 140–51.

Poole, A.J., and Cord-Ruwisch, R. (2004). Treatment of strongflow wool scouring effluent by biological emulsion destabilisation. Water Res. *38*, 1419–1426.

Probian, C., Wülfing, A., and Harder, J. (2003). Anaerobic mineralization of quaternary carbon atoms: Isolation of denitrifying bacteria on pivalic acid (2,2-dimethylpropionic acid). Appl. Environ. Microbiol. *69*, 1866–1870.

Pruneda-Paz, J.L., Linares, M., Cabrera, J.E., and Genti-Raimondi, S. (2004). TeiR, a LuxR-type transcription factor required for testosterone degradation in *Comamonas testosteroni*. J. Bacteriol. *186*, 1430–1437.

Raman, D.R., Williams, E.L., Layton, A.C., Burns, R.T., Easter, J.P., Daugherty, A.S., Mullen, M.D., and Sayler, G.S. (2004). Estrogen content of dairy and swine wastes. Environ. Sci. Technol. *38*, 3567–3573.

Ren, Y.X., Nakano, K., Nomura, M., Chiba, N., and Nishimura, O. (2007). Effects of bacterial activity on estrogen removal in nitrifying activated sludge. Water. Res. *41*, 3089–3096.

Rengarajan, J., Bloom, B.R., and Rubin, E.J. (2005). Genome-wide requirements for *Mycobacterium tuberculosis* adaptation and survival in macrophages. Proc. Natl. Acad. Sci. U.S.A. *102*, 8327–8332.

Ridlon, J.M., Kang, D.J., and Hylemon, P.B. (2006). Bile salt biotransformations by human intestinal bacteria. J. Lipid Res. *47*, 241–259.

Rosłoniec, K.Z., Wilbrink, M., Capyk, J.K., Mohn, W.W., Ostendorf, M., van der Geize, R., Dijkhuizen, L., and Eltis, L.D. (2009). Cytochrome P450 125 (CYP125) catalyzes C26-hydroxylation to initiate sterol side chain degradation in *Rhodococcus jostii* RHA1. Mol. Microbiol. *74*, 1031–1043.

Routledge, E.J., Sheahan, D., Desbrow, C., Brighty, G.C., Waldock, M., and Sumpter, J.P. (1998). Identification of estrogenic chemicals in STW Effluent. 2. *In vivo* responses in trout and roach. Environ. Sci. Technol. 32, 1559–1565.

Schnappinger, D., Ehrt, S., Voskuil, M.I., Liu, Y., Mangan, J.A., Monahan, I.M., Dolganov, G., Efron, B., Butcher, P.D., Nathan, C., *et al.* (2003). Transcriptional adaptation of *Mycobacterium tuberculosis* within macrophages: insights into the phagosomal environment. J. Exp. Med. *198*, 693–704.

Schoemer, U., and Martin, C.K.A. (1980). Microbial transformation of sterols. Biotechnol. Bioeng. *22*, 11–25.

Schubert, K., Ritter, F., Sorkina, T., Böhme, K.H., and Hörhold, C. (1969). Abbau von Steroiden – VIII. Bildung von [1,4–14C] Bernsteinsäure aus [1,3α –14C] 7α-methyl-5,6,7,7 α-tetrahydroindan-1,5-dion-4-(3-propionsäure) durch *Nocardia opaca*. J. Steroid Biochem. *1*, 1–7.

Sedlaczek, L., and Smith, L.L. (1988). Biotransformations of steroids. Crit. Rev. Biotechnol. 7, 187–236.

Sih, C.J. (1962). Mechanism of steroid oxidation by microorganisms. Biochim. Biophys. Acta *62*, 541–547.

Sih, C.J., and Wang, K.C. (1963). Mechanisms of steroid oxidation by microorganisms. II. Isolation and characterization of 3aα-H-4α-[3′-propionic acid]-7aß methyl-hexahydro-1,5-indanedione. J. Am. Chem. Soc. 85, 2135–2137.

Sih, C.J., and Whitlock, H.W. (1968). Biochemistry of steroids. Annu. Rev. Biochem. *37*, 661–694.

Sih, C.J., Wang, K.C., Gibson, D.T., and Whitlock, H.W. (1965). On the mechanism of ring A clavage in the degradation of 9,10-seco steroids by microorganisms. J. Am. Chem. Soc. *87*, 1386–1387.

Skowasch, D., Möbus, E., and Maser, E. (2002). Identification of a novel *Comamonas testosteroni* gene encoding a steroid inducible extradiol dioxygenase. Biophys. Res. Commun. *294*, 560–566.

Silva, E., Rajapakse, N., and Kortenkamp, A. (2002). Something from 'nothing' – Eight weak estrogenic chemicals combined at concentrations below NOECs produce significant mixture effects. Environ. Sci. Technol. *36*, 1751–1756.

Slaytor, M., and Bloch, K. (1965). Metabolic transformation of cholestenediols. J. Biol. Chem. *240*, 4598–4602.

Söhngen, N.L. (1913). Benzin, petroleum, paraffinöl und paraffin als kohlenstoff- und energiequelle für mikroben. Zentr. Bacteriol. Parasitenk. Abt II. *27*, 595–609.

Szentirmai, A. (1990). Microbial physiology of sidechain degradation of sterols. J. Ind. Microbiol. Biotechnol. *6*, 101–115.

Tak, J.D. (1942). On bacteria decomposing cholesterol. Antonie van Leeuwenhoek J. Microbiol. Serol. *8*, 32–40.

Talalay, P., Dobson, M.M., and Tapley, D.F. (1952). Oxidative degradation of testosterone by adaptive enzymes. Nature *170*, 620–621.

Tamaoka, J., Duk-Mo, H., and Komagata, K. (1987). Reclassification of *Pseudomonas acidovorans* den Dooren de Jong 1926 and *Pseudomonas testosteroni* Marcus and Talalay 1956 as *Comamonas acidovorans* comb. nov. and *Comamonas testosterona* comb. nov., with an amended description of the genus *Comamonas*. Int. J. Syst. Bacteriol. *37*, 52–59.

Tarlera, S., and Denner, E.B.M. (2003). *Sterolibacterium denitrificans* gen. nov., sp. nov., a novel cholesterol-oxidizing, denitrifying member of the β-Proteobacteria. Int. J. Syst. Evol. Microbiol. 53, 1085–1091.

Taylor, C.D., Smith, S.O., and Gagosian, R.B. (1981). Use of microbial enrichments for the study of the anaerobic degradation of cholesterol. Geochim. Cosmochim. Acta *45*, 2161–2168.

Thomas, J.E., Caroll, R., Sy, L.P., and Watanabe, M. (1989). Isolation and characterization of a 50 kDa testosterone-binding protein from *Pseudomonas testosteroni*. J. Steroid. Biochem. 32, 27–34.

Turfitt, G.E. (1944). Microbiological agencies in the degradation of steroids: I. The cholesterol-decomposing organisms of soils. J. Bacteriol. *47*, 487–493.

Turfitt, G.E. (1948). The microbiological degradation of steroids 4. Fission of the steroid molecule. Biochem. J. *42*, 376–383.

Vader, J.S., van Ginkel, C.G., Sperling, F.M.G.M., de Jong, J., de Boer, W., de Graaf, J. S., van der Most, M., and Stokman, P.G.W. (2000). Degradation of ethinyl estradiol by nitrifying activated sludge. Chemosphere *41*, 1239–1243.

Vannini, A., Volpari, C., Gargioli, C., Muraglia, E., Cortese, R., De Francseco, R., Neddermann, P., and Di Marco, S. (2002). The crystal structure of the quorum sensing protein TraR bound to its autoinducer and target DNA. EMBO J. *21*, 4393–4401

Veiga, P., Juste, C., Lepercq, P., Saunier, K., Béguet, F., and Gérard, P. (2005). Correlation between faecal microbial community structure and cholesterol-to-coprostanol conversion in the human gut. FEMS Microbiol. Lett. *242*, 81–86.

Vrielink, A., Lloyd, L.F., and Blow, D.M. (1991). Crystal structure of cholesterol oxidase from *Brevibacterium sterolicum* refined at 1.8 Å resolution. J. Mol. Biol. *219*, 533–554.

Watanabe, M., and Po, L. (1974). Testosterone uptake by membrane vesicles of *Pseudomonas testosteroni*. Biochim. Biophys. Acta *345*, 419–429.

Whitmarsh, J.M. (1964). Intermediates of microbiological metabolism of cholesterol. Biochem. J. *90*, 23.

Xiong, G., and Maser, E. (2001). Regulation of the steroid-inducible 3α-hydroxysteroid dehydrogenase/carbonyl reductase gene in *Comamonas testosteroni*. J. Biol. Chem. *276*, 9961–9970.

Xiong, G., Markowetz, S., and Maser, E. (2003a). Regulation of 3α-hydroxysteroid dehydrogenase/carbonyl reductase in *Comamonas testosteroni*: function and relationship of two operators. Chem. Biol. Interact. 143–144, 411–423.

Xiong, G., Martin, H.J., and Maser, E. (2003b). Identification and characterization of a novel translational regulator of the steroid-inducible 3α-hydroxysteroid dehydrogenase/carbonyl reductase gene in *Comamonas testosteroni*. J. Biol. Chem. *278*, 47400–47407.

Yam, K.C., D'Angelo, I., Kalscheuer, R., Zhu, H., Wang, J.X., Snieckus, V., Ly, L.H., Converse, P.J., Jacobs, W.R., Jr., Strynadka, N., *et al.* (2009). Studies of a ring-cleaving dioxygenase illuminate the role of cholesterol metabolism in the pathogenesis of *Mycobacterium tuberculosis*. PLoS Pathog. *5*, e1000344.

Yang, X., Dubnau, E., Smith, I., and Sampson, N.S. (2007) Rv1106c from *Mycobacterium tuberculosis* is a 3β-hydroxysteroid dehydrogenase. Biochemistry *46*, 9058–9067.

Yi, T., and Harper, W.F., Jr. (2007). The effect of biomass characteristics on the partitioning and sorption hysteresis of 17α-ethinylestradiol. Water Res. *41*, 1543–1553.

Yoshimoto, T., Nagai, F., Fujimoto, J., Watanabe, K., Mizukoshi, H., Makino, T., Kimura, K., Saino, H., Sawada, H., and Omura, H. (2004). Degradation of estrogens by *Rhodococcus zopfii* and *Rhodococcus equi* isolates from activated sludge in wastewater treatment plants. Appl. Environ. Microbiol. *70*, 5283–5289.

Yue, Q.K., Kass, I.J., Sampson, N.S., and Vrielink, A. (1999). Crystal structure determination of cholesterol oxidase from *Streptomyces* and structural characterization of key active Site mutants. Biochemistry 38, 4277–4286.

Organophosphorus Compound-degrading Bacteria and Enzymes, and their Application

3

Katsumasa Abe, Shouji Takahashi and Yoshio Kera

Abstract

Organophosphorus compounds are widely used as pesticides, flame retardants and plasticizers. Owing to their large-scale use, they have been detected in various environments, including surface and ground water, soil and air in the world. Since many of these compounds are toxic for many organisms including human, their widespread contamination has become a serious problem for various organisms in the environment. Microbial degradation of organophosphorus compounds has received attention, because it is thought to be more environmentally friendly way than chemical methods. Therefore, many microorganisms capable of degrading organophosphorus compounds have been isolated and characterized. In this chapter, we describe organophosphorus compound-degrading bacteria and enzymes, and also their application for bioremediation purpose.

Introduction

Organophosphorus compounds are widely used as pesticides, flame retardants and plasticizers. These compounds, as well as chemical warfare agents, irreversibly inhibit the enzyme acetylcholinesterase, which hydrolyses the neurotransmitter acetylcholine to choline and acetic acid at postsynaptic membranes of central cholinergic synapses, by covalently binding to its active-site serine residue. The phosphorylation of acetylcholinesterase results in accumulation of acetylcholine at synapses and death of many organisms (Ragnarsdottir, 2000).

The use of organophosphorus compounds has increased dramatically after world war II. The structure, mammalian toxicity and half-life of well-known organophosphorus compounds are summarized in Fig. 3.1 and Table 3.1. Organophosphorus pesticides, such as parathion, chlorpyrifos, dimethoate, marathion and fenitrothion, are approximately account for approximately 38% of total pesticide usage (Singh, 2009). Owing to their large-scale use, they have been detected in various environments, including surface and ground waters, soil and air in the world. The widespread contamination has resulted in death of numerous non-target species in the environment.

Microbial degradation of organophosphorus compounds has received attention, because it is thought to be more environmentally friendly way than chemical methods. Therefore, many microorganisms capable of degrading the compounds have been isolated and characterized. Some reviews, which describe well the toxicity and microbial degradation of organophosphorus compounds, have been published (Raushel, 2002; Singh and Walker, 2006; Singh, 2009). In this chapter, we describe organophosphorus compound-degrading bacteria and enzymes, and also their application for bioremediation purpose.

Organophosphorus compound-degrading bacteria

The first organophosphorus compound-degrading bacterium was isolated from the Philippines and identified as *Fravobacterium* ATCC 27551

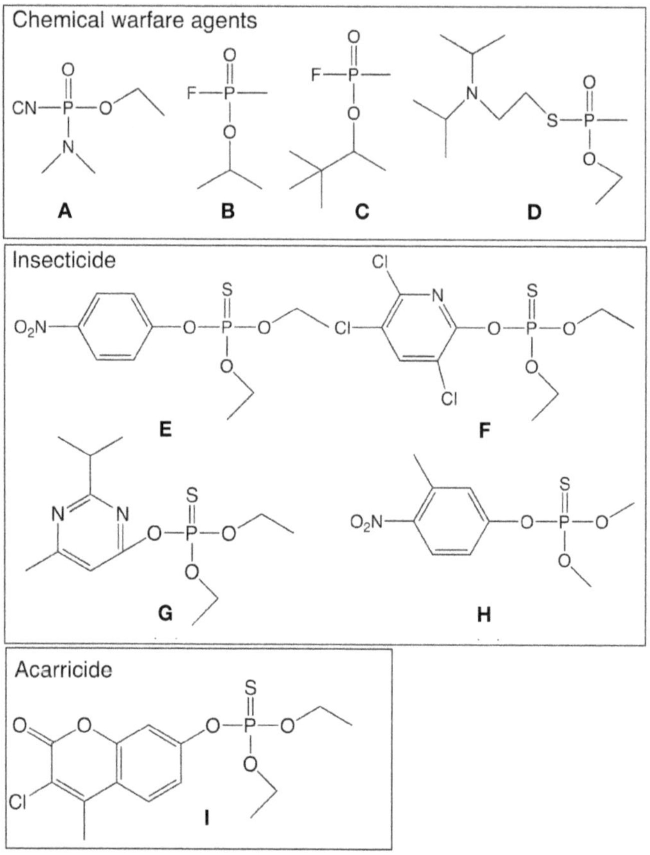

Figure 3.1 Structures of selected organophosphorus compounds: (A) Tabun; (B) Sarin; (C) Soman; (D) VX; (E) parathion; (F) chlorpyrifos; (G) diazinon; (H) fenitrothion; (I) coumaphos.

Table 3.1 Toxicity and half-life of the well-known organophosphorus compounds

Compound name	Mammalian toxicity	Half-life (days)
Chemical warfare agent[1]		
Tabun (GA)	150–400	1.5–2.5
Sarin (GB)	75–100	1.5–2.5
Soman (GD)	35–50	1.5–2.5
VX	10	4–42
Insecticide[2]		
Parathion	2–10	30–180
Chlorpyrifos	135–163	16–120
Diazinon	80–300	80–300
Fenitrothion	1700	12–28
Acaricide[2]		
Coumaphos	16–41	24–1400

[1]Toxicity of chemical warfare agents is expressed as the lethal dose by inhalation (mg/min·m^3).

[2]Toxicity of insecticides and acaricide is expressed as the LD_{50} value (mg/min·kg).

Some of data in Tables 3.1 and 3.3 in the previous review (Singh and Walker, 2006) are rearranged and given.

(Sethunathan and Yoshida, 1973). Then, various bacteria and fungi capable of degrading organophosphorus pesticides and insecticides have been isolated from the environments contaminated with organophosphorus compounds, using media containing various organophosphorus compounds as a source of carbon, nitrogen or phosphorus (Singh, 2009). *Flavobacterium* ATCC 27551 and *Brevundimonas diminuta* (formerly *Pseudomonas diminuta*) can degrade various organophosphorus triesters and thioesters, and fluorophosphonate compounds, such as coumaphos, methyl parathion, chlorpyrifos, VX, Soman and Sarin (Caldwell and Raushel, 1991). These two bacteria, therefore, are the most well-characterized microorganisms for the degradation of those compounds.

While a large number of studies have been conducted with bacteria degrading aryldialkyl phosphate, such as paraoxon and chlorpyrifos, there is little known about bacteria degrading trihaloalkyl phosphate, such as tris(1,3-dichloro-2-propyl) phosphate (TDCPP), tris(2-chloroethyl) phosphate (TCEP) and tris(2,3-dibromopropyl) phosphate (TBPP) (Fig. 3.2).

Trihaloalkyl phosphates have been used as flame retardants and plasticizers in a wide variety of applications. However, chlorinated organophosphorus flame retardants, such as TDCPP and TCEP, have received little attention with regard to human exposure and potential adverse effects, in comparison with polybrominated diphenyl ethers and other brominated flame retardants including TBPP (Reemtsma *et al.*, 2008).

The widespread use of TDCPP and TCEP has led to their contamination in various environments, including indoor air and house dust (Marklund *et al.*, 2005; Stapleton *et al.*, 2009), surface water (Andresen *et al.*, 2004), ground water (Fries and Puttmann, 2003), rain and snow (Julia and Wilhelm, 2009), and sediment (Kawagoshi *et al.*, 1999) in concentrations ranging from ng/l up to several μg/l. Particularly, a higher contamination level was observed in leachates and raw water of waste disposal sites (Kawagoshi *et al.*, 2002). Furthermore, they were also detected in drinking water (World Health Organization, 1998), because they are hardly eliminated not only in the environment but also in sewage treatment plants (Andresen and Bester, 2006).

Although the toxicity of trihaloalkyl phosphates is considerably lower than organophosphorus pesticides and chemical warfare agents, many studies have shown several toxic effects of the compounds. TCEP has been shown to cause adverse effects on brain, liver and kidney, and

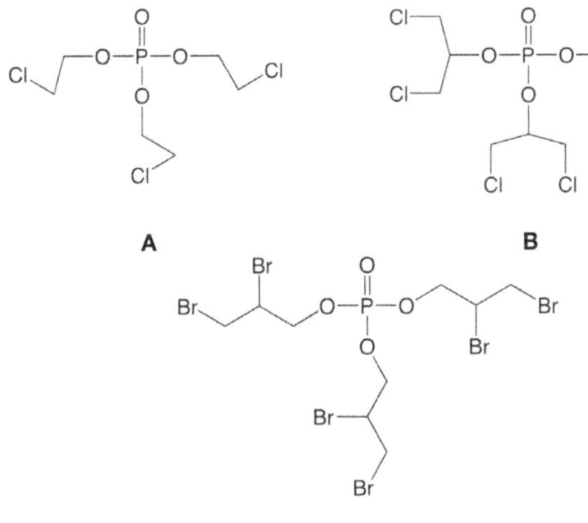

Figure 3.2 Structures of trihaloalkylphosphate used as a flame retardants; (A) tris(2-chloroethyl) phosphate; (B) tris(1,3-dichloro-2-propyl) phosphate; (C) tris(2,3-dibromopropyl) phosphate.

the fertility of male rats and mice (World Health Organization, 1998). In addition, it produced tumours at various organ sites (U.S. National Toxicology Program, 1991) and has been also suspected to possess carcinogenicity (Matthews *et al.*, 1993) and shown to inhibit the expression of cell cycle regulatory proteins, DNA synthesis and cell numbers (Ren *et al.*, 2008), although there are contradictory results on its carcinogenicity (Follmann and Wober, 2006). TDCPP has exhibited genotoxicity in several *in vitro* assays conducted in prokaryotic and eukaryotic cells (Soderlund *et al.*, 1985) and produced some indications of carcinogenicity (World Health Organization, 1998). These situations have prompted the recognition of potential ecological and human health concerns.

Recently, novel bacteria degrading trihaloalkyl phosphates, *Shingomonas* sp. strain TDK1 and *Sphingobium* sp. strain TCM1, were isolated by growth on TDCPP or TCEP as a sole phosphorus source, respectively (Fig. 3.3) (Takahashi *et al.*, 2008, 2010).

Both strains were also able to grow on another trihaloalkyl phosphate, tris(2,3-dibromopropyl) phosphate, and on triaryl phosphates, tricresyl and triphenyl phosphates (Table 3.2). In addition, strain TCM1 was able to grow moderately on tributyl phosphate and slightly on tris(2-butoxyethyl) phosphate, triethyl phosphate and trimethyl phosphate. However, strain TDK1 was not able to grow on all trialkyl phosphates tested. These results showed that the strains can degrade and assimilate not only the chlorinated organophosphorus flame retardants but also several other organophosphorus flame retardants and plasticizers, and that the strains have different substrate specificity for trialkyl phosphates.

Detoxification of hydrolysis products of organophosphorus compounds

The hydrolysis products of organophosphorus compounds are often toxic. Since *p*-nitrophenol, a hydrolysis product of parathion and methyl parathion, is known to be very toxic and has often been detected in soil, it has been listed as the priority pollutant by the United States Environmental Protection Agency (Letzel *et al.*, 2003). There are two different pathways for *p*-nitrophenol degradation via 4-nitrocatechol or hydroquinone (Fig. 3.4A). Some bacteria capable of hydrolysing parathion and methyl parathion can further degrade *p*-nitrophenol. For example, *Pseudomonas putida* that utilizes methyl parathion as sole carbon and/or phosphorus source can degrades *p*-nitrophenol to maleyl acetate via hydroquinone and 1,2,4-benzenetriol (Rani and Lalithakumari, 1994).

Bacteria that degrade only *p*-nitrophenol have been also isolated. *Arthrobacter aurescens* TW17 (Hanne *et al.*, 1993), *Pseudomonas* sp. (Sudhakar *et al.*, 1978), *Rhodococcus* sp. (Schafer *et al.*, 1996) and *Serratia* sp. (Pakala *et al.*, 2007) degrade *p*-nitrophenol via formation of 4-nitrocatechol, 1,2,4-benzenetriol and maleylacetate as intermediate products. On the other hand, bacteria, such as *Moraxella* sp. (Spain and Gibson, 1991), *Arthrobacter protophormiae* RKJ100 (Chauhan *et al.*, 2000) and *Pseudomonas* subsp. (Prakash *et al.*, 1996), use hydroquinone pathway which form of

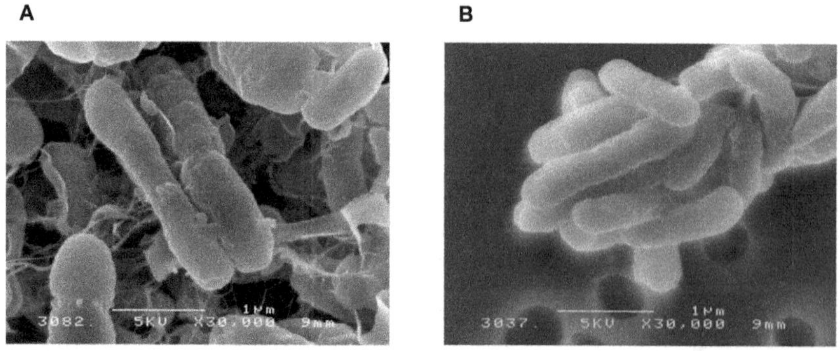

Figure 3.3 SEM micrographs of *Sphingobium* sp. TCM1 (A) and *Sphingomonas* sp. TDK1 (B).

Table 3.2 Substrate specificity of strains TCM1 and TDK1

	Relative growth of indicated strain[1]	
Substrate	TCM1	TDK1
None[2]	1.0 ± 0.3	1.0 ± 0.3
NaH_2PO_4	22 ± 4.2	6.0 ± 1.6
Trihaloalkyl		
Tris(2-chloroethyl) phosphate	18 ± 1.1	4.2 ± 1.4
Tris(1,3-dichloro-2-propyl) phosphate	12 ± 5.3	5.5 ± 0.80
Tris(2,3-dibromopropyl) phosphate	14 ± 1.6	5.9 ± 0.66
Triaryl		
Tricresyl phosphate	18 ± 1.5	5.3 ± 0.80
Triphenyl phosphate	17 ± 1.1	5.8 ± 0.71
Trialkyl		
Tributyl phosphate	7.1 ± 1.4	1.0 ± 0.32
Tris(2-butoxyethyl) phosphate	2.8 ± 0.32	0.98 ± 0.17
Triethyl phosphate	2.5 ± 0.27	0.81 ± 0.25
Tris(2-ethylhexyl) phosphate	1.3 ± 0.78	1.0 ± 0.22
Trimethyl phosphate	2.5 ± 0.14	0.85 ± 0.27

[1]The cell suspension was inoculated at a final optical density at 660 nm (OD_{660}) of approximately 0.05, and it was incubated at 30°C with shaking, with or without substrate. Growth (OD_{660}) was measured after 96 h. Relative values to the growth without substrate in each strain are expressed as means ± standard deviation of three independent experiments.

[2]OD_{660} was 0.041 ± 0.011 in TCM1 and 0.059 ± 0.016 in TDK1.

Data are modified from our previous study (Takahashi *et al.*, 2010).

1,4-benzoquinone and hydroquinone as intermediate products.

Shingomonas sp. TCM1 and *Sphingobium* sp. TDK1 can degrade TCEP and TDCPP, but not degrade their hydrolysis products, 2-chloroethanol and 1,3-dichloro-2-propanol. Bacteria that degrade the chloroalchols have been also isolated. *Xanthobacter autotrophicus* GJ10 isolated on 2-chloroethane as a sole carbon and energy source degrade 2-chloroethanol via several enzymatic reactions (Fig. 3.4B) (Janssen *et al.*, 1985) and *Arthrobacter* sp. PY1 can metabolize 1,3-dichloro-2-propanol to glycerol using a haloalchol dehalogenase (Yonetani *et al.*, 2004).

Thus, even if organophosphorus compound-degrading bacteria cannot further degrade of toxic metabolites, the complete detoxification of organophosphorus compounds could be achieved by using consortium of organophosphorus compound-degrading bacteria and some bacteria that subsequently degrade those toxic metabolites.

Bacterial phosphotriesterases and their genetic engineering

In microorganisms, organophosphorus compounds are degraded by phosphotriesterases, which catalyse the first step of their degradation. Many bacterial phosphotriesterases have been purified and characterized, and the encoding genes have been cloned. According to their primary structure similarities, well-characterized bacterial phosphotriesterases are mainly classified into three groups, organophosphorus hydrolase (OPH), organophosphorus acid anhydrolase (OPAA) and methyl parathion hydrolase (MPH) (Singh, 2009). The properties of selected bacterial phosphotriesterases are shown in Table 3.3. Although their primary structure is different, all of these enzymes require divalent cation for catalytic activity.

OPH encoded by *opd* gene belongs to amidohydrolase superfamily and exhibits broad substrate specificity. OPH from *Brevundimonas*

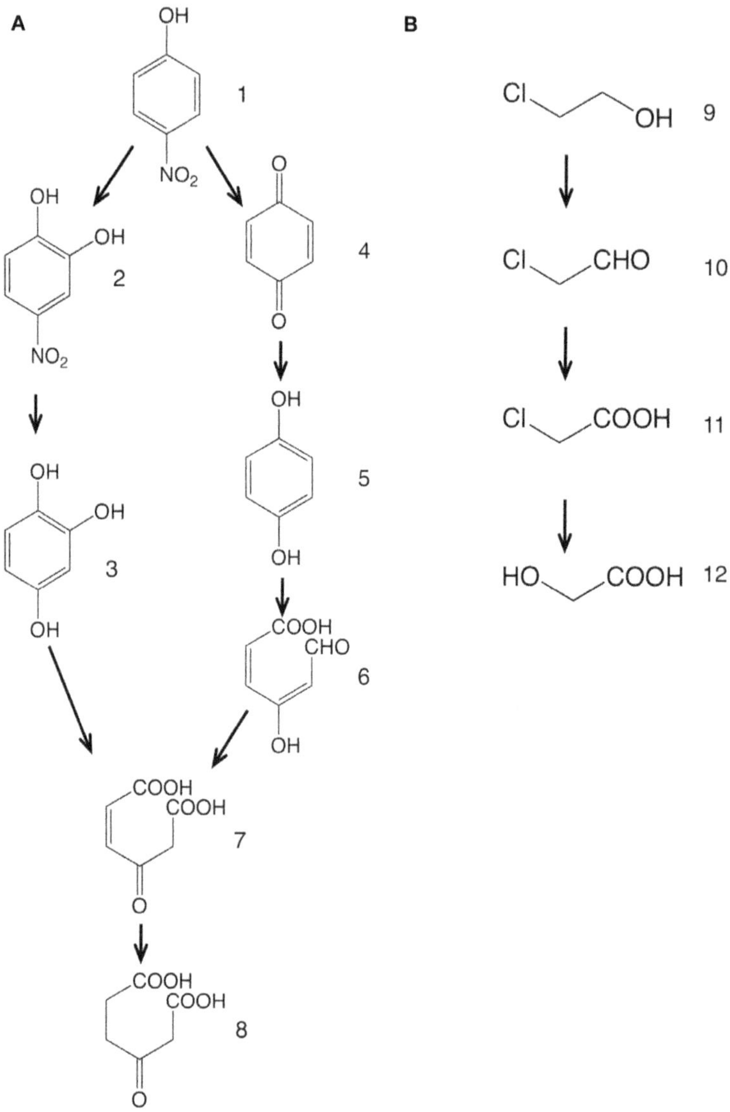

Figure 3.4 Proposed pathways of degradation of *p*-nitrophenol (A) (Chauhan *et al.*, 2000) and 2-chloroethanol (B) (Janssen *et al.*, 1985). Compounds: 1, *p*-nitrophenol; 2, 4-nitrocatechol; 3, 1,2,4-benzenetriol; 4, 1,4-benzoquinone; 5, hydroquinone; 6, *cis, trans*-4-hydroxymuconic semialdehyde; 7, maleylactate; 8, 3-oxoadipate; 9, 2-chloroetanol; 10, chloroacetaldehyde; 11, chloroacetic acid; 12, glycolic acid.

diminuta has extensively characterized. Its X-ray structure and proposed reaction mechanism have been reported (Benning *et al.*, 1994; Aubert *et al.*, 2004). This enzyme is homodimer and folds as a $(\alpha/\beta)8$ barrel and a binuclear metal centre with two transition metal ion such as zinc is found at the active site. OPDA encoded by *opdA*, a homologue of OPH, has been isolated from *Agrobacterium radiobacter* and characterized (Horne *et al.*, 2002).

OPAA encoded by *opaA* belongs to dipeptidase superfamily. Recently, the crystal structure of OPAA from *Altermonas* sp. strain JD6.5 has been resolved (Vyas *et al.*, 2010). OPH, OPDA and MPH are homodimer, whereas OPAA is monomer in solution. The subunit of OPAA is composed of two domains, amino and carboxy domains, with the latter exhibiting a 'pita bread' architecture and harbouring the active site with the binuclear ions. OPH has the hydrolysis

Table 3.3 Bacterial phosphotriesterases

Organism	Enzyme (gene)	Subunit MW	Quaternary structure	Substrate	Accession number	PDB code	Reference
Brevundimonas diminuta	OPH (*opd*)	39,000	Dimer	Parathion, paraoxon, diazinon, coumaphos, chlorpyrifos, diisopropyl fluorophosphates (DFP), Sarin, Soman, Tabun, VX	M20392	1PTA	Dumas *et al.* (1989)
Agrobacterium radiobacter	OPDA (*opdA*)	35,000	Dimer	Parathion, paraoxon, diazinon, phosmet, fenthion	AY043245	2D2J	Horne *et al.* (2002)
Alteromonas sp. JD6.5	OPAA (*opaA*)	59,000	Monomer	DFP, paraoxon, Sarin, Soman, cyclosarin	ASU29240	3L24	DeFrank and Cheng (1991)
Pseudomonas sp. WBC-3	MPH (*mpd*)	35,000	Dimer	Parathion, fenitrothion	AY251554	1P9E	Dong *et al.* (2005)
Burkholderia sp. NF100	FedA (*fedA*)	55,000	N.D.	Fenitrothion, parathion, diazinon	AB235840	N.D.	Tago *et al.* (2006a)

ND, not determined.

activity against P–O, P–C, P–F and P–S bonds. On the other hand, OPAA cannot hydrolyse P–S bond,s and the hydrolysis activity of the enzyme is higher against P–F bonds than P–O and P–C bonds (DeFrank and Cheng, 1991).

MPH encoded by mpd belongs to β-lactamase superfamily. The crystal structure of MPH from *Pseudomonas* sp. WBC-3 has been reported (Dong *et al.*, 2005). MPH is homodimer and each subunit has a mixed hybrid binuclear zinc centre.

Fenitrothion hydrolase (FedA) has been isolated only from *Burkhorderia* sp. NF100 and does not share amino acid or gene-sequence similarity with OPH, OPDA, OPAA and MPH. While OPH, MPD and OPAA act as homodimer or monomer, FedA requires a putative membrane protein FedB for full enzymatic activity. Therefore, it is considered that FedA forms a complex with FedB and that the complex locates in the cell membrane fraction (Tago *et al.*, 2006b).

Since there are little known about the enzymatic properties of MPH and FedA, further studies are required to obtain more detailed information on their substrate specificity and kinetic property of both enzymes, as well as quaternary structure of FedA.

Recently, the improvements of catalytic activity and substrate specificity of phosphotriesterases by using genetic engineering techniques have been reported. Cho *et al.* (2002, 2004) reported that an OPH variant finally obtained by DNA shuffling exhibited a 725-fold increased activity towards a poor substrate, chlorpyrifos. Yang *et al.* (2010) co-expressed MPH and OPH to the periplasm and cell surface of *Escherichia coli*, respectively, and the engineered *E. coli* strain acquired a broad-spectrum degradation activity against various organophosphorus compounds.

In future, genetically engineered organophosphorus compound-degrading enzymes and microorganisms with high activity and stability may provide efficient tools for biodegradation.

Bioremediation of environments contaminated with organophosphorus compounds by microorganisms

Some success in the use of microorganisms for the degradation of organophosphorus compounds-contaminated soil and water has been reported.

Barles *et al.* (1979) have demonstrated that the inoculation of an acclimated culture capable of utilizing parathion as a sole carbon and energy source into soil resulted in the complete degradation of up to 1250 kg/ha parathion within 35 days. Walter *et al.* (1996) have reported that a biofilter with microorganism consortium capable of degrading coumaphos degraded 1200 mg/l coumaphos in a waste dip to a level below 0.10 mg/l in 7–10 days.

Bioremediation of soil contaminated with organophosphorus compounds have also been examined by using isolated bacterial strains. Cycoń *et al.* (2009) have described that isolated strains designated DDS-1, DDS-2 and DDS-3 were identified as *Serratia liquefaciens, Serratia marcescens* and *Pseudomonas* sp., respectively, and each strain alone or their consortium efficiently degraded diazinon (100 mg/kg soil) in sterilized soil. It was also reported the addition of isolated strain *Enterobacter* strain B-14, which can mineralize chlorpyrifos, to soil with a low indigenous population of chlorpyrifos-degrading bacteria resulted in a higher degradation rate than was observed in non-inoculated soil (Singh *et al.*, 2004).

In general, it is easily postulated that the function of organophosphorus compound-degrading bacteria may be affected by the presence of other microorganisms and various chemical compounds in the soil and water. Further characterization of organophosphorus compound-degrading bacteria and analysis of biochemical properties and transcriptional regulations of phosphotriesterases are required for developing more useful systems with microorganisms to degrade organophosphorus compounds.

References

Andresen, J., and Bester, K. (2006). Elimination of organophosphate ester flame retardants and plasticizers in drinking water purification. Water Res. *40*, 621–629.

Andresen, J.A., Grundmann, A., and Bester, K. (2004). Organophosphorus flame retardants and plasticisers in surface waters. Sci. Total Environ. 332, 155–166.

Aubert, S.D., Li, Y., and Raushel, F.M. (2004). Mechanism for the hydrolysis of organophosphates by the bacterial phosphotriesterase. Biochemistry *43*, 5707–5715.

Barles, R.W., Daughton, C.G., and Hsieh, D.P. (1979). Accelerated parathion degradation in soil inoculated with acclimated bacteria under field conditions. Arch. Environ. Contam. Toxicol. *8*, 647–660.

Benning, M.M., Kuo, J.M., Raushel, F.M., and Holden, H.M. (1994). Three-dimensional structure of phosphotriesterase: an enzyme capable of detoxifying organophosphate nerve agents. Biochemistry *33*, 15001–15007.

Caldwell, S.R., and Raushel, F.M. (1991). Detoxification of organophosphate pesticides using a nylon based immobilized phosphotriesterase from *Pseudomonas diminuta*. Appl. Biochem. Biotechnol. *31*, 59–73.

Chauhan, A., Chakraborti, A.K., and Jain, R.K. (2000). Plasmid-encoded degradation of p-nitrophenol and 4-nitrocatechol by *Arthrobacter* protophormiae. Biochem. Biophys. Res. Commun. *270*, 733–740.

Cho, C.M., Mulchandani, A., and Chen, W. (2002). Bacterial cell surface display of organophosphorus hydrolase for selective screening of improved hydrolysis of organophosphate nerve agents. Appl. Environ. Microbiol. *68*, 2026–2030.

Cho, C.M., Mulchandani, A., and Chen, W. (2004). Altering the substrate specificity of organophosphorus hydrolase for enhanced hydrolysis of chlorpyrifos. Appl. Environ. Microbiol. *70*, 4681–4685.

Cycon, M., Wojcik, M., and Piotrowska-Seget, Z. (2009). Biodegradation of the organophosphorus insecticide diazinon by *Serratia* sp. and *Pseudomonas* sp. and their use in bioremediation of contaminated soil. Chemosphere *76*, 494–501.

DeFrank, J.J., and Cheng, T.C. (1991). Purification and properties of an organophosphorus acid anhydrase from a halophilic bacterial isolate. J. Bacteriol. *173*, 1938–1943.

Dong, Y.J., Bartlam, M., Sun, L., Zhou, Y.F., Zhang, Z.P., Zhang, C.G., Rao, Z., and Zhang, X.E. (2005). Crystal structure of methyl parathion hydrolase from *Pseudomonas* sp. WBC-3. J. Mol. Biol. *353*, 655–663.

Dumas, D.P., Caldwell, S.R., Wild, J.R., and Raushel, F.M. (1989). Purification and properties of the phosphotriesterase from *Pseudomonas diminuta*. J. Biol. Chem. *264*, 19659–19665.

Follmann, W., and Wober, J. (2006). Investigation of cytotoxic, genotoxic, mutagenic, and estrogenic effects of the flame retardants tris-(2-chloroethyl)-phosphate (TCEP) and tris-(2-chloropropyl)-phosphate (TCPP) *in vitro*. Toxicol. Lett. *161*, 124–134.

Fries, E., and Puttmann, W. (2003). Monitoring of the three organophosphate esters TBP, TCEP and TBEP in river water and ground water (Oder, Germany). J. Environ. Monit. *5*, 346–352.

Hanne, L.F., Kirk, L.L., Appel, S.M., Narayan, A.D., and Bains, K.K. (1993). Degradation and induction specificity in actinomycetes that degrade p-nitrophenol. Appl. Environ. Microbiol. *59*, 3505–3508.

Horne, I., Sutherland, T.D., Harcourt, R.L., Russell, R.J., and Oakeshott, J.G. (2002). Identification of an opd (organophosphate degradation) gene in an *Agrobacterium* isolate. Appl. Environ. Microbiol. *68*, 3371–3376.

Janssen, D.B., Scheper, A., Dijkhuizen, L., and Witholt, B. (1985). Degradation of halogenated aliphatic compounds by *Xanthobacter autotrophicus* GJ10. Appl. Environ. Microbiol. *49*, 673–677.

Kawagoshi, Y., Fukunaga, I., and Itoh, H. (1999). Distribution of organophosphoric acid triesters between water and sediment at a sea-based solid waste disposal site. J. Mater. Cycle Waste Manage. *1*, 53–61.

Kawagoshi, Y., Nakamura, S., and Fukunaga, I. (2002). Degradation of organophosphoric esters in leachate from a sea-based solid waste disposal site. Chemosphere *48*, 219–225.

Letzel, S., Goen, T., Bader, M., Angerer, J., and Kraus, T. (2003). Exposure to nitroaromatic explosives and health effects during disposal of military waste. Occup. Environ. Med. *60*, 483–488.

Marklund, A., Andersson, B., and Haglund, P. (2005). Organophosphorus flame retardants and plasticizers in air from various indoor environments. J. Environ. Monit. 7, 814–819.

Matthews, H.B., Eustis, S.L., and Haseman, J. (1993). Toxicity and carcinogenicity of chronic exposure to tris(2-chloroethyl)phosphate. Fundam. Appl. Toxicol. *20*, 477–485.

Pakala, S.B., Gorla, P., Pinjari, A.B., Krovidi, R.K., Baru, R., Yanamandra, M., Merrick, M., and Siddavattam, D. (2007). Biodegradation of methyl parathion and p-nitrophenol: evidence for the presence of a p-nitrophenol 2-hydroxylase in a Gram-negative *Serratia* sp. strain DS001. Appl. Microbiol. Biotechnol. *73*, 1452–1462.

Prakash, D., Chauhan, A., and Jain, R.K. (1996). Plasmid-encoded degradation of p-nitrophenol by *Pseudomonas cepacia*. Biochem. Biophys. Res. Commun. *224*, 375–381.

Ragnarsdottir, K.V. (2000). Environmental fate and toxicology of organophosphate pesticides. J. Geol. Soc. London *157*, 859–876.

Rani, N.L., and Lalithakumari, D. (1994). Degradation of methyl parathion by *Pseudomonas putida*. Can. J. Microbiol. *40*, 1000–1006.

Raushel, F.M. (2002). Bacterial detoxification of organophosphate nerve agents. Curr. Opin. Microbiol. *5*, 288–295.

Reemtsma, T., Quintana, J.B., Rodil, R., Garcia-López, M., and Rodríguez, I. (2008). Organophosphorus flame retardants and plasticizers in water and air. I. Occurrence and fate. Trends Analyt. Chem. *27*, 727–737.

Regnery, J., and Puttman, W. (2009). Organophosphorus flame retardants and plasticizers in rain and snow from middle Germany. CLEAN – Soil, Air, Water 37, 334–342.

Ren, X., Lee, Y.J., Han, H.J., and Kim, I.S. (2008). Effect of tris-(2-chloroethyl)-phosphate (TCEP) at environmental concentration on the levels of cell cycle regulatory protein expression in primary cultured rabbit renal proximal tubule cells. Chemosphere *74*, 84–88.

Schafer, A., Harms, H., and Zehnder, A.J. (1996). Biodegradation of 4-nitroanisole by two *Rhodococcus* spp. Biodegradation 7, 249–255.

Sethunathan, N., and Yoshida, T. (1973). A *Flavobacterium* sp. that degrades diazinon and parathion. Can. J. Microbiol. *19*, 873–875.

Singh, B.K. (2009). Organophosphorus-degrading bacteria: ecology and industrial applications. Nat. Rev. Microbiol. 7, 156–164.

Singh, B.K., and Walker, A. (2006). Microbial degradation of organophosphorus compounds. FEMS Microbiol. Rev. *30*, 428–471.

Singh, B.K., Walker, A., Morgan, J.A., and Wright, D.J. (2004). Biodegradation of chlorpyrifos by *Enterobacter* strain B-14 and its use in bioremediation of contaminated soils. Appl. Environ. Microbiol. *70*, 4855–4863.

Soderlund, E.J., Dybing, E., Holme, J.A., Hongslo, J.K., Rivedal, E., Sanner, T., and Nelson, S.D. (1985). Comparative genotoxicity and nephrotoxicity studies of the two halogenated flame retardants tris(1,3-dichloro-2-propyl)phosphate and tris(2,3-dibromopropyl)phosphate. Acta Pharmacol. Toxicol. (Copenhagen) *56*, 20–29.

Spain, J.C., and Gibson, D.T. (1991). Pathway for biodegradation of p-nitrophenol in a *Moraxella* sp. Appl. Environ. Microbiol. 57, 812–819.

Stapleton, H.M., Klosterhaus, S., Eagle, S., Fuh, J., Meeker, J.D., Blum, A., and Webster, T.F. (2009). Detection of organophosphate flame retardants in furniture foam and U.S. house dust. Environ. Sci. Technol. *43*, 7490–7495.

Sudhakar, B., Siddaramappa, R., Wahid, P.A., and Sethunathan, N. (1978). Conversion of p-nitrophenol to 4-nitrocatechol by a *Pseudomonas* sp. Antonie Van Leeuwenhoek *44*, 171–176.

Tago, K., Yonezawa, S., Ohkouchi, T., Hashimoto, M., and Hayatsu, M. (2006a). Purification and characterization of fenitrothion hydrolase from *Burkholderia* sp. NF100. J. Biosci. Bioeng. *101*, 80–82.

Tago, K., Yonezawa, S., Ohkouchi, T., Ninomiya, T., Hashimoto, M., and Hayatsu, M. (2006b). A novel organophosphorus pesticide hydrolase gene encoded on a plasmid in *Burkholderia* sp. strain NF100. Microbes Environ. *21*, 53–57.

Takahashi, S., Kawashima, K., Kawasaki, M., Kamito, J., Endo, Y., Akatsu, K., Horino, S., Yamada, R.H., and Kera, Y. (2008). Enrichment and characterization of chlorinated organophosphate ester-degrading mixed bacterial cultures. J. Biosci. Bioeng. *106*, 27–32.

Takahashi, S., Satake, I., Konuma, I., Kawashima, K., Kawasaki, M., Mori, S., Morino, J., Mori, J., Xu, H., Abe, K., *et al.* (2010). Isolation and identification of persistent chlorinated organophosphorus flame retardant-degrading bacteria. Appl. Environ. Microbiol. 76, 5292–5296.

U.S. National Toxicology Program. (1991). NTP toxicology and carcinogenesis studies of tris(2-chloroethyl) phosphate (CAS No. 115-96-8) in F344/N rats and B6C3F1 mice (Gavage Studies). Natl. Toxicol Program Tech. Rep. Ser. *391*, 1–233.

Vyas, N.K., Nickitenko, A., Rastogi, V.K., Shah, S.S., and Quiocho, F.A. (2010). Structural insights into the dual activities of the nerve agent degrading organophosphate anhydrolase/prolidase. Biochemistry *49*, 547–559.

Walter, W.M., Pedro, L.D.V., and Jeffrey, S.K. (1996). Biodegradation of the organophosphate insecticide coumaphos in highly contaminated soils and in liquid wastes. Pestic. Sci. *48*, 149–155.

World Health Organization. (1998). Environmental health criteria 209. Flame retardants: tris(chloropropyl) phosphate and tris(2-chloroethyl) phosphate. (World Health Organization: Geneva, Switzerland).

Yang, C., Freudl, R., Qiao, C., and Mulchandani, A. (2010). Cotranslocation of methyl parathion hydrolase to the periplasm and of organophosphorus hydrolase to the cell surface of *Escherichia coli* by the Tat pathway and ice nucleation protein display system. Appl. Environ. Microbiol. *76*, 434–440.

Yonetani, R., Ikatsu, H., Miyake-Nakayama, C., Fujiwara, E., Maehara, Y., Miyoshi, S.-i., Matsuoka, H., and Shinoda, S. (2004). Isolation and characterization of a 1,3-dichloro-2-propanol-degrading bacterium. J. Health Sci. *50*, 605–612.

Biodegradation of Hydrocarbons in Bioturbated Marine Sediments

4

Philippe Cuny, Cristiana Cravo-Laureau, Vincent Grossi, Franck Gilbert and Cécile Militon

Abstract

Sediments can serve as sinks for hydrocarbon contaminants in marine ecosystems. Once settled, hydrocarbons fate will be dominated by several abiotic and biotic processes that will result in either their partial or total degradation or in a selective preservation when buried within the sediment. Biodegradation of hydrocarbons in marine sediments is mainly due to the existence of prokaryotes harbouring specific catabolic genes enabling the degradation of these compounds under oxic, suboxic and anoxic conditions. The interplay of the various factors that govern hydrocarbons biodegradation in marine sediments is highly complex as illustrated by bioturbation processes carried out by macrofaunal organisms. For instance, the redox oscillation regimes generated by macrofaunal organisms, and the efficiency of metabolic coupling between functional groups associated to these specific redox regimes, are probably determinant factors controlling the biodegradation rates of hydrocarbons in marine sediments. From the understanding of how these natural occurring factors may modulate the rates of hydrocarbons biodegradation, innovative bioremediation strategies may emerge.

Introduction

Oil has been increasingly used by man not only as an energy source (e.g. fuel for transportation and heating system) but also, since the beginning of the 20th century, as a raw material for many chemical products. Even if oil transportation safety has been greatly improved over the last decades, significantly lowering accidental oil spills, hundreds of millions of litres of petroleum still enter the environment from both natural and anthropogenic sources every year (Head and Swannell, 1999; Head *et al.*, 2006). Furthermore, as the more accessible oil reservoirs are being emptied, marine and particularly deep-sea reservoirs are becoming the 'new Eldorado for black gold', generating new threats to marine ecosystems as shown by the recent and dramatic BP Deepwater Horizon Gulf of Mexico oil spill. Hydrocarbons will still be one of the most important organic contaminant in marine ecosystems for decades. This is attested by the increasing occurrence of several polycyclic aromatic hydrocarbons (PAHs) species worldwide, especially in marine sediments, and mostly in areas close to intense industrial activities (Nikolaou *et al.*, 2009).

As highlighted by Borch *et al.* (2010), despite recent development, we still face major challenges in unravelling and understanding the intricate and coupled environmental processes that control contaminant fate and transport in the environment. This is particularly true in marine sediments where ends up part of spilled oil or hydrocarbons like PAHs.

Oil hydrocarbons biodegradation by microorganisms, and in particular by bacteria, is known to be one of the prevalent processes acting in the removal of these contaminants from marine ecosystems (Van Hamme *et al.*, 2003; Head *et al.*, 2006). Since the first discovery of hydrocarbon-degrading organisms almost a century ago (Söhngen, 1913), most work has been conducted on single bacteria, bacterial consortium or contaminated microcosms in laboratory conditions. Field studies in marine areas submitted to an oil spill have revealed *in situ* changes in oil and bacterial communities (McNaughton, *et al.*, 1999; Delille and Delille, 2000; Kasai *et al.*, 2001; Röling *et al.*, 2002; Grossi *et al.*, 2002; Maruyama *et al.*, 2003; Röling *et al.*, 2004; Miralles *et al.*, 2007a,b; Alonso-Gutiérrez *et al.*, 2009; Miralles *et al.*, 2010). Oil degradation involves progressive or sequential reactions, in which the hydrocarbon-degrading microorganisms (primary degraders) carry out the initial attack on the oil constituents; this produces intermediate compounds that are subsequently utilized by different groups of organisms, resulting in further degradation. These reactions lead to an apparent sequential removal of hydrocarbons from alkanes to polycyclic aromatic compounds and according to molecular weight (from low to high molecular weight; Head *et al.*, 2006). Overall, the fate of oil appears closely dependent on the ecosystem characteristics and it is often hard to predict exact removal rate of hydrocarbons even for a specific area. This is perhaps not as surprising as marine ecosystems are complex systems characterized by heterogeneity and multiple-scale interactions (direct and indirect) between organisms and their environment. The comprehension of the fate and effects of an oil spill in the real world implies the understanding of the functioning of the ecological network operating in these systems. Of particular interest is the understanding of the link between bioturbation (i.e. mixing of sediment particles and solutes by benthic fauna), microorganisms-degraders and fate of deposited oil in marine sediments.

Sedimentation of oil is mainly due to physical interactions between oil droplets and suspended particulate matter (SPM). In coastal regions characterized by high concentrations of SPM, about 20–30% of the total spill may sediment to the seabed (Muschenheim and Lee, 2002; Payne *et al.*, 2003). Oil or weathered oil can also reach the intertidal sediments under the action of waves or currents. Once deposited on the sediment, contaminating oil can prevent ventilation of upper layers and stimulate microbial consumption of oxygen, resulting in the development of extensive anoxic zones (Miralles *et al.*, 2007b). Similarly, owing to their hydrophobic and lipophilic character, PAHs are strongly bound to particles and tend to accumulate in the sediment. Sediment-living organisms can also affect the fate (biodegradation) and transport of hydrocarbons through activities like feeding and burrowing, collectively generating bioturbation. In particular, macrofauna, through their feeding, burrowing and ventilation activities, have an important influence on microbial activity and sediment metabolism in marine ecosystems (Aller and Aller, 1998; Kristensen, 2001). In non-bioturbated sediments, the aerobic degradation of hydrocarbons will only take place in the upper part (sometimes less than a few mm) of the sediment layer, most of the biodegradation being carried out under suboxic or anoxic conditions. If the aerobic and the anaerobic biodegradation of these compounds are well documented, little is still known about degradation in the transition zone from oxic to anoxic conditions (Berthe-Corti and Fetzner, 2002) and far less under the oscillating redox conditions which prevail in bioturbated sediments (Aller, 1994).

After a brief review of the overall actors in hydrocarbon biodegradation and of the aerobic and anaerobic hydrocarbon activation mechanisms, we will address the question of the impact of bioturbation on hydrocarbon dynamics in marine sediments. The last part of this chapter will be dedicated to the evaluation of using oscillating oxic-anoxic conditions as a potential strategy for bioremediation.

Role of oxygen in hydrocarbons biodegradation: main factor?

The influence of oxygen on organic matter (OM) preservation is under debate and studies comparing aerobic and anaerobic degradation reached so far contrasting conclusions. Some studies suggest that oxygen has no significant impact on the rate of OM degradation, which may rather be influenced by the age, origin, chemical composition or structure of OM (Kristensen and Blackburn, 1987; Lee, 1992; Kristensen *et al.*, 1995, Hulthe *et al.*, 1998; Sun *et al.*, 2000). Other studies show that the degradation of a variety of organic components, including lipids, is largely controlled by oxygen and that aerobic degradation is generally faster than anaerobic one (Harvey and Macko, 1997; Sun *et al.*, 1997, 2002; Bianchi *et al.*, 2000; Grossi *et al.*, 2003), indicating a dominant control of oxygen on OM fate. For instance, oxygen is a key variable in microbial PAHs degradation and mineralization, determining their degradation rates (McNally *et al.*, 1999, Lei *et al.*, 2005), pathways (Meckenstock *et al.*, 2004), and the composition of bacteria involved in PAHs degradation (Roslev *et al.*, 1998; Eriksson *et al.*, 2003). Aerobic PAHs degradation rates are generally orders of magnitude higher than those under nitrate- or sulfate-reducing conditions (Coates *et al.*, 1997; Rockne and Strand, 1998). However, in some environments, the role of molecular oxygen on alkane or even (low molecular weight) PAHs biodegradation may not be as obvious as previously thought (see section 'Bioremediation strategies under oscillating conditions').

Although hydrocarbon-polluted marine sites are often deficient in oxygen or completely anoxic, aerobic hydrocarbon-degrading bacteria (HCB) have been the subject of intensive investigations whereas efforts in studying anaerobic HCB are more recent. This may partly be explained by the oxygen paradigm aforementioned and by the relative easiness to study aerobic biodegradation processes. Furthermore, while the contribution of *Bacteria* to the fate of hydrocarbons in environment is well established, the role of *Archaea*, and of micro- and macroeukaryotes is still poorly understood. Potential role of micro- and macroeukaryotes will not be developed in this chapter.

Aerobic hydrocarbon-degrading bacteria in marine sediments

During the last two decades, studies of benthic microbial communities inhabiting surface sediments accidentally or experimentally contaminated by organic pollutants, have been performed either *in situ* or in laboratory microcosms. Such experiments allow isolating HCB, characterizing their catabolic abilities and understanding their role in oil-spilled marine sediments (McNaughton *et al.*, 1999; Delille *et al.*, 2002; Röling *et al.*, 2004; Yakimov *et al.*, 2005; Alonso-Gutiérrez *et al.*, 2009). The use of biomolecular approaches is giving us the opportunity to follow the succession of bacterial communities after hydrocarbon inputs, and hence to identify autochthonous bacterial populations resistant to organic pollutants or potentially involved in bioremediation processes. Following the addition of oil or pure model hydrocarbons, several trends are reported through the analysis of marine bacterial communities. After the *Prestige* or the *Nakhoda* oil spills, bacterial abundance remained stable in the studied marine samples (Kasai *et al.*, 2001; Maruyama *et al.*, 2003); on the other hand, Taketani *et al.* (2009) detected a positive response of bacterial abundance to the input of hydrocarbons in both pristine and polluted mangrove sediments. The assessment of rRNA/rDNA ratio and counts of heterotrophic and hydrocarbonoclastic populations are often higher in contaminated marine samples, suggesting increased activity of some bacterial populations due to hydrocarbons (Alonso-Gutiérrez *et al.*, 2009; Taketani *et al.*, 2009). Another recurrent trend is the rapid and significant reduction of bacterial diversity following hydrocarbons addition (Röling *et al*, 2002; Alonso-Gutiérrez *et al.*, 2009). This decrease is often correlated with blooms of several bacterial phylotypes belonging to Proteobacteria (e.g. *Alcanivorax*, *Cycloclasticus*, *Marinobacter*, *Neptunomonas*, and *Thalassolituus*) and Bacteroidetes, which could transiently contribute strongly to the microbial community in contaminated sites (Röling *et al.*, 2004; Yakimov *et al.*, 2005). Gammaproteobacteria are generally dominant during the first weeks of the bioremediation, particularly after the addition of nutrients acting as bio-stimulating agents. Then within less than one month, they are either replaced by Alphaproteobacteria or associated with members of this class (Röling *et al*, 2002). Actinobacteria may appear several months later suggesting a role in the biodegradation of recalcitrant fractions of the spilled-oil (Alonso-Gutiérrez *et al.*, 2009). Some authors have observed that the resilience of the bacterial diversity can be coupled with a decrease in hydrocarbons content after an oiling event suggesting a restoration of impacted ecosystems (Delille *et al*, 2002). However, even if the bacterial diversity has recovered its pre-oil level, the bacterial community structure is significantly different from that before the spill and toxic daughter products still remain even after a long period (Röling *et al*, 2002). These culture-independent studies give a first glance at potential hydrocarbon degraders' candidates inhabiting contaminated marine areas; nevertheless they are unable to establish a formal link between bacterial populations and their role in hydrocarbon biodegradation.

Catabolic abilities of marine benthic bacteria could be investigated following the successful isolation of diverse HCB from contaminated or non-contaminated marine sediments. Isolation approaches use both conventional enrichment techniques using petroleum hydrocarbons as substrates for growth (Yakimov *et al.*, 1998; Tapilatu *et al.*, 2010) and dilution-to-extinction techniques with sterilized and hydrocarbon-amended synthetic seawater as media (Dykesterhouse *et al.*, 1995; Staley, 2010). More than 100 aerobic and facultative anaerobic bacteria able to act on various hydrocarbon classes such as straight or branched aliphatic hydrocarbons, alicyclic hydrocarbons and/or aromatic hydrocarbons have been isolated from benthic environments (e.g. shoreline, deep-sea and estuarine sediments; Table 4.1). According to their ability to use petroleum hydrocarbons for growth, these bacteria could be potentially involved in the bioremediation of oil-spilled marine sediments. However, only some of them play a pivotal role in the restoration of these environments. As highlighted by bacterial successions observed in oil-spilled benthic area, the bacterial populations were rapidly dominated by only few phylotypes, namely *Alcanivorax* spp., *Marinobacter* spp., *Cycloclasticus* spp., *Neptunomonas* spp. and *Thalassolituus* spp. (Yakimov *et al.*, 2005; Hedlund and Costa, 2010). This selection may be partly explained by the high competitiveness of these marine bacteria for petroleum-derived aliphatic and aromatic hydrocarbons as carbon and energy sources. This feature, carried out by the so called 'obligate' hydrocarbonoclastic bacteria (OHCB), was discovered while characterizing the phenotype of *Alcanivorax borkumensis* isolated from seawater/sediment samples collected in the North Sea (Yakimov *et al.*, 1998). Some OHCB exhibit other special features which improve their fitness in polluted sediments, such as the production of powerful biosurfactants which increase the availability of hydrophobic hydrocarbons, or the development of efficient nutrients (Mn, Mg, N and P) uptake systems overcoming typical unbalanced C:N:P ratio of oil polluted marine sites (Schneiker *et al.*, 2006).

Recent advances in molecular and '-omics' technologies have been helpful to identify genes and mechanisms involved in hydrocarbons degradation by marine bacteria (Schneiker *et al.*, 2006; Desai *et al.*, 2010). Under aerobic conditions, molecular oxygen plays a crucial role by acting as electron acceptor for the microbial respiration and as co-substrate for the activation of hydrocarbons via oxygenation reactions. In this section, a brief overview of (non-methane) *n*-alkanes and aromatic hydrocarbons activation mechanisms will be presented, with emphasis on genes involved. More detailed descriptions of overall hydrocarbon degradation mechanisms have been recently reviewed (Van Hamme *et al.*, 2003; Fetzner, 2010; Pieper *et al.*, 2010; Pérez-Pantoja *et al.*, 2010; Rojo, 2010; Smith *et al.*, 2010). *n*-Alkanes are generally poorly reactive compounds owing to a lack of functional groups, thus initial activation is critical. Alkane-degrading microbes generally perform stepwise oxidations starting from the terminal methyl group (Fig. 4.1A, pathway I) which leads to the formation of fatty alcohol, aldehyde and fatty acid (subterminal oxidation has been observed in some degraders; Fig. 4.1A, pathway II). Produced carboxylic acid is then combined with CoA and processed by β-oxidation to yield acetyl-CoA that enters the tricarboxylic acid (Krebs) cycle.

Enzymes involved in initial oxidation are classified according to the chain length of the substrate. For short-chain length (C_1–C_4) *n*-alkanes, activation is performed by enzymes that are related to methane monooxygenase (MMO) produced by aerobic methane oxidizing bacteria (Methenotrophs). This enzyme family contains two distinct forms of MMO: a soluble di-iron methane monooxygenase (sMMO) and a membrane-bound copper-containing methane monooxygenase (pMMO), which are encoded by *mmoX* and *pmoA* genes, respectively. MMO are involved in methane oxidation but also in the co-oxidation of various hydrocarbons. sMMO catalyse the co-oxidation of saturated, unsaturated, linear, branched and cyclic hydrocarbons with up to eight carbons, whereas pMMO act on alkanes and alkenes with up to five carbons

Table 4.1 Aerobic hydrocarbon-degrading strains isolated from marine sediments

Phylum/class/strain	Hydrocarbons utilized	Origin	Reference
Actinobacteria			
Gordonia sp. I37	Tetradecane and isoprene	Estuarine sediments	Alvarez *et al*. (2009)
Leifsonia sp. I47	Tetradecane and isoprene	Estuarine sediments	Alvarez *et al*. (2009)
Microbacterium oxydans MP-PYR1	Phenanthrene and pyrene	Mangrove sediments	Ke *et al*. (2009)
Mycobacterium parafortuitum CK1	Phenanthrene, fluorene, fluoranthene and pyrene	Mangrove sediments	Zhou *et al*. (2006)
Mycobacterium gilvum HH1	Phenanthrene, fluorene, fluoranthene, anthracene and pyrene	Mangrove sediments	Zhou *et al*. (2006)
Mycobacterium gilvum HH2	Phenanthrene, fluorene, fluoranthene and pyrene	Mangrove sediments	Zhou *et al*. (2006)
Mycobacterium psychrotolerans HH3	Phenanthrene, fluorene, fluoranthene and pyrene	Mangrove sediments	Zhou *et al*. (2006)
Mycobacterium gilvum MC	Phenanthrene, fluorene, fluoranthene and pyrene	Mangrove sediments	Zhou *et al*. (2006)
Mycobacterium chubuense MD	Phenanthrene, fluorene, fluoranthene, anthracene and pyrene	Mangrove sediments	Zhou *et al*. (2006)
Mycobacterium vanbaalenii MPHC	Phenanthrene, fluorene, fluoranthene, anthracene and pyrene	Mangrove sediments	Zhou *et al*. (2006)
Mycobacterium gilvum TE	Phenanthrene, fluorene, fluoranthene and pyrene	Mangrove sediments	Zhou *et al*. (2006)
Mycobacterium gilvum TI	Phenanthrene, fluorene and fluoranthene	Mangrove sediments	Zhou *et al*. (2006)
Mycobacterium fortuitum TM	Phenanthrene, fluorene and fluoranthene	Mangrove sediments	Zhou *et al*. (2006)
Mycobacterium farcinogenes TU	Phenanthrene, fluorene, fluoranthene and pyrene	Mangrove sediments	Zhou *et al*. (2006)
Rhodococcus ruber SKOO	Phenanthrene, fluoranthene and pyrene	Mangrove sediments	Guo *et al*. (2005)
Rhodococcus ruber HH6	Phenanthrene, fluorene and fluoranthene	Mangrove sediments	Guo *et al*. (2005)
Rhodococcus opacus CCS	Phenanthrene, fluoranthene and pyrene	Mangrove sediments	Guo *et al*. (2005)
Terrabacter sp. FLO	Phenanthrene, fluorene, anthracene and fluoranthene	Mangrove sediments	Guo *et al*. (2005)
Terrabacter sp. HH4	Phenanthrene, fluorene, anthracene and fluoranthene	Mangrove sediments	Guo *et al*. (2005)
Terrabacter sp.HH5	Phenanthrene, fluorene, anthracene and fluoranthene	Mangrove sediments	Guo *et al*. (2005)
Terrabacter sp. HF	Phenanthrene, fluorene, anthracene and fluoranthene	Mangrove sediments	Guo *et al*. (2005)
Firmicutes			
Bacillus cereus EJB8	*p*-Xylene	Deep-sea sediment	Wang *et al*. (2008)
Bacillus cereus Py5	Pyrene	Mangrove sediments	Lin and Cai (2008)
Bacillus fusiformis EJB1	Benzene, toluene and p-xylene	Deep-sea sediment	Wang *et al*. (2008)
Bacillus fusiformis EJB5	*p*-Xylene	Deep-sea sediment	Wang *et al*. (2008)
Bacillus megaterium Py6	Pyrene	Mangrove sediments	Lin and Cai (2008)
Bacillus mojavensis JB5	Benzene and toluene	Deep-sea sediment	Wang *et al*. (2008)
Bacillus mojavensis JB7	Toluene	Deep-sea sediment	Wang *et al*. (2008)
Bacillus sp. *SFI 13*	Fluoranthene	Lagoon sediments	Ben Said *et al*. (2008)

Table 4.1 continued

Phylum/class/strain	Hydrocarbons utilized	Origin	Reference
Bacillus indicus strain 10	Naphthalene	Coastal sediment	Nair *et al.* (2008)
Bacillus simplex strain 4 (11, 12)	Naphthalene	Coastal sediment	Nair *et al.* (2008)
Paenibacillus sp. ORNaP1	Naphthalene, anthracene and diesel fuel	Highly polluted lagoon sediments	Pepi *et al.* (2009)
Paenibacillus sp. *Strain 5*	Naphthalene	Coastal sediment	Nair *et al.* (2008)
Planomicrobium alkanoclasticum MAE2	Branched and *n*-alkanes (C_{11} to C_{33})	Intertidal beach sediment	Engelhardt *et al.* (2001)
Staphylococcus sp. *SFI 17*	Fluoranthene	Lagoon sediments	Ben Said *et al.* (2008)
Bacteroidetes			
Dyadobacter sp. I8	Tetradecane, isoprene and pristane	Estuarine sediments	Alvarez *et al.* (2009)
Yeosuana aromativorans GW1–1	Pyrene and benzo[a]pyrene	Estuarine sediments	Kwon *et al.* (2006)
Alphaproteobacteria			
Agrobacterium tumefaciens SFI 6	Fluoranthene	Lagoon sediments	Ben Said *et al.* (2008)
Kordiimonas gwangyangensis GW14–5	Benzo[a]pyrene, fluorene, phenanthrene, anthracene and pyrene	Marine sediments	Kwon *et al.* (2005)
Lutibacterium anuloederans LC8	Hexadecane, naphthalene and phenanthrene	Marine macrofaunal burrow sediments	Chung and King (2001)
Novosphingobium pentaromativorans	Hexadecane, naphthalene and phenanthrene	Marine macrofaunal burrow sediments	Chung and King (2001)
US6-1	Fluorene, phenanthrene, fluoranthene, anthracene, pyrene, benz[a] anthracene, chrysene, benz[b]fluoranthene and benzo[a]pyrene	Estuarine muddy sediment	Sohn *et al.* (2004)
Paracoccus versutus SPNT	Phenanthrene, fluoranthene and pyrene	Mangrove sediments	Guo *et al.* (2005)
Rhodococcus sp. i24	Tetradecane and isoprene	Estuarine sediments	Alvarez *et al.* (2009)
Sphingobium yanoikuyae TD4	Phenanthrene and fluorene	Mangrove sediments	Guo *et al.* (2005)
Sphingomonas herbicidovorans FCQ	Phenanthrene, fluorene and fluoranthene	Mangrove sediments	Guo *et al.* (2005)
Sphingomonas herbicidovorans MB	Phenanthrene and fluorene	Mangrove sediments	Guo *et al.* (2005)
Sphingomonas herbicidovorans MPHA	Phenanthrene, fluorene, fluoranthene and pyrene	Mangrove sediments	Guo *et al.* (2005)
Sphingomonas paucimobilis AC3	Phenanthrene and fluorene	Mangrove sediments	Guo *et al.* (2005)
Sphingomonas paucimobilis FCIB	Phenanthrene, fluorene, fluoranthene, pyrene and anthracene	Mangrove sediments	Guo *et al.* (2005)

Sphingomonas paucimobilis FCL	Phenanthrene, fluorene and fluoranthene	Mangrove sediments	Guo *et al*. (2005)
Sphingomonas paucimobilis HL2	Phenanthrene, fluorene and anthracene	Mangrove sediments	Guo *et al*. (2005)
Sphingomonas paucimobilis HL5	Fluorene	Mangrove sediments	Guo *et al*. (2005)
Sphingomonas sp. 2MPII	2-Methylphenanthrene	Coastal hydrocarbon-polluted sediment	Gilewicz *et al*. (1997)
Sphingomonas sp. HL3	Phenanthrene and fluorene	Mangrove sediments	Guo *et al*. (2005)
Sphingomonas sp. HL4	Phenanthrene, fluorene, fluoranthene and anthracene	Mangrove sediments	Guo *et al*. (2005)
Sphingomonas sp. MW4	Phenanthrene, fluorene and fluoranthene	Mangrove sediments	Guo *et al*. (2005)
Sphingomonas sp. MWFG	Phenanthrene, fluoranthene and pyrene	Mangrove sediments	Guo *et al*. (2005)
Sphingomonas sp. MWW	Phenanthrene	Mangrove sediments	Guo *et al*. (2005)
Sphingomonas sp. SKNG	Phenanthrene	Mangrove sediments	Guo *et al*. (2005)
Sphingomonas sp. SPNW	Phenanthrene	Mangrove sediments	Guo *et al*. (2005)
Sphingomonas wittichii HL1	Phenanthrene, fluorene, fluoranthene, anthracene	Mangrove sediments	Guo *et al*. (2005)
Sphingopyxis composta C7	Phenanthrene, fluorene, fluoranthene, anthracene, pyrene	Mangrove sediments	Guo *et al*. (2005)
Betaproteobacteria			
Achromobacter xylosoxidans SFI 5	Fluoranthene	Lagoon sediments	Ben Said *et al*. (2008)
Achromobacter xylosoxidans SFI 20	Fluoranthene	Lagoon sediments	Ben Said *et al*. (2008)
Gammaproteobacteria			
Acinetobacter sp. SFI 18	Fluoranthene	Lagoon sediments	Ben Said *et al*. (2008)
Acinetobacter sp. SFI 19	Fluoranthene	Lagoon sediments	Ben Said *et al*. (2008)
Alcanivorax borkumensis SK2	Straight-chain alkanes up to C_{32} in length, long-chain isoprenoids, phytane and pristane and alkyl-aromatic hydrocarbons	Sea water/sediment samples	Yakimov *et al*. (1998)
Alcanivorax jadensis T9	Tetradecane, hexadecane and pristane	Intertidal sediment	Bruns and Berthe-Corti (1999)
Cycloclasticus pugetti PS-1	Naphthalene, phenanthrene, anthracene and toluene	Marine sediments	Dyksterhouse *et al*. (1995)
Cycloclasticus sp. P1	Naphthalene, 2-methylnaphthalene, 2,6-dimethylnaphthalene, biphenyl, fluorene, acenaphthene, DBF, DBT, phenanthrene, anthracene, and pyrene	Deep-sea sediment	Wang *et al*. (2008)
Cycloclasticus spirillensus M4–6	Naphthalene and phenanthrene	Marine macrofaunal burrow sediments	Chung and King (2001)
Marinobacter alkaliphilus ODP1200D-1.5	Crude oil, *n*-pentadecane, *n*-tridecane, *n*-undecane, *n*-decane, *n*-nonane, butane, and kerosene	Sub-seafloor alkaline serpentine mud volcano	Takai *et al*. (2005)

Table 4.1 continued

Phylum/class/strain	Hydrocarbons utilized	Origin	Reference
Marinobacter hydrocarbonoclasticus SP17	Tetradecane, hexadecane, phenyldecane, phenanthrene, eicosane, heneicosane and pristane	Coastal hydrocarbon-polluted sediment	Gauthier *et al.* (1992)
Marinobacter sp. GPM2507	Fluoranthene	Mangrove sediments	Brito *et al.* (2006)
Marinobacter sp. GPM2508	Pristane	Mangrove sediments	Brito *et al.* (2006)
Marinobacter sp. GPM2512	Octadecane	Mangrove sediments	Brito *et al.* (2006)
Marinobacter sp. NCE312	Naphthalene and 2-methylnaphthalene	Creosote-contaminated marine sediment	Hedlund *et al.* (2001)
Marinobacter squalenivorans 2Asq64	Benzothiophene, tetradecane, 1-tetradecene, perhydrofluorene and squalene	Marine environment	Rontani *et al.* (2003)
Neptunomonas naphthovorans NAG-2N-126	Naphthalene and 2-methylnaphthalene	Contaminated marine sediment	Hedlund *et al.* (1999)
Oleiphilus messinensis ME102	Aliphatic hydrocarbons (C_{11} to C_{20}) and their oxidized derivatives	Seawater/sediment samples	Golyshin *et al.* (2002)
Pseudomonas mendocina SFI 8	Fluoranthene	Lagoon sediments	Ben Said *et al.* (2008)
Pseudomonas nitroreducens SFI 15	Fluoranthene	Lagoon sediments	Ben Said *et al.* (2008)
Pseudomonas pohangensis strain 9	Naphthalene	Coastal sediment	Nair *et al.* (2008)
Pseudomonas putida SFI 2	Fluoranthene	Lagoon sediments	Ben Said *et al.* (2008)
Pseudomonas sp. SFI 10	Fluoranthene	Lagoon sediments	Ben Said *et al.* (2008)
Pseudomonas sp. ORNaP2	Naphthalene	Highly polluted lagoon sediments	Pepi *et al.* (2009)
Pseudomonas sp. SFI 1	Fluoranthene	Lagoon sediments	Ben Said *et al.* (2008)
Pseudomonas sp. SFI 11	Fluoranthene	Lagoon sediments	Ben Said *et al.* (2008)
Pseudomonas sp. SFI 12	Fluoranthene	Lagoon sediments	Ben Said *et al.* (2008)
Pseudomonas sp. SFI 3	Fluoranthene	Lagoon sediments	Ben Said *et al.* (2008)

Pseudomonas sp. SFI 7	Fluoranthene	Lagoon sediments	Ben Said *et al*. (2008)
Shewanella oneidensis SFI 4	Fluoranthene	Lagoon sediments	Ben Said *et al*. (2008)
Shigella boydii SFI 16	Fluoranthene	Lagoon sediments	Ben Said *et al*. (2008)
Stenotrophomonas maltophilia			
SFI 14	Fluoranthene	Lagoon sediments	Ben Said *et al*. (2008)
Thalassolituus oleivorans MIL-1	Aliphatic hydrocarbons (C7 to C20) and their oxidized derivatives	Seawater/sediment samples	Yakimov *et al*. (2004)
Uncultured *Pseudomonas* sp. SFI 9	Fluoranthene	Lagoon sediments	Ben Said *et al*. (2008)
Unknown			
Strain HCCW	Phenanthrene	Mangrove sediments	Guo *et al*. (2005)
Strain YOWG	Phenanthrene	Mangrove sediments	Guo *et al*. (2005)
Strain KLHS	Phenanthrene	Mangrove sediments	Guo *et al*. (2005)
Strain MPSS	Phenanthrene, fluoranthene and pyrene	Mangrove sediments	Guo *et al*. (2005)

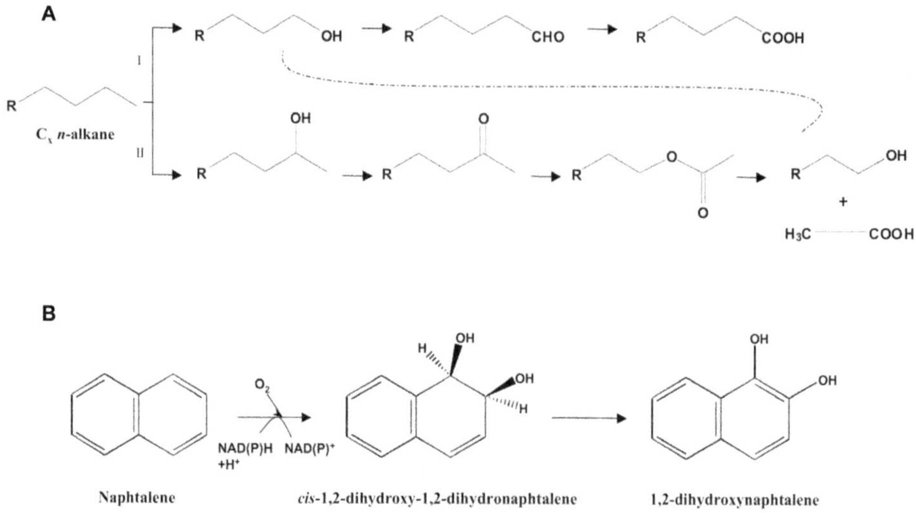

Figure 4.1 Most frequent pathways for the aerobic activation of alkanes (A) and naphthalene (B).

(Berthe-Corti and Bruns, 2001). Methanotrophs are known to be active in environments with high methane content like seafloor hydrocarbon seeps and methane vents along the continental margins (Suess *et al.*, 1999). Thus, the fate of deposited or seeped oil hydrocarbons in such environments may be adventitiously affected by methanotroph activities.

Medium-chain length (C_5–C_{16}) alkanes may be oxidized by two main classes of enzymes: integral membrane non-heme iron alkane hydroxylases (the *alk* system) and soluble cytochromes P450. The former is an enzymatic system often found in alkane-degrading bacteria, which has been deeply investigated in *Pseudomonas putida* GPo1 (Smits *et al.*, 2002). Genes related to alkane hydroxylase genes of *P. putida* GPo1 (*alkB*) are mined in different genomes of HCB (Golyshin *et al.*, 2002; Sabirova *et al.*, 2006) and in various environments such as soils and aquifers (Sotsky *et al.*, 1994; Guo *et al.*, 1997; Whyte *et al.*, 2002). In the case of marine sediments, only few studies report on the diversity and the distribution of *alkB* genes (Kuhn *et al.*, 2009; Wasmund *et al.*, 2009). Nevertheless, numerous alkane hydroxylase genes (*alkB*), distantly related to previously described AlkB sequences and with original transcription directions, have been discovered in deep-sea sediments suggesting that these marine environments may be a reservoir of novel catalytic enzymes (Xu *et al.*, 2008). In most studied environments, alkane exposure induced a higher copy number of *alkB* genes (Whyte *et al.*, 2002; Heiss-Blanquet *et al.*, 2005) but a decrease of *alkB* genes diversity (Kuhn *et al.*, 2009) suggesting a bloom of specialized hydrocarbonoclastic organisms. The second class of hydroxylases, mediating the terminal hydroxylation of medium-chain *n*-alkanes, gathers enzymes related to the soluble cytochrome P450 CYP153A1 from *Acinetobacter* sp. EB104 (Maier *et al.*, 2001). These enzymes have been isolated from several bacterial strains that degrade medium-chain alkanes (e.g. *Alcanivorax borkumensis* SK2) and from various environments such as surface seawater and contaminated soil or groundwater (Kubota *et al.*, 2005; Schneiker *et al.*, 2006; Wang *et al.*, 2010). The prevalence of such enzymes in marine sediments is still poorly documented (Kuhn *et al.*, 2009). For long-chain length ($>C_{20}$) alkanes, enzymes responsible for the initial oxidization are identified but not well characterized thus far (Maeng *et al.*, 1996; Tani *et al.*, 2001; Throne-Holst *et al.*, 2007; Rojo, 2010).

It has been highlighted that some HCB may carry different alkane hydroxylase systems (Sabirova *et al.*, 2006; Wang *et al.*, 2010). Therefore, the genome analysis of *A. borkumensis* SK2 shows the presence of genes encoding two AlkB systems (AlkB1 and AlkB2) and three cytochromes P450 (Schneiker *et al.*, 2006). Transcriptomics and proteomics experiments show upregulation of *alkB1* genes on C_5–C_{12} *n*-alkanes, *alkB2* genes on C_8–C_{16} *n*-alkanes and P450 genes on isoprenoid hydrocarbons. Metabolic versatility may be then advantageous for microbes, allowing the use of a wide range of resources in oil-perturbed environments. *A. borkumensis* SK2 can also degrade *n*-alkanes up to C_{32}, branched alkanes, alkylarenes and alkyl-cycloalkanes suggesting activity of other enzymatic systems yet undiscovered (Schneiker *et al.*, 2006).

Aromatic hydrocarbons can be used by microorganisms through different catabolic routes than alkanes. Aromatic hydrocarbons are first transformed into few intermediates via a number of peripheral pathways; they are then channelled by a few central pathways to the central cellular metabolism. The initial activation step of aromatic hydrocarbons generally consists in the introduction of molecular oxygen at two adjacent carbon atoms of the aromatic nucleus (Fig. 4.1B). This reaction features a multicomponent aromatic ring-hydroxylating-dioxygenase enzyme system leading to the formation of *cis*-dihydrodiol. These compounds are re aromatized to yield dihydroxylated substrates which can be cleaved and funnelled into the Krebs cycle (Habe and Homori, 2003). Alternatively, peripheral reactions preparing aromatic hydrocarbons for ring cleavage can be performed by enzymatic systems like soluble di-iron monooxygenases or like single or two components flavoprotein monooxygenases (Pérez-Pantoja *et al.*, 2010). The most studied group of oxygenases contain sequences related to naphthalene 1,2-dioxygenase from *P. putida* NCIB 9816-4. Genes encoding the alpha subunit of this dioxygenase (*nahAc* genes) are used as functional biomarkers to directly investigate microbes with specific biodegradation potential in the environment (Widada *et al.*, 2002). Recently, novel PAHs catabolic genes, evolutionarily different from *nah*-like genes, have been characterized in aromatic HCB and in marine sediments (Hedlund *et al.*, 1999; Saito *et al.*, 2000; Gomes *et al.*, 2007; Lozada *et al.*, 2008; Marcos *et al.*, 2009).

It is clear that knowledge on aerobic hydrocarbon-degrading microorganisms and their catabolic routes is constantly improving, but there are still important gaps on hydrocarbons biodegradation under hypoxic or oxygen-oscillating conditions. Under micro-aerophilic conditions, hydrocarbons co-metabolism by ammonium-oxidizing bacteria (AOB) could play an important role in bioremediation. AOB are ubiquitous in sediments and particularly abundant at oxic/anoxic interfaces exhibiting high fluxes of ammonia coming from the anoxic areas and low molecular oxygen concentrations (Berthe-Corti *et al.*, 2002). These nitrifying bacteria possess the ammonia monooxygenase enzyme which sequentially oxidizes ammonia to nitrite and then nitrite to nitrate. This enzyme exhibits broad substrate specificity and may contribute to the initial activation of short-chain aliphatic hydrocarbons and of numerous aromatics. Under aerobic conditions, aromatic hydrocarbons can be subject to non-oxygenolytic ring cleavage. The involvement of CoA-dependent pathways in aromatic hydrocarbons biodegradation is known for decades but only under anaerobic conditions. As suggested by Pérez-Pantoja *et al.* (2010), CoA dependent pathways could play an important role in hydrocarbons degradation under oxygen oscillating conditions which prevail in the burrows of numerous benthic macro-organisms (see section 'Fate of hydrocarbons in bioturbated sediments: back to the real world').

Anaerobic hydrocarbon-degrading bacteria in marine sediments: previously unsuspected actors

Crucial role of molecular oxygen in hydrocarbon activation (see above) has long favoured the view that microbes could not act anaerobically on hydrocarbons, and that hydrocarbons buried in anoxic subsurface sediments are likely to be preserved for a long period of time. Convincing evidence for anaerobic bacterial utilization of hydrocarbons emerged by the end of the twentieth century (for reviews see Spormann and Widdel, 2000; Widdel and Rabus, 2001; Widdel *et al.*, 2010). Research on and detection of anaerobic hydrocarbon-degrading microorganisms have been investigated either by *in situ* observations pointing at the disappearance of targeted hydrocarbons in anoxic habitats (Hanson *et al.*, 1999; Grossi *et al.*, 2002; Jones *et al.*, 2008), and/or by the classical use of enrichment cultures on specific hydrocarbons (e.g. Hanson *et al.*, 1999). Both approaches yielded to the detection of metabolically and phylogenetically distinct anaerobic hydrocarbon degraders which could either be isolated in pure cultures and described in detail, or be enriched in cultures and characterized by molecular analysis. This has provided fundamental information on anaerobic biodegradation pathways for compounds such as alkanes, BTEXs and PAHs, and on the microorganisms able to oxidize hydrocarbons in the absence of oxygen with nitrate, sulfate or other electron acceptors. A general view is that anaerobic microorganisms utilizing hydrocarbons always exhibit slower growth than their aerobic counterparts. Utilization of only a distinct, narrow range of *n*-alkanes by individual strains under anoxic conditions seems to be a common feature. Anaerobic bacteria that degrade *n*-alkanes with six or more carbon with sulfate or nitrate as electron acceptor are the most frequently isolated, whereas anaerobic utilization of shorter-chain *n*-alkanes has been observed far less frequently, and this process is thought to be limited to gas seep areas. Knowledge about anaerobic degraders of branched and cyclic alkanes is also scarce and oxidation of such hydrocarbons has been shown only in enriched communities (Bregnard *et al.*, 1997; Rios-Hernandez *et al.*, 2003) or in a pure denitrifying bacterium (Wilkes *et al.*, 2003).

Degraders of non-methane hydrocarbons are phylogenetically diverse and many sedimentary hydrocarbonoclastic (nitrate or sulfate reducing) Proteobacteria have been isolated in pure cultures (Table 4.2). Sulfate reduction in anoxic marine sediments has been shown to play a predominant role in the mineralization of a wide variety of aliphatic and aromatic hydrocarbons (Coates *et al.*, 1997) and sulfate-reducing hydrocarbonoclastic *Bacteria* (SRB) belonging to the *Desulfobacteraceae* (e.g. *Desulfosarcina, Desulfococcus, Desulfatibacillum, Desulfatiferula*) have been isolated (for a review see Widdel *et al.*, 2010). These results are supported by the observation of dominant *Bacteria* closely related to *Desulfococcus–Desulfonema–Desulfosarcina*-like and to *Desulfovibrio–Desulfomicrobium*-like groups during *in situ* oil biodegradation experiments in infralittoral sediments (Miralles *et al.*, 2007a). Several hydrocarbonoclastic SRB may also grow in syntrophic association with methane-oxidizing *Archaea* (Teske, 2010). In such consortia, bacterial strains might reduce sulfate with scavenged methane-derived product(s) produced by archaeal partner (Schink, 2002). Several *Archaea* potentially involved in these processes in sediments or in enrichment cultures belong to anaerobic methanotrophs groups (ANME-1, ANME-2 and ANME-3) mainly placed within the *Methanosarcinales* (*Euryarchaeota*). Long-chain *n*-alkane degradation by methanogenic bacterial communities has also been shown using hydrocarbon-rich sediments (Zengler *et al.*, 1999; Alain *et al.*, 2006). Some members of these consortia have been identified and belong to *Methanosaeta, Methanoculleus* and *Synthrophus*. These synthrophic associations might have a significant role in hydrocarbons degradation in marine sediment as suggested by recent *in situ* experiments (Miralles *et al.*, 2010).

The anaerobic activation of alkanes is of particular interest, as they are very unreactive compounds. Two mechanisms of anaerobic *n*-alkane degradation have been described to date. They involve unprecedented biochemical reactions that differ completely from those used in aerobic hydrocarbons metabolism and yield to a series of specific metabolites which can be used to assess the *in situ* biodegradation of these contaminants in anoxic environments. Field samples collected at different sites show the presence of these metabolites and demonstrate that microbial degradation of hydrocarbons is actively taking place within these anoxic environments (Aitken *et al.*, 2004; Young and Phelps, 2005). The most common and precisely described mechanism of anaerobic alkane oxidation involves the activation at the subterminal carbon of the alkane and the addition to a molecule of fumarate yielding an alkylsuccinate. This molecule of alkylsuccinate then rearranges (following an assumed activation to the thioester with free CoA or via a CoA-transferase) into methyl-branched fatty acids

Table 4.2 Anaerobic hydrocarbon degraders in pure cultures or highly enriched cultured from marine sediments

Phylum/class/strain	Electron acceptor	Hydrocarbons utilized	Origin	Reference
Firmicutes				
Enrichment (Propane60-GuB)	SO_4^{2-}	Propane	Sediment from hydrocarbons seep area	Kniemeyer *et al*. (2007)
Enrichment (Firmicutes and Proteobacteria)	SO_4^{2-}	Benzene	Guaymas Basin sediment	Phelps *et al*. (1998)
Deltaproteobacteria				
Desulfurothermus naphthae TD3	SO_4^{2-}	*n*-alkanes (C_6–C_{16}	Guaymas Basin sediment	Rueter *et al*. (1994)
Desulfatibacillum aliphaticivorans CV2803	SO_4^{2-}	*n*-alkanes (C_{13}–C_{18}), *n*-1-alkenes (C_7–C_{23})	Hydrocarbon polluted marine sediments	Cravo-Laureau *et al*. (2004a)
Strain BuS5	SO_4^{2-}	Propane, *n*-butane	Sediment from hydrocarbons seep area	Kniemeyer *et al*. (2007)
Strain PL12	SO_4^{2-}	*n*-alkanes (C_{16}, C_{10})	Coastal sediments	Higashioka *et al*. (2009)
Strain Pnd3	SO_4^{2-}	*n*-alkanes (C_{14}–C_{17}), *n*-1-hexadecene (C_{16})	Marine sediment	Aeckersberg *et al*. (1991)
Strain AK-01	SO_4^{-}	*n*-alkanes (C_{13}–C_{18}), *n-1*-alkenes (C_{15}, C_{16})	Chronically contaminated sediments	So and Young (1999)
Desulfatibacillum alkenivorans PF2803	SO_4^{2-}	*n-1*-Alkenes (C_8–C_{23})	Oil-polluted sediments	Cravo-Laureau *et al*. (2004b)
Desulfatiferula olefinivorans LM2801	SO_4^{2-}	*n*-1-Alkenes (C_{14}–C_{23})	Brackish sediment of a wastewater decantation facility of an oil refinery	Cravo-Laureau *et al*. (2007)
Strains GhAcy 1, GhAcy 3	Fermentation	Acetylene	Marine sediment	Schink (1985)
Strains NaphS2, NaphS3, NaphS6	SO_4^{2-}	Naphthalene, 2-methylnaphthalene	Marine sediment	Galushko *et al*. (1999)
Desulfobacula toluolica Tol2	SO_4^{2-}	Toluene	Marine sediment	Rabus *et al*. (1993)
Desulfobacula phenolica PhO1	SO_4^{2-}	Toluene	Marine mud	Rabus *et al*. (1993)
Strain EbS7	SO_4^{2-}	Ethylbenzene	Marine sediment	Kniemeyer *et al*. (2003)

Table 4.2 continued

Phylum/class/strain	Electron acceptor	Hydrocarbons utilized	Origin	Reference
Enrichment (Protobacteria, Firmicutes, Chloroflexi)	SO_4^{2-}	*n*-Propane, *n*-Pentane	Hydrocarbon seep	Savage *et al.* (2010)
Enrichment (Butane12-Gme)	SO_4^{2-}	*n*-Butane	Sediment from hydrocarbons seep area	Kniemeyer *et al.* (2007)
Enrichment	SO_4^{-}	*n*-Butane	Marine sediment	Musat and Widdel (2008)
Enrichment	SO_4^{-}	Phenanthrene	Contaminated marine sediment	Davidova *et al.* (2007)
Enrichment	SO_4^{2-}	*p*-Xylene	Marine sediment	Nakagawa *et al.* (2008)
Gammaproteobacteria				
Pseudomonas sp. AI1	NO_3^-	*n*-1-Heptadecene (C_{17})	Hydrocarbon-polluted marine sediment	Gilewicz *et al.* (1991)
Pseudomonas balearica (BerOc6)	NO_3^-	*n*-Alkanes (C_{15}–C_{18})	Estuarine sediment	Grossi *et al.* (2008)
Marinobacter sp. (2sq31;2Asq64)	NO_3^-	Squalene	Marine sediments	Bonin *et al.* (2002), Rontani *et al.* (2002)
Pseudomonas stuzeri NAP-3–1	NO_3^-	Naphthalene	Contaminated marine sediment	Rockne *et al.* (2000)
Vibrio pelagius NAP-4	NO_3^-	Naphthalene	Contaminated marine sediment	Rockne *et al.* (2000)
Unknown				
Strain NAP-3–2	NO_3^-	Naphthalene	Contaminated marine sediment	Rockne *et al.* (2000)

which can be further metabolized via β-oxidation sequences (Fig. 4.2A, pathway I; Wilkes *et al.*, 2002; Cravo-Laureau *et al.*, 2005). In the unique case of propane, the fumarate molecule is added to one of the terminal carbon atoms of the alkane (Kniemeyer *et al.*, 2007). The second pathway is based on the addition of inorganic carbon at C-3 of the alkane (Fig. 4.2A, pathway II) but the precise mechanism remained incompletely understood (So *et al.*, 2003). Both pathways, which seem to occur simultaneously in mixed sulfate-reducing communities (Callaghan *et al.*, 2006), oppose an hypothetical mechanism presented in previous literature involving the dehydrogenation of alkanes to the corresponding alk-1-enes (Aeckersberg *et al.*, 1998; Grossi *et al.*, 2008). A third pathway has been suggested in a denitrifying bacterium *Pseudomonas balearica* strain BerOc6 isolated from a brackish lagoon but the mechanism involved is unknown (Grossi *et al.*, 2008).

Both addition to fumarate and carboxylation have been observed previously as initial reactions during the anaerobic degradation of some aromatic hydrocarbons. The reaction of alkyl chains with a fumarate co-substrate yielding an alkylsuccinate adduct has been demonstrated for compounds such as toluene, xylenes, ethylbenzene, *p*-cresol, or methylnaphthalene (Biegert *et al.*, 1996; Kniemeyer *et al.*, 2003; Heider, 2007). Carboxylation reactions have been proposed as an activation mechanism in the anaerobic oxidation of unsubstituted compounds such as benzene, naphthalene or phenanthrene (Annweiler *et al.*, 2002; Musat *et al.*, 2008). In the case of naphthalene, the methylated intermediate is then further converted by addition to fumarate in analogy to anaerobic toluene degradation, (Fig. 4.2B; Safinowski and Meckenstock, 2004). Some denitrifying and Fe(III)-reducing hydrocarbon degrading

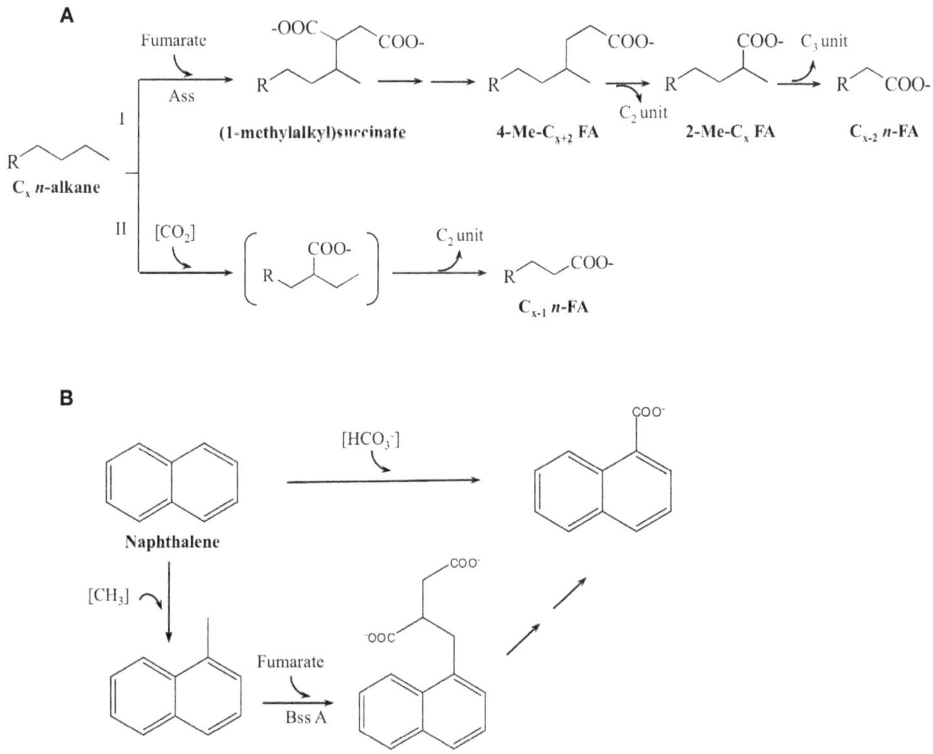

Figure 4.2 Initial activation reactions involved in anaerobic degradation of alkanes (A) and naphthalene (B).

bacteria can also activate toluene, ethylbenzene and *p*-cresol via hydroxylation (see review by Boll and Heider, 2010).

The finding of benzylsuccinate as a metabolite in toluene-degrading anaerobic cultures and subsequent enzymatic measurements (Biegert *et al.*, 1996; Beller and Spormann, 1997) have allowed a detailed description of the toluene-activating benzylsuccinate synthase (mostly termed BssA). This glycyl radical enzyme has since been demonstrated to be involved in the addition to fumarate of other aromatic hydrocarbons (Heider, 2007). More recently, an enzyme involved in 2-methylnaphthalene activation by methyl group addition to fumarate (Nms) has been identified from SRB enriched from Mediterranean sediment (Musat *et al.*, 2009). These enzymes belonging to the glycyl radical enzymes group are analogous to the large subunit of BssA. The enzyme involved in the methylation reaction of unsubstituted hydrocarbons is unknown; a likely candidate might be a B_{12}-containing methyltransferase.

A related radical-catalysed alkane addition to fumarate, similar to toluene activation, has been suggested by spectroscopic (EPR) measurement of an organic radical in a denitrifying bacterium grown on *n*-hexane (Rabus *et al.*, 2001). This assumption is further corroborated by the analysis of genes and relationships of deduced proteins in anaerobic alkane degraders (Callaghan *et al.*, 2008; Grundmann *et al.*, 2008) and hydrocarbon-impacted environments (Callaghan *et al.*, 2010). The apparent glycyl radical proteins from the anaerobic alkane degraders represent mainly the large subunits of the alkane-activating enzyme, alkylsuccinyl synthase (Ass; Callaghan *et al.*, 2008) or (1-methylalkyl)succinate synthase (Mas; Grundmann *et al.*, 2008), but smaller proteins with some similarity to those found in BssA may be part of the native Ass or Mas enzymes.

Undoubtedly, anaerobic bacterial processes play a significant role in the fate of hydrocarbons in marine sediments, in parallel to the aforementioned aerobic biodegradation. Consequently, both aerobic and anaerobic processes are likely to be strongly complementary in the remediation of hydrocarbon-polluted marine sediments.

Fate of hydrocarbons in bioturbated sediments: back to the real world

Aquatic communities are structured by complex interactions between biotic and abiotic processes. These interactions are particularly relevant in benthic communities. One of the central ideas to emerge recently has been the concept of 'ecosystem engineers' – species that change the availability of resources to other species by physical alteration of the environment. Jones *et al.* (1994) distinguished two categories: 'autogenic' engineers that affect the habitat by their physical presence, and 'allogenic' engineers that alter the environment via their behaviour and activities. As a subset of allogenic engineering, bioturbation – mixing of sediment particles and solutes by benthic fauna – is recognized as one of the major processes that influence the structure and function of sediment environments (Palmer *et al.*, 1997; Lohrer *et al.*, 2004). Particularly, these and other number of macrofaunal activities like particle manipulation, grazing, excretion/secretion, burrow/tube construction are known or have been inferred to stimulate or modulate microbial metabolism and capacity for organic matter degradation in bioturbated sediments (Aller and Aller, 1998; Kristensen and Kostka, 2005).

Bioturbation, the ever-changing sea floor

Through bioturbation dwelling macroorganisms introduce temporal and spatial heterogeneity in the sediment system (Rhoads, 1974; Aller, 1982; Berkenbusch and Rowden, 1999; Pischedda *et al.*, 2008). This is partly linked to the diverse sediment mixing modes which have been classified into five functional groups: biodiffusors, upwards and downwards conveyors, regenerators and gallery-diffusors (Gardner *et al.*, 1987; François *et al.*, 1997; Gérino *et al.*, 2003). Another example is the organisms' active irrigation activities. Renewal of burrow water (i.e. bioirrigation) with oxygen-rich overlying water allows important functions for infaunal animals, such as supply of oxygen, removal of toxic metabolites and providing suspended food (Kristensen, 2001). Bioirrigation is intermittent with active and quiet periods which follow a more or less cyclic rhythm. Temporal patterns of irrigation are species-specific and may vary with physiological state and feeding behaviour of the animal (Christensen *et al.*, 1999). For instance, the polychaete *Nereis virens* is active for about 20% of the time with 5–10 min irrigation periods followed by 20–30 min of inactivity (Kristensen, 1989) whereas, the polychaete *Arenicola marina* proceeds in cycles of 40–60 min duration with about 10 min of intense irrigation followed by a progressive decrease in the activity (Kristensen, 2001).

Transport of reactants across redox boundaries due to bioturbation may considerably affect rates and pathways of organic matter mineralization (Aller, 1994; Gilbert *et al.*, 1996; Hulthe *et al.*, 1998; Sun *et al.*, 1999; Grossi *et al.*, 2003). In marine sediments, oxygen-deficient and anoxic conditions are common, and bacterial metabolisms utilizing alternative electron acceptors other than molecular oxygen for organic matter oxidation dominate. There is a successive utilization of inorganic electron acceptors following, in principle, the free energy yield ($\Delta G^{\circ\prime}$) per mole of organic carbon oxidized by each of these compounds, resulting in a biogeochemical zonation (Fig. 4.3A). Owing to synthrophic associations (Sieber *et al.*, 2010) and different linkages between chemolithotrophic and suboxic/anoxic processes, this zonation may be in fact more complex (Burdrige, 2006; Hulth *et al.*, 2005). Anyway, owing to the reworking activity, the biological mixing alters aerobic/anaerobic boundaries and biogeochemical zonation (Fig. 4.3B) generating geometrically and temporally complex redox mosaics (Aller, 1994). This results in the modification of the expression of the main bacterial activities, which in turns influences the degradation of OM (Kristensen and Blackburn, 1987; Aller, 1994; Aller, 2001). For instance, sulfate reduction rates are up to three orders of magnitude higher (140–790 nmol SO_4^{2-}/cm^3 per day) in reduced microniches associated with ghost shrimp (*Neotrypaea californiensis*) burrows compared with the surrounding sediments (Bertics and Ziebis, 2010). Thus, one major consequence of macrofaunal bioturbation is the establishment of a highly heterogeneous system (Fig. 4.3C) associated with dynamic biogeochemical variables constantly changing with wide amplitudes, on both small spatial (mm, cm) and short temporal (s, min) scales (Glud *et al.*, 1998; Aller, 2001; Wenzhöfer and Glud, 2004; Glud, 2008; Pischedda *et al.*, 2008, 2009). The relative frequency and duration of redox changes was suggested to be critical for microbial activities with the existence of rapid switches in dominant metabolic processes (Aller, 1994).

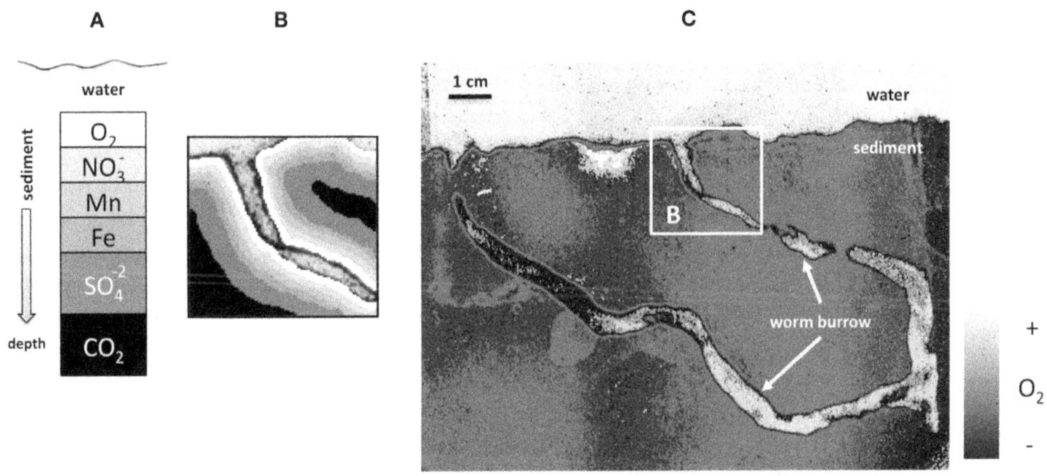

Figure 4.3 Theoretical biogeochemical redox zonation of electron acceptors used in the sediments in (A) a non-bioturbated sediment and (B) the microenvironment surrounding the entrance of an irrigated macrofaunal burrow. (C) Two-dimensional oxygen distribution pattern around a burrow of the polychaete *Nereis virens* obtained with a planar optode. The intermittent irrigation of the burrow allows deeper oxygen penetration within the mostly anoxic sediment matrix.

Last but not least, densities of benthic organisms carrying bioturbation are often high and could reach up to thousands individuals per square metre (e.g. >3500 *Amphiura filiformis* ind./m^2 in the south-eastern deep trench of Skagerrak, Sweden; Rosenberg, 1995), potentially leading to a preponderant control of biogeochemical processes within the sediments. However, both experimental and field measurements have demonstrated that mineralization rates (more particularly denitrification rates) could also vary inversely with bioirrigation intensity when reaching highest organism densities and irrigation rates (Berelson *et al.*, 1998; Gilbert *et al.*, 2003).

Overall, bioturbation has a pivotal role in the transport, biodegradation, accumulation and bioavailability of hydrocarbons in marine sediments. An understanding of the effects of redox oscillating conditions prevailing in bioturbated sediments may lead to further relevant bioremediation strategies, as will be discussed in the last part of this chapter.

The bioturbation-driven hydrocarbons transport

The fate and transport of hydrocarbons in sediments often involve very complex processes that depend on the physical/chemical properties of the hydrocarbons and the sediments, as well as the type of benthic communities inhabiting the sediment. Owing to their hydrophobic and lipophilic characters, hydrocarbons (particularly those of high molecular weight like PAHs having a high octanol–water partition coefficient K_{ow}) tend to be strongly associated with the particulate fraction of the sediment, thus partitioning only weakly into the porewaters and overlying water. Once settled they can be passively buried more or less rapidly, according to the sedimentation rate. They can also be actively buried under the action of macrobenthic organisms. In a field study of simulated oil spill in a coastal area (deposition of Arabian light crude oil at the surface of sediment cores), Grossi *et al.* (2002) showed a continuous macrobenthic burial of hydrocarbons under the form of accumulation zones moving deeper with time (at depths of 10–20 mm, 20–30 mm and 20–55 mm after 2, 6 and 12 months of *in situ* incubation respectively). Similarly, 10 days of laboratory experiments with four benthic species (the bivalve *Abra nitida*, the polychaete *Scalibregma inflatum*, and the echinoderms *Amphiura filiformis* and *Echinocardium cordatum*) showed a transfer of hydrocarbons from the surface sediment down to approximately 5 cm depth by both continuous (biodiffusion) and non-continuous (biotransport) transport (see Caradec *et al.*, 2004a). Interestingly, the hydrocarbons repartition was species-specific, the four studied species having different effects on particulate matter redistribution in the upper part of the sediments.

Conversely, bioturbation can also significantly increase the release of previously buried hydrocarbons or corresponding metabolites (see next section) from the sediment into the water column. Indeed, in a microcosm experiment, Schaffner *et al.* (1997) showed that macrofauna activities such as tube and burrow irrigation enhanced the short-term (a few days) diffusive loss (desorption) of pyrene and benzo[a]pyrene. A similar observation is provided by Christensen *et al.* (2002b), who demonstrated that bioturbation activities of the marine polychaetes *A. marina* and *Nereis diversicolor* enhanced the flux of sediment-associated pyrene and its metabolites into overlying water. More recently, Qin *et al.* (2010) have shown an enhanced PAHs desorption in the burrow of the crab *Helice tiensinensis*, which could be attributed to the increase in dissolved organic matter in pore water as well as to the mechanical mixing by the crab. Overall, it has been proposed that bioturbation may be responsible for a large, if not dominant, fraction of the chemical quantities released from the sediment into the water column. The bioturbation-driven release of contaminants from the sediment involves initial bioturbation-generated transport of particle-bound contaminants into the sediment–water interface, followed by rapid desorption and subsequent diffusion through the benthic boundary layer into overlying waters (Thibodeaux and Bierman, 2003). For instance, it has been demonstrated that the continuous activity of the invasive polychaete *Marenzelleria* spp. can be equivalent or even more important than physical disturbance in the remobilization of sediment-bound contaminants in the Baltic Sea (Hedman *et al.*, 2010). It was further shown that *Marenzelleria neglecta* accumulates low amounts of contaminants but also significantly stimulates aquatic release of bioavailable sequestered contaminants. Bioturbation by this polychaete at realistic (localized maxima of 30–40,000 ind./m^2, as reported by Kube *et al.*, 1996) or even low densities (2350 ind./m^2), increases the average flux of released dissolved polychlorobiphenyls (PCBs) by a factor of 0.7–3.0 (Granberg *et al.*, 2008). As it will be further discussed in the section dedicated to bioavailability and bioturbation, hydrocarbons desorption from particles also occurs in deposit feeders guts allowing dissolved hydrocarbons to be available to degrading organisms.

Other factors like the quality and quantity of OM present within or deposited on the surface of the sediment may also strongly influence the fate of hydrocarbons in bioturbated sediments. Indeed, OM quality affects organism activity and feeding behaviour, and is therefore also likely to affect hydrocarbons fate in benthic systems. Sediment organic enrichment may lead to the enhancement of the biomass of organisms and of the bioturbation potential that in turn will influence hydrocarbons cycling (Konovalov *et al.*, 2010). Furthermore, benthic assemblages are suggested to preferentially select high-quality food particles. This results in quality-dependent mixing intensities in both coastal and deep areas (Smith *et al.*, 1993; Miller *et al.*, 2000; Fornes *et al.*, 2001; Green *et al.*, 2002). Food-rich particles like phytodetritus are mixed faster into sediment than older food-poor particles (Josefson *et al.*, 2002) slowing down their degradation by removing it from the oxidized layer (Blair *et al.*, 1996; Osinga *et al.*, 1997; Van de Bund *et al.*, 2001). Yet, it has been shown that phytoplankton plays a key role in the transport and distribution of PAHs in the southern Baltic environment. PAHs are selectively sorbed from water by phytoplankton leading to maximum concentrations (sum of 12 PAHs) as high as 16,000 ng/g, values about twice as high as the maximum levels in sediments of the southern Baltic (Kowalewska and Konat, 1997). Once deposited, sorbed PAHs may then be rapidly buried and sequestered in anoxic sediments. However, as the intensity of enrichment increases, enhanced microbial and faunal activities deplete the oxygen level at the sediment–water interface, thereby reducing the depth of biological reworking (Pearson and Rosenberg, 1978). Oxygen deficits also lead to the inhibition of irrigation by macrobenthos, at least those that are not sulfide tolerant (Gamenick *et al.*, 1996). This results in reduced cycling/remobilization and consequently increased preservation of hydrocarbons owing to bioturbation decrease. Extended periods of eutrophication-induced hypoxia within bottom water layers may also lead to complete depletion of bioturbating macrofauna, subsequent spreading of laminated sediments and enhanced storage of contaminants, as observed in the Baltic sea (Granberg *et al.*, 2008).

It is likely that there are seasonal differences in the mobilization of organic contaminants in the benthic ecosystem linked to the variability of organic matter inputs, the type, the abundance and the level of activity of the organisms (e.g. high or low bioturbation in summer and winter, respectively) (Schaffner *et al.*, 1997). Because of bioturbation, sediment-bound hydrocarbons may thus either be transferred to oxic layers were they could be biodegraded aerobically or buried into anoxic layers and subjected to anaerobic biodegradation. By remobilizing old and refractory buried organic particles, deep reworking can for instance re-expose hydrocarbons to oxygen and enhance their degradation.

Accumulation, biotransformation and transfer to higher trophic levels of hydrocarbons

Macrobenthic infauna can be important vectors in the transfer of sediment-associated contaminants to higher trophic levels. Accumulation of contaminants from one level to the next along the food chain depends on feeding rate and assimilation efficiency; two factors that are species-dependent and closely related to the food quality and contaminant type (Palmqvist *et al.*, 2006). Experimental enrichment with high quality OM has been reported to stimulate the accumulation of sediment-associated PAHs in various marine invertebrates (e.g. Granberg and Forbes, 2006; Granberg and Selck, 2007). Furthermore, it is suggested that bioaccumulation and trophic transfer of sediment-associated PAHs should increase following inputs of fresh organic matter, for example after sedimentation of phytoplankton blooms

(Granberg and Selck, 2007). Besides accumulation in fatty tissues and transfer to highest trophic levels, elimination of water-insoluble and water-soluble metabolites by benthic invertebrates could also be a critical factor governing the fate of hydrocarbons. The biotransformation process generally involves both cytochrome P450 monooxygenase or mixed-function oxygenase and one more subsequent enzymatic conjugation to an organic anion, rendering the metabolite water-soluble and thereby greatly facilitating urinary excretion of this compound (Klaassen, 1996; Granberg and Selck, 2007). Biotransformation may not only change the bioavailability of hydrocarbons but also may change their toxicity. Produced metabolites are indeed often more toxic than the parent compounds (Klaassen, 1996). The cascading effect is that the whole benthic communities, from bioturbating macroorganisms to microbial-degraders, may then be affected, resulting in drastic changes in biodegradation of these compounds. Differences in potential trophic transfer between parent compounds and their respective metabolites have also been observed (Palmqvist *et al.*, 2006). Granberg and Selck (2007) stress the importance of considering behavioural characteristics of the macrofauna present, along with the trophic situation within the system. In particular, selective feeding by deposit feeders may result in different patterns of particle selection on sediments with varying levels of contamination. And yet, concentrations of organic contaminants such as PAHs or PCBs are often much higher in smaller size and low-density fractions of sediments (Prahl and Carpenter, 1983; Piérard *et al.*, 1996; Horng, 1998). For instance, microcosm experiments run with the subsurface deposit feeder polychaete *Capitella* sp. (genus indicator of heavily polluted sediments) demonstrate that this worm feeds selectively on the finest particles (median size of minerals particles ingested about 4 μm) of a sediment contaminated with PAHs (about 31.1 μg/g), potentially carrying the greatest mass-specific concentration of contaminants (Horng and Taghon, 1999). The ingested particles were subsequently incorporated into long-lived faecal pellets whose estimated 'half-life' is from 5 to 33 years (Taghon *et al.*, 1984). This incorporation did not, however, affect the loss rates of PAHs. Desorption and biodegradation rate constants are similar for faecal pellets and sediment (Horng and Taghon, 2001).

Bioturbation and biodegradation of hydrocarbons

As explained above, bioturbation can have strong effects on the transport of hydrocarbons which can be transformed and accumulated by bioturbating macroorganisms. But, to what extend does bioturbation enhance hydrocarbons biodegradation? Field experiments have shown that the activity of benthic organisms can have a significant influence on the qualitative and quantitative fate of acyclic hydrocarbons following a petroleum contamination in marine coastal sediments (Gilbert *et al.*, 1996; Grossi *et al.*, 2002; Miralles *et al.*, 2007b). These experiments are carried out in muddy sand sediments rich in OM characterized by a thin permanently oxygenated uppermost layer of less than 3 mm. Rates of *n*-alkanes biodegradation observed in the layer where highest bioturbation activity occurs (2–4 cm) are closed to those in the oxic superficial layer (0–1 cm) (Miralles *et al.*, 2007b). Biodegradation was much lower in the non-bioturbated permanently anoxic layer (8–10 cm). Indeed, after 503 days of *in situ* (water column depth of 20 m) incubation of oiled sedimentary cores (Arabian light oil, 20 g/kg dry weight), total losses of *n*-alkanes and of isoprenoid alkanes (n-C_{16}, n-C_{17}, n-C_{18}, n-C_{20}, n-C_{25}, n-C_{30}, pristane and phytane) range from 56.6% to 91.4%, from 51.2% to 84.5% and from 17.4% to 34.2% in the 0–1 cm, 2–4 cm and 8–10 cm layers, respectively. Intermittent oxygen penetration into the mostly anoxic sediment associated to the activity of macrobenthic organisms may explain the extent of biodegradation down to the 2–4 cm layer. It is concluded, however, that part of the biodegradation activity was associated to sulfate-reducing bacteria (SRB). In particular, clone libraries analyses constructed for oiled sediment, show that the most abundant clones were closely related to strains capable of hydrocarbons degradation coupled with sulfate-reduction (e.g. strains Hxd3, AK01 and *Desulfatibacillum aliphaticivorans*). In marine reduced and heavily contaminated sediments, hydrocarbons degradation/sulfate-reduction coupling seems to be the most relevant among the different potential anaerobic processes. Indeed, sulfate is abundant in coastal and estuarine seawater, whereas nitrate concentrations are typically low and Fe(III) is often only sparsely available (Rothermich *et al.*, 2002). It is therefore likely that oscillating redox conditions generated by bioturbation may favour the re-oxidation of sulfides refuelling the SRB. But the implication of aerobic degraders cannot be ruled out and may be determinant when PAHs are considered.

Because PAHs are persistent, toxic and many of them mutagenic, carcinogenic and/or teratogenic (e.g. Belkin *et al.*, 1994) their fate in bioturbated sediments have been the focus of several laboratory studies. Different works demonstrate that polychaete species like *Capitella* sp., *A. marina* and *N. virens* significantly enhance degradation of different PAHs (Bauer *et al.*, 1988; McElroy *et al.*, 1990; Christensen *et al.*, 2002b; Timmermann *et al.*, 2008). For instance, compared with defaunated sediments, the presence of these two latter worms increases the microbial pyrene minerlization 1.8 to 2-fold (Timmermann *et al.*, 2008). ^{14}C-Pyrene mineralization potential, measured as the initial rate of $^{14}CO_2$ production, is equally high in surface and burrow sediments and is significantly lower in reduced and presumably anoxic sediments (Timmermann *et al.*, 2008). These findings are in accordance with other studies which demonstrate that PAHs degradation rates of burrow sediments of different taxa are generally greater than those for non-burrow sediments (and sometimes even higher than surface sediments) (Chung and King, 1999; Selck *et al.*, 2005; Qin *et al.*, 2010). Taken together, these results highlight the central role of oxygenation due to burrow construction and irrigation leading to an increase activity of aerobic bacterial degraders. This is further verified by Montgomery *et al.* (2008) who observe an increased capacity for PAHs mineralization in bioirrigated coastal sediments (rates up to ten times higher) compared with poorly mixed sulfidic sediments. Indeed, anaerobic PAHs degradation is lower under both nitrate- and sulfate-reducing conditions (Coates *et al.*, 1997; Rockne and Strand, 1998). Chung and King (1999) compare the effects of available electron acceptors (oxygen, nitrate, ferric iron and sulfate) on PAHs degradation. They observe that in the burrow sediments from the polychaetes *N. virens* and *Clymenella torquata*, significant degradation of PAHs, hexadecane and benzene only occur in the presence of molecular oxygen. Interestingly, the capacity for oxic phenanthrene degradation is unaffected by alternating oxic and anoxic conditions, suggesting significant anoxia tolerance of microbial hydrocarbon-degrading communities. In fact, the construction of tubes and burrows and their intermittent ventilation by benthic inhabitants allow the circulation of oxygenated water into anoxic parts of the sediment, which are thus subjected to oscillating conditions of oxygenation (Forster and Graf, 1995; Timmermann *et al.*, 2006; Pischedda, 2009). At a smaller scale, the mucus layer that stabilizes the burrow wall is a barrier for solute diffusion that establishes step chemical gradients between anoxic sediment and burrow lumen (Aller and Yingst, 1978, Boudreau and Marinelli, 1994). The periodical character of the irrigation and the step gradients also induce pH and redox oscillations in the structure, at frequencies depending of the macrofaunal organisms (Aller, 1988; Kristensen, 2000). The frequency and duration of these processes may notably modulate the degradation rates of hydrocarbons (Selck *et al.*, 2005). Therefore, burrow walls should not be considered simply as an extension of the sediment–water interface as they are characterized by unique physical and chemical properties as well as by specific microbial communities (Matsui *et al.*, 2004; Papaspyrou *et al.*, 2005, 2006; Pischedda, 2009). Particularly, it is suggested that different burrow sediments may harbour unique assemblages of PAHs degraders (Chung and King, 1999). Indeed, two aerobic, obligatory marine PAH-degrading bacteria, *Lutibacterium anuloederans*, and *Cycloclasticus spirillensus*, were isolated from burrows of the polychaete *N. virens* and the mollusc *Mya arenaria*, respectively (Table 4.1). Both strains use naphthalene and phenanthrene aerobically as their sole carbon sources (Chung and King, 2001). Similarly, aerobic PAH-degrading bacteria associated with the rhizosphere of salt marsh plants have been isolated (Daane *et al*, 2001). Yet, the rhizospheres of marine plants are known to enhance oxygen availability in sediments (Calhoun and King, 1997; Azzoni *et al.*, 2001). Likewise, the presence of the marine polychaete *N. diversicolor* in oil contaminated sediments is shown to favour the development of strictly aerobic alkane-degrading bacteria of the genus *Alcanivorax* (Cuny *et al.*, 2007) and to enhance oil biodegradation (Christensen *et al.*, 2002a).

However, as stressed by Banta and Andersen (2003), different mechanisms by which infauna affect contaminants can interact in complicated ('the real world') and, at times, contradictory ways. For instance, fluoranthene (added as a contaminated top layer) mineralization was shown to decrease in the presence of the polychaete *A. marina*. This was explained by the rapid transport of fluoranthene from the surface into deeper anoxic zones due to the conveyor belt feeding behaviour of this worm (Kure and Forbes, 1997). In the same way, increased desorption and resulting bioavailability of hydrocarbons in the burrow walls caused by bioirrigation are not systematically the rule as it will be discussed in the next section.

While it has been demonstrated that bioturbation could have a significant impact on the burial and the degradation of hydrocarbons, the influence of redox oscillation on their degradation has been so far poorly studied. Furthermore, only little attention has been paid to pathways and enzymes that are active in bacteria which utilize hydrocarbons at suboxic or even quasi-anoxic conditions. Yet, in suboxic to quasi-anoxic habitats, where alkane, ammonium and methane oxidizing bacteria as well as other organotrophic microorganisms live in close vicinity, a cooperative effect with respect to *n*-alkane degradation may occur (Berthe-Corti and Fetzner, 2002). Using a diffusively 'open' incubation sedimentary system that mimics varying conditions of oxygenation in bioturbated sediments, Sun *et al.* (1993) show that redox changes influence the diagenetic pathways of chloropigments. Later on, using a similar experimental approach, these authors demonstrated that the frequency of oxic–anoxic oscillation could also alter the rates and pathways of degradation of individual cell-associated lipids (Sun *et al.*, 2002). Undoubtedly, the effects of redox oscillations on bacterial coupling and hydrocarbons biodegradation requires further investigation. Bioremediation strategies under these oscillating conditions will be discussed in the last section of this chapter.

Bioturbation and bioavailability of hydrocarbons: role of digestive fluids and mucus

Bioavailability refers to the extent of exposure of living organisms to chemicals in the environment (Ehlers and Luthy, 2003). Bioavailability corresponds, on one hand, to the accessible quantity (i.e. mass of contaminants) which can become available to, for example, biodegradation and bio-uptake and, on the other hand, to the chemical activity of the contaminant that quantifies the potential for spontaneous physicochemical processes, such as diffusion, sorption, and partitioning (Reichenberg and Mayer, 2006). It is therefore a complex and dynamic variable that is influenced by the interactions and physical properties of the contaminant(s), the sediment matrix, and the organisms present within a given environment (Parales *et al.*, 2008). The bioavailability not only provides a measurement of the potential toxicity of a contaminant or of its metabolites, but also an estimate of the potential effectiveness of microbial bioremediation processes. When degradation rates exceed rates of hydrocarbons transfer, the growth of microorganisms-degraders becomes constrained by the rate of pollutant transfer (Bosma *et al.*, 1997). The degree of sorption of hydrocarbons onto particles depends on several characteristics of the particles (e.g. grain size, chemical composition and charge) as well as on the type of hydrocarbon (e.g. organic carbon normalized sorption coefficient K_{oc}). Particularly, the sorption of organic contaminants has been shown to be greatly affected by the type and amount of carbonaceous materials present within the sedimentary matrix, which in turn affect contaminants transport and fate (Jeong *et al.*, 2008). For instance, black carbon (carbonaceous geosorbent), which is a byproduct of incomplete combustion or pyrolysis of biomass and fossil fuels, limits the availability of PAHs in pore water, as well as the availability in the intestinal tract of the organisms (Koelmans *et al.*, 2006). As shown by Koelmans *et al.* (2006) the presence of black carbon lowers exposure and toxicological risks of persistent organic pollutants (POPs, PAHs,

PCBs) in a none proportional way: depending on the black carbon content (about 9% of total organic carbon in aquatic sediments), a 10- to 100-fold decrease of POPs concentration may be observed in the organisms. Because of this strong effect on contaminants sorption and bioavailability, Koelmans *et al.* (2006) suggest that the presence of this geosorbent has a positive implication for sediment remediation by lowering the exposure of organisms to these toxic compounds.

Extended exposure of contaminants to sediment matrix further reduces bioavailability (a process known as ageing) by strengthening the binding between hydrophobic contaminants and the sediment particles and/or OM, therefore decreasing the rate of desorption into water. Besides chemical changes, changes in the sediment pore size and structure as well as shifts in the spatial distribution of the hydrophobic organic compounds due to diffusive transfer do occur. Combined with the exclusion of microorganisms from certain parts of the geomaterial, this may also lead to an increase of the 'non-bioaccessible' or 'irreversibly bound' pool of contaminants (Harms *et al.*, 2010). Over time, contaminants may become therefore less susceptible to microbial biodegradation. As aforementioned (see section 'Aerobic hydrocarbon-degrading bacteria in marine sediments') hydrocarbon-degrading microorganisms may however produce biosurfactants (surface-active agents) of diverse chemical natures and molecular sizes to enhance the uptake of hydrocarbons. In particular, some of them are able to produce high molecular weight biosurfactants, bound tightly to surfaces, which increase the surface area of hydrophobic water-insoluble substrates and increase their bioavailability. Growth of bacteria and rate of bioremediation will then be enhanced (Ron, 2000; Ron and Rosenberg, 2002).

Bioturbation may affect bioavailability in several ways. First, as previously mentioned, hydrocarbons transport and biotransformation by marine invertebrates may reduce bioavailability. Granberg and Selck (2007) demonstrate that the activities of the polychaete *N. diversicolor* and of the dwelling brittle star *Amphiura filiformis* lead to the subduction of sediment-associated pyrene and produced metabolites. This transfer is associated to the increase with depth of the fraction of pyrene covalently bound to sediment OM, resulting in the removal of pyrene from the bioavailable pool.

In contrast, it is also hypothesized that the enhanced remediation of hydrocarbon-contaminated sediments in the presence of sediment-reworking invertebrates can be attributed to the increased bioavailability of PAHs and other hydrocarbons to HCB owing to solubilizers within the animal gut (Weston and Mayer, 1998). Indeed, the gut of some marine invertebrates (particularly of deposit feeders), contain high levels of biosurfactants (Mayer *et al.*, 1997) that solubilize organic compounds at many times their seawater, and equilibrium partitioning-predicted, solubility (Voparil *et al.*, 2003). The solubilization capacity appears to be taxa-dependent. Digestive fluids of polychaetes and echiurans are relatively strong, those of sipunculan and mollusc have intermediate strength and echinoderms and a cnidarian have very weak ones (Mayer *et al.*, 2001). For instance, in the guts of the deposit-feeding polychaete *A. marina*, surfactant micelles solubilize phenanthrene and benzo[a]pyrene at concentrations 10 (4.6 μg/ml) and 1000 (2.0 μg/ml) times greater than their seawater solubility, respectively (Voparil and Mayer, 2000). The source of these surfactants remains unclear (Ahrens *et al.*, 2001) even if, surfactant-resistant and biosurfactant-producing bacteria were isolated from the guts of two polychaetes deposit feeders, *N. succinea* and *Amphitrite ornate* (Plante *et al.*, 2008). On the contrary, even if *N. diversicolor* and *A. marina*, by oxygenating the sediments, greatly enhance pyrene mineralization (see previous section), they also reduce its bioavailablility (Timmermann *et al.*, 2008). The authors suggest that gut fluids of these polychaetes may catalyse oxidative coupling of pyrene metabolites to OM, lowering its bioavailability as described by Giessing and Mayer (2004). They further hypothesize that a similar phenomenon may occur in the organic rich burrow walls of these worms owing to the presence of burrow-stabilizing mucus.

Finally, the feeding activity of macrofaunal organisms (ingestion of sediment/egestion of faecal material) may have more than a direct effect (digestive enzymatic processes, assimilation, and solubilization) on sediment-bound hydrocarbons. As a matter of fact, shifts in the composition of the bacterial assemblages occur during passage through the gut of deposit-feeders (Dobbs and Guckert, 1988; Findlay *et al.*, 1990; Grossmann and Reichardt, 1991; Wilde and Plante, 2002; Grossi *et al.*, 2006). These changes can be attributed to the digestion of a subset of ingested bacteria (bacteriolytic activity) with the concomitant stimulation or inoculation of other groups. One has to keep in mind that in order to meet their nutritional needs, deposit feeders ingest massive volumes of sediment by bulk feeding (Andersen and Kristensen, 2002). How these taxa-specific changes influence the capacity of the microbial communities to degrade hydrocarbons is still not known.

Bioremediation strategies under oscillating conditions

Oxic/anoxic switches are widespread in environments (e.g. Wakeham and Canuel, 2006). At the water/sediment interface, redox properties in sediment can exhibit daily to seasonal changes, owing to variations of dissolved oxygen content in bottom waters (Parker and O'Reilly, 1991). Switches between oxic and anoxic states can occur during bioturbation activities in sediments deposits (Aller, 1994; Brune *et al.*, 2000). Environments submitted to oxic/anoxic switches such as the coastal marine sediments, thus constitute particular functional systems impacting the bacterial diversity and its role on organic compounds degradation (e.g. Kogure and Wada, 2005).

Effect of oxic/anoxic oscillations on OM and hydrocarbons biodegradation

The fate of OM submitted to oxic/anoxic switches has been studied, with special focus on the impact of oscillations on organic carbon decomposition and remineralization (e.g. Sun *et al.*, 1993, 2002; Aller, 1994; Caradec *et al.*, 2004b; Abril *et al.*, 2010). Most of these studies have addressed phytoplankton degradation, in particular lipids from microalgal cells that constitute a major source of organic detritus for marine sediments. The degradation of these compounds is often followed under three oxygenation conditions: permanently oxic, strictly anoxic and oscillating, most often in bioreactors. As already mentioned, Aller (1994) shows that redox oscillations stimulate bacterial metabolisms, promoting decomposition and minimizing burial of organic carbon. Oscillations, frequency and duration determine the adaptation of the microorganisms involved (consortia), their respiration response, decomposition potential, and biomass synthesis or destruction. In a complementary way, Sun *et al.* (2002) demonstrate that the frequency of oscillations also alters the rates and pathways of lipid degradation. Furthermore, oscillation of redox conditions regulates microbial activity and lipid degradation, and minimizes the accumulation of intermediates (Sun *et al.*, 2002). The oxic-anoxic-oxic redox transitions promote degradation of organic material in continental margin sediments (Hulthe *et al.*, 1998). By monitoring the mineralization of OM in freshwater, Bastviken *et al.* (2004) show that a sequential change in the oxygen regime does not result in greater degradation, but that initially anoxic conditions favour subsequent oxic mineralization. OM mineralization is also enhanced during re-oxidation periods (Abril *et al.*, 2010). C and N mass balance analyses suggest an increase of OM mineralization under redox oscillation on short time scale (day to week). Despite these promising conclusions, microbial diversity and biogeochemical mechanisms involved in the fate of pollutants in oxic/anoxic 'oscillating' environments still remain poorly documented.

Coastal marine ecosystems remain susceptible to pollution and are regularly contaminated by hydrocarbons. Most studies done on hydrocarbons degradation processes are restricted to strict oxic or anoxic conditions (e.g. Eriksson *et al.*, 1999; Boopathy, 2004; Gafarov *et al.*, 2006; Head *et al.*, 2006). Although marine sediments act as pollutant reservoirs and are submitted to oxic/anoxic oscillations, little attention has been paid to the influence of such conditions on hydrocarbons degradation. Indeed, if several studies report on the effect of bioturbation on hydrocarbons degradation (see above), they only represent indirect approaches in regard to oscillating conditions. Only two studies focus on the degradation of hydrocarbons under oscillating conditions. Löser *et al.* (1998) study hydrocarbons degradation in a pilot-scale plant with diesel fuel contaminated sand submitted to aerobic/anaerobic switching compared with purely aerobic conditions. In both experimental conditions (switching and permanently aerobic) less than 50% of the diesel fuel is degraded after 27 days of incubation. Against all expectations, aerobic/anaerobic changes neither accelerate the degradation rate nor reduce the residual hydrocarbon content. Nevertheless, the masses of carbon dioxide produced and oxygen consumed are greater during aerobic/anaerobic changes than under continuous aerobic conditions. Conversely, Vieira *et al.* (2009a) show that intermittent aeration during the biodegradation of effluents contaminated with diesel oil and gasoline is better than under continuous aeration. These authors analyse the removal of hydrocarbons present in a synthetic effluent (containing diesel oil and gasoline) bioaugmented with a mixed bacterial culture using intermittent aeration. They demonstrate that 90% of total petroleum hydrocarbons (TPH) are removed after 49 days of processing (Vieira *et al.*, 2009a). Following an optimization of operational conditions (aeration interval and agitation speed) a comparative study between intermittent aeration, constant aeration and without aeration, shows a stabilization of the TPH removal 18 days after the beginning of the experiment, with a reduction of 90%, 84%, and 70%, respectively (Vieira *et al.*, 2009b). These studies demonstrate that intermittent aeration conditions thus result in greater TPH removal and constitute an alternative to continuous aeration processes in the biodegradation of contaminated effluents. Diverging conclusions obtained by Löser *et al.* (1998) and Vieira *et al.* (2009a,b) could be explained by the maladaptation of microorganisms to pollution (i.e. lack of enzymatic equipment for hydrocarbons degradation) in the artificially contaminated pristine sand (Löser *et al.*, 1998). In such conditions, hydrocarbons could constitute a critical inhibition stress for bacterial communities lacking specific biodegradation capacities. Although these studies carefully address the fate of hydrocarbons and their degradation, they are not focused on the behaviour and evolution of bacterial communities.

Recent studies are investigating the impact of oxic/anoxic switches on both endogenous hydrocarbons biodegradation and autochtonous bacterial communities from a coastal contaminated environment (Vitte, 2010; Vitte *et al.*, 2011). Results demonstrate that autochthonous bacteria present strong adaptation ability to changes of environmental parameters (aeration *versus* anoxia). As found by Löser *et al.* (1998), the efficiency of hydrocarbons removal is similar in both switching and permanent oxic conditions. Interestingly, these findings are similar to those obtained by Chung and King (1999) for PAHs in laboratory experiments with sediment slurries prepared using *N. virens* burrow wall sediments (see section 'Bioturbation and biodegredation of hydrocarbons'). Thus, a bioremediation methodology of oxic/anoxic switches could be a useful process for the treatment of hydrocarbons contaminated sludges or effluents, as already mentioned by Vieira *et al.* (2009b). This methodology presents a low cost processing, as its efficiency is similar to a permanent oxic incubation that requires more expensive continuous oxygen supply. Moreover, switching conditions provide less biomass during anoxic periods and consequently less sludge to be eliminated after treatment.

Hydrocarbons bioremediation under oscillating oxic–anoxic conditions: a new original strategy?

Physical, chemical and biological processes have been developed and applied in removing pollutants from marine contaminated sites. Since microorganisms harbour an important catabolic diversity that allows them to effectively use hydrocarbons, bioremediation represents an attractive option. Techniques of bioremediation using microorganisms degrading capacities emerged a few decades ago in order to attenuate toxics in the environments in a natural way (e.g. Bamforth and Singleton, 2005). These strategies show diverse results with a general good efficiency in the laboratory, but with quite difficult applications in the field (Prince *et al*, 2003). One factor limiting the implementation of bioremediation is the lack of knowledge about the factors controlling the growth, metabolism and response of bacteria in contaminated natural environments (Lovley, 2003).

The common strategy to accelerate the natural microbial processes involved in oil degradation is *in situ* application of the treatments. These methods include addition of oil degrading bacteria (bioaugmentation) and addition of nutrients stimulating the growth of hydrocarbon indigenous degraders (biostimulation) (e.g. Perelo, 2010). This latter appears as an excellent strategy for oil bioremediation in marine environments (for review see Nikolopoulou and Kalogerakis, 2010). Nevertheless, in environments treated in such way, some major drawbacks exist. Indeed, no enhancement of oil biodegradation should be expected when fertilizers are added to sediments (Nikolopoulou and Kalogerakis, 2010). Although the aeration (simple tilling) of the top layer of sediments in coastal zones during low tide could be done, this is not applicable to all marine environments where oxygen delivery seems to be very difficult.

In the light of these data, the following developments have to be considered:

- There is a pressing need to understand the role of microorganisms and their behaviour in biodegradation processes, and the knowledge of their responses to environmental parameter fluctuations is necessary to obtain tools to assist in mitigation. Although studies have demonstrated hydrocarbons biodegradation capacities in coastal sediments, our understanding of the microorganisms involved, their genetic and enzymatic capacities, their interactions, as well as their functioning under oscillating redox conditions, is largely unknown. As highlighted by Borch *et al.* (2010), research addressing the biogeochemical functioning of polluted environments characterized by fluctuating redox conditions is scarce and warrants further attention.
- Besides the presence of oxic/anoxic switches in natural environments like the burrows of numerous macrofaunal organisms, such conditions may be used for a new strategy of bioremediation, by using different biodegradation capacities of autochthonous bacteria. The promising results obtained using bioreactors subjected to oxic/anoxic switches (Vitte, 2010; Vitte *et al.*, 2011) reinforce this new insight into hydrocarbons biodepollution, even though this seems more adapted to the treatment of contaminated sludge or effluents than to marine coastal sediments. Further research developments are essential in this area with the aim of implementing such technology.

Conclusion

Once settled, hydrocarbons fate is dominated by several abiotic and biotic processes that result in either their partial or total degradation or to a selective preservation within the buried sediments. Biodegradation of petroleum or pyrogenic hydrocarbons in marine sediments is mainly due to the existence of specific prokaryotes having the capacity to degrade these compounds (or at least to carry out the first step of activation) under oxic, suboxic and anoxic conditions. Much work is needed on the mechanisms and the microbial actors responsible for hydrocarbons biodegradation in suboxic and anoxic sediments. In these widespread sedimentary environments, the overall mineralization process often relies on diverse and dynamic communities that require hierarchical cooperation (e.g. synthrophic) among several different metabolic groups. Yet, the role of these consortia in the bioremediation of contaminated environments is poorly studied. The redox oscillation regimes generated by the macrofaunal organisms, and the efficiency of metabolic coupling between functional groups associated to these specific redox regimes, are probably determinant factors which control the biodegradation of hydrocarbons in marine sediments. From the understanding of how these natural occurring factors modulate the rates of hydrocarbons biodegradation, innovative bioremediation strategies may emerge. Bioturbation can also affect the distribution and bioavailability of hydrocarbons resulting either in increased losses (e.g. desorption and diffusion of contaminants from bed sediments back into the water column, transfer through the food chains) or biodegradation, but may also lead to an enhanced sequestration. Macrofaunal organisms may also transform hydrocarbons into more soluble and bioavailable intermediate compounds but sometimes more toxic than the parent molecules.

As highlighted in this chapter, the fate of hydrocarbons in the real world is pretty much more complex than in pure cultures or in microcosm's experiments from which most of our knowledge comes from. The interplay of the various factors that govern hydrocarbons biodegradation in marine sediments is diverse and intricate and many details are not yet understood. Apart from the presence of specific interconnected communities of hydrocarbon-degraders, the benthic assemblages' composition (species, functional group, feeding behaviours, density of organisms, bioturbation activity), the physico-chemical characteristics of specific hydrocarbon contaminants and of sedimentary matrix, other ecological considerations are of relevant interest. Direct or indirect contaminant-induced changes in behaviour, physiology, competitive interactions, and predation/grazing rate can alter species abundances or community composition (Fleeger *et al.*, 2003). This can in turn modify the overall ecosystem functioning and the natural remediation rate of hydrocarbon-contaminated sediments. For instance, higher abundances of meiofauna can significantly decrease by a top-down control the microbial mineralization of PAHs such as naphthalene, and significantly modify the bacterial community composition in natural marine sediments (Näslund *et al.*, 2010). Yet, macrobenthic burrows may have promotive effects resulting in increased densities of meiobenthic organisms (Dittmann, 1996).

Going down to smaller scales, what are the cross-linked effects of an oil spill or a contamination by toxic PAHs, on the microbial food webs that control bacterial communities having a pivotal role in biodegradation processes (Head *et al.*, 2006)? Marine sediments can harbour up to 1 million virus species per kg of marine sediment (Breitbart *et al.*, 2004). How viruses manipulate microbial communities as well as catabolic genes transfer in contaminated marine sediments (Rohwer and Thurber, 2009)?

As stressed by Berthe-Corti and Nachtkamp (2010), the biggest challenge for the future is to understand the network of these diverse influences which require future research efforts in different fields. A more holistic and integrative ecological approach is definitely needed in order to understand and predict the fate of hydrocarbons in marine bioturbated sediments.

References

Abril, G., Commarieu, M.V., Etcheber, H., Deborde, J., Deflandre, B., Zivadinovic, M.K., Chaillou, G., and Anschutz, P. (2010). *In vitro* simulation of oxic/suboxic diagenesis in an estuarine fluid mud subjected to redox oscillations. Estuar. Coast. Shelf. Sci. *88*, 279–291.

Aeckersberg, F., Bak, F., and Widdel, F. (1991). Anaerobic oxidation of saturated hydrocarbons to CO2 by a new type of sulfate-reducing bacterium. Arch. Microbiol. *156*, 5–14.

Aeckersberg, F., Rainey, F.A., and Widdel, F., (1998). Growth, natural relationships, cellular fatty acids and metabolic adaptation of sulfate-reducing bacteria that utilize long-chain alkanes under anoxic conditions. Arch. Microbiol. *170*, 361–369.

Ahrens, M.J., Hertz, J.H., Lamoureux, E.M., Lopez, G.R., McElroy, A.E., and Brownawell B.J. (2001). The role of digestive surfactants in determining bioavailability of sediment-bound hydrophobic organic contaminants to 2 deposit-feeding polychaetes. Mar. Ecol. Prog. Ser. *212*, 145–157.

Aitken, C.M., Jones, D.M., and Larter, S.R. (2004). Anaerobic hydrocarbon biodegradation in deep subsurface oil reservoirs. Nature *431*, 291–294.

Alain, K., Holler, T., Musat, F., Elvert, M., Treude, T., and Krüger, M. (2006). Microbiological investigation of methaneand hydrocarbon-discharging mud volcanoes in the Carpathian Mountains, Romania. Environ. Microbiol. *8*, 574–590.

Aller, R.C. (1982). The effects of macrobenthos on chemical properties of marine sediments and overlying waters. In Animal–Sediment Relations: The Biogenic Alteration of Sediments, McCall, P.L., and Tevesz, M.J.S., eds. (Plenum Press: New York), pp. 53–102.

Aller, R.C. (1988). Benthic fauna and biogeochemical processes in marine sediments: the role of burrow structures. In Nitrogen Cycling in Coastal Marine Environments, Blackburn, T.H., Sorensen, J., eds. (John Wiley & Sons: Chichester, UK), pp. 301–338.

Aller, R.C. (1994). Bioturbation and remineralization of sedimentary organic matter: effects of redox oscillation. Chem. Geol. *114*, 331–345.

Aller, R.C. (2001). Transport and reactions in the bioirrigated zone. In The Benthic Boundary Layer: Transport Processes and Biogeochemistry, Boudreau, B.P., Jørgensen, B.B., eds. (Oxford Press: Oxford, UK), pp. 269–301.

Aller, R.C., and Aller, J.Y. (1998). The effect of biogenic irrigation intensity and solute exchange on diagenetic reaction rates in marine sediments. J. Mar. Res. *56*, 905–936

Aller, R.C., and Yingst, J.Y. (1978). Biogeochemistry of tube-dwellings – Study of sedentary polychaete *Amphitrite ornata* (Leidy) J. Mar. Res. *36*, 201–254.

Alonso-Gutiérrez, J., Figueras, A., Albaigés, J., Jiménez, N., Viñas, M., Solanas, A.M., and Novoa, B. (2009). Bacterial communities from shoreline environments (Costa da Morte, northwestern Spain) affected by the prestige oil spill. Appl. Environ. Microbiol. *75*, 3407–3418.

Alvarez, L.A., Exton, D.A., Timmis, K.N., Suggett, D.J., and McGenity, T.J. (2009). Characterization of marine isoprene-degrading communities. Environ. Microbiol. *11*, 3280–3291.

Andersen, M., and Kristensen, E. (2002). The importance of bacteria and microalgae in the diet of the deposit-feeding polychaete *Arenicola marina*. Ophelia *56*, 179–196.

Annweiler, E., Michaelis, W., and Meckenstock, R.U. (2002). Identical ring cleavage products during anaerobic degradation of naphthalene, 2-methylnaphthalene, and tetralin indicate a new metabolic pathway. Appl. Environ. Microbiol. 68, 852–858.

Azzoni, R., Giordani, G., Bartoli, M., Welsh, D.T, and Viaroli, P. (2001). Iron, sulphur and phosphorus cycling in the rhizosphere sediments of a eutrophic *Ruppia cirrhosa* meadow (Valle Smarlacca, Italy). J. Sea Res. *45*, 15–26.

Bamforth, S.M., and Singleton, I. (2005). Bioremediation of polycyclic aromatic hydrocarbons: current knowledge and future directions. J. Chem. Technol. Biotechnol. *80*, 723–736.

Banta, G.T., and Andersen, O. (2003). Bioturbation and the fate of sediment pollutants – experimental case studies of selected infauna species. Vie Milieu *53*, 233–248.

Bastviken, D., Persson, L., Odham, G., and Tranvik, L. (2004). Degradation of dissolved organic matter in oxic and anoxic lake water. Limnol. Oceanogr. *49*, 109–116.

Bauer, J.E., Kerr, R.P., Bautista, M.F., Decker, C.J., and Capone, D.G. (1988). Stimulation of microbial activities and polycyclic aromatic hydrocarbon degradation in marine sediments inhabited by *Capitella capitata*. Mar. Environ. Res. *25*, 63–84.

Belkin, S., Stieber, M., Tiehm, A., Frimmel, F.H., Abeliovich, A., Werner, P., and Ulitzur, S. (1994). Toxicity and genotoxicity enhancement during polycyclic aromatic hydrocarbons' biodegradation. Environ. Toxicol. Water Qual. *9*, 303–309.

Beller, H.R., and Spormann, A.M. (1997). Anaerobic activation of toluene and o-xylene by addition to fumarate in denitrifying strain T. J. Bacteriol. 179, 670–676.

Ben Said, O., Goñi-Urriza, M.S., El Bour, M., Dellali, M., Aissa, P., and Duran, R. (2008). Characterization of aerobic polycyclic aromatic hydrocarbon-degrading bacteria from Bizerte lagoon sediments, Tunisia. J. Appl. Microbiol. *104*, 987–997.

Berelson, W.M., Heggie D., Longmore, A., Kilgore, T., Nicholson, G., and Skyring, G. (1998). Benthic nutrient recycling in Port Phillip Bay, Australia Estuar. Coast. Shelf Sci. *46*, 917–934.

Berkenbusch, K., and Rowden, A.A. (1999). Factors influencing sediment turnover by the burrowing ghost shrimp *Callianassa filholi* (Decapoda: Thalassinidea). J. Exp. Mar. Biol. Ecol. *238*, 283–292

Berthe-Corti, L., and Bruns, A. (2001). Composition and activity of marine alkane-degrading bacterial communities in the transition from suboxic to anoxic conditions. Microb. Ecol. *42*, 46–55.

Berthe-Corti, L., and Fetzner, S. (2002). Bacterial metabolism of *n*-alkanes and ammonia under oxic, suboxic and anoxic conditions. Acta Biotechnol. *22*, 299–336.

Berthe-Corti, L., and Nachtkamp, M. (2010). Bacterial communities in hydrocarbon-contaminated marine coastal environments. In Handbook of Hydrocarbon and Lipid Microbiology, Timmis, K.N., ed. (Springer-Verlag: Berlin, Germany), pp. 2349–2360.

Bertics V.J., and Wiebke Ziebis, W. (2010). Bioturbation and the role of microniches for sulfate reduction in coastal marine sediments. Environ. Microbiol. *12*, 3022–3034.

Bianchi, T.S., Johansson, B., and Elmgren, R. (2000). Breakdown of phytoplankton pigments in Baltic sediments: effects of anoxia and loss of deposit-feeding macrofauna. J. Exp. Mar. Biol. Ecol. *251*, 161–183.

Biegert, T., Fuchs, G., and Heider, J. (1996). Evidence that anaerobic oxidation of toluene in the denitrifying bacterium *Thauera aromatica* is initiated by formation of benzylsuccinate from toluene and fumarate. Eur. J. Biochem. 238, 661–668.

Blair, N.E., Levin, L.A., DeMaster, D.J., and Plaia, G. (1996). The short-term fate of fresh algal carbon in continental slope sediments. Limnol. Oceanogr. *41*, 1208–1219.

Boll, M., and Heider, J. (2010). Anaerobic degradation of hydrocarbons: Mechanisms of C–H bond activation in the absence of oxygen. In Handbook of Hydrocarbon and Lipid Microbiology, Timmis, K. N., ed. (Springer-Verlag: Heidelberg, Germany), pp. 1011–1024.

Bonin, P.C., Michotey, V.D., Mouzdahir, A., and Rontani, J.F. (2002). Anaerobic biodegradation of squalene: using DGGE to monitor the isolation of denitrifying bacteria taken from enrichment cultures. FEMS Microbiol. Ecol. *42*, 37–49.

Boopathy, R. (2004). Anaerobic biodegradation of no. 2 diesel fuel in soil: a soil column study. Biores. Technol. *94*, 143–151.

Borch, T., Kretzschmar, R., Kappler, A., Cappellen, P.V., Ginder-Vogel, M., Voegelin, A., and Campbell, K. (2010). Biogeochemical redox processes and their impact on contaminant dynamics. Environ. Sci. Technol. *44*, 15–23.

Bosma, T.N.P., Middeldorp, P.J.M., Schraa, G., and Zehnder, A.J.B. (1997). Mass transfer limitation of biotransformation: quantifying bioavailability. Environ. Sci. Technol. *31*, 248–252.

Boudreau, B.P., and Marinelli, R.L. (1994). A modeling study of discontinuous biological irrigation. J. Mar. Res. *52*, 947–968.

Bregnard, T.P., Haner, A., Hohener, P., and Zeyer, J. (1997). Anaerobic degradation of pristane in nitrate-reducing microcosms and enrichment cultures. Appl. Environ. Microbiol. *63*, 2077–2081.

Breitbart, M., Felts, B., Kelley, S., Mahaffy, J.M., Nulton, J., Salamon, P., and Rohwer, F. (2004). Diversity and population structure of a near-shore marine-sediment viral community. Proc. Biol. Sci. *271*, 565–574.

Brito, E.M., Guyoneaud, R., Goni-Urriza, M., Ranchou-Peyruse, A., Verbaere, A., Crapez, M.A., Wasserman, J.C., and Duran, R. (2006). Characterization of hydrocarbonoclastic bacterial communities from mangrove sediments in Guanabara Bay, Brazil. Res. Microbiol. *157*, 752–762.

Brune, A., Frenzel, P., and Cypionka, H. (2000). Life at the oxic–anoxic interface: microbial activities and adaptations. FEMS Microbiol. Rev. *24*, 691–710.

Bruns, A., and Berthe-Corti, L. (1999). *Fundibacter jadensis* gen. nov., sp. nov., a new slightly halophilic bacterium, isolated from intertidal sediment. Int. J. Syst. Bacteriol. *2*, 441–448.

Burdige, D.J. (2006). Geochemistry of Marine Sediments (Princeton University Press: Princeton, USA).

Calhoun, A., and King, G.M. (1997). Regulation of root-associated methanotrophy by oxygen availability in the rhizosphere of two aquatic macrophytes. Appl. Environ. Microbiol. *63*, 3051–3058.

Callaghan, A.V., Gieg, L., Kropp, K., Suflita, J., and Young, L. (2006). Comparison of mechanisms of alkane metabolism under sulfate-reducing conditions among two bacterial isolates and a bacterial consortium. Appl. Environ. Microbiol. *72*, 4274–4282.

Callaghan, A.V., Wawrik, B., Nı′ Chadhain, S.M., Young, L.Y., and Zylstra, G.J. (2008). Anaerobic alkane-degrading strain AK-01 contains two alkylsuccinate synthase genes. Biochem. Biophys. Res. Commun. *366*, 142–148.

Callaghan, A.V., Davidova, I.A., Savage-Ashlock, K., Parisi, V.A., Gieg, L.M., Suflita, J.M., Kukor, J.J., and Wawrik, B. (2010). Diversity of benyzl- and alkylsuccinate synthase genes in hydrocarbon-impacted environments and enrichment cultures. Environ. Sci. Technol. *44*, 7287–7294.

Caradec, S., Grossi, V., Hulth, S., Stora, G., and Gilbert, F. (2004a). Macrofaunal reworking activities and hydrocarbon redistribution in an experimental sediment system. J. Sea. Res. *52*, 199–210.

Caradec, S., Grossi, V., Gilbert, F., Guigue, C., and Goutx, M. (2004b). Influence of various redox conditions on the degradation of microalgal triacylglycerols and fatty acids in marine sediments. Org. Geochem. *35*, 277–287.

Christensen, B., Vedel, A., and Kristensen, E. (1999). Carbon and nitrogen fluxes in sediment inhabited by suspension-feeding (*Nereis diversicolor*) and non suspension-feeding (*Nereis virens*) polychaetes. Mar. Ecol. Prog. Ser. *192*, 203–217.

Christensen, J.H., Hansen, A.B., Mortensen, J., and Andersen, O. (2002a). A preliminary study on fate of oil in marine sediments inhabited by the polychaete *Nereis diversicolor*. In Oil and Hydrocarbon Spills III: Modelling, Analysis and Control, Brebbia, C.A., ed. (Wessex Institute of Technology Press: Southampton, UK), pp. 289–300.

Christensen, M., Andersen, O., and Banta, G.T. (2002b). Metabolism of pyrene by the polychaetes *Nereis diversicolor* and *Arenicola marina*. Aquat. Toxicol. *58*, 15–25.

Chung, W.K., and King, G.M. (1999). Biogeochemical transformations and potential polyaromatic hydrocarbon degradation in macrofaunal burrow sediments. Aquat. Microb. Ecol. *19*, 285–295.

Chung, W.K., and King G.M. (2001). Isolation, characterization and polyaromatic hydrocarbon degradation potential of aerobic bacteria from marine macrofaunal burrow sediments and description of *Lutibacterium anuloederans* gen. nov., sp. nov., and *Cycloclasticus spirillensis* sp. nov. Appl. Environ. Microbiol. *67*, 5585–5592.

Coates, J.D., Woodward, J., Allen, J., Philp, P., and Lovley, D.R., (1997). Anaerobic degradation of polycyclic aromatic hydrocarbons and alkanes in petroleum-contaminated marine harbor sediments. Appl. Environ. Microbiol. *63*, 3589–3593.

Cravo-Laureau, C., Matheron, R., Cayol, J.-L., Joulian, C., and Hirschler-Réa, A. (2004a). *Desulfatibacillum aliphaticivorans* gen. nov., sp. nov., an *n*-alkane- and *n*-alkene-degrading, sulfate-reducing bacterium. Int. J. System. Evol. Microbiol. 54, 77–83.

Cravo-Laureau, C., Matheron, R., Joulian, C., Cayol, J.L., and Hirschler-Réa, A. (2004b). *Desulfatibacillum alkenivorans* sp. nov., a novel *n*-alkene-degrading, sulfate-reducing bacterium, and emended description of the genus *Desulfatibacillum*. Int. J. Syst. Evol. Microbiol. *54*, 1639–1642.

Cravo-Laureau, C., Grossi, V., Raphel, D., Matheron, R., and Hirschler-Réa, A. (2005). Anaerobic *n*-alkane metabolism by a sulfate-reducing bacterium, *Desulfatibacillum aliphaticivorans* strain CV2803T. Appl. Environ. Microbiol. *71*, 3458–3467.

Cravo-Laureau, C., Labat, C., Joulian, C., Matheron, R., and Hirschler-Réa, A. (2007). *Desulfatiferula olefinivorans* gen. nov., sp. nov., a long-chain *n*-alkene-degrading, sulfate-reducing bacterium. Int. J. Syst. Evol. Microbiol. 57, 2699–2702.

Cuny, P., Miralles, G., Cornet-Barthaux, V., Acquaviva, M., Stora, G., Grossi, V., and Gilbert, F. (2007). Influence of bioturbation by the polychaete *Nereis diversicolor* on the structure of bacterial communities in oil contaminated coastal sediments. Mar. Poll. Bull. *54*, 452–459.

Daane, L.L., Harjono, I., Zylstra, G.J., and Häggblom, M.M. (2001). Isolation and characterization of polycyclic aromatic hydrocarbon-degrading bacteria associated with the rhizosphere of salt marsh plants. Appl. Environ. Microbiol. *67*, 2683–2691.

Davidova, I.A., Gieg, L.M., Duncan, K.E., and Suflita, J.M. (2007). Anaerobic phenanthrene mineralization by a carboxylating sulfate-reducing bacterial enrichment. ISME J. *1*, 436–442.

Delille, D., and Delille, B. (2000). Field observations on the variability of crude oil impact on indigenous hydrocarbon-degrading bacteria from sub-Antarctic intertidal sediments. Mar. Environ. Res. *49*, 403–417.

Delille, D., Delille, B, and Pelletier, E. (2002). Effectiveness of bioremediation of crude oil contaminated subantarctic intertidal sediment: the microbial response. Microb. Ecol. *44*, 118–126.

Desai, C., Pathak, H., and Madamwar, D. (2010). Advances in molecular and '-omics' technologies to gauge microbial communities and bioremediation at xenobiotic/anthropogen contaminated sites. Bioresource Technol. *101*, 1558–1569.

Dittmann, S., (1996). Effects of macrobenthic burrows on infaunal communities in tropical tidal flats. Mar. Ecol. Prog. Ser. *134*, 119–130.

Dobbs, F.C., and Guckert, J.B. (1988). Microbial food resources of the macrofaunal-deposit feeder *Ptychodera bahamensis* (Hemicordata: Enteropneusta). Mar. Ecol. Prog. Ser. *45*, 127–136.

Dyksterhouse, S.E., Gray, J.P., Herwig, R.P., Lara, J.C., and Staley, J.T. (1995). *Cycloclasticus pugetii* gen. nov., sp. nov., an aromatic hydrocarbon-degrading bacterium from marine sediments. Int. J. Syst. Bacteriol. *45*, 116–123.

Ehlers, L. J., and Luthy, R.G. (2003). Contaminant bioavailability in soil and sediment. Environ. Sci. Technol. 37, 295 302.

Engelhardt, M.A., Daly, K., Swannell, R.P., and Head, I.M. (2001). Isolation and characterization of a novel hydrocarbon-degrading, Gram-positive bacterium, isolated from intertidal beach sediment, and description of *Planococcus alkanoclasticus* sp. nov. J. Appl. Microbiol. *90*, 237–247.

Eriksson, M., Dalhammar, G., and Borg-Karlson, A.K. (1999). Aerobic degradation of a hydrocarbon mixture in natural uncontaminated potting soil by indigenous microorganisms at 20°C and 6°C. Appl. Microbiol. Biotechnol. *51*, 532–535.

Eriksson, M., Sodersten, E., Yu, Z.T., Dalhammar, G., and Mohn, W.W. (2003) Degradation of polycyclic aromatic hydrocarbons at low temperature under aerobic and nitrate-reducing conditions in enrichment cultures from northern soils. Appl. Environ. Microbiol. *69*, 275–284.

Fetzner, S. (2010). Aerobic degradation of halogenated aliphatics. In Handbook of Hydrocarbon and Lipid Microbiology, Timmis, K.N., ed. (: Springer-Verlag: Berlin, Heidelberg, Germany), pp.865–886.

Findlay, R.H., Trexler, M.B., and White, D.C. (1990). Response of a benthic microbial community to biotic disturbance. Mar. Ecol. Prog. Ser. *62*, 135–148.

Fleeger, J.W., Carman, K.R., and Nisbet, R.M. (2003). Indirect effects of contaminants in aquatic ecosystems. Sci. Total Environ. *317*, 207–233.

Fornes, W.L., DeMaster, D.J., and Smith, C.R. (2001). A particle introduction experiment in Santa Catalina Basin sediments: Testing the age-dependant mixing hypothesis. J. Mar. Res. *59*, 97–112.

Forster, S., and Graf, G. (1995). Impact of irrigation on oxygen flux into the sediment – intermittent pumping by *Callianassa subterranea* and piston pumping by *Lanice conchilega*. Mar. Biol. *123*, 335–346.

François, F., Poggiale, J.C., Durbec, J.P., and Stora, G. (1997). A new approach for the modelling of sediment reworking induced by a macrobenthic community. Acta Biotheor. *45*, 295–319.

Gafarov, A.B., Panov, A.V., Filonov, A.E., and Boronin, A.M. (2006). Change in the composition of a bacterial association degrading aromatic compounds during oil sludge detoxification in a continuous-flow microbial reactor. Appl. Biochem. Microbiol. *42*, 160–165.

Galushko, A., Minz, D., Schink, B., and Widdel, F. (1999). Anaerobic degradation of naphthalene by a pure culture of a novel type of marine sulphate-reducing bacterium. Environ. Microbiol. *1*, 415–420.

Gamenick, I., Jahn, A., Vopel, K., and Giere, O. (1996). Hypoxia and sulfide as structuring factors in a macrozoobenthic community on the Baltic Sea shore: colonisation studies and tolerance experiments. Mar. Ecol. Prog. Ser. *144*, 73–85.

Gardner, L.R., Sharma, P., and Moore, W.S. (1987). A regeneration model for the effect of bioturbation by fiddler crabs on 210Pb profiles in salt-marsh sediments. J. Environ. Radioact. *5*, 25–36.

Gauthier, M.J., Lafay, B., Christen, R., Fernandez, L., Acquaviva, M., Bonin, P., and Bertrand, J.C. (1992). *Marinobacter hydrocarbonoclasticus* gen. nov., sp. nov., a new extremely halotolerant, hydrocarbon degrading marine bacterium. Int. J. Syst. Bacteriol. *42*, 568–576.

Gérino, M., Stora, G., François-Carcaillet, F., Gilbert, F., Poggiale, J.C., Mermillod-Blondin, F., Desrosiers, G., and Vervier, P. (2003). Macro-invertebrate functional groups in freshwater and marine sediments: A common mechanistic classification. Life Environ. *53*, 221–231.

Giessing, A.M.B., and Mayer, L.M. (2004). Oxidative coupling during gut passage in marine deposit-feeding invertebrates. Limnol. Oceanogr., *49*, 716–726.

Gilbert, F., Stora, G., and Bertrand, J.C. (1996). In situ bioturbation and fate in an experimental contaminated Mediterranean coastal ecosystem. Chemosphere *33*, 1449–1458.

Gilbert, F., Stora, G., Desrosiers, G., Deflandre, B., Bertrand, J-C., Poggiale, J-C., and Gagné J-P. (2001). Alteration and release of aliphatic compounds by the polychaete *Nereis virens* (Sars) experimentally fed with hydrocarbons. J. Exp. Mar. Biol. Ecol. *256*, 199–213.

Gilbert, F., Hulth, S., and Aller, R.C. (2003). The influence of macrofaunal burrow spacing and diffusive scaling on sedimentary nitrification and denitrification: An experimental and model approach. J. Mar. Res. *61*, 101–125.

Gilewicz, M., Monpert, G., Acquaviva, M., Mille, G., and Bertand, J-C. (1991). Anaerobic oxidation of 1-*n*-heptadecene by a marine denitrifying bacterium. Appl. Microbiol. Biotechnol. *36*, 252–256.

Gilewicz, M., Ni'matuzahroh, Nadalig, T., Budzinski H., Doumenq, P., Michotey, V., and Bertrand, J.C. (1997). Isolation and characterization of a marine bacterium capable of utilizing 2-methylphenanthrene. Appl. Microbiol. Biotechnol. *48*, 528–533.

Glud, R.N. (2008). Oxygen dynamics of marine sediments. Mar. Biol. Res. *4*, 243–289.

Glud, R.N., Santegoeds, C.M., Beer, D.D., Kohls, O., and Ramsing, N.B. (1998). Oxygen dynamics at the base of a biofilm studied with planar optodes. Aquat. Microb. Ecol. *14*, 223–233.

Golyshin, P.N., Chernikova, T.N., Abraham, W.R., Lünsdorf, H., Timmis, K.N., and Yakimov, M.M. (2002). *Oleiphilaceae* fam. nov., to include *Oleiphilus messinensis* gen. nov., sp. nov., a novel marine bacterium that obligately utilizes hydrocarbons. Int. J. Syst. Evol. Microbiol. *52*, 901–911.

Gomes, N.C., Borges, L.R., Paranhos, R., Pinto, F.N., Krögerrecklenfort, E., Mendonça-Hagler, L.C., and Smalla, K. (2007). Diversity of ndo genes in mangrove sediments exposed to different sources of polycyclic aromatic hydrocarbon pollution. Appl. Environ. Microbiol. *7*, 7392–7399.

Granberg, M.E., and Forbes, T.L. (2006). Role of sediment organic matter quality and feeding history in dietary absorption and accumulation of pyrene in the mud snail (*Hydrobia ulvae*). Environ. Toxicol. Chem. *25*, 995–1006.

Granberg, M.E., and Selck, H. (2007). Effects of sediment organic matter quality on bioaccumulation, degradation, and distribution of pyrene in two macrofaunal species and their surrounding sediment. Mar. Environ. Res. *64*, 313–35.

Granberg, M.E., Gunnarsson, J.S., Hedman, J.E., Rosenberg, R., Jonsson, P. (2008). Bioturbation-driven release of organic contaminants from Baltic Sea sediments mediated by the invading polychaete *Marenzelleria neglecta*. Environ. Sci. Technol. *42*, 1058–1065.

Green, M.A., Aller, R.C., Cochran, J.K., Lee, C., and Aller, J.Y. (2002). Bioturbation in shelf/slope sediments off Cape Hatteras, North Carolina: the use of 234Th, Chl-a, and Br- to evaluate rates of particle and solute transport. Deep-Sea Res. II *49*, 4627–4644.

Grossi, V., Massias, D., Stora, G., and Bertrand, J.C. (2002). Burial, exportation and degradation of acyclic petroleum hydrocarbons following a simulated oil spill in bioturbated Mediterranean coastal sediments. Chemosphere *48*, 947–954.

Grossi, V., Caradec, S., Gilbert, F. (2003). Burial and reactivity of sedimentary microalgal lipids in bioturbated Mediterranean coastal sediments. Mar. Chem. *81*, 57–69.

Grossi, V., Cuny, P., Caradec, S., Nérini, D., Pancost, R., and Gilbert, F. (2006). Impact of feeding by *Arenicola marina* (L.) and ageing of faecal material on fatty acid distribution and bacterial community structure in marine sediments: an experimental approach. J. Exp. Mar. Biol. Ecol. *336*, 54–64.

Grossi, V., Cravo-Laureau, C., Guyoneaud, R., Ranchou-Peyruse, A., and Hirschler-Réa, A. (2008). Metabolism of *n*-alkanes and *n*-alkenes by anaerobic bacteria: a summary. Org. Geochem. *39*, 1197–1203.

Grossmann, S., and Reichardt, W. (1991). Impact of *Arenicola marina* on bacteria in intertidal sediments. Mar. Ecol. Prog. Ser. *77*, 85–96.

Grundmann, O., Behrends, A., Rabus, R., Amann, J., Halder, T., Heider, J., and Widdel, F. (2008). Genes encoding the candidate enzyme for anaerobic activation of *n*-alkanes in the denitrifying bacterium, strain HxN1. Environ. Microbiol. 10, 376–385.

Gunnarsson, J.S., Granberg, M.E., Nilsson, H.C., Rosenberg, R., and Hellman, B. (1999). Influence of sediment organic matter quality on growth and polychlorobiphenyl bioavailability in *Echinodermata* (*Amphiura filiformis*). Environ. Toxicol. Chem. *18*, 1534–1543.

Guo, C., Sun, W., Harsh, J.B., and Ogram, A. (1997). Hybridization analysis of microbial DNA from fuel oil-contaminated and noncontaminated soil. Microb. Ecol. *34*, 178–187.

Guo, C.L., Zhou, H.W., Wong, Y.S., and Tam, N.F. (2005). Isolation of PAH-degrading bacteria from mangrove sediments and their biodegradation potential. Mar. Pollut. Bull. *51*, 1054–1061.

Habe, H., and Omori, T. (2003). Genetics of polycyclic aromatic hydrocarbon metabolism in diverse aerobic bacteria. Biosci. Biotech. Bioch. *67*, 225–243.

Hanson, J.R., Macalady, J.L., Harris, D., and Scow, K.M. (1999). Linking toluene degradation with specific microbial populations in soil. Appl. Envir. Microbiol. *65*, 5403–5408.

Harms, H., Wick, L.Y., Smith, K.E.C. (2010). Matrix-hydrophobic compound interactions. In Handbook of Hydrocarbon and Lipid Microbiology, Timmis, K.N., ed. (Springer Verlag: Berlin, Germany), pp. 1467–1478.

Harvey, H.R., and Macko, S.A. (1997). Kinetics of phytoplankton decay during simulated sedimentation: changes in lipids under oxic and anoxic conditions. Org. Geochem. *27*, 129–140.

Head I.M., and Swannell, R.P. (1999). Bioremediation of petroleum hydrocarbon contaminants in marine habitats. Curr. Opin. Biotechnol. *10*, 234–239.

Head I.M., Jones D.M., and Roling, W.F.M. (2006). Marine microorganisms make a meal of oil. Nat. Rev. Microbiol. *4*, 173–182.

Hedlund, B.P., and Costa, K.C. (2010). Neptunomonas. In Handbook of Hydrocarbon and Lipid Microbiology, Timmis, K.N., ed. (Springer-Verlag: Berlin Heidelberg, Germany), pp. 1773–1780.

Hedlund, B.P., Geiselbrecht, A.D., Bair, T.J., and Staley, J.T. (1999). Polycyclic aromatic hydrocarbon degradation by a new marine bacterium, *Neptunomonas naphthovorans* gen. nov., sp. nov. Appl. Environ. Microbiol. *65*, 251–259.

Hedlund, B.P., Geiselbrecht, A.D., and Staley, J.T. (2001). *Marinobacter* strain NCE312 has a *Pseudomonas*-like naphthalene dioxygenase. FEMS Microbiol. Lett. *201*, 47–51.

Hedman, J.E., Stempa Tocca, J., and Gunnarsson, J.S. (2010). Remobilization of polychlorinated biphenyl from baltic sea sediment: Comparing the roles of bioturbation and physical resuspension. Environ. Toxicol. Chem. *28*, 2241 – 2249.

Heider, J. (2007). Adding handles to unhandy substrates: anaerobic hydrocarbon activation mechanisms. Curr. Opin. Chem. Biol. *11*, 188–194.

Heiss-Blanquet, S., Benoit, Y., Marechaux, C., and Monot, F. (2005). Assessing the role of alkane hydroxylase genotypes in environmental samples by competitive PCR. J. Appl. Microbiol. 99, 1392–1403.

Higashioka, Y., Kojima, H., Nakagawa, T., Sato, S., and Fukui, M. (2009). A novel n-alkane-degrading bacterium as a minor member of p-xylene-degrading sulfate-reducing consortium. Biodegradation 20, 383–390.

Horng, C.-Y. (1998). Influence of the marine polychaete, *Capitella* sp. I, on the fate of sediment-bound polycyclic aromatic hydrocarbons: the role of feeding activity. PhD thesis, Rutgers University, New Brunswick.

Horng, C.Y., and Taghon, G.L. (1999). Effects of contaminated sediments on particle size selection by the polychaete *Capitella* sp. I. J. Exp. Mar. Biol. Ecol. 242, 41–57.

Horng, C.-Y., and Taghon, G.L. (2001). Loss rates for polycyclic hydrocarbons from sediment and deposit-feeder fecal pellets. Environ. Toxicol. Chem. 20, 1465–1473.

Hulth, S., Aller, R.C., Canfield, D.E., Dalsgaard, T., Engström, P., Gilbert, F., Sundbäck, K., and Thamdrup, B. (2005). Nitrogen removal in marine environments: recent findings and future research challenges. Mar. Chem. 94, 125–145.

Hulthe, G., Hulth S., and Hall, P.O.J. (1998). Effect of oxygen on degradation rate of refractory and labile organic matter in continental margin sediments, Geochim. Cosmochim. Acta 62, 1319–1328.

Jeong, S., Wander M.M., Kleineidam, S., Grathwohl, P., Ligouis, B., and Werth, C.J. (2008). The role of condensed carbonaceous materials on the sorption of hydrophobic organic contaminants in subsurface sediments. Environ. Sci. Technol. 42, 1458–1464.

Jones, C.G., Lawton, J.H., and Shachak, M. (1994). Organisms as ecosystem engineers. Oikos 69, 373–386.

Jones, D.M., Head, I.M., Gray, N.D., Adams, J.J., Rowan, A.K., Aitken, C.M., Bennett, B., Huang, H., Brown, A., Bowler, B.F.J., *et al.* (2008). Crude oil biodegradation via methanogenesis in subsurface petroleum reservoirs. Nature 451, 176–180.

Josefson, A.B., Forbes, T.L., and Rosenberg, R. (2002). Fate of phytodetritus in marine sediments: functional importance of macrofaunal activity. Mar. Ecol. Prog. Ser. 230, 71–85.

Kasai, Y., Kishira, H., Syutsubo, K., and Harayama, S. (2001). Molecular detection of marine bacterial populations on beaches contaminated by the Nakhodka tanker oil-spill accident. Environ. Microbiol. 3, 246–255.

Ke, L., Bao, W., Chen, L., Wong, Y.S., and Tam, N.F. (2009). Effects of humic acid on solubility and biodegradation of polycyclic aromatic hydrocarbons in liquid media and mangrove sediment slurries. Chemosphere 76, 1102–1108.

Klaassen, C.D. (1996). Casarett & Doull's Toxicology – the Basic Science of Poisons. McGraw-Hill: New York.

Kniemeyer, O., Fischer, T., Wilkes, H., Glöckner, F-O., and Widdel, F. (2003). Anaerobic degradation of ethylbenzene by a new type of marine sulfate-reducing bacterium. Appl. Environ. Microbiol. 69, 760–768.

Kniemeyer, O., Musat, F., Sievert, S., Knittel, K., Wilkes, H., Blumenberg, M., Michaelis, W., Classen, A., Bolm, C., Joye, S., *et al.* (2007). Anaerobic oxidation of short-chain hydrocarbons by marine sulphate-reducing bacteria. Nature 449, 898–902.

Koelmans, A.A., Jonker, M.T.O., Cornelissen, G., Bucheli, T.D., Van Noort, P.C.M., and Gustafsson, Ö. (2006) Black carbon: The reverse of its dark side. Chemosphere 63, 365–377.

Kogure, K., and Wada, M. (2005). Impacts of macrobenthic bioturbation in marine sediment on bacterial metabolic activity. Microb. Environ. 20, 191–199.

Konovalov, D., Renaud, P.E., Berge, J., Voronkov, A.Y., and Cochrane, S.K.J. (2010). Contaminants, benthic communities, and bioturbation: potential for PAH mobilization in Artic sediments. Chem. Ecol. 26, 197–208.

Kowalewska, G., Konat, J. (1997). The role of phytoplankton in the transport and distribution of polynuclear aromatic hydrocarbons (PAHs) in the southern Baltic environment. Oceanologia 39, 267–277.

Kristensen, E. (1989). Oxygen and carbon dioxide exchange in the polychaete *Nereis virens*: influence of ventilation activity and starvation. Mar. Biol. 101, 381–388.

Kristensen, E. (2000). Organic matter diagenesis at the oxic/anoxic interface in coastal marine sediments, with emphasis on the role of burrowing animals. Hydrobiologia 426, 1–24.

Kristensen, E. (2001). Impact of polychaetes (*Nereis* spp. and *Arenicola marina*) on carbon biogeochemistry in coastal marine sediments. Geochem. Trans. 2, 92.

Kristensen, E., and Blackburn, T.H. (1987). The fate of organic carbon and nitrogen in experimental marine sediment systems: Influence of bioturbation and anoxia. J. Mar. Res. 45, 231–257.

Kristensen, E., and Kostka, J.E. (2005). Macrofaunal burrows and irrigation in marine sediment: microbiological and biogeochemical interactions. In Interactions between Macro- and Microorganisms in Marine Sediments, Kristensen, E., Haese, R.R., Kostka, J.E., eds (American Geophysical Union: Washington, DC, USA), pp. 125–158.

Kristensen, E., Ahmed, S.I., and Devol, A.H. (1995). Aerobic and anaerobic decomposition of organic matter in marine sediment: which is fastest? Limnol. Oceanogr. 40, 1430–1437.

Kube, J., Zettler, M.L., Gosselck, F., Ossig, S., and Powilleit, M. (1996). Distribution of *Marenzelleria viridis* (Polychaeta, Spionidae) in the southwestern Baltic Sea in 1993/94 – ten years after introduction. Sarsia 81, 131–142.

Kubota, M., Nadate, M., Yasumoto-Hirose, M., Uchiyama, T., Kagami, O., Shizuri, Y., and Misawa, N. (2005). Isolation and functional analysis of cytochrome P450 CYP153A genes from various environments. Biosci. Biotechnol. Biochem. 69, 2421–2430.

Kuhn, E., Bellicanta, G.S., and Pellizari, V.H. (2009). New alk genes detected in Antarctic marine sediments. Environ. Microbiol. *11*, 669–673.

Kure, L.K, and Forbes, T.L. (1997). Impact of bioturbation by *Arenicola marina* on the fate of particle-bound fluoranthene. Mar. Ecol. Progr. Ser. *156*, 157–166.

Kwon, K.K., Lee, H.S., Jung, H.B., Kang, J.H., and Kim, S.J. (2006). Yeosuana aromativorans gen. nov., sp. nov., a mesophilic marine bacterium belonging to the family Flavobacteriaceae, isolated from estuarine sediment of the South Sea, Korea. Int. J. Syst. Evol. Microbiol. *56*, 727–732.

Kwon, K.K., Lee, H.S., Yang, S.H., and Kim, S.J. (2005). *Kordiimonas gwangyangensis* gen. nov., sp. nov., a marine bacterium isolated from marine sediments that forms a distinct phyletic lineage (Kordiimonadales ord. nov.) in the '*Alphaproteobacteria*'. Int. J. Syst. Evol. Microbiol. *55*, 2033–2037.

Lee, C. (1992). Controls on organic preservation: The use of stratified water bodies to compare intrinsic rates of decomposition in oxic and anoxic systems. Geochim. Cosmochim. Acta *56*, 3323–3335.

Lei, L., Khodadoust, A.P., Suidan, M.T., and Tabak, H.H. (2005). Biodegradation of sediment-bound PAHs in field contaminated sediment. Water. Res. *39*, 349–361.

Lin, Y., and Cai, L.X. (2008). PAH-degrading microbial consortium and its pyrene-degrading plasmids from mangrove sediment samples in Huian, China. Mar. Pollut. Bull. *57*, 703–706.

Lohrer, A.M., Thrush, S.F., and Gibbs, M.M. (2004). Bioturbators enhance ecosystem function through complex biogeochemical interactions. Nature *431*, 1092–1095.

Löser, C., Seidel, H., Zehnsdorf, A., and Stottmeister, U. (1998). Microbial degradation of hydrocarbons in soil during aerobic/anaerobic changes and under purely aerobic conditions. Appl. Microbiol. Biotechnol. *49*, 631–636.

Lovley, D.R. (2003). Cleaning up with genomics: applying molecular biology to bioremediation. Nature Rev. Microbiol. *1*, 35–44.

Lozada, M., Riva Mercadal, J.P., Guerrero, L.D., Di Marzio, W.D., Ferrero, M.A., and Dionisi, H.M. (2008). Novel aromatic ring-hydroxylating dioxygenase genes from coastal marine sediments of Patagonia. BMC Microbiol. 25, 8–50.

Maeng, J.H., Sakai, Y., Ishige, T., Tani, Y., and Kato, N. (1996). Diversity of dioxygenases that catalyze the first step of oxidation of long-chain *n*-alkanes in *Acinetobacter* sp. Strain M-1. FEMS Microbiol. Lett. *141*, 177–182.

Maier, T., Forster, H.H., Asperger, O., and Hahn, U. (2001). Molecular characterization of the 56-kDA CYP153 from *Acinetobacter* sp. EB104. Biochem. Biophys. Res. Commun. *286*, 652–658.

Marcos, M.S., Lozada, M., and Dionisi, H.M. (2009). Aromatic hydrocarbon degradation genes from chronically polluted Subantarctic marine sediments. Lett. Appl. Microbiol. *49*, 602–608.

Maruyama, A., Ishiwata, H., Kitamura, K., Sunamura, M., Fujita, T., Matsuo, M., and Higashihara, T. (2003). Dynamics of microbial populations and strong selection for *Cycloclasticus pugetii* following the Nakhodka oil spill. Microb. Ecol. *46*, 442–453.

Matsui, G.Y., Ringelberg, D.B., and Lovell, C.R. (2004). Sulfate-reducing bacteria in tubes constructed by the marine infaunal polychaete *Diopatra cuprea*. Appl. Environ. Microbiol. *70*, 7053–7065.

Mayer, L.M., Schick, L.L., Self, R.F.L., Jumars, P.A. (1997). Digestive environments of benthic macroinvertebrate guts: enzymes, surfactants and dissolved organic matter. J. Mar. Res. *55*, 785–812.

Mayer, L.M., Weston, D.P., and Bock M.J. (2001). Benzo[a]pyrene and zinc solubilization by digestive fluids of benthic invertebrates – a cross-phyletic study. Environ. Toxicol. Chem. *20*, 1890–1900.

McElroy, A.E., Farrington, J.W., and Teal, J.M. (1990). Influence of mode of exposure and the presence of a tubiculous polychaete on the fate of benz[a]anthracene in the benthos. Environ. Sci. Technol. *24*, 1648–1655.

McNally, D.L., Mihelcic, J.R., and Lueking, D.R. (1999). Biodegradation of mixtures of polycyclic aromatic hydrocarbons under aerobic and nitrate-reducing conditions. Chemosphere *38*, 1313–1321.

McNaughton, S.J., Stephen, J.R., Venosa, A.D., Davis, G.A., Chang, Y.J., and White, D.C. (1999). Microbial population changes during bioremediation of an experimental oil spill. Appl. Environ. Microbiol. *65*, 3566–3574.

Meckenstock, R.U., Safinowski, M., and Griebler, C. (2004). Anaerobic degradation of polycyclic aromatic hydrocarbons. FEMS Microbiol. Ecol. *49*, 27–36.

Miller, R.J., Smith, C.R., DeMaster, D.J., and Fornes, W.L. (2000). Feeding selectivity and rapid particle processing by deep-sea megafaunal deposit feeders: A 234Th tracer approach. J. Mar. Res. *58*, 653–673.

Miralles, G., Acquaviva, M., Bertrand, J.C., and Cuny, P. (2010). Response of an archaeal community from anoxic coastal marine sediments to experimental petroleum contamination. Aquat. Microb. Ecol. *59*, 25–31.

Miralles, G., Grossi, V.,Acquaviva, M.,Duran, R.,Bertrand, J.C., and Cuny, P. (2007a) Alkane biodegradation and dynamics of phylogenetic subgroups of sulfate reducing bacteria in an anoxic coastal marine sediment artificially contaminated with oil. Chemosphere *68*, 1327–1334.

Miralles, G., Nérini, D., Manté, C., Acquaviva, M., Doumenq, P., Michotey, V., Nazaret, S., Bertrand, J.C., and Cuny, P. (2007b). Effects of spilled oil on bacterial communities of mediterranean coastal anoxic sediments chronically subjected to oil hydrocarbon contamination. Microb. Ecol. *54*, 646–661.

Montgomery M.T., Osburn C.L., Furukawa, Y., and Gieskes J.M. (2008). Increased capacity for polycyclic aromatic hydrocarbon mineralization in bioirrigated coastal marine sediments. Biorem. J. *12*, 98–110.

Musat F., Galushko A., Jacob J., Widdel F., Kube M., Reinhardt R., Wilkes H., Schink B., and Rabus, R. (2009). Anaerobic degradation of naphthalene and 2-methylnaphthalene by strains of marine sulfate-reducing bacteria. Environ. Microbiol. *11*, 209–219.

Musat, F., and Widdel, F. (2008). Anaerobic degradation of benzene by a marine sulfate-reducing enrichment culture, and cell hybridization of the dominant phylotype. Environ. Microbiol. *10*, 10–19.

Muschenheim, D.K, and Lee, K. (2002). Removal of oil from the sea surface through particulate interactions: review and prospectus. Spill Sci. Technol. Bull. *8*, 9–18.

Nair, D., Fernández-Acero, F.J., García-Luque, E., Riba, I., and Del Valls, T.A. (2008). Isolation and characterization of naphthalene-degrading bacteria from sediments of Cadiz area (SW Spain). Environ. Toxicol. *23*, 576–582.

Nakagawa, T., Sato, S., and Fukui, M. (2008). Anaerobic degradation of *p*-xylene in sediment-free sulfate-reducing enrichment culture. Biodegradation *19*, 909–913.

Näslund, J., Nascimento, F.J.A., and Gunnarsson, J.S. (2010). Meiofauna reduces bacterial mineralization of naphthalene in marine sediment. ISME J., 1–10.

Nikolaou, A., Kostopoulou, M., Lofrano, G., Meric, S., Petsas, A., and Vagi, M. (2009). Levels and toxicity of polycyclic aromatic hydrocarbons in marine sediments. Trends Anal. Chem. *28*, 653–664.

Nikolopoulou, M., and Kalogerakis, N. (2010). Biostimulation strategies for enhanced bioremediation of marine oil spills including chronic pollution. In Handbook of Hydrocarbon and Lipid Microbiology, Timmis, K.N., ed. (Springer-Verlag: Berlin Heidelberg, Germany), pp. 2521–2529.

Osinga, R., Kop, A.J., Malschaert, J.F.P., and Van Duyl, F.C. (1997). Effects of the sea urchin *Echinocardium cordatum* on bacterial production and carbon flow in experimental benthic systems under increasing organic loading. J. Sea Res. 37, 109–121.

Palmer, M.A., Covich, A., Finlay, B.J., Gibert, J., Hyde, K.D., Johnson R.K., Kairesalo, T., Lake, S., Lovelle, C.R., Naiman, R.J., *et al.* (1997). Biodiversity and ecosystem processes in freshwater sediments. Ambio *26*, 571–577.

Palmqvist, A., Rasmussen, L.J., and Forbes, V. (2006). Influence of biotransformation on trophic transfer of the PAH, fluoranthene. Aquat. Toxicol. *80*, 309–319.

Papaspyrou, S., Gregersen, T., Cox, R.P., Thessalou-Legaki, M., and Kristensen, E. (2005). Sediment properties and bacterial community in burrows of the ghost shrimp *Pestarella tyrrhena* (Decapoda: Thalassinidea). Aquat. Microb. Ecol. *38*, 181–190.

Papaspyrou, S., Gregersen, T., Kristensen, E., Christensen B., and Cox, R.P. (2006). Microbial reaction rates and bacterial communities in sediment surrounding burrows of two nereidid polychaetes (*Nereis diversicolor* and *N. virens*). Mar. Biol. *148*, 541–550.

Parales, R.E., Ju, K.S., Rollefson, J., and Ditty, J.L. (2008). Bioavailability, transport and chemotaxis of organic pollutants. In Microbial Bioremediation, Diaz, E., ed. (Caister Academic Press: Norfolk, UK), pp. 145–187.

Parker, C.A., and O'Reilly, J.E. (1991). Oxygen depletion in Long Island Sound: A historical perspective. Estuaries *14*, 248–264.

Payne, J.R., Clayton, J., John, R., and Kirstein, B.E. (2003). Oil/suspended particulate material interactions and sedimentation. Spill Sci. Technol. Bull. *8*, 201–221.

Pearson, T.H., and Rosenberg, R. (1978). Macrobenthic succession in relation to organic enrichment and pollution of the marine environment. Oceanogr. Mar. Biol. Ann. Rev. *16*, 229–311.

Pepi, M., Lobianco, A., Renzi, M., Perra, G., Bernardini, E., Marvasi, M., Gasperini, S., Volterrani, M., Franchi, E., Heipieper, H.J., *et al.* (2009). Two naphthalene degrading bacteria belonging to the genera *PaeniBacillus* and *Pseudomonas* isolated from a highly polluted lagoon perform different sensitivities to the organic and heavy metal contaminants. Extremophiles *13*, 839–848.

Perelo L. W. (2010). Review: *In situ* and bioremediation of organic pollutants in aquatic sediments. J. Hazard. Mater. *177*, 81–89.

Pérez-Pantoja, D., Gonzalez, B., and Pieper, D.H. (2010). Aerobic degradation of aromatic hydrocarbons. In Handbook of Hydrocarbon and Lipid Microbiology, Timmis, K.N., ed. (Springer-Verlag: Berlin Heidelberg, Germany), pp. 799–838.

Phelps, C.D., Kerkhof, L.J., and Young, L.Y. (1998). Molecular characterization of a sulfate-reducing consortium which mineralizes benzene. FEMS Microbiol. Ecol. *27*, 269–279.

Pieper, D.H., Gonzalez, B., Camara, B., Pérez-Pantoja, D., and Reineke, W. (2010). Aerobic degradation of chloroaromatics. In Handbook of Hydrocarbon and Lipid Microbiology, Timmis, K.N., ed. (Springer-Verlag: Berlin Heidelberg, Germany), pp.839–864.

Pierard, C., Budzinski, H., and Garrigues, P. (1996). Grain-size distribution of polychlorobiphenyls in coastal sediments. Environ. Sci. Technol. *30*, 2776–2783.

Pischedda, L. (2009). Bioturbation et hétérogénéité de l'oxygène dans les sédiments marins appréhendées par une méthode optique de quantification 2D. Incidence sur les communautés microbiennes sédimentaires. PhD thesis, Université de la Méditérranée, Marseille, p. 297.

Pischedda, L., Poggiale, J.C., Cuny, P., and Gilbert, F. (2008). Imaging oxygen distribution in marine sediments. The importance of bioturbation and sediment heterogeneity. Acta Biotheor. *56*, 123–135.

Plante, C. J., Coe, K.M., and Plante, R.G. (2008). Isolation of surfactant-resistant bacteria from natural, surfactant-rich marine habitats. Appl. Environ. Microbiol. *74*, 5093–5099.

Prahl, F.G., and Carpenter, R. (1983). Polycyclic aromatic hydrocarbon (PAH) – phase associations in Washington coastal sediment. Geochim. Cosmochim. Acta *47*, 1013–1023

Prince, R.C., Lessard, R.R., and Clark, J.R. (2003). Bioremediation of marine oil spills. Oil Gas Sci. Technol. Rev. IFP *58*, 463–468.

Qin, X., Sun, H., Wang, C., Yu, Y., and Sun, T. (2010). Impacts of crab bioturbation on the fate of polycyclic aromatic hydrocarbons in sediment from the Beitang estuary of Tianjin, China. Environ. Toxicol. Chem. *29*, 1248–1255.

Rabus, R., Nordhaus, R., Ludwig, W., and Widdel, F. (1993). Complete oxidation of toluene under strictly anoxic conditions by a new sulfate-reducing bacterium. Appl. Environ. Microbiol. *59*, 1444–1451.

Rabus, R., Wilkes, H., Behrends, A., Armstroff, A., Fischer, T., Pierik, A.J., and Widdel, F. (2001). Anaerobic initial reaction of *n*-alkanes in a denitrifying bacterium: Evidence for (1-methylpentyl) succinate as initial product and for involvement of an organic radical in *n*-hexane metabolism. J. Bacteriol. *183*, 1707–1715.

Reichenberg, F., and Mayer, P. (2006). Two complementary sides of bioavailability: accessibility and chemical activity of organic contaminants in sediments and soils. Environ. Toxicol. Chem. *25*, 1239–1245.

Rhoads, D.C. (1974). Organism-sediment relations on the muddy sea floor. Oceanogr. Mar. Biol. Annu. Rev. *12*, 263–300.

Riisgard, H.U. (1991). Suspension feeding in the polychaete *Nereis diversicolor*. Mar. Ecol. Progr. Ser. *70*, 29–37.

Rios-Hernandez, L.A., Gieg, L.M., and Suflita, J.M. (2003). Biodegradation of an alicyclic hydrocarbon by a sulfate-reducing enrichment from a gas condensate-contaminated aquifer. Appl. Environ. Microbiol. *69*, 434–443.

Rockne, K.J., and Strand, S.E. (1998). Biodegradation of bicyclic and polycyclic aromatic hydrocarbons in anaerobic enrichments. Environ. Sci. Technol. *32*, 3962–3967.

Rockne, K.J., Chee-Sanford, J.C., Sanford, R.A., Hedlund, B.P., Staley, J.T., and Strand, S.E. (2000). Anaerobic naphthalene degradation by microbial pure cultures under nitrate-reducing conditions. Appl. Environ. Microbiol. *66*, 1595–1601.

Rohwer, F., and Thurber, R.V. (2009). Marine viruses: Manipulating the marine environment. Nature *459*, 207–212.

Rojo, F. (2010). Enzymes for aerobic degradation of alkanes. In Handbook of Hydrocarbon and Lipid Microbiology, Timmis, K.N., ed. (Springer-Verlag: Berlin Heidelberg, Germany), pp. 781–798.

Röling, W.F., Milner, M.G., Jones, D.M., Lee, K., Daniel, F., Swannell, R.J., and Head, I.M. (2002). Robust hydrocarbon degradation and dynamics of bacterial communities during nutrient-enhanced oil spill bioremediation. Appl. Environ. Microbiol. *68*, 5537–5548.

Röling, W.F., Milner, M.G., Jones, D.M., Fratepietro, F., Swannell, R.P., Daniel, F., and Head, I.M. (2004). Bacterial community dynamics and hydrocarbon degradation during a field-scale evaluation of bioremediation on a mudflat beach contaminated with buried oil. Appl. Environ. Microbiol. *70*, 2603–2613.

Ron, E.Z. (2000). Microbial life on petroleum. In Journey to Diverse Microbial Worlds. Seckbach, J., ed. (Kluwer Academic Publishers: Dordrecht, The Netherlands), pp. 303–315.

Ron, E.Z., and Rosenberg, E. (2002). Biosurfactants and oil bioremediation. Curr. Opin. Biotechnol. *13*, 249–252.

Rontani, J.F., Mouzdahir, A., Michotey, V., and Bonin, P. (2002). Aerobic and anaerobic metabolism of squalene by a denitrifying bacterium isolated from marine sediment. Arch. Microbiol. *178*, 279–287.

Rontani, J.F., Mouzdahir, A., Michotey, V., Caumette, P., and Bonin, P. (2003). Production of a polyunsaturated isoprenoid wax ester during aerobic metabolism of squalene by *Marinobacter squalenivorans* sp. nov. Appl. Environ. Microbiol. *69*, 4167–4176.

Rosenberg, R. (1995). Benthic marine fauna structured by hydrodynamic processes and food availability. Neth. J. Sea Res. 34, 303–317.

Roslev, P., Iversen, N., and Henriksen, K. (1998). Direct fingerprinting of metabolically active bacteria in environmental samples by substrate specific radiolabelling and lipid analysis. J. Microbiol. Methods *31*, 99–111.

Rothermich, M.M., Hayes, L.A., and Lovley, D.R. (2002). Anaerobic, sulfatedependent degradation of polycyclic aromatic hydrocarbons in petroleum-contaminated harbor sediment. Environ. Sci. Technol. *36*, 4811–4817.

Rueter, P., Rabus, R., Wilkes, H., Aeckersberg, F., Rainey, F.A., Jannasch, H.W., and Widdel, F. (1994). Anaerobic oxidation of hydrocarbons in crude oil by new types of sulphate-reducing bacteria. Nature 372, 455–458.

Sabirova, J.S., Ferrer, M., Regenhardt, D., Timmis, K.N., and Golyshin, P.N. (2006). Proteomic insights into metabolic adaptations in *Alcanivorax borkumensis* induced by alkane utilization. J. Bacteriol. *188*, 3763–73.

Safinowski, M., and Meckenstock, R.U. (2004). Enzymatic reactions in anaerobic 2-methylnaphthalene degradation by the sulphate-reducing enrichment culture N 47. FEMS Microbiol. Lett. *240*, 99–104.

Saito, A., Iwabuchi, T., and Harayama, S. (2000). A novel phenanthrene dioxygenase from *Nocardioides* sp. Strain KP7: expression in *Escherichia coli*. J. Bacteriol. *182*, 2134–2141.

Schaffner, L.C., Dickhut, R.M., Mitra, S., Lay, P.W., and Brouwer-Riel, C. (1997). Effects of physical chemistry and bioturbation by estuarine macrofauna on the transport of hydrophobic organic contaminants in the benthos. Environ. Sci. Technol. *31*, 3120–3125.

Schink, B. (1985). Fermentation of acetylene by an obligate anaerobe, *Pelobacter acetylenicus*. Arch. Microbiol. *142*, 295–301.

Schink, B. (2002). Synergistic interactions in the microbial world. Antonie van Leeuwenhoek *81*, 257–261.

Schneiker, S., Martins dos Santos, V.A., Bartels, D., Bekel, T., Brecht, M., Buhrmester, J., Chernikova, T.N., Denaro, R., Ferrer, M., Gertler, C., *et al.* (2006). Genome sequence of the ubiquitous hydrocarbon-degrading marine bacterium *Alcanivorax borkumensis*. Nat. Biotechnol. *24*, 997–1004.

Selck, H., Granberg, M.E., and Forbes, V.E. (2005). Impact of sediment organic matter quality on the fate and effects of fluoranthene in the infaunal brittle star Amphiura filiformis. Mar. Environ. Res. *59*, 19–45.

Sieber, J.R., McInerney, M.J., Plugge, C.M., Schink, B., and Gunsalus, R.P. (2010). Methanogenesis: syntrophic

metabolism. In Handbook of Hydrocarbon and Lipid Microbiology, Timmis, K.N., ed. (Berlin, Germany: Springer-Verlag), pp. 337–355.
Smith, C.R., Pope, R.H., DeMaster, D.J., and Magaard, L. (1993). Age-dependant mixing of deep-sea sediments. Geochim. Cosmochim. Acta *57*, 1473–1488.
Smith, T.J., Trotsenko, Y.A., and Murrell, J.C. (2010). Physiology and biochemistry of the aerobic methane oxidizing bacteria. In Handbook of Hydrocarbon and Lipid Microbiology, Timmis, K.N., ed. (Springer-Verlag: Berlin Heidelberg, Germany), pp. 767–780.
Smits, T.H.M., Balada, S.B., Witholt, B., and van Beilen, J.B. (2002). Functional analysis of alkane hydroxylases from Gram-negative and Gram-positive bacteria. J. Bacteriol. *184*, 1733–1742.
So, C.M., and Young, L.Y. (1999). Isolation and characterization of a sulfate-reducing bacterium that anaerobically degrades alkanes. Appl. Environ. Microbiol. *65*, 2969–2976.
So, C.M., Phelps, C.D., and Young, L.Y. (2003). Anaerobic transformation of alkanes to fatty acids by a sulfate-reducing bacterium, strain Hxd3. Appl. Environ. Microbiol. *69*, 3892–3900.
Sohn, J.H., Kwon, K.K., Kang, J.H., Jung, H.B., and Kim, S.J. (2004). *Novosphingobium pentaromativorans* sp. nov., a high-molecular-mass polycyclic aromatic hydrocarbon-degrading bacterium isolated from estuarine sediment. Int. J. Syst. Evol. Microbiol. *54*, 1483–1487.
Söhngen, N. L. (1913). Benzin, petroleum, paraffinöl und paraffin als kohlenstoff und energiequelle für mikroben. Zentralbl. Bakteriol. Parasitenkd. Infektionskr. Hyg. Abt. *237*, 595–609.
Sotsky, J.B., Greer, C.W., and Atlas, R.M. (1994). Frequency of genes in aromatic and aliphatic hydrocarbon biodegradation pathways within bacterial populations from Alaskan sediments. Can. J. Microbiol. *40*, 981–985.
Spormann, A.M., and Widdel, F. (2000). Metabolism of alkylbenzenes, alkanes, and other hydrocarbons in anaerobic bacteria. Biodegradation *11*, 85–105.
Staley, J.T. (2010). Cycloclasticus: a genus of marine polycyclic aromatic hydrocarbon degrading bacteria. In Handbook of Hydrocarbon and Lipid Microbiology, Timmis, K.N., ed. (Springer-Verlag: Berlin Heidelberg, Germany), pp. 1781–1786.
Suess, E., Collier, R.W., Greinert, J., Linke, P., Rehder, G., Trehu, A., Wallmann, K., Winckler, G., Zuleger, E., and Torres, M.E. (1999). Gas hydrate destabilization: enhanced dewatering, benthic material turnover and large methane plumes at the Cascadia convergent margin. Earth Planet. Sci. Lett. *170*, 1–15.
Sun, M., Lee, C., and Aller, R.C. (1993). Laboratory studies and anoxic degradation of chlorophyll-a in Long Island Sound sediments. Geochim. Cosmochim. Acta *57*, 147–157.
Sun, M.Y., Lee, C., and Aller, R.C. (1993). Laboratory studies of oxic and anoxic degradation of chlorophyll-a in Long Island Sound sediments. Geochim. Cosmochim. Acta *57*, 147–157.
Sun, M.Y., Aller, R.C., Lee, C., and Wakeham, S.G. (1999). Enhanced degradation of algal lipids by benthic macrofaunal activity: Effect of *Yoldia limatula*. J. Mar. Res. *57*, 775–804.
Sun, M.Y., Aller, R.C., Lee, C., and Wakeham S.G. (2002). Effects of oxygen and redox oscillation on degradation of cell-associated lipids in surficial marine sediments. Geochim. Cosmochim. Acta *66*, 2003–2012.
Sun, M.-Y., Wakeham, S.G., and Lee, C. (1997). Rates and mechanisms of fatty acids degradation in oxic and anoxic coastal marine sediments of Long Island Sound, New York, USA. Geochim. Cosmochim. Acta *61*, 341–355.
Sun, M.-Y., Shi, W., and Lee, R. F. (2000). Lipid-degrading enzyme activities associated with distribution and degradation of fatty acids in the mixing zone of Altamaha estuarine sediments. Org. Geochem. *31*, 889–902.
Taghon, G.L., Nowell, A.R.M., and Jumars, P.A. (1984). Transport and breakdown of fecal pellets: biological and sedimentological consequences. Limnol. Oceanogr. *29*, 64–72.
Takai, K., Moyer, C.L., Miyazaki, M., Nogi, Y., Hirayama, H., Nealson, K.H., and Horikoshi, K. (2005). *Marinobacter alkaliphilus* sp. nov., a novel alkaliphilic bacterium isolated from subseafloor alkaline serpentine mud from Ocean Drilling Program Site 1200 at South Chamorro Seamount, Mariana Forearc. Extremophiles *9*, 17–27.
Taketani, R.G., dos Santos, H.F., van Elsas, J.D., and Rosado, A.S. (2009). Characterisation of the effect of a simulated hydrocarbon spill on diazotrophs in mangrove sediment mesocosm. Antonie Van Leeuwenhoek *96*, 343–354.
Tani, A., Ishige, T., Sakai, Y., and Kato, N. (2001). Gene structures and regulation of the alkane hydroxylase complex in *Acinetobacter* sp. Strain M-1. J. Bacteriol. *183*, 1819–1823
Tapilatu, Y., Acquaviva, M., Guigue, C., Miralles, G., Bertrand, J.C., and Cuny, P. (2010). Isolation of alkane-degrading bacteria from deep-sea Mediterranean sediments. Lett. Appl. Microbiol. *50*, 234–236.
Teske, A. (2010). Sulfate-reducing and methanogenic hydrocarbon-oxidizing microbial communities in the marine environment. In Handbook of Hydrocarbon and Lipid Microbiology, Timmis, K.N., ed. (Springer-Verlag: Berlin Heidelberg, Germany), pp. 2203–2224.
Thibodeaux, L.J., and Bierman, V.J. (2003). The bioturbation-driven chemical release process. Environ. Sci. Technol. *37*, 252A–258A.
Throne-Holst, M., Wentzel, A., Ellingsen, T.E., Kotlar, H.K., and Zotchev, S.B. (2007). Identification of novel genes involved in long-chain *n*-alkane degradation by *Acinetobacter* sp. strain DSM *n*-17874. Appl. Environ. Microbiol. *73*, 3327–3332.
Timmermann, K., Banta, G.T., and Glud, R.N. (2006). Linking *Arenicola marina* irrigation behaviour to oxygen transport and dynamics in sandy sediments. J. Mar. Res. *64*, 915–938.
Timmermann, K., Banta, G.T., Johnsen, A.R., and Andersen, O. (2008). Effects of the polychaetes

Arenicola marina and *Nereis diversicolor* on microbial pyrene mineralization. Aquat. Microb. Ecol. *50*, 197–207.

Van de Bund, W.J., Olafsson, E., Modig, H., and Elmgren, R. (2001). Effects of the coexisting Baltic amphipods *Monoporeia affinis* and *Pontoporeia femorata* on the fate of a simulated spring diatom bloom. Mar. Ecol. Prog. Ser. *22*, 107–115.

Van Hamme, J.D., Singh, A., and Ward, O.P. (2003). Recent advances in petroleum microbiology. Microbiol. Mol. Biol. Rev. *67*, 503–549.

Vieira, P.A., Faria, S., Vieira, R.B., De França, F.P., and Cardoso, V.L. (2009a). Statistical analysis and optimization of nitrogen, phosphorus, and inoculum concentrations for the biodegradation of petroleum hydrocarbons by response surface methodology. World J. Microb. Biot. *25*, 427–438.

Vieira, P.A., Vieira, R.B., Faria, S., Ribeiro, E.J., and Cardoso, V.L. (2009b). Biodegradation of diesel oil and gasoline contaminated effluent employing intermittent aeration. J. Hazard. Mater. *168*, 1366–1372.

Vitte, I. (2010). Etude de communautés bactériennes impliquées dans la dégradation des hydrocarbures en conditions d'oscillations contrôlées: oxie/anoxie. PhD Thesis. University of Pau, France, p. 178.

Vitte, I., Duran, R., Jézéquel, R., Caumette, P., and Cravo-Laureau, C. (2011). Effect of oxic/anoxic switches on bacterial communities and PAHs biodegradation in an oil contaminated sludge. Environ. Sci. Pol. Res. In press.

Voparil, I.M., and Mayer, L.M. (2000). Dissolution of sedimentary polycyclic aromatic hydrocarbons into the lugworm's (*Arenicola marina*) digestive fluids. Environ. Sci. Technol. *34*, 1221–1228.

Voparil, I.M., Mayer, L.M., and Place, A.R. (2003). Interactions among contaminants and nutritional lipids during mobilization by digestive fluids of marine invertebrates. Environ. Sci. Technol. *37*, 3117–22.

Wakeham, S.G., and Canuel, E.A. (2006). Degradation and preservation of organic matter in marine sediments. In Handbook of Environmental Chemistry, Vol. 2 Reactions and Processes part N, Hutzinger, O., ed. (Springer-Verlag: Berlin Heidelberg, Germany), pp. 295–321.

Wang, B., Lai, Q., Cui, Z., Tan, T., and Shao, Z. (2008a). A pyrene-degrading consortium from deep-sea sediment of the West Pacific and its key member *Cycloclasticus* sp. P1. Environ. Microbiol. *10*, 1948–1963.

Wang, L., Qiao, N., Sun, F., and Shao, Z. (2008b). Isolation, gene detection and solvent tolerance of benzene, toluene and xylene degrading bacteria from nearshore surface water and Pacific Ocean sediment. Extremophiles *12*, 335–342.

Wang, L., Wang, W., Lai, Q., and Shao, Z. (2010). Gene diversity of CYP153A and AlkB alkane hydroxylases in oil-degrading bacteria isolated from the Atlantic Ocean. Environ. Microbiol. *12*, 1230–1242.

Wasmund, K., Burns, K.A., Kurtböke, D.I., and Bourne, D.G. (2009). Novel alkane hydroxylase gene (alkB) diversity in sediments associated with hydrocarbon seeps in the Timor Sea, Australia. Appl. Environ. Microbiol. *75*, 7391–7398.

Wenzhöfer, F., and Glud, R.N. (2004). Small-scale spatial and temporal variability in coastal benthic O2 dynamics: Effects of fauna activity. Limnol. Oceanogr. *49*, 1471–1481.

Weston, D.P., and Mayer, L.M. (1998). *In vitro* digestive fluid extraction as a measure of the bioavailability of sediment-associated polycyclic aromatic hydrocarbons: sources of variation and implications of partitioning models. Environ. Toxicol. Chem. *17*, 820–829.

Whyte, L.G., Schultz, A., van Beilen, J.B., Luz, A.P., Pellizari, V., and Labbe, D. (2002). Prevalence of alkane monooxygenase genes in Arctic and Antarctic hydrocarbon-contaminated and pristine soils. FEMS Microbiol. Ecol. *41*, 141–150.

Widada, J., Nojiri, H., Kasuga, K., Yoshida, T., Habe, H., and Omori, T. (2002). Molecular detection and diversity of polycyclic aromatic hydrocarbon-degrading bacteria isolated from geographically diverse sites. Appl. Microbiol. Biot. *58*, 202–209.

Widdel, F., and Rabus, R (2001). Anaerobic biodegradation of saturated and aromatic hydrocarbons. Curr. Opin. Biotech. *12*, 259–276.

Widdel, F., Knittel, K., and Galushko, A. (2010). Anaerobic hydrocarbon-degrading microorganisms: an overview. In Handbook of Hydrocarbon and Lipid Microbiology, Timmis, K.N., ed. (Springer-Verlag: Berlin Heidelberg, Germany), pp. 1997–2020.

Wilde, S.B., and Plante, C.J. (2002). Spatial heterogeneity of bacterial assemblages inmarine sediments: the influence of deposit feeding by *Balanoglossus aurantiacus*. Estuar. Coast. Shelf Sci. *55*, 97–107.

Wilkes, H., Kühner, S., Bolm, C., Fischer, T., Classen, A., Widdel, F., and Rabus, A. (2003). Formation of n-alkane and cycloalkane-derived organic acids during anaerobic growth of a denitrifying bacterium with crude oil. Org. Geochem. *34*, 1313–1323.

Wilkes, H., Rabus, R., Fischer, T., Armstroff, A., Behrends, A., and Widdel, F. (2002). Anaerobic degradation of n-hexane in a denitrifying bacterium: further degradation of the initial intermediate (1-methylpentyl)succinate via C-skeleton rearrangement. Archiv. Microbiol. *177*, 235–243.

Xu, M., Xiao, X., and Wang, F. (2008). Isolation and characterization of alkane hydroxylases from a metagenomic library of Pacific deep-sea sediment. Extremophiles *12*, 255–262.

Yakimov, M.M., Golyshin, P.N., Lang, S., Moore, E.R., Abraham, W.R., Lunsdorf, H., and Timmis, K.N. (1998). *Alcanivorax borkumensis* gen., nov., sp. nov., a new hydrocarbon-degrading and surfactant-producing marine bacterium. Int. J. Syst. Bacteriol. *48*, 339–348.

Yakimov, M.M., Giuliano, L., Denaro, R., Crisafi, E., Chernikova, T.N., Abraham, W.R., Luensdorf, H., Timmis, K.N., and Golyshin, P.N. (2004). *Thalassolituus oleivorans* gen. nov., sp. nov., a novel marine bacterium that obligately utilizes hydrocarbons. Int. J. Syst. Evol. Microbiol. *54*, 141–148.

Yakimov, M.M., Denaro, R., Genovese, M., Cappello, S., D'Auria, G., Chernikova, T.N., Timmis, K.N., Golyshin, P.N., and Giluliano, L. (2005). Natural microbial diversity in superficial sediments of Milazzo Harbor (Sicily) and community successions during microcosm enrichment with various hydrocarbons. Environ. Microbiol. 7, 1426–1441.

Young, L.Y., and Phelps, C.D. (2005). Metabolic biomarkers for monitoring *in situ* anaerobic hydrocarbon degradation. Environ. Health Persp. 113, 62–67.

Zengler, K., Richnow, H.H., Rossello-Mora, R., Michaelis, W., and Widdel, F. (1999). Methane formation from long-chain alkanes by anaerobic microorganisms. Nature *401*, 266–269.

Zhou, H.W., Guo, C.L., Wong, Y.S., and Tam, N.F. (2006). Genetic diversity of dioxygenase genes in polycyclic aromatic hydrocarbon-degrading bacteria isolated from mangrove sediments. FEMS Microbiol. Lett. *262*, 148–157.

Hydrocarbon-degrading Soil Bacteria: Current Research

5

Anna-Irini Koukkou and Elpiniki Vandera

Abstract

Hydrocarbons are the major representatives of non-metal pollutants found in many contaminated soils by natural or industrial and social activities. Their removal from polluted environmental niches depends to a great extent on microbial degradation, which can also be applied on several technological applications. The extended microbial diversity in soil has served as a rich source for the isolation of efficient PAH-degrading strains. Bacterial isolates with the ability to use PAHs as an alternative source of carbon and energy facilitate their mineralization to harmless products. Culture-based approaches have resulted in the isolation of a range of soil hydrocarbon degrading bacteria, which primarily are members of different subdivisions of Proteobacteria as well as of the high G+C Gram-positive bacteria. Generally, in polluted soils Gram-negative bacteria such as *Pseudomonas, Burkholderia* and *Sphingomonas* seem to degrade preferentially lower molecular weight PAHs such as naphthalene and phenanthrene, while Gram-positive isolates are more specialized in the degradation of high molecular weight PAHs such as pyrene.

Introduction

Hydrocarbons are widespread in the environment owing to anthropogenic as well as various natural processes, whereas their high concentrations in soils is mainly due to industrial and social activities, such as fuel combustion, pyrolytic processes, spillage of petroleum products, waste incinerators and domestic heaters (Wakeham *et al.*, 1980; Freeman and Cattell, 1990). The risk of oil spillage, involved in many activities of the petroleum industry, poses a serious environmental problem, because of the potential pollution of air, water and soil. Petrol is a complex mixture of various components including aliphatic (i.e. alkanes) and aromatic hydrocarbons such as polycyclic aromatic hydrocarbons (PAHs). Biodegradation is considered the most appropriate way to dispose of hydrocarbons from contaminated soils (Alexander, 1981).

Biodegradation of alkanes by soil bacteria

Alkanes are saturated hydrocarbons of different sizes and structures. Numerous bacterial strains have been isolated and characterized so far, which are capable to degrade alkanes as a source of carbon and energy. Most of them belong to the α-, β- and γ-Proteobacteria and some are members of the Gram-positives G+C rich Actinobacteria. The most common soil bacteria that degrade petroleum hydrocarbons belong to the genus *Pseudomonas* (Ko and Lebeault, 1999; Van Hamme *et al.*, 2000; Yuste *et al.*, 2000; Barathi and Vasudevan, 2001, 2003; Belhaj *et al.*, 2002; Chaerun *et al.*, 2004; Wongsa *et al.*, 2004; Karamalidis *et al.*, 2010). References for alkane degraders also exist for other genera, such as *Bacillus* (Rahman *et al.*, 2002; Chaerun *et al.*, 2004; Arvanitis *et al.*, 2008), *Paracoccus* (Chaerun *et al.*, 2004), *Serratia* (Wongsa *et al.*, 2004), *Micrococcus* (Rahman *et al.*, 2002), *Corynebacterium* (Bouchez-Naïtali *et al.*, 1999; Rahman *et al.*, 2002), *Flavobacterium* (Van Hamme *et al.*, 2000; Rahman *et al.*, 2002),

Acinetobacter (Sakai *et al.*, 1994; Van Hamme *et al.*, 2000; Koma *et al.*, 2001; Margesin *et al.*, 2003; Throne-Holst *et al.*, 2006, 2007), *Rhodococcus* (Ko and Lebeault, 1999; Whyte *et al.*, 2002; Margesin *et al.*, 2003; Larkin *et al.*, 2005; Hamamura *et al.*, 2006; Auffret *et al.*, 2009; Lee *et al.*, 2010), *Nocardia* (Nhi-Chong *et al.*, 2010), *Alcaligenes* (Van Hamme *et al.*, 2000; Pepi *et al.*, 2003), *Mycobacterium* (Sakai *et al.*, 2004), *Ochrobactrum* (Van Hamme *et al.*, 2000; Katsivela *et al.*, 2005), *Stenotrophomonas* (Van Hamme *et al.*, 2000), *Yersinia* (Van Hamme *et al.*, 2000), *Dietzia* (von der Weid *et al.*, 2007), *Gordonia* (Franzetti *et al.*, 2009) and *Microbacterium* (Zhang *et al.*, 2004).

Several excellent reviews are also available describing enzymes, pathways and regulation of genes responsible for the degradation of alkanes (van Beilen *et al.*, 2003; van Hamme *et al.*, 2003; van Beilen and Funhoff, 2007; Wentzel *et al.*, 2007; Rojo, 2009).

Generally, alkanes are more easily degraded in the environment than PAHs. Stroud *et al.* (2007) proved that the catabolic potential of soil was hexadecane > naphthalene > phenanthrene > pyrene >B[a]P, determined by the decrease in lag time (defined by the time taken for the production of 5% $^{14}CO_2$ as a result of mineralization of ^{14}C-labelled hydrocarbons added in the soil). In spite of this, short chain alkanes are not feasibly biotransformed owing to their strong volatility and low solubility (Lee *et al.*, 2010).

Very recently Nhi-Chong, *et al.* (2010) reported the oxidation of aliphatic, branched chain, as well as aromatic hydrocarbons by the soil bacterium *Nocardia cyriacigeorgica* isolated from oil-polluted sand samples collected in the Saudi Arabian desert.

Alkanes biodegradation under aerobic and anaerobic conditions are reviewed elsewhere in this book. In this chapter we will focus on PAH-degrading soil bacteria, as well as on factors, such as temperature and pH, involved in biodegradation of petroleum hydrocarbons in soil.

Biodegradation of PAHs by soil bacteria

Polycyclic aromatic hydrocarbons (PAHs) are ubiquitous compounds detected in soils and sediments as a result of both natural and anthropogenic activities (Agency for Toxic Substances and Disease Registry 1995). They constitute an important class of organic compounds which are generally characterized as molecules consisting of two or more aromatic rings in various configurations (Blumer, 1976). Their formation often results from incomplete combustion of organic materials, gas production, wood treatment facilities and waste incineration (Mueller *et al.*, 1991; IARC, 1983, Cerniglia, 1992, Kanaly and Harayama, 2000, Kim *et al.*, 2003a). Soils contaminated with polycyclic aromatic hydrocarbons are a major environmental problem worldwide. In industrial countries, improper disposal of industrial waste or accidental spillage of crude or refined oil, as well as anthropogenic combustion activities are a principal source of PAHs in soils. This has led to a significant increase of PAHs concentration in the soil over the last 100–150 years (Juhasz and Naidu, 2000).

The US Environmental Protection Agency and other environmental authorities have listed PAHs as priority pollutants because they are toxic, carcinogenic, teratogenic and recalcitrant in the environment (Kanaly and Harayama, 2000; Samanta *et al.*, 2002; Dandie *et al.*, 2004). Most recently, a series of 15 PAHs were reported as carcinogens by the Department of Public Health of America (Public Health Service, 2005). PAHs are highly lipid-soluble, rapidly absorbed from the gastrointestinal tract of mammals and accumulated in the body fat (Cerniglia, 1984). Carcinogenicity has been demonstrated for several PAHs, such as benzo[a]pyrene, benzo[a]anthracene, benzo[a]fluoranthene, dibenzo[a,h]anthracene, as well as for their metabolites in the human body, like epoxides produced by the oxidation of PAHs via the cytochrome P450-mediated mixed function system (Samanta *et al.*, 2002, and references therein). Interest for the environmental fate of PAHs and mechanisms for their degradation is motivated by the risks threatening public health as a result of their huge dispersion, of their low bioavailability due to strong adsorption on soil and their consequent bioaccumulation through the food chain (Phillips, 1983; Kanaly and Harayama, 2000). Biodegradation is considered the most appropriate way to remove substances such as

PAHs from contaminated soils (Alexander, 1981; Cerniglia, 1993; Daane *et al.*, 2001).

The microbial degradation of PAHs depends on cellular uptake rate and metabolism. Little information is available about how PAHs travel across bacterial membranes to reach the cytoplasmic catabolic enzymes. Although an undisputed molecular mechanism for active hydrocarbon uptake is not known so far, some descriptions of active PAH uptake have been published. Whitman *et al.* (1998) reported that the first step in the utilization of naphthalene by *Pseudomonas fluorescens* is its transfer into the cell by a specific energy-linked transport system. In contrast, Bateman and coworkers were the first who studied the uptake of naphthalene by *Pseudomonas putida* strain PpG1064, a process that was not found to be associated with an active transport system (Bateman *et al.*, 1986). The first conclusive evidence provided for bacterial PAH uptake was based on a passive diffusion mechanism (Sikkema *et al.*, 1995). Bugg *et al.* (2000) demonstrated that *Pseudomonas fluorescens* LP6a excreted PAHs by an active efflux mechanism, but, in agreement with the previous findings, their influx depended on passive diffusion. Ortega-Calvo and Alexander (1994) observed that the rate of degradation of naphthalene by a strain of *Arthrobacter* sp. was faster than the rate of dispersion of naphthalene in the aqueous phase. They suggested that degradation of naphthalene in this case involved two physiologically distinct bacterial populations, one suspended in the aqueous phase and the second growing at the interface of non-aqueous liquids and water where concentration of naphthalene was much higher. The first population initiated biodegradation of naphthalene which was dispersed in the water phase and it depended on the rate of partitioning of the substrate to water. The second population transformed naphthalene faster apparently because cells attached to the interface acquired their substrate at a higher rate. The authors proposed that the higher availability of naphthalene to the interface attached cells may involve hydrophobic interactions or membrane surfactants facilitating substrate uptake (Ortega-Calvo and Alexander, 1994). Rodrigues *et al.* (2005) studied the uptake of fluorene and phenanthrene by *Pseudomonas putida* ATCC 17514 and showed that these cells were able to overcome the problem of lower aqueous solubility via their attachment to the solid PAH substrate with the help of extracellular polymeric substances. Apparently, changes in cell surface electric charge, triggered by these interactions, promoted the bio-availability and uptake of such hydrophobic compounds (Rodrigues *et al.*, 2005). The mechanism of uptake of phenanthrene by *Mycobacterium* sp. strain RJGII-135 was examined with induced and uninduced cells (Miyata *et al.*, 2004). The results were consistent with the presence of a saturable energy-dependent mechanism for transport of phenanthrene in induced cells only. Similar results have been reported for *Arthrobacter* sp. strain Sphe3 that appeared to internalize phenanthrene by two mechanisms: a passive diffusion when cells were grown on glucose, as well as an induced active transport system when cells were grown in the presence of phenanthrene (Kallimanis *et al.*, 2007). Very recently, Neher and Lueking (2009) demonstrated also the requirement for OmpW, a porin for the transport of hydrophobic molecules, for the growth of *P. fluorescens* on naphthalene as sole source of carbon and energy, supporting the existence of a mechanism transporting naphthalene. The relevant gene was isolated, cloned, sequenced and shown to be exclusively localized on an indigenous large catabolic plasmid. Specific mutations in this gene abolished growth on naphthalene and other PAHs (Neher and Lueking, 2009).

The bacterial cell envelope is the first and major front of defence against environmental threats (Jordan *et al.*, 2008). Maintenance of optimal fluidity, stability and permeability of bacterial membranes is required, as a result of adaptation to environmental stimuli for the selective uptake and excretion of substrates and metabolites by protein carriers embedded in the membrane lipids bilayer (Tsitko *et al.*, 1999; Brown *et al.*, 2000). Previous studies have demonstrated that hydrocarbons influence the function of biological membranes by affecting permeability as well as the activity of membrane enzymes (Sikkema *et al.*, 1994). Many studies have shown that bacteria able to survive in the presence of toxic aromatic compounds change their membrane fluidity by modifying their fatty acids (Heipieper *et al.*, 1992; Sikkema *et al.*, 1995;

Weber and de Bont, 1996). Effects of naphthalene on the whole cell-derived fatty acid composition of *Pseudomonas* strains during naphthalene degradation have been investigated by Mrozik *et al.* (2005). Their results showed that cells reacted to the addition of naphthalene with an increase in the saturated/unsaturated ratio as well as with alterations in the percentage of hydroxy, cyclopropane and branched fatty acids. Wick *et al.* (2003b) have reported that growth on anthracene may lead to an overall increase of the membrane fluidity of *Mycobacterium* sp. LB501T as opposed to exponential growth on glucose. Kallimanis *et al.* (2007) have reported that *Arthrobacter* sp. strain Sphe3 responded to the presence of phenanthrene by altering the composition of membrane phospholipids and fatty acids thus increasing its membrane fluidity. Alterations of fatty acid profiles as an adaptation response of microorganisms to the toxicity effects of hydrocarbons has been studied extensively and several reports postulate that fatty acid profiles are reliable stress markers for monitoring a biodegradation process (Heipieper *et al.*, 2003 and references there in).

PAH-degrading soil bacteria

PAH-degrading bacteria have been isolated from various natural environments. Especially the high microbial diversity in soil has served as a rich source of efficient PAH-degrading strains. Bacterial strains with the ability to use PAHs as an alternative source of essential nutrients, facilitate the mineralization of PAHs to harmless products. Several PAHs are known to be converted to carcinogenic or potentially genotoxic products (White, 1986). Strains able to use PAHs as sole source of carbon and energy, leading to their complete conversion to carbon dioxide and water, possess a substantial potential for the bioremediation of PAH-polluted environments. A list of soil and sediment bacterial isolates capable to use PAHs as sole source of carbon and energy is presented in Table 5.1.

Although numerous bacterial strains that can use PAHs such as naphthalene (two rings), phenanthrene (three rings), anthracene (three rings) and pyrene (four rings) as sole source of carbon and energy have been isolated and characterized, strains able to use high-molecular-weight PAHs (five or more rings) have not been found (Table 5.1). Several bacterial strains such as *Sphingomonas paucimobilis* EPA 505, *Mycobacterium* sp. RJGII-135, *Mycobacterium vanbaalenii* PYR-1 *Rhodococcus* sp. UW1, *Burkholderia cepacia* strains, *Stenotrophomonas maltophila* strains and *Sphingomonas yanoikuyae* JAR02 can degrade benzo[a]pyrene (a five-ring PAH) in the presence of other energy sources (Schneider *et al.*, 1996; Ye *et al.*, 1996; Juhasz *et al.*, 1996, 1997, 1998; Walter *et al.*, 1991; Aitken *et al.*, 1998; Moody *et al.*, 2004; Rentz *et al.*, 2008). The high molecular weight PAHs are resistant to biodegradation because of their even lower solubility and their high tendency to interact with non-aqueous phases and soil organic matter. Consequently, they become potentially unavailable for microbial degradation, since bacteria are known to degrade chemicals only when they are dissolved in water (Johnsen *et al.*, 2005). Bioavailability of PAHs decreases logarithmically in relation to the increase of molecular weight (Johnsen *et al.*, 2005).

Almost all PAH-degrading isolates are aerobic and able to use PAHs as sole source of carbon and energy. Aerobic degradation pathways of PAHs have been studied extensively. Usually, the initial step in PAH metabolism is to introduce oxygen atoms into an aromatic ring, leading to the production of dihydrodiol compounds. This reaction is catalysed by multicomponent enzymes, called ring hydroxylating dioxygenases (RHDs). An overview of the current knowledge on the genetics, structure, catalytic mechanism and diversity of RHDs is presented in Chapter 7.

The dihydrodiol compounds formed by the initial dioxygenation are usually transformed to the central intermediates of the aromatic compounds degradation, such as catechols and protocatechuates, which are incorporated in the TCA cycle.

Description of the biodegradation pathways known so far is beyond the scope of this chapter and readers are referred to a number of excellent reviews on this topic such as those by Habe and Omori (2003), Peng *et al.* (2008) and Haritash and Kaushik (2009). The major pathways that yield carbon and energy from degradation of PAHs are summarized in Fig. 5.1.

Table 5.1 Partial list of soil bacterial isolates that grow on PAHs as sole source of carbon and energy

Lineage	Organism	PAH (as sole carbon and energy source)	Source of Isolation	Reference
Gram-negative				
Alphaproteobacteria	*Paracoccus* sp. Ophe1	Anthracene, chrysene, fluorene, fluoranthene, pyrene	PAH-contaminated soil (oil refinery site, Greece).	Zhang *et al*. (2004)
	Sphingomonas paucimobilis EPA 505	Fluoranthene	Soil highly contaminated with coal tar creosote	Mueller *et al*. (1990), Ye *et al*. (1996)
	Sphingomonas paucimobilis TNE12	Fluoranthene, phenanthrene	Creosote-contaminated soil from Tennessee	Shuttleworth and Cerniglia (1996); Shuttleworth *et al*. (2000)
	Sphingomonas paucimobilis var. EPA 505	Fluoranthene, naphthalene, phenanthrene	Soil highly contaminated with coal tar creosote from a creosote waste site in Pensacola, Fla.	Mueller *et al*. (1989), Story *et al*. (2000)
	Sphingomonas paucimobilis BA2	Anthracene	Soil samples collected from different sites contaminated with either fuel or coal tar oil in Hamburg	Kästner *et al*. (1994)
	Sphingomonas paucimobilis ZX4		Oil-contaminated soil near factory	Xia *et al*. (2005)
	Sphingomonas (sphingobium) yanoikuyae B1(formerly *Beijerinckia* sp. B1)	Naphthalene, phenanthrene	Polluted stream	Gibson *et al*. (1973), Khan *et al*. (1996), Ferraro *et al*. (2007)
	Sphingomonas yanoikuyae R1	Fluoranthene	PAH- contaminated soils and sediments	Kazunga *et al*. (2001)
	Sphingomonas yanoikuyae JAR02	Phenanthrene	soil collected at the Iowa City oil disposal facility	Rentz *et al*. (2008)
	Sphingomonas sp. A4	Acenaphthene, acenaphthylene	Soil	Komatsu *et al*. (1993), Pinyakong *et al*. (2004)
	Sphingomonas sp. CHY-1	Anthracene, crysene, naphthalene, phenanthrene	Soil heavily contaminated with PAHs from a coal gasification site in the Rhône-Alpes region, France	Willison *et al*. (2004)
	Sphingomonas sp. GY2B	Naphthalene, phenanthrene	Soils contaminated by crude oil at a site near Guangzhou Petrifaction Company, Guangzhou, China	Tao *et al*. (2007)
	Sphingomonas sp. KS14	Naphthalene, phenanthrene	Soil from waste oil disposal site	Cho and Kim (2001)
	Sphingomonas sp. LB126	Fluorene	Soil and sludge	Bastiaens *et al*. (2000)

Table 5.1 continued

Lineage	Organism	PAH (as sole carbon and energy source)	Source of Isolation	Reference
	Sphingomonas sp. P2	Phenanthrene	Petroleum-contaminated soil in Thailand	Grosser *et al*. (1991) Pinyakong *et al*. (2000)
	Sphingomonas sp. ZL5	Phenanthrene	Crude oil-contaminated soil of Liaohe oil field in China	Liu *et al*. (2004)
	Sphingomonas sp. ZP1	Phenanthrene	PAH-contaminated soil in oil refinery fields in Shanghai	Zhao *et al*. (2008)
	Sinorhizobium sp. C4	Phenanthrene	PAH-contaminated soil, Hili, HI, USA	Keum *et al*. (2006)
Betaproteobacteria	*Acidovorax* sp. NA3	Phenanthrene	Contaminated soil from the site of a former manufactured gas plant.	Singleton *et al*. (2009)
	Alcaligenes denitrificans WW1	Fluoranthene	Soil	Weissenfels *et al*. (1990)
	Alcaligenes faecalis AFK2	Phenanthrene	Soil	Kiyohara *et al*. (1982)
	Burkholderia (formerly *Pseudomonas*) sp. JS150	Naphthalene	Mixture of sewage from Tyndall Air Force Base and Panama City, FL	Haigler *et al*. (1992); Spain and Nishino, (1987)
	Burkholderia sp. C3	Phenanthrene	PAHs contaminated soil from Hilo, Hawaii	Seo *et al*. (2007)
	Burkholderia sp. RP007	Anthracene, naphthalene, phenanthrene	PAH-contaminated site in New Zealand	Laurie and Lloyd-Jones (1999)
	Burkholderia sp. TNFYE-5	Phenanthrene	Crude oil-contaminated site in Tennessee, USA	Kang *et al*. (2003)
	Comamonas testosteroni GZ38A	Anthracene, phenanthrene	River sediment	Goyal and Zylstra (1996)
	Comamonas testosteroni GZ39	Naphthalene, phenanthrene	River sediment	Goyal and Zylstra (1996)
	Comamonas testosteroni GZ42	Naphthalene, phenanthrene	River sediment	Goyal and Zylstra (1996)
	Ralstonia (formerly Pseudomonas) sp. *U2*	Napthalene	Oil-contaminated soil in Venezuela	Fuenmayor *et al*. (1998), Zhou *et al*. (2001)
Gammaproteobacteria	*Pseudomonas Alcaligenes* PA-10	Fluoranthene	From a bioreactor treating PAH contaminated waste	Gordon and Dobson (2001)
	Pseudomonas saccharophila P-15	Naphthalene, methylnaphthalenes, phenanthrene	Creosote-contaminated soil	Stringfellow and Aitken (1994, 1995)
	Pseudomonas stutzeri P-16	Naphthalene, methylnaphthalenes, phenanthrene	Creosote-contaminated soil	Stringfellow and Aitken (1994, 1995)

	Pseudomonas stutzeri P16	Fluoranthene	Soil	Kazunga *et al*. (2001)
	Pseudomonas stutzeri ZP2	Phenanthrene	PAH-contaminated soil containing waste from an oil refinery field in Shanghai, China	Zhao *et al*. (2009)
	Pseudomonas sp. Cam-1	Naphthalene	PCB-contaminated Arctic Soil from a Vancouver location	Master and Mohn (1998)
	Pseudomonas sp. PP2	Phenanthrene	Gas station soil in Ahmednagar, India.	Prabhu and Phale (2003)
	Pseudomonas sp. Wphe1	Naphthalene, phenanthrene	PAH-contaminated soil contaminated, wood processing site, Greece	Zhang *et al*. (2004)
	Stenotrophomonas maltophila VUN 10,010	Pyrene	PAH-contaminated soil from an abandoned factory site located near Port Melbourne, Victoria, Australia	Boonchan *et al*. (1998, 2000)
	Stenotrophomonas maltophila (formerly *Burkholderia cepacia*) VUN 10,003	Benzo[a]anthracene, benzo[a] pyrene, fluoranthene, pyrene	Soil from an abandoned factory site located near Port Melbourne, Australia	Juhasz *et al*. (1997, 2000)
Gram-positive bacteria				
Actinobacteria	*Arthrobacter phenanthrenivorans*	Phenanthrene	Creosote-contaminated soil in Epirus, Greece	Kallimanis *et al*. (2007, 2009)
	Arthrobacter sulphureus RKJ4	Phenanthrene	Soil sample taken from Gujarat oil fields in India	Samanta *et al*. (1999)
	Arthrobacter sp. F101	Fluorene	Sludge from an oil refinery wastewater treatment plant	Grifoll *et al*. (1992)
	Arthrobacter sp. P1–1	Phenanthrene	PAH-contaminated soil from Hilo, HI, USA	Seo *et al*. (2006a)
	Arthrobacter sp. RP17	Phenanthrene	Organic soil in the Sacramento river delta	Schwartz and Scow (2001)
	Brevibacterium sp. DPO-1361	Fluorene	NA	Trenz *et al*. (1994)
	Gordona sp. BP9	Fluoranthene, naphthalene, phenanthrene, pyrene	Soil from different contaminated sites in the Hamburg area	Kästner *et al*. (1994)
	Janibacter anophelis JY11	Anthracene, naphthalene, phenanthrene, pyrene	Polluted soil in Jinan oil refinery factory, in Shandong province, China	Zhang *et al*. (2009)
	Janibacter sp. YY-1	Fluorene	Soil of an agricultural field in Japan	Yamazoe *et al*. (2004)
	Microbacterium sp. F10a	Phenanthrene	Oil-polluted soil in Nanjing, China	Sheng *et al*. (2009)
	Microbacterium sp.	Phenanthrene	PAH-contaminated soil, steel industry site, Greece	Zhang *et al*. (2004)
	Nocardia otitidiscaviarum TSH1	Anthracene, naphthalene, phenanthrene, pyrene	Contaminated soil, petroindustrial sites	Zeinali *et al*. (2007)

Table 5.1 continued

Lineage	Organism	PAH (as sole carbon and energy source)	Source of Isolation	Reference
	Mycobacterium austroafricanum GTI-23	Fluoranthene, phenanthrene, pyrene	Soil from a former manufactured gas plant site in Iowa	Bogan *et al*. (2003)
	Mycobacterium flavescens	Pyrene	Sediment from the Grand Calumet River in north-western Indiana	Dean-Ross and Cerniglia (1996)
	Mycobacterium fluoranthenivorans	Fluoranthene	PAH polluted soils from a former coal gas plant in Saarbruecken-Burbach, Germany	Hormisch *et al*. (2004)
	Mycobacterium frederiksbergense FAn9	Fluoranthene, phenanthrene, pyrene	Coal tar-contaminated soil on the site of a former gas works at Frederiksberg, Denmark	Willumsen *et al*. (2001)
	Mycobacterium gilvum BBI	Fluorene, fluoranthene, phenanthrene, pyrene	PAH-contaminated soil from a former coal gasification site at Karlsruhe, Germany	Boldrin *et al*. (1993)
	Mycobacterium pyrenivorans 17A3	Pyrene	PAH polluted soil from a former coking plant in Übach-Palenberg, Germany	Derz *et al*. (2004)
	Mycobacterium vanbaalenii, PYR-1	Phenanthrene, pyrene	Sediment samples collected near an oil field near Port Aransas, Texas	Heitkamp *et al*. (1998), Khan *et al*. (2002)
	Mycobacterium sp. BG1	Phenanthrene	Sediment from the Great Bay Estuary, NH	Guerin *et al*. (1988)
	Mycobacterium sp., strains JLS, KMS, MCS	Phenanthrene, pyrene	PAH-contaminated soil from the land treatment unit (LTU2) at the Champion International Superfund Site in Libby, MT	Miller *et al*. (2004)
	Mycobacterium sp. KR20	Fluoranthene	PAH- contaminated soil of a former gasworks plant site in Kassel, Germany	Rehmann *et al*. (1999)
	Mycobacterium sp. CH1	Fluoranthene, phenanthrene, pyrene	(PAH)-contaminated freshwater sediments	Churchill *et al*. (1999)
	Mycobacterium sp. 1B	Fluoranthene, phenanthrene, pyrene	Manufactured gas plant-contaminated soil	Dandie *et al*. (2004)
	Mycobacterium sp. MHP-1	Pyrene	Soil at Chiba Prefecture	Habe *et al*. (2004)
	Mycobacterium sp. 6PY1	Phenanthrene, pyrene	PAH-contaminated soil	Krivobok *et al*. (2003)
	Mycobacterium sp. LB501T	Anthracene	PAH-contaminated soil	Bastiaens *et al*. (2000); Van Herwijnen *et al*. (2003)

	Mycobacterium sp. LP1	Phenanthrene, pyrene	Agricultural topsoil collected in Uppsala, Sweden	Pizzul *et al*. (2007)
	Mycobacterium sp. RJGII-135	Pyrene	Soils from three abandoned coal gasification plants in southern Illinois	Grosser *et al*. (1991, Schneider *et al*. (1996)
	Mycobacterium sp. SNP 11	Fluoranthene, fluorene, phenanthrene, pyrene	Grass-land soil	Pagnout *et al*. (2007)
	Mycobacterium sp. Spyr1	Anthracene, fluoranthene, phenanthrene, pyrene	Soil from a creosote polluted site in Epirus, Greece	Karabika *et al*. (2009)
	Mycobacterium sp. VF1	Anthracene, fluoranthene, naphthalene, phenanthrene, pyrene	Soil from different contaminated sites in the Hamburg area	Kästner *et al*. (1994)
	Rhodococcus aetherivorans	Naphthalene	From a bacterial consortium, including bacteria from soil at a gas station polluted by leaking tanks	Auffret *et al*. (2009)
	Rhodococcus wratislaviensis	Naphthalene	From a bacterial consortium, including bacteria from soil at a gas station polluted by leaking tanks	Auffret *et al*. (2009)
	Rhodococcus opacus M213	Naphthalene	Fuel oil-contaminated soil in Idaho, USA	Uz *et al*. (2000)
	Rhodococcus sp. UW1	Anthracene, fluoranthene, phenanthrene	Contaminated soil	Walter *et al*. (1991)
Bacilli	*Bacillus cereus* P21	Fluoranthene	PAH-contaminated soil	Kazunga *et al*. (2001)
	Bacillus sp. JF8	Naphthalene	Compost	Shimura *et al*. (1999)
	Bacillus sp.	Anthracene, naphthalene, phenanthrene	Soil	Hanumanthanaik *et al*. (2000)
	Staphylococcus sp. PN/Y	Phenanthrene	Petroleum-contaminated soil (Noonmati refinery site, India)	Mallick *et al*. (2007)

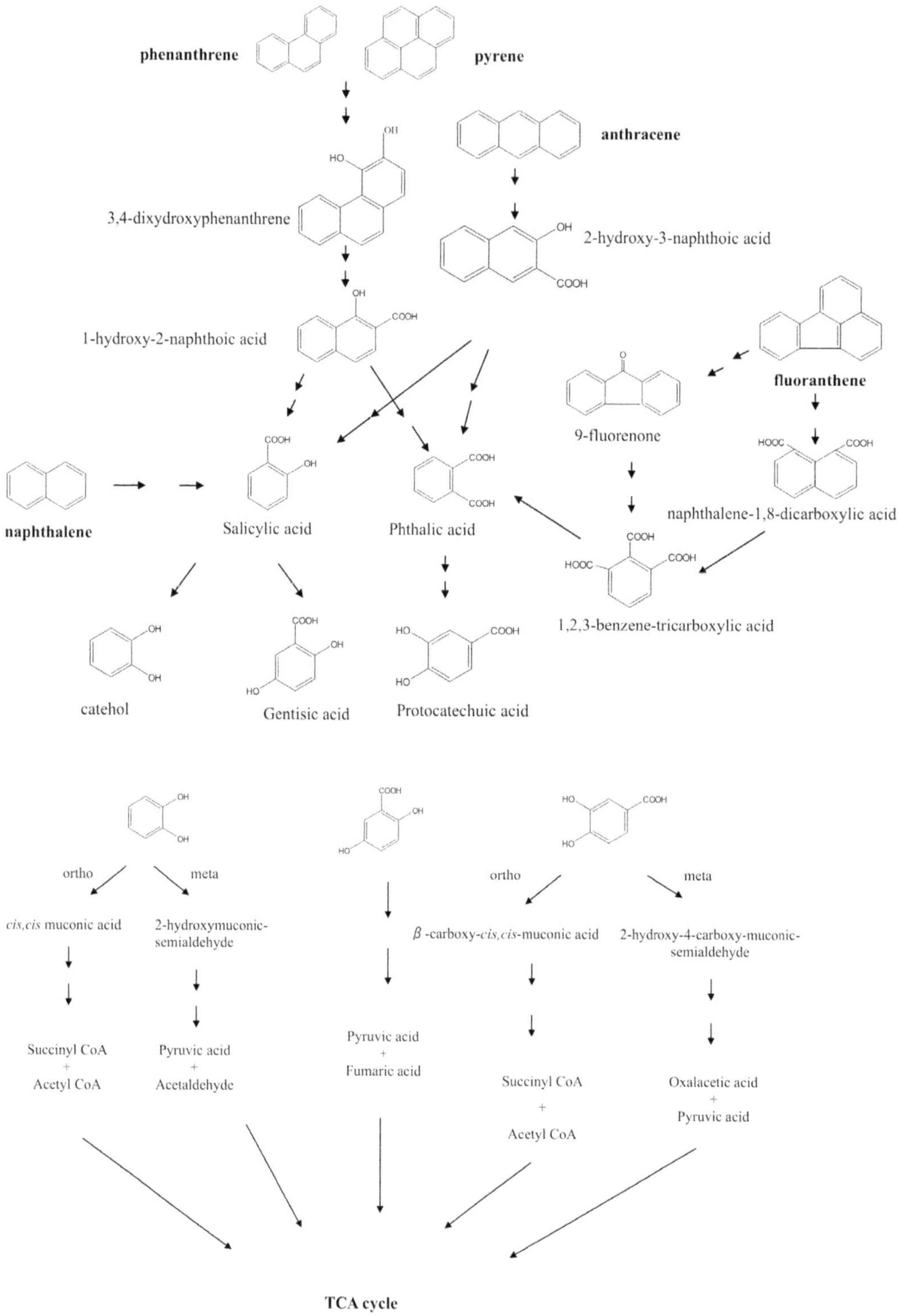

Figure 5.1 Proposed pathways for bacterial degradation of polycyclic aromatic hydrocarbons (PAHs). Single arrows indicate one-step reactions, double arrows indicate transformations of two or more steps.

Although detailed characterization of anthracene, phenanthrene and naphthalene biodegradation pathways is available, only limited information is available to describe the biodegradation pathways of PAHs with more than three rings (Kanaly and Harayama, 2000; Moody *et al.*, 2004, Rentz *et al.*, 2008). Several genes that encode enzymes responsible for PAH catabolism have been identified. The fundamental knowledge of catabolic genes in different bacterial species can give useful information on the evolution of structure–function relationships as well as the evolution and diversity of catabolic genes through horizontal transfer, DNA rearrangements, mutations etc., contributing to the elucidation of molecular mechanisms by which bacteria adapt to the presence of xenobiotic compounds. Although the last two decades genetic studies of PAH degradation had focused on naphthalene catabolic genes in cells of *Pseudomonas* sp., recently knowledge started to accumulate about PAH catabolic genes from both Gram-negative and Gram-positive bacteria (Demanèche *et al.*, 2004; Schuler *et al.*, 2008, 2009). A very well conserved group of PAH catabolic genes that has been studied over the past two decades is the group of *nah*-like genes (*nah, ndo, pah,* and *dox*), isolated from species of *Pseudomonas* (Habe and Omori, 2003). Recently, PAH catabolic genes have been identified from other non-*Pseudomonas* Gram-negative bacteria, as well as from Gram-positive bacteria, which are evolutionarily distinct from *nah*-like genes (Saito *et al.*, 2000; Kim *et al.*, 2007).

Many genes involved in the degradation of PAHs are often located on large plasmids as reviewed in Chapter 9.

The degradation of naphthalene by soil *Pseudomonas* was first reported by Davies and Evans (1964), while the first bacterium able to extensively metabolize a PAH with four aromatic rings, such as pyrene was isolated in the late 1980s from Heitkamp *et al.* (1988). Generally, it seems that PAH mineralization in soil is dominated by bacterial strains belonging to very limited number of Phyla. This may be due to the fact that the majority of microorganisms from environmental samples cannot be cultivated under laboratory conditions (Janssen *et al.*, 2002). The genetic diversity of soil bacteria is extremely high (>10 billion prokaryotes per gram of soil) but current cultivation methods are efficient for only 100- to 1000-fold lower number of cells (Rosselló-Mora and Amann, 2001). Culture independent approaches have revealed a detailed picture of microbial communities in the environment demonstrating that uncultivated species represent the bulk of the microbial diversity (Cardenas and Tiedje, 2008; Handelsman, 2004). Cultivation methods are biased towards fast growing species belonging mostly to Proteobacteria, Actinobacteria, Bacteroidetes and Firmicutes (Hugenholtz, 2002). Johnsen *et al.* (2002) characterized the phenanthrene-degrading microbial communities by examining incorporation of [^{13}C]phenanthrene-derived carbon into sterols and polar lipid fatty acids (PLFAs). By using the ^{13}C approach to identify bacteria that metabolized phenanthrene *in situ*, the authors avoided the possible biases associated with enrichment and isolation. Stach and Burns (2002) showed that biofilm culture methods are more appropriate for the study of community–level interactions in PAH-degrading microbial communities. Their study compared conventional batch enrichment with a biofilm culture method for the isolation of PAH-degrading communities from a PAH-contaminated soil. Their results showed that the biofilm community possesses considerably greater number of species and gene diversity than that of the enriched community.

Several cultivation methods have been developed to increase microbial recovery from the environment (Bollmann *et al.*, 2007; Cardenas and Tiedje, 2008). Bastiaens *et al.* (2000) compared two different procedures for the isolation of PAH-utilizing bacteria from PAH-contaminated soil and sludge samples: (a) a traditional liquid enrichment method in which PAHs were supplied as crystals and (b) a new method in which PAH degraders were enriched on and recovered from hydrophobic membranes containing sorbed PAHs. Although both methods were successful, different bacterial strains able to grow on PAHs as the sole source of carbon and energy were selected from the same source. The liquid enrichment mainly selected for *Sphingomonas* sp., whereas the membrane method exclusively led to the selection of *Mycobacterium* sp.

Very recently, Jacques *et al.* (2009) have presented a modification of the classical enrichment method of isolation of PAH-degrading bacteria employing a pre-enrichment respirometric analysis, to rapidly obtain isolates with necessary genetic and biochemical elements for improved soil colonization and PAH degradation. The modified enrichment of PAH-degrading microorganisms reduced the enrichment period without imposing new adaptations to *in vitro* biodegradation conditions and thus maintaining the original genetic, physiological and morphological traits for efficient PAH degradation in natural soils.

Many studies indicate that certain groups of soil bacteria are specialized in colonization of PAH-contaminated soil and play the main role in the biodegradation process. As shown in Table 5.1, PAH-degrading isolates mostly belong to the genera *Sphingomonas* and *Mycobacterium*. It has also been reported that initial degradation of PAHs in soil is done by sphingomonads and other fast-growing bacteria, whereas in later stages the sphingomonads are outcompeted by relatively slow-growing mycobacteria (Johnsen *et al.*, 2007).

Generally in polluted-soils, Gram-negative bacteria such as *Pseudomonas*, *Burkholderia* and *Sphingomonas* seem to degrade preferentially lower molecular weight PAHs such as naphthalene and phenanthrene, while Gram-positive isolates are more specialized in the degradation of high molecular weight PAHs such as pyrene (Heitkamp *et al.*, 1988; Grosser *et al.*, 1991; Boldrin *et al.*, 1993; Dean-Ross and Cerniglia, 1996).

The phylum Actinobacteria

Among Gram-positive PAH-degrading isolates, actinobacteria oxidize the largest variety of PAHs. Members of the genus *Mycobacterium*, are considered among the most important representatives of this group. Recent reports on the physiology and ecology of PAH-degrading *Mycobacteria* suggest that strains of this genus are specialists in adjusting and proliferating in the oligotrophic environment of PAH-polluted soils (Uyttebroek *et al.*, 2006).

Mycobacterium species are widespread organisms belonging to the phylum of actinobacteria with typical characteristics of high G+C content (>61%) and long chain mycolic acids (Lévy-Frébault and Portaels, 1992). Fast-growing *Mycobacterium* strains are an important group of polycyclic aromatic hydrocarbon (PAH)-degrading bacteria, with very different biochemical characteristics including non-pathogenic, pathogenic and facultative pathogenic species.

Kim *et al.* (2005a) presented the numerical and genetic analysis of PAH-degrading mycobacteria isolated from heavily contaminated sites and uncontaminated humus soil from the Natural Park, Schwaebische Alb, Germany. Their studies demonstrated that PAH-degrading strains were diverse from fast-growing and thermosensitive mycobacteria that could be responsible for the degradation of a variety of PAHs, including fluoranthene, phenanthrene and pyrene, in various environments. The uncontaminated humus soil-originating PAH-degrading mycobacteria showed the typical traits for PAH-degrading mycobacteria isolated from heavily contaminated sites. Although significant differences between the two subgroups were observed in the utilization of polyalcoholic sugars, they were phylogenetically related to each other (Kim *et al.*, 2005a).

Mycobacterial wall, rich in mycolic acids, provides resistance to toxic hydrocarbons (Sokolovska *et al.*, 2003; Nicolau *et al.*, 2009). Presence of mycolic acids in the cell wall is very critical for the mineralization of recalcitrant PAHs, because they facilitate the attachment of the cells to the hydrophobic compounds (Bastiaens *et al.*, 2000; Wick *et al.*, 2002b, 2003a; Pizzul *et al.*, 2007). Mycolic acids are high-molecular-mass a-alkyl b-hydroxy fatty acids containing 70–90 carbon atoms which form a highly ordered, close-packed quasicrystalline layer of abnormal thickness, perpendicular to peptidoglycan (Nicaido *et al.*, 1993). Wick *et al.* (2002a) have examined physiological responses of anthracene-degrading *Mycobacterium frederiksbergense* LB501T to poorly water-soluble anthracene, using solid anthracene as sole carbon source. Their results showed that cells growing on anthracene were significantly more hydrophobic and more negatively charged than those growing on glucose. They have also showed that *Mycobacterium* sp. LB501T exhibits a high specific affinity for anthracene, combined with very low substrate requirements for cell maintenance forming biofilms on the surface of anthracene crystals in shaken liquid cultures (Wick *et al.*, 2001, 2002a).

Mycobacterium sp. LB501T cells grown on mixed-substrates with different molar ratios of glucose and anthracene exhibited different composition of cell-wall fatty acids. A higher proportion of anthracene in the carbon source mixture led to higher cell surface hydrophobicities and more hydrophobic mycolic acids (Wick *et al.*, 2003a).

Krivobok *et al.* (2003), using a combination of *in vivo* protein-labelling experiments and 2D-PAGE analysis, have investigated the enzymes involved in PAH degradation by *Mycobacterium* sp. strain 6PY1 able to use pyrene as sole source of carbon and energy. Kim *et al.* (2004) have also identified the proteins induced by PAH in *M. vanbaalenii* PYR-1 using 2D-PAGE electrophoresis as well as *de novo* sequencing methods. Strain PYR-1, originally isolated in 1986 from oil-contaminated estuarine sediment, is well known for its ability to degrade a wide range of HMW-PAHs (Heitkamp and Cerniglia, 1988, 1989; Kelley *et al.*, 1991; Kelley and Cerniglia, 1995; Moody *et al.*, 2003, *2004*, 2005).

Very recently, a genomic analysis of PAH degradation in *Mycobacterium vanbaalenii* PYR-1 has been reported (Kim *et al.*, 2008). The authors showed that sequences and arrangements of aromatic catabolic genes of strain PYR-1 appear to have diverged considerably from those of other bacterial species (Kim *et al.*, 2008).

The complete genome sequence of several PAH degrading *Mycobacterium* strains is now available at the Joint Genome Institute (JGI) (http://img.jgi.doe.gov) (Kim *et al.*, 2008; Kallimanis *et al.*, manuscript in preparation), which are expected to facilitate functional genomic studies in this genus and provide a holistic view on the mechanisms of PAH metabolism.

Arthrobacter species are members of the actinomycete-coryneform bacteria characterized by a high G+C content (Loveland-Curtze *et al.*, 1999). *Arthrobacter* spp. lack mycolic acids and contain predominantly straight-chain saturated, iso- and anteiso-methyl-branched long-chain fatty acids (Collins, 1987). The genus *Arthrobacter* was first proposed by Conn and Dimmick (1947) and its description was finally emended by Koch *et al.* (1995). *Arthrobacter* spp. are ubiquitous in all soils, having considerable nutritional versatility utilizing a wide and varied range of natural as well as xenobiotic compounds. A large number of bacterial species are isolated and characterized as efficient degraders of persistent organic pollutants such as pesticides, phenols, chlorophenols, polychlorinated biphenyls (PCB), nitrobenzoate, chlorobenzoate, haloalkanes, dibenzothiophene (DBT) and carbazole (CBZ) (Engelhardt *et al.*, 1982; Marks *et al.*, 1984; Scholtz *et al.*, 1988; Lin and Wang, 1991; Athiel *et al.*, 1995; Bae *et al.*, 1996; Kar *et al.*, 1997; Turnbull *et al.*, 2001; Daane *et al.*, 2001; Negrete-Raymond *et al.*, 2003; Pandey *et al.*, 2003; Paul *et al.*, 2004; Karigar *et al.*, 2006; Seo *et al.*, 2006a; Unell *et al.*, 2007; Ferreira *et al.*, 2008).

Despite the fact that *Arthrobacter* spp. have been found to be among the predominant members of culturable communities from several terrestrial environments (Crocker *et al.*, 2000), and although high G+C coryneform bacteria are considered important members of PAH-degrading microbial population, so far little information on *Arthrobacter* strains able to degrade PAHs is available and detailed research on the catabolism of PAHs is still limited within this genus (Grifoll *et al.*, 1995; Samanta *et al.*, 1999; Seo *et al.*, 2006b; Kallimanis *et al.*, 2007).

Very recently, the genome of the novel species *A. phenanthrenivorans* strain Sphe3 was sequenced by the DOE-Joint Genome Institute (Kallimanis *et al.*, 2011). Strain Sphe3 is able to metabolize phenanthrene as sole source of carbon and energy at rates faster than those reported for other *Arthrobacter* species (Kallimanis *et al.*, 2007, 2009). The genome analysis of strain Sphe3 in combination with proteomic studies will allow us to understand the molecular basis of phenanthrene metabolism in this genus.

Factors influencing petroleum hydrocarbon biodegradation

A number of limiting factors such as the composition of polluted soils, microbial community, temperature and pH have been recognized to affect the biodegradation of petroleum hydrocarbons. For example, hydrocarbons with 5–10 carbon atoms have been shown to be highly toxic for the majority of hydrocarbon-degrading microorganisms, while *n*-alkanes in the C_{10}–C_{20} range are the most biodegradable petroleum hydrocarbons. It has been shown that hydrocarbons with C_5–C_{10} tend to disrupt lipid membrane structures of microorganisms (Okoh, 2006).

In addition, as microbial communities play a major role in the biodegradation rates in soil, several reports have described the biodegradation of petroleum hydrocarbons by adding soil microorganisms able to degrade petroleum hydrocarbons. Such an approach, known as bioaugmentation, provides certain advantages over biostimulation in cases where pollutant toxicity or lack of appropriate microorganisms is important (Richard and Vogel, 1999; Barathi and Vasudevan, 2001; Bento *et al.*, 2005). However, several researchers have raised doubts about the efficiency of this technique for soil bioremediation (Vogel 1996; Johnsen *et al.*, 2007). Indigenous predators, parasites and toxicants are known to severely restrict biodegradation (Mallory *et al.*, 1983; Murakami and Alexander, 1989). Many reports have described the use of immobilized cells for degradation of toxic compounds in order to overcome adverse environmental conditions that threaten microbial survival (Wiesel *et al.*, 1993; Weir *et al.*, 1995; Jeřábková *et al.*, 1997; Manohar and Karegoudar, 1998; Diaz *et al.*, 2002; Grötzschel *et al.*, 2002; Li *et al.*, 2005). Furthermore, immobilized microbial cells can be handled more easily, because contamination during transport, application and storage can be avoided (Cassidy *et al.*, 1996; Park and Chang, 2000). Karabika *et al.* (2009) studied the ability of alginate/starch entrapped cells of a *Mycobacterium* sp. strain Spyr1 to degrade pyrene as well as mixtures of PAHs in soil microcosms. Although alginate/starch immobilized cells had slower biodegradation rates than free cells in the removal of pyrene and other PAHs from contaminated spiked soil, the long-term preservation of immobilized Spyr1 cells without decrease of their PAH-degrading activity provides an advantage in soil bioremediation. Recently, Karamalides *et al.* (2010) compared the biodegradation of petroleum-polluted soil by biostimulation of indigenous microorganisms with the biodegradation by biostimulation and bioaugmentation with free or alginate/starch-encapsulated cells of *P. aeruginosa* strain Spet, isolated from a petroleum polluted soil. Their results showed that bioaugmentation with free or encapsulated *P. aeruginosa* cells and/or soil dilution had no significant effect on biodegradation.

Biodegradation of petroleum hydrocarbons in extreme soil environment

Temperature

Temperature plays a very important role on the rate of petroleum hydrocarbon biodegradation. It affects the population and the activity of microorganisms, as well as the solubility of hydrocarbons (Foght *et al.*, 1996; Venosa and Zhu, 2003). The source of organisms may be critical for PAH biodegradation applications at low temperature because mesophilic PAH degraders are relatively ubiquitous in soils whereas psychrotolerant PAH degraders may be less ubiquitous or may require a very long time to multiply (Eriksson *et al.*, 2003).

The biodegradation of hydrocarbons has been shown to occur in various extreme habitats, such as in low and high temperature environments. Cold-tolerant bacteria, belonging to the genus *Rhodococcus* and capable to degrade *n*-alkanes, have been isolated from oil-contaminated soils in Antarctica and Arctic areas (Whyte *et al.*, 1998; Bej *et al.*, 2000). Margesin *et al.* (2003) using cultural-independent analysis in oil-contaminated European Alpine soils have reported that strains belonging to the genera *Pseudomonas, Acinetobacter* and *Rhodococcus* were involved in the degradation of *n*-alkanes in these cold environments.

Bacterial strains, identified as *Pseudomonas* spp., were isolated from a petroleum-contaminated soil from Pangnirtung, Baffin Island, Canada, and proved to degrade C5–C12 n-alkanes, toluene, and naphthalene at both 5 and 25°C (Whyte *et al.*, 1997). These strains possessed both the alk catabolic pathway for alkane and the nah catabolic pathway for polycyclic aromatic hydrocarbon biodegradation (Whyte *et al.*, 1997). The psychrotrophic *Rhodococcus* sp. strain Q15 mineralized a broad range of aliphatics to a greater extent at 5°C than 0°C (Whyte *et al.*, 1998). The psychrotolerant bacterial strain *Sphingomonas* sp. Ant. 17, isolated from Antarctic fuel-contaminated soil, degrades the aromatic fraction of several different crude oils at low temperatures even without the addition of nutrients (Baraniecki *et al.*, 2002). *Rhodococcus* sp. strain EH831, isolated from an oil-contaminated soil, able to degrade hexane and a wide range of other hydrocarbons, maintained its activity at low temperatures (Lee *et al.*, 2010).

Aislabie *et al.* (2000) have reported the isolation of hydrocarbon-degrading bacteria from soils of the Ross Sea region in Antarctica. They found that the alkane degraders were predominantly *Rhodococcus* sp., whereas the aromatic degraders were identified as *Sphingomonas* sp. or *Pseudomonas* sp. Very recently, a *Microbacterium* sp. strain F10 isolated from oil-polluted soil and able to degrade phenanthrene and pyrene in low-temperature environment was also characterized (Sheng *et al.*, 2009).

Elevated temperature increases the bioavailability and biodegradation of PAHs because it facilitates their solubility and mass transfer rates in aqueous solutions (Annweiler *et al.*, 2000; Margesin and Schinner, 2001; Feitkenhauer and Markl, 2003; Feitkenhauer *et al.*, 2003).

Indigenous thermophilic hydrocarbon degraders are of special significance as they could be efficiently used for bioremediation of hydrocarbon-polluted soils (Margesin and Schinner, 2001). Many *Bacillus*, *GeoBacillus* and *Thermus* species have been isolated from high-temperature environments (Sorokhoh *et al.*, 1993; Feitkenhauer *et al.*, 2003; Hao *et al.*, 2004; Lei *et al.*, 2006; Marchant *et al.*, 2006; Meintanis *et al.*, 2006; Liu *et al.*, 2009).

Viamajala *et al.* reported the isolation of three thermophilic Geobacilli from compost that grew on phenantrene at 60°C and exhibiting a higher degradation rate as compared with that known for mesophiles. By comparing the phenanthrene solubilization rates with degradation rates at ambient and elevated temperatures, they showed that while solubilization rates of PAHs were significantly enhanced at elevated temperatures, the biodegradation of PAHs was probably mass transfer limited even at elevated temperatures (Viamajala *et al.*, 2007).

Annweiler *et al.* (2000) have reported the isolation of a thermophilic aerobic bacterium *Bacillus thermoleovorans* Hamburg 2 from a contaminated compost consisting of lignite-treated wooden ties which grows at 60°C on naphthalene as the sole source of carbon and energy. The moderate thermophilic bacterial strain TSH1 isolated from a petro-industrial wastewater-contaminated soil identified as *Nocardia otitidiscaviarum* was capable of degrading straight-chain aliphatic hydrocarbons as well as polycyclic aromatic hydrocarbons such as naphthalene, phenanthrene, anthracene and pyrene (Zeinali *et al.*, 2008).

The effect of temperature as a determinant for selecting microbial populations associated with alkane-degradation has been examined in crude oil-amended soil microcosms by Hamamura *et al.* (2008). The growth of a *Pseudomonas* sp. isolate was examined at different temperatures with hexadecane or glucose as a growth substrate. The generation time of the culture increased at lower temperatures regardless if the carbon source was hexadecane or glucose, indicating that the decrease in growth rates at low temperatures was not caused by a decrease of the hexadecane solubility (Hamamura *et al.*, 2008).

pH

The pH of culture media affected the levels of PAH *o*-quinone reductases in a pyrene degrading *Mycobacterium* sp. strain PYR 100 (Kim *et al.*, 2003b). Kim *et al.* (2005b) also reported that the pH of the media and biotransformation buffer affected the rates of growth of *M. vanbaalenii* PYR-1 and its degradation ability of phenanthrene and pyrene (Kim *et al.*, 2005b). They reported that an acidic pH seemed to render the mycobacterial cells more permeable to hydrophobic substrates owing to the possible disruption of the mycolic acid layer under acidic conditions. Even though acidification of the medium significantly reduced the growth rate, the disruption of the cell permeability barrier increased the substrate concentration in the cytosol. Similar results were reported by Grosser *et al.* (1991) for a pyrene-degrading *Mycobacterium* sp. that mineralized pyrene more rapidly in acidic soil compared to that in slightly alkaline soil, although in long-term experiments they found that pyrene degradation was more rapid in the alkaline soil.

Uyttebroek *et al.* (2007) have compared the microbial communities of cultures able to utilize pyrene as a sole source of carbon and energy enriched from acidic PAH-contaminated soil with those enriched from a PAH-contaminated soil sampled at the same site but with a slightly alkaline pH. The above studies indicated that the cultures enriched from the acidic PAH-contaminated soil were dominated by a single bacterium closely related to *Mycobacterium montefiorense*, belonging to the slow-growing *Mycobacterium* species, whereas the culture enriched on pyrene at pH 7 from the slightly alkaline soil was dominated by *Pseudomonas putida* and a fast-growing *Mycobacterium* species. They also reported that bacteria able to degrade high-molecular-weight PAHs at low pH were present in the acidic PAH-contaminated soil (Uyttebroek *et al.*, 2007).

Salt

Although many reports have described the biodegradation of PAHs using marine halophilic bacteria, the degradation of phenanthrene under hypersaline conditions by a halophilic soil bacterial consortium was only recently published (Zhao *et al.*, 2009).

Conclusions

Culture-based approaches have resulted in the isolation of a range of soil hydrocarbon-degrading bacteria. PAH-degrading bacteria are members of the different subdivisions of the Proteobacteria as well as of the high G+C Gram-positive bacteria. These bacteria have been used to elucidate the metabolism of PAHs providing information on the types and diversity of metabolic pathways. Microbial genomic analysis will provide a basis to decipher the evolution of PAH catabolic genes and may clarify the mechanisms of how these genes are disseminated among bacteria within the same phylogenetic group and beyond. This information will assist to exploit microorganisms with a deeper profiling of microbial activity at the metabolic level in bioremediation of hydrocarbon-contaminated soils.

Acknowledgements

This work was supported by the programme 'Pythagoras II' of EPEAEK with 25% National Funds and 75% European Social Funds (ESF) and by the Greek Secretariat for Research and Technology (Programme PENED-01ED547).

References

Aislabie, J., Foght, J., and Saul, D. (2000). Aromatic hydrocarbon-degrading bacteria from soil near Scott base, Antarctica. Polar Biol. *23*, 183–188.

Aitken, M.D., Stringfellow, W.T., Nagel, R.D., Kazunga, C., and Chen, S.-H. (1998). Characteristics of phenanthrene-degrading bacteria isolated from soils contaminated with polycyclic aromatic hydro-carbons. Can. J. Microbiol. *44*, 743–752.

Alexander, M. (1981). Biodegradation of chemicals of environmental concern. Science *211*, 132–138.

Annweiler, E., Richnow, H.H., Antranikian, G., Hebenbrock, S., Garms, C., Franke, S., Francke, W., and Michaelis, W. (2000). Naphthalene degradation and incorporation of naphthalene-derived carbon into biomass by the thermophile *Bacillus thermoleovorans*. Appl. Environ. Microbiol. *68*, 518–523.

Arvanitis, N., Katsifas, E.A., Chalkou, K.I., Meintanis, C., and Karagouni, A.D. (2008). A refinery sludge deposition site: presence of nahH and alkJ genes and crude oil biodegradation ability of bacterial isolates. Biotechnol. Lett. *30*, 2105–2110.

Athiel, P., Alfizar, Mercadier, C., Vega, D., Bastide, J., Davet, P., Brunel, B., and Cleyet-Marel, J.-C. (1995). Degradation of iprodione by a soil *Arthrobacter*-like strain. Appl. Environ. Microbiol. *61*, 3216–3220.

Auffret, M., Labbé, D., Thouand, G., Greer, C.W., and Fayolle-Guichard, F. (2009). Degradation of a Mixture of hydrocarbons, gasoline, and diesel oil additives by *Rhodococcus aetherivorans* and *Rhodococcus wratislaviensis*. Appl. Environ. Microbiol. *75*, 7774–7782.

Bae, H.S., Lee, J.M., and Lee, S.-T. (1996). Biodegradation of 4-chlorophenol via a hydroquinone pathway by *Arthrobacter ureafaciens* CPR706. FEMS Microbiol. Lett. *145*, 125–129.

Baraniecki, C.A., Aislabie, J., and Foght, J.M. (2002). Characterization of *Sphingomonas* sp. Ant 17, an aromatic hydrocarbon-degrading bacterium isolated from antarctic soil. Microb. Ecol. *43*, 44–54.

Barathi, S., and Vasudevan, N. (2001). Utilization of petroleum hydrocarbons by *Pseudomonas fluorescens* isolated from a petroleum-contaminated soil. Environ. Int. *26*, 413–416.

Barathi, S., and Vasudevan, N. (2003). Bioremediation of crude oil contaminated soil by augmentation of *Pseudomonas fluorescens* NS1. J. Environ. Sci. Heal. A. *38*, 1857–1866.

Bastiaens, L., Springael, D., Wattiau, P., Harms, H., de Wachter, R., Verachtert, H., and Diels, L. (2000). Isolation of adherent polycyclic aromatic hydrocarbon (PAH) degrading bacteria using PAH sorbing carriers. Appl. Envriron. Microbiol. *66*, 1834–1843.

Bateman, J.N., Speer, B., Feduik, L., and Hartline, R.A. (1986). Naphthalene association and uptake in *Pseudomonas putida*. Appl. Environ. Microbiol. *166*, 155–161.

Bej, A.K., Saul D., and Aislabie, J. (2000). Cold-tolerant alkane-degrading *Rhodococcus* species from Antarctica. Polar Biol. *23*, 100–105

Belhaj, A., Desnoues, N., and Elmerich, C. (2002). Alkane biodegradation in *Pseudomonas aeruginosa* strains isolated from a polluted zone: Identification of alkB and alkB-related genes. Res. Microbiol. *153*, 339–344.

Bento, F.M., Camargo, F.A.O., Okeke, B.C., Frankenberger, W.T. (2005). Comparative bioremediation of soils contaminated with diesel oil by natural attenuation, biostimulation and bioaugmentation. Biores. Technol. *96*, 1049–1055.

Blumer, M. (1976). Polycyclic aromatic compounds in nature. Sci. Am. *234*, 35–45.

Bogan, B.W., Lahner, L.M., Sullivan, W.R., and Paterek, J.R. (2003). Degradation of straight-chain aliphatic and high-molecular-weight polycyclic aromatic hydrocarbons by a strain of *Mycobacterium austroafricanum*. J. Appl. Microbiol. *94*, 230–239.

Boldrin, B., Tiehm, A., and Fritzsche, C. (1993). Degradation of phenanthrene, fluorene, fluoranthene and pyrene by a *Mycobacterium* sp. Appl. Environ. Microbiol. *59*, 1927–1930.

Bollmann, A., Lewis, K., and Epstein, S.S. (2007). Incubation of environmental samples in a diffusion chamber increases the diversity of recovered isolates. Appl. Environ. Microbiol. *73*, 6386–6390.

Boonchan, S., Britz, M.L., and Stanley, G.A. (1998). Surfactant-enhanced biodegradation of high molecular weight polycyclic aromatic hydrocarbons by *Stenotrophomonas maltophilia*. Biotechnol. Bioeng. *59*, 482–494.

Boonchan, S., Britz, M., and Stanley, G. (2000). Degradation and mineralization of high-molecular-weight polycyclic aromatic hydrocarbons by defined fungal-bacterial cocultures. Appl. Environ. Microbiol. *66*, 1007–1019.

Bouchez-Naïtali, M., Rakatozafy, H., Marchal, R., Leveau, J.-Y., and Vandecasteele, J.-P. (1999). Diversity of bacterial strains degrading hexadecane in relation to the mode of substrate uptake. J. Appl. Microbiol. *86*, 421–428.

Brown, G.R., Sutcliffe, I.C., Bendell, D., and Cummings, S.P. (2000). The modification of the membrane of *Oceanomonas baumannii* when subjected to both osmotic and organic solvent stress. FEMS Microbiol. Lett. *189*, 149–154.

Bugg, T., Foght, J.M., Pickard, M.A., and Gray, M.R. (2000). Uptake and active efflux of polycyclic aromatic hydrocarbons by *Pseudomonas fluorescens* LP6a. Appl. Environ. Microbiol. *66*, 5387–5392.

Cardenas, E., and Tiedje, J.M. (2008). New tools for discovering and characterizing microbial diversity. Curr. Opin. Biotechnol. *19*, 544–549.

Cassidy, M.B., Lee, H., and Trevors, J.T. (1996). Environmental applications of immobilized microbial cells: a review. J. Ind. Microbiol. *16*, 79–101.

Cerniglia, C.E. (1984). Microbial metabolism of polycyclic aromatic hydrocarbons. Adv. Appl. Microbiol. *30*, 31–71.

Cerniglia, C.E. (1992). Biodegradation of polycyclic aromatic hydrocarbons. Biodegradation *3*, 351–368.

Cerniglia, C.E. (1993). Biodegradation of polycyclic aromatic hydrocarbons. Curr. Opin. Biotech. *4*, 331–338.

Chaerun, S.K., Tazaki, K., Asada, R., and Kogure, K. (2004). Bioremediation of coastal areas 5 years after the Nakhodka oil spill in the Sea of Japan: isolation and characterization of hydrocarbon-degrading bacteria. Environ. Int. *30*, 911–922.

Cho, J.-C., and Kim, S.-J. (2001). Detection of mega plasmid from polycyclic aromatic hydrocarbon-degrading *Sphingomonas* sp. strain KS14. J. Mol. Microb. Biotech. 3, 503–506.

Churchill, S.A., Harper, J.P., and Churchill, P.F. (1999). Isolation and characterization of a *Mycobacterium* species capable of degrading three- and four-ring aromatic and aliphatic hydrocarbons. Appl. Environ. Microbiol. *65*, 549–552.

Collins, M.D. (1987). Transfer of *Arthrobacter variabilis* (Muller) to the genus *Corynebacterium*, as *Corynebacterium variabilis* comb. Nov. Int. J. Syst. Bacteriol. *37*, 287–288.

Conn, H.J., and Dimmick, I. (1947). Soil bacteria similar in morphology to *Mycobacterium* and *Corynebacterium*. J. Bacteriol. *54*, 291–303.

Crocker, F.H., Fredrickson, J.K., White, D.C., Ringelberg, D.B., and Balkwill, D.L. (2000). Phylogenetic and physiological diversity of *Arthrobacter* strains isolated from unconsolidated subsurface sediments. Microbiology *146*, 1295–1310.

Daane, L.L., Harjono, I., Zylstra, G.J., and Häggblom, M.M. (2001). Isolation and characterization of polycyclic aromatic hydrocarbon-degrading bacteria associated with rhizosphere of salt marsh plants. Appl. Environ. Microbiol. *67*, 2683–2691.

Dandie, C.E., Thomas, S.M., Bentham, R.H., and McClure, N.C. (2004). Physiological characterization of *Mycobacterium* sp. strain 1B isolated from a bacterial culture able to degrade high-molecular-weight polycyclic aromatic hydrocarbons. J. Appl. Microbiol. *97*, 246–255.

Davies, J.I., and Evans, W.C. (1964). Oxidative metabolism of naphthalene by soil pseudomonads: the ring-fission mechanism. Biochem. J. *91*, 251–261.

Dean-Ross, D., and Cerniglia, C.E. (1996). Degradation of pyrene by *Mycobacterium flavescens*. Appl. Microbiol. Biot. *46*, 307–312.

Demanèche, S., Meyer, C,M., Micoud, J., Louwagie, M., Willison, J.C., and Jouanneau, Y. (2004). Identification and functional analysis of two aromatic-ring-hydroxylating dioxygenases from a *Sphingomonas* sp. strain that degrades various polycyclic aromatic hydrocarbons. Appl. Environ. Microbiol. *70*, 6714–6725.

Derz, K., Klinner, U., Schuphan, I., Stackebrandt, E., and Kroppenstedt, R.M. (2004). *Mycobacterium pyrenivorans* sp. nov., a novel polycyclic-aromatic-hydrocarbon-degrading species. Int. J. Syst. Evol. Micr. *54*, 2313–2317.

Diaz, M.P., Boyd, K.G., Grigson, S.J.W., and Burgess, J.B. (2002). Biodegradation of crude oil across a wide range of salinities by an extremely halotolerant bacterial consortium MPD-M, immobilised onto polypropylene fibers. Biotechnol. Bioeng. 79, 145–153.

Engelhardt, G., Ziegler, W., Wallnofer, P.R., Jarczyk, H.J., and Oehlmann, L. (1982). Degradation of the triazinone herbicide metamitron by *Arthrobacter* sp. DSM 20389. J. Agr. Food Chem. *30*, 278–282.

Eriksson, M., Sodersten, E., Yu, Z., Dalhammar, G., and Mohn, W.W. (2003). Degradation of polycyclic aromatic hydrocarbons at low temperature under aerobic and nitrate-reducing conditions in enrichment cultures from northern soils. Appl. Environ. Microbiol. *69*, 275–284.

Feitkenhauer, H., and Markl, H. (2003). Biodegradation of aliphatic and aromatic hydrocarbons at high temperatures. Water Sci. Technol. *47*, 123–130.

Feitkenhauer, H., Müller, R., and Märkl, H. (2003). Degradation of polycyclic aromatic hydrocarbons and long chain alkanes at 60–70 °C by *Thermus* and *Bacillus* spp. Biodegradation *14*, 367–372.

Ferraro, D.J., Brown, E.N., Yu, C.-L., Parales, R.E., Gibson, D.T., and Ramaswamy, S. (2007). Structural investigations of the ferredoxin and terminal oxygenase components of the biphenyl 2,3-dioxygenase from *Sphingobium yanoikuyae* B1. BMC Struct. Biol. *7*, 10.

Ferreira, M.I.M., Marchesi, J.R., and Janssen, D.B. (2008). Degradation of 4-fluorophenol by *Arthrobacter* sp. strain IF1. Appl. Microbiol. Biot. *78*, 709–717.

Foght, J.M., Westlake, D., Johnson, W.M., and Ridgway, H.F. (1996). Environmental gasoline-utilising isolates and clinical isolates of *Pseudomonas aeruginosa* are taxonomically indistinguishable by chemotaxonomic and molecular techniques. Microbiology *142*, 1333–2340.

Franzetti, A., Caredda, P., Ruggeri, C., La Colla, P., Tamburini, E., Papacchini, M., and Bestetti, G. (2009). Potential applications of surface active compounds by *Gordonia* sp. strain BS29 in soil remediation technologies. Chemosphere 75, 801–807.

Freeman, D.J., and Cattell, F.C.R. (1990). Woodburning as a source of atmospheric polycyclic aromatic hydrocarbons. Environ. Sci. Technol. *24*, 1581–1585.

Fuenmayor, S.L., Wild, M., Boyes, A.L., and Williams, P.A. (1998). A gene cluster encoding steps in conversion of naphthalene to gentisate in *Pseudomonas* sp. strain U2. J. Bacteriol. *180*, 2522–2530.

Gibson, D.T., Roberts, R.L., Wells, H.C., and Kobaf, V.M. (1973). Oxidation of biphenyl by a Beijerinckia species. Biochem. Bioph. Res. Co. *50*, 211–219.

Gordon, L., and Dobson, A.D.W. (2001). Fluoranthene degradation in *Pseudomonas alcaligenes* PA-10. Biodegradation *12*, 393–400.

Goyal, A.K., and Zylstra, G.J. (1996). Molecular cloning of novel genes for polycyclic aromatic hydrocarbon degradation from *Comamonas testosteroni* GZ39. Appl. Environ. Microbiol. *62*, 230–236.

Grifoll, M., Casellas, M., Bayona, J.M., and Solanas, A.M. (1992). Isolation and characterization of a fluorene-degrading bacterium: identification of ring oxidation and ring fission products. Appl. Environ. Microbiol. *58*, 2910–2917.

Grifoll, M., Selifonov, S.A., Gatlin, C.V., and Chapman, P.J. (1995). Actions of a versatile fluorene-degrading bacterial isolate on polycyclic aromatic compounds. Appl. Environ. Microbiol. *61*, 3711–3723.

Grosser, R.J., Warshawsky, D., and Vestal, J.R. (1991). Indigenous and enhanced mineralization of pyrene, benzo[a]pyrene, and carbazole in soils. Appl. Environ. Microbiol. 57,b3462–3469.

Grötzschel, S., Köster, J., Abed, R.M.M., and de Beer, D. (2002). Degradation of petroleum model compounds immobilised on clay by a hypersaline microbial mat. Biodegradation *13*, 273–283.

Guerin, W.F., and Jones, G.E. (1988). Mineralization of phenanthrene by a *Mycobacterium* sp. Appl. Environ. Microbiol. *54*, 937–944.

Habe, H., and Omori, T. (2003). Genetics of polycyclic aromatic hydrocarbon metabolism in diverse aerobic bacteria. Biosci. Biotech. Biochim. *67*, 225–243.

Habe, H., Kanemitsu, M., Nomura, M., Takemura, T., Iwata, K., Nojiri, H., Yamane, H., and Omori, T. (2004). Isolation and characterization of an alkaliphilic bacterium utilizing pyrene as a carbon source. J. Biosci. Bioeng. *98*, 306–308.

Haigler, B.E., Pettigrew, C.A., and Spain, J.C. (1992). Biodegradation of mixtures of substituted benzenes by *Pseudomonas* sp. strain JS150. Appl. Environ. Microbiol. *58*, 2237–2244.

Hamamura, N., Olson, S.H., Ward, D.M., and Inskeep, W.P. (2006). Microbial population dynamics associated with crude-oil biodegradation in diverse soils. Appl. Environ. Microbiol. *72*, 6316–6324.

Hamamura, N., Fukui, M., Ward, D.M., and Inskeep, W.P. (2008). Assessing soil microbial populations responding to crude-oil amendment at different temperatures using phylogenetic, functional gene (alkB) and physiological analyses. Environ. Sci. Technol. *42*, 7580–7586.

Handelsman, J. (2004). Metagenomics: application of genomics to uncultured microorganisms. Microbiol. Mol. Biol. R. *68*, 669–685.

Hanumanthanaik, P.D., and Ninnekar, H.Z. (2000). Biodegradation of phenanthrene by a *Bacillus* species. Curr. Microbiol. *41*, 11–14.

Hao, R., Lu, A., and Wang, G. (2004). Crude-oil-degrading thermophilic bacterium isolated from an oil field. Can. J. Microbiol. *50*, 175–182.

Haritash, A.K., and Kaushik, C.P. (2009). Biodegradation aspects of polycyclic aromatic hydrocarbons (PAHs): A review. J. Hazard. Mater. *169*, 1–15.

Heipieper, H.-J., Diefenbach, R., and Keweloh, H. (1992). Conversion of cis unsaturated fatty acids to trans, a possible mechanism for the protection of phenol-degrading *Pseudomonas putida* P8 from substrate toxicity. Appl. Environ. Microbiol. *58*, 1847–1852.

Heipieper, H.J., Meinhardt, F., and Segura, A. (2003). The cis-trans isomerase of unsaturated fatty acids in *Pseudomonas* and *Vibrio*: biochemistry, molecular biology and physiological function of a unique stress adaptive mechanism. FEMS Microbiol. Lett. *229*, 1–7.

Heitkamp, M.A., and Cerniglia, C.E. (1989). Polycyclic aromatic hydrocarbon degradation by a *Mycobacterium* sp. in microcosms containing sediment and water from a pristine ecosystem. Appl. Environ. Microbiol. *55*, 1968–1973.

Heitkamp, M.A., Franklin, W., and Cerniglia, C.E. (1988). Microbial metabolism of polycyclic aromatic hydrocarbons: isolation and characterization of a pyrene-degrading bacterium. Appl. Environ. Microbiol. *54*, 2549–2555.

Heitkamp, M.A., Freeman, J.P., Miller, D.W., and Cerniglia, C.E. (1998). Pyrene degradation by a *Mycobacterium* sp.: Identification of ring oxidation and ring fission products. Apll. Environ. Microb. *54*, 2556–2565.

Heitkamp, M.M., and Cerniglia, C.E. (1988). Mineralization of polycyclic aromatic hydrocarbons by a bacterium isolated from sediment below an oil field. Appl. Environ. Microbiol. *54*, 1612–1614.

Hormisch, D., Brost, I., Kohring, G.W., Giffhorn, F., Kroppenstedt, R.M., Stackebrandt, E., Färber, P., and Holzapfel, W.H. (2004). *Mycobacterium fluoranthenivorans* sp. nov., a fluoranthene and aflatoxin B1 degrading bacterium from contaminated soil of a former coal gas plant. Syst. Appl. Microbiol. *27*, 653–660.

Hugenholtz, P. (2002). Exploring prokaryotic diversity in the genomic era. Genome Biol. *3*, reviews 0003.1–0003.8.

Jacques, R.J.S., Okeke, B.C., Bento, F.M., Peralba, M.C.R., and Camargo, F.A.O. (2009). Improved enrichment and isolation of polycyclic aromatic hydrocarbons (PAH)-degrading microorganisms in soil using anthracene as a model PAH. Curr. Microbiol. *58*, 628–634.

Janssen, P.H., Yates, P.S., Grinton, B.E., Taylor, P.M., and Sait, M. (2002). Improved culturability of soil bacteria and isolation in pure culture of novel members of the divisions Acidobacteria, Proteobacteria and *Verrucomicrobia*. Appl. Environ. Microbiol. *68*, 2391–2396.

Jeřábková, H., Králová, B., Krejèů, V., Sanchez, J.L.I., and Roig, M.G. (1997). Use of polyurethane foam for the biodegradation of n-alkanes by immobilised cells of *Pseudomonas*. Biotechnol. Techniques. *11*, 391–394.

Johnsen, A., Wick, L.Y, and Harms, H. (2005). Principles of microbial PAH-degradation in soil. Environ. Pollut. *133*, 71–84.

Johnsen, A.R., Winding, A., Karlson, U., and Roslev, P. (2002). Linking of microorganisms to phenanthrene metabolism in soil by analysis of 13C-labeled cell lipids. Appl. Environ. Microbiol. *68*, 6106–6113.

Johnsen, A.R., Schmidt, S., Hybholt, T.K., Henriksen, S., Jacobsen, C.S., and Andersen, O. (2007). Strong impact on the polycyclic aromatic hydrocarbon (PAH)-degrading community of a PAH-polluted soil but marginal effect on PAH degradation when priming with bioremediated soil dominated by *Mycobacteria*. Appl. Environ. Microbiol. 73, 1474–1480.

Jordan, S., Hutchings, M.I., and Mascher, T. (2008). Cell envelope stress response in Gram-positive bacteria. FEMS Microbiol. R. *32*, 107–146.

Juhasz, A.L. (1998). Microbial degradation of high molecular weight polycyclic aromatic hydrocarbons. PhD Thesis, Victoria University of Technology, Melbourne, Australia.

Juhasz, A.L., and Naidu, R. (2000). Bioremediation of high molecular weight polycyclic aromatic hydrocarbons: a review of the microbial degradation of benzo[a] pyrene. Int. Biodeter. Biodegradation. *45*, 57–88.

Juhasz, A.L., Britz, M.L., and Stanley, G.A. (1996). Degradation of high molecular weight polycyclic aromatic hydrocarbons by *Pseudomonas cepacia*. Biotechnol. Lett. *18*, 577–582.

Juhasz, A.L., Britz, M.L., and Stanley, G.A. (1997). Degradation of fluoranthene, pyrene, benz[a] anthracene and dibenz[a,h]anthracene by *Burkholderia cepacia*. J. Appl. Microbiol. *83*, 189–198.

Juhasz, A.L., Stanley, G.A., and Britz, M.L. (2000). Microbial degradation and detoxification of high molecular weight polycyclic aromatic hydrocarbons by *Stenotrophomonas maltophilia* strain VUN 10,003. Lett. Appl. Microbiol. *30*, 396–401.

Kallimanis, A., Frillingos S., Drainas C., and Koukkou A. I. (2007). Taxonomic identification, phenanthrene uptake activity, and membrane lipid alterations of the PAH degrading *Arthrobacter* sp. strain Sphe3. Appl. Microbiol. Biot. *76*, 709–17.

Kallimanis, A., Kavakiotis, K., Perisynakis, A., Spröer, C., Pukall, R., Drainas, C., and Koukkou, A. I. (2009). *Arthrobacter phenanthrenivorans* sp. nov., to accommodate the phenanthrene-degrading bacterium *Arthrobacter* sp. strain Sphe3. Int. J. Syst. Evol. Micr. *59*, 275–279.

Kallimanis, A., La Butti, K.M., Lapidus, A., Clum, A., Lykidis, A., Mavromatis, K., Pagani, I., Liolios, K., Ivanova, N., Goodwin, L., *et al.* (2010). Complete genome sequence of *Anthrobacter phenanthrenivorans type* strain (Sphe3). Stand. Genomic Sci. (In press).

Kanaly, R.A., and Harayama, S. (2000). Biodegradation of high-molecular-weight polycyclic aromatic hydrocarbons by bacteria. J. Bacteriol. *182*, 2059–2067.

Kang, H., Hwang, S.Y., Kim, Y.M., Kim, E., Kim, Y.-S., Kim, S.-K., Kim, S.W., Cerniglia, C.E., Skuttleworth, K.L., and Zylstra, G.J. (2003). Degradation of phenanthrene and naphthalene by *Burkholderia* species strain. Can. J. Microbiol. *49*, 139–144.

Kar, S., Swaminathan, T., and Baradarajan, A. (1997). Biodegradation of phenol and cresol isomermixtures by *Arthrobacter*. World J. Microb. Biot. *13*, 659–663.

Karabika, E., Kallimanis, A., Dados, A., Pilidis, G., Drainas, C., and Koukkou, A.I. (2009). Taxonomic identification and use of free and entrapped cells of a new *Mycobacterium* sp., strain Spyr1 for degradation of polycyclic aromatic hydrocarbons (PAHs). Appl. Biochem. Biotech. *159*, 155–167.

Karamalidis, A.K., Evangelou, A.C, Karabika, E., Koukkou, A.I., Drainas, C., and Voudrias, E.A. (2010). Laboratory scale bioremediation of petroleum-contaminated soil by indigenous microorganisms and added *Pseudomonas aeruginosa* strain Spet. Biores. Technol. *101*, 6546–6552.

Karigar, C., Mahesh, A., Nagenahalli, M., and Yun, D.J. (2006). Phenol degradation by immobilized cells of *Arthrobacter citreus*. Biodegradation *17*, 47–55.

Kästner, M., Breuer-Jammali, M., and Mahro, B. (1994). Enumeration and characterization of the soil microflora from hydrocarbon-contaminated soil sites able to mineralize polycyclic aromatic hydrocarbons (PAH). Appl. Microbiol. Biot. *41*, 267–273.

Katsivela, E., Moore, E.R.B., Maroukli, D., Strömpl, C., Pieper, D., and Kalogerakis, N. (2005). Bacterial community dynamics during in situ bioremediation of petroleum waste sludge in landfarming sites. Biodegradation *16*, 169–180.

Kazunga, C., Aitken. M.D., Gold, A., and Sangaiah, R. (2001). Fluoranthene-2,3- and -1,5-diones are novel products from the bacterial transformation of fluoranthene. Environ. Sci. Technol. *35*, 917–922.

Kelley, I., and Cerniglia, C.E. (1995). Degradation of a mixture of highmolecular-weight polycyclic aromatic hydrocarbons by a *Mycobacterium* strain PYR-1. J. Soil Contam. *4*, 77–91.

Kelley, I., Freeman, J.P., Evans, F.E., and Cerniglia, C.E. (1991). Identification of a carboxylic acid metabolite from the catabolism of fluoranthene by a *Mycobacterium* sp. Appl. Environ. Microbiol. *57*, 636–641.

Keum, Y.-S., Seo, J.-S., Hu, Y., and Li, Q.X. (2006). Degradation pathways of phenanthrene by *Sinorhizobium* sp. C4. Appl. Microbiol. Biot. *71*, 935–941.

Khan, A.A., Wang, R.F., Cao, W.W., Franklin, W., and Cerniglia, C.E. (1996). Reclassification of a polycyclic aromatic hydrocarbon-metabolising bacterium *Beijerinckia* sp. strain B1, as *Sphingomonas yanoikuyae* by fatty acid analysis, protein pattern analysis, DNA-DNA hybridisation, and 16S ribosomal DNA sequencing. Int. J. Syst. Bacteriol. *46*, 466–469.

Khan, A.A., Kim, S.-J., Pain, D.D., and Cerniglia, C.E. (2002). Classification of a polycyclic aromatic hydrocarbon-metabolizing bacterium, *Mycobacterium* sp. strain PYR-1, as *Mycobacterium vanbaalenii* sp. nov. Int. J. Syst. Evol. Micr. *52*, 1997–2002.

Kim, S.-J., Kweon, O., Jones, R.C., Freeman, J.P., Edmondson, R.D., and Cerniglia, C.E. (2007). Complete and integrated pyrene degradation pathway in *Mycobacterium vanbaalenii* PYR-1 based on systems biology. J. Bacteriol. *189*, 464–472.

Kim, S.-J., Kweon, Z.O., Jones, R.C., Edmondson, R.D., and Cerniglia, C.E. (2008). Genomic analysis of polycyclic aromatic hydrocarbon degradation in *Mycobacterium vanbaalenii* PYR-1. Biodegradation *19*, 859–881.

Kim, T.J., Lee, E.Y., Kim, Y.J., Cho, K.-S., and Ryu, H.W. (2003a). Degradation of polyaromatic hydrocarbons by *Burkholderia cepacia* 2A-12. World J. Microb. Biot. *19*, 411–417.

Kim, Y.-H., Engesser, K.-H., and Cerniglia, C.E. (2003b). Two polycyclic aromatic hydrocarbon *o*-quinone reductases from a pyrene-degrading *Mycobacterium*. Arch. Biochem. Biophys. *416*, 209–217.

Kim, Y.H., Moody, J.D., Freeman, J.P., Brezna, B., Engesser, K.H., and Cerniglia, C.E. (2004). Evidence for the existence of PAH-quinone reductase and catechol-o-methyltransferase in *Mycobacterium vanbaalenii* PYR-1. J. Ind. Microbiol. Biot. *31*, 507–516.

Kim, Y.-H., Engesser, K.-H., and Cerniglia, C.E. (2005a). Numerical and genetic analysis of polycyclic aromatic hydrocarbon-degrading mycobacteria. Microbial Ecol. *50*, 110–119.

Kim, Y.-H., Freeman, J.P., Moody, J.D., Engesser, K.-H., and Cerniglia, C.E. (2005b). Effects of pH on the degradation of phenathrene and pyrene by *Mycobacterium vanbaalenii* PYR-1. Appl. Microbiol. Biot. *67*, 275–285.

Kiyohara, H., Nagao, K., Kouno, K., and Yano, K. (1982). Phenanthrene-degrading phenotype of *Alcaligens faecalis* AFK2. Appl. Environ. Microbiol. *43*, 458–461.

Ko, S.H., and Lebeault, J.M. (1999). Effect of a mixed culture on co-oxidation during the degradation of saturated hydrocarbon mixture. J. Appl. Microbiol. *87*, 72–79.

Koch, C., Schumann, P., and Stackebrandt, E. (1995). Reclassification of *Micrococcus agilis* (Ali-Cohen 1889) to the genus *Arthrobacter* as *Arthrobacter agilis* comb. nov. and emendation of the genus *Arthrobacter*. Int. J. Syst. Bacteriol. *45*, 837–839.

Koma, D., Hasumi, F., Yamamoto, E., Ohta, T., Chung, S.-Y., and Kubo, M. (2001). Biodegradation of long-chain n-paraffins from waste oil of car engine by *Acinetobacter* sp. J. Biosci. Bioeng. *91*, 94–96.

Komatsu, T., Omori, T., and Kodama, T. (1993). Microbial degradation of the polycyclic aromatic hydrocarbons acenaphthene and acenaphthylene by a pure bacterial culture. Biosci. Biotech. Bioch. *57*, 864–865.

Krivobok, S., Kuony, S., Meyer, C., Louwagie, M., Willison, J.C., and Jouanneau, Y. (2003). Identification of pyrene-induced proteins in *Mycobacterium* sp. strain 6YP1: evidence for two ring-hydroxylating dioxygenases. J. Bacteriol. *185*, 3828–3841.

Larkin, M.J., Kulakov, L.A., and Allen C.C.R. (2005). Biodegradation and *Rhodococcus* – masters of catabolic versatility. Curr. Opin. Biotech. *16*, 282–290.

Laurie, A.D., and Lloyd-Jones, G. (1999). The *phn* genes of *Burkholderia* sp. strain RP007 constitute a divergent gene cluster for polycyclic aromatic hydrocarbon catabolism. J. Bacteriol. *181*, 531–540.

Lee, E.-H., Kim, J., Cho, K.-S., Ahn, Y.G., and Hwang, G.-S. (2010). Degradation of hexane and other recalcitrant hydrocarbons by a novel isolate, *Rhodococcus* sp. EH831. Environ. Sci. Pollut. R. *17*, 64–77.

Lei, W., Yun, T., Shuo, W., Rulin, L., Muzhi, L., Yan, Z., Fenglai, L., and Lu, F. (2006). Isolation and characterization of a novel thermophilic *Bacillus* strain degrading long-chain n-alkanes. Extremophiles *10*, 347–356.

Lévy-Frébault, V.V., and Portaels, F. (1992). Proposed minimal standards for the genus *Mycobacterium* and for description of new slowly growing *Mycobacterium* species. Int. J. Syst. Bacteriol. *42*, 315–323.

Li, P., Wang, X., Stagnitti, F., Li, L., Gong, Z., Zhang, H., Xiong, X., and Austin, C. (2005). Degradation of phenanthrene and pyrene in soil slurry reactors with immobilized bacteria *Zoogloea* sp. Environ. Eng. Sci. *22*, 390–399.

Lin, J.-E., and Wang, H.Y. (1991). Degradation of pentachlorophenol by non-immobilized, immobilized and co-immobilized *Arthrobacter* cells. J. Ferment. Bioeng. 72, 311–314.

Liu, Y., Zhang, J., and Zhang, Z. (2004). Isolation and characterization of polycyclic aromatic hydrocarbons-degrading *Sphingomonas* sp. strain ZL5. Biodegradation *15*, 205–212.

Liu, Y.-C., Zhou, T.-T., Zhang, J., Xu, L., Zhang, Z.-H., Shen, Q.-R., and Shen, B. (2009). Molecular characterization of the alkB gene in the thermophilic *GeoBacillus* sp. strain MH-1. Res. Microbiol. *160*, 560–566.

Loveland-Curtze, J., Sheridan, P.P., Gutshall, K.R., and Brenchley, J.E. (1999). Biochemical and phylogenetic analyses of phychrophilic isolates belonging to the *Arthrobacter* subgroup and description of *Arthrobacter psychrolactophilus*, sp. nov. Arch. Microbiol. *171*, 355–363.

Mallick, S., Chatterjee, S., and Dutta, T.K. (2007). A novel degradation pathway in the assimilation of phenanthrene by *Staphylococcus* sp. strain PN/Y via *meta*-cleavage of 2-Hydroxy-1-naphthoic acid: formation of Trans-2,3-dioxo-5-(2′-hydroxyphenyl)-pent-4-enoic acid Microbiology *153*, 2104–2115.

Mallory, L.M., Yuk, C.-S., Liang, L.-N., and Alexander, M. (1983). Alternative prey: a mechanism for elimination of bacterial species by protozoa. Appl. Environ. Microbiol. *46*, 1073–1079.

Manohar, S., and Karegoudar, T.B. (1998). Degradation of naphthalene by cells of *Pseudomonas* sp. strain NGK 1 immobilized in Alginate, Agar and polyacrylamide. Appl. Microbiol. Biotech. *49*, 785–792.

Marchant, R., Sharkey, F.H., Banat, I.M., Rahman, T.J., and Perfumo, A. (2006). The degradation of n-hexadecane in soil by thermophilic *Geobacilli*. FEMS Microbiol. Ecol. *56*, 44–54.

Margesin, R., and Schinner, F. (2001). Biodegradation and bioremediation of hydrocarbons in extreme environments. Appl. Microbiol. Biot. *56*, 650–663.

Margesin, R., Labbé, D., Schinner, F., Greer, C.W., and Whyte, L.G. (2003). Characterization of hydrocarbon-degrading microbial populations in contaminated and pristine alpine soils. Appl. Environ. Microbiol. *69*, 3085–3092.

Marks, T.S., Smith, A.R.W., and Quirk, A.V. (1984). Degradation of 4-chlorobenzoic acid by *Arthrobacter* sp. Appl. Environ. Microbiol. *48*, 1020–1025.

Master, E.R., and Mohn, W.W. (1998). Psychrotolerant Bacteria isolated from arctic soil that degrade polychlorinated biphenyls at low temperatures. Appl. Environ. Microbiol. *64*, 4823–4829.

Meintanis, C., Chalkou, K.I., Kormas, K.Ar., and Karagouni, A.D. (2006). Biodegradation of crude oil by thermophilic bacteria isolated from a volcano island. Biodegradation *17*, 3–9.

Miller, C.D., Hall, K., Liang, Y.N., Nieman, K., Sorensen, D., Issa, B., Anderson, A.J., and Sims, R.C. (2004). Isolation and characterization of polycyclic aromatic hydrocarbon-degrading *Mycobacterium* isolates from soil. Microbial. Ecol. *48*, 230–238.

Miyata, N., Iwahori, K., Foght, J.M., and Gray, M.R. (2004). Saturable, energy-dependent uptake of phenanthrene in aqueous phase by *Mycobacterium* sp. strain RJGII-135. Appl. Environ. Microbiol. *70*, 363–369.

Moody, J.D., Freeman, J.P., and Cerniglia, C.E. (2005). Degradation of benz[a]anthracene by *Mycobacterium vanbaalenii* strain PYR-1. Biodegradation *16*, 513–526.

Moody, J.D., Fu, P.P., Freeman, J.P., and Cerniglia, C.E. (2003). Regio- and stereoselective metabolism of 7,12-dimethylbenz[a]anthracene by *Mycobacterium vanbaalenii* PYR-1. Appl. Environ. Microbiol. *69*, 3924–3931.

Moody, J.D., Freeman, J.P., Fu, P.P., and Cerniglia, C.E. (2004). Degradation of benzo[a]pyrene by *Mycobacterium vanbaalenii* PYR-1. Appl. Environ. Microbiol. *70*, 340–345.

Mrozik, A., Labuzek, S., and Piotrowska-Seget, Z. (2005). Changes in fatty acid composition in *Pseudomonas putida* and *Pseudomonas stutzeri* during naphthalene degradation. Microbiol. Res. *160*, 149–157.

Mueller, J.G., Chapman, P.J., and Pritchard, P.H. (1989). Action of a fluoranthene-utilizing bacterial community on polycyclic aromatic hydrocarbon components of creosote. Appl. Environ. Microbiol. *55*, 3085–3090.

Mueller, J.G., Chapman, P.J., Blattmann, B.O., and Pritchard, P.H. (1990). Isoaltion and characterization of a fluoranthene-utilizing strain of *Pseudomonas paucimobilis*. Appl. Environ. Microbiol. *56*, 1079–1086.

Mueller, J.G., Lantz, S.E., Blattmann, B.O., and Chapman, P.J. (1991). Bench-scale evaluation of alternative biological treatment process for the remediation of pentachlorophenol and creosote contaminated materials: solid phase bioremediation. Environ. Sci. Technol. *25*, 1045–1055.

Murakami, Y., and Alexander, M. (1989). Destruction and formation of toxins by one bacterial species affect biodegradation by a second species. Biotechnol. Bioeng. *33*, 832–838.

Negrete-Raymond, A.C., Weder, B., and Wackett, L.P. (2003). Catabolism of arylboronic acids by *Arthrobacter nicotinovorans* strain PBA. Appl. Environ. Microbiol. *69*, 4263–4267.

Neher, T.M., and Lueking, D.R. (2009). *Pseudomonas fluorescens* ompW: plasmid localization and requirement for naphthalene uptake. Can. J. Microbiol. *55*, 553–563.

Nhi-Chong, L.T., Mikolasch, A., Awe, S., Sheikhany, H., Klenk, H.-P., and Schauer, F. (2010). Oxidation of aliphatic, branched chain, and aromatic hydrocarbons by *Nocardia cyriacigeorgica* isolated from oil-polluted sand samples collected in the saudi arabian desert. J. Bas. Microb. *50*, 241–253.

Nicaido, H., Kim, S.-H., and Rosenberg, E.Y. (1993). Physical organization of lipids in the cell wall of *Mycobacterium chelonae*. Mol. Microbiol. *8*, 1025–1030.

Nicolau, E., Kuhn, L., Marchal, R., and Jouanneau, Y. (2009). Proteomic investigation of enzymes involved in 2-ethylhexyl nitrate biodegradation in *Mycobacterium austroafricanum* IFP 2173. Res. Microbiol. *160*, 838–847.

Okoh, A.I. (2006). Biodegradation alternative in the cleanup of petroleum hydrocarbon pollutants. Biotechnol. Mol. Biol. Rev. *1*, 38–50.

Ortega-Calvo, J.-J., and Alexander, M. (1994). Roles of Bacterial attachment and spontaneous partitioning in the biodegradation of naphthalene initially present in nonaqueous-phase lipids. Appl. Environ. Microbiol. *60*, 2643–2646.

Pagnout, C., Frache, G., Poupin, P., Maunit, B., Muller, J.-F., and Férard, J.-F. (2007). Isolation and characterization of a gene cluster involved in pah degradation in *Mycobacterium* sp. strain SNP11: expression in *Mycobacterium smegmatis* mc2155. Res. Microbiol. *158*, 175–186.

Pandey, G., Paul, D., and Jain, R.K. (2003). Branching of o-nitrobenzoate degradation pathway in *Arthrobacter protophormiae* RKJ100: identification of new intermadiates. FEMS Microbiol. Lett. *229*, 231–236.

Park, J.K., and Chang, H.N. (2000). Microencapsulation of microbial cells. Biotechnol. Adv. *18*, 303–319.

Paul, D., Chauhan, A., Pandey, G., and Jain, R.K. (2004). Degradation of p-hydroxybenzoate via protocatechuate in *Arthrobacter protophormiae* RKJ100 and *Burkholderia cepacia* RKJ200. Curr. Sci. *87*, 1263–1268.

Peng, R.-H., Xiong, A.-S., Xue, Y., Fu, X.-Y., Gao, F., Zhao, W., Tian, Y.-S., and Yao, Q.-H. (2008). Microbial biodegradation of polyaromatic hydrocarbons. FEMS Microbiol. Rev. *32*, 927–955.

Pepi, M., Minacci, A., Di Cello, F., Baldi, F., and Fani, R. (2003). Long-term analysis of diesel fuel consumption in a co-culture of *Acinetobacter venetianus*, *Pseudomonas putida* and *Alcaligenes faecalis*. Antonie van Leeuwenhoek *83*, 3–9.

Phillips, D.H. (1983). Fifty years of benzo[a]pyrene. Nature *303*, 468–472.

Pinyakong, O., Habe, H., Kouzuma, A., Nojiri, H., Yamane, H., and Omori, T. (2004). Isolation and characterization of genes encoding polycyclic aromatic hydrocarbon dioxygenase from acenaphthene and acenaphthylene degrading *Sphingomonas* sp. strain A4. FEMS Microbiol. Lett. *238*, 297–305.

Pinyakong, O., Habe, H., Supaka, N., Pinpanichkarn, P., Juntongjin, K., Yoshida, T., Furihata, K., Nojiri, H., Yamane, H., and Omori, T. (2000). Identification of novel metabolites in the degradation of phenanthrene by *Sphingomonas* sp. strain P2. FEMS Microbiol. Lett. *191*, 115–121.

Pizzul, L., Sjögren, A., del Pilar Castillo, M., and Stenström, J. (2007). Degradation of polycyclic aromatic hydrocarbons in soil by a two-step sequential treatment. Biodegradation *18*, 607–616.

Prabhu, Y., and Phale, P.S. (2003). Biodegradation of phenanthrene by *Pseudomonas* sp. strain PP2: novel metabolic pathway, role of biosurfactant and cell surface hydrophobicity in hydrocarbon assimilation. Appl. Microbiol. Biotech. *61*, 342–351.

Rahman, K.S.M., Lakshmanaperumalsamy, R.T., and Banat, I.M. (2002). Occurence of crude oil degrading bacteria in gasoline and diesel station soils. J. Basic Microb. *42*, 284–291.

Rehman, K., Hertkorn, N., and Kettrup, A.A. (1999). Bacterial fluoranthene degradation-indication for a novel degradation pathway. In Novel Approaches for Bioremediation of Organic Pollution, Fass, R., Flashner, Y., and Reuvery, S, eds. Proceedings of the 42nd Oholo Conference Eilat, Israel (3.5–7.5.1998) (Kluwer Academic/Plenum: New York, Boston, Dordrecht, London, Moscow), pp.39–46.

Rentz, J.A., Alvarez, P.J.J., and Schnoor, J.L. (2008). Benzo[a]pyrene degradation by *Sphingomonas yanoikuyae* JAR02. Environ. Pollut. *151*, 669–677.

Richard, J.Y., and Vogel, T.M. (1999). Characterization of a soil bacterial consortium capable of degrading diesel fuel. Int. Biodeter. Biodegr. *44*, 93–100.

Rodrigues, A.C., Wuertz, S., Brito, A.G., and Melo, L.F. (2005). Fluorene and phenanthrene uptake by *Pseudomonas putida* ATCC 17514: kinetics and physiological aspects. Biotechnol. Bioeng. *90*, 281–289.

Rojo, F. (2009). Degradation of alkanes by bacteria. Environ. Microbiol. *11*, 2477–2490.

Rosselló-Mora, R., and Amann, R. (2001). The species concept for Procaryotes. FEMS Microbiol. Rev. *25*, 39–67.

Saito, A., Iwabuchi, T., and Harayama, S. (2000). A novel phenanthrene dioxygenase from *Nocardioides* sp. strain KP7: Expression in *Escherichia coli*. J. Bacteriol. *182*, 2134–2141.

Sakai, Y., Maeng, J.H., Tani, Y., and Kato, N. (1994). Use of long-chain *n*-alkanes (C13-C44) by an isolate, *Acinetobacter* sp. M-1. Biosci. Biotech. Biochim. *58*, 2128–2130.

Sakai, Y., Takahashi, H., Wakasa, Y., Kotani, T., Yurimoto, H., Miyachi, N., VanVeldhoven, P.P., and Kato, N. (2004). Role of α-methylacyl coenzyme a racemase in the degradation of methyl-branched alkanes by *Mycobacterium* sp. strain P101. J. Bacteriol. *186*, 7214–7220.

Samanta, S.K., Chakraborti, A.K., and Jain, R.K. (1999). Degradation of phenanthrene by different bacteria: evidence for novel transformation sequences involving the formation of 1-naphthol. Appl. Microbiol. Biot. *53*, 98–107.

Samanta, S.K., Singh, O.V., and Jain, R.K. (2002). Polycyclic aromatic hydrocarbons: environmental pollution and bioremediation. Trends Biotechnol. *20*, 243–248.

Schneider, J., Grosser, R., Jayasimhulu, K., Xue, W., and Warshawsky, D. (1996). Degradation of pyrene, benz[a]anthracene and benzo[a]pyrene by *Mycobacterium* sp., strain RJGII-135, isolated from a former coal gasification site. Appl. Environ. Microbiol. *62*, 13–19.

Scholtz, R., Messi, F., Leisinger, T., and Cook, A.M. (1988). Three dehalogenases and physiological restraints in the biodegradation of haloalkanes by *Arthrobacter* sp. strain HA1. Appl. Environ. Microbiol. *54*, 3034–3038.

Schuler, L., Jouanneau, Y., Ní Chadhain, S.M., Meyer, C., Pouli, M., Zylstra, G.J., Hols, P., and Agathos, S.N. (2009). Characterization of a ring-hydroxylating dioxygenase from phenanthrene-degrading *Sphingomonas* sp. strain LH128 able to oxidize benz[a] anthracene. Appl. Microbiol. Biot. *83*, 465–475.

Schuler, L., Ní Chadhain, S.M., Jouanneau, Y., Meyer, C., Zylstra, G.J., Hols, P., and Agathos, S.N. (2008). Characterization of a novel angular dioxygenase from fluorene-degrading *Sphingomonas* sp. strain LB126. Appl. Environ. Microbiol. *74*, 1050–1057.

Schwartz, E., and Scow, K.M. (2001). Repeated inoculation as a strategy for the remediation of low concentrations of phenanthrene in soil. Biodegradation *12*, 201–207.

Seo, J.-S., Keum, Y.-S., Cho, I.K., and Li, Q.X. (2006a). Degradation of dibenzothiophene and carbazole by *Arthrobacter* sp. P1-1. Int. Biodeter. Biodegradation *58*, 36–43.

Seo, J-S., Keum, Y-S., Hu, Y., Lee, S-E., and Li, Q.X. (2006b). Phenanthrene degradation in *Arthrobacter* sp. P1-1: initial 1,2-, 3,4- and 9,10-dioxygenation, and *meta*- and *ortho*-cleavages of naphthalene-1,2-diol after its formation from naphthalene-1,2-dicarboxylic acid and hydroxyl naphthoic acids. Chemosphere *65*, 2388–2394.

Seo, J.-S., Keum, Y.-S., Hu, Y., Lee, S.-E., and Li, Q.X. (2007). Degradation of phenanthrene by *Burkholderia* sp. C3: initial 1,2- and 3,4-dioxygenation and *Meta*- and *Ortho*-cleavage of naphthalene-1,2-diol. Biodegradation *18*, 123–131.

Sheng, X.F., He, L.Y., Zhou, L., and Shen, Y.Y. (2009). Characterization of *Microbacterium* sp. F10a and its role in polycyclic aromatic hydrocarbon removal in low-temperature soil. Can. J. Microbiol. *55*, 529–535.

Shimura, M., Mukerjee-Dhar, G., Kimbara, K., Nagato, H., Kiyohara, H., and Hatta, T. (1999). Isolation and characterization of a thermophilic *Bacillus* sp. JF8 capable of degrading polychlorinated biphenyls and naphthalene. FEMS Microbiol. Lett. *178*, 87–93.

Shuttleworth, K.L., and Cerniglia, C.E. (1996). Bacterial degradation of low concentrations of phenanthrene and inhibition by naphthalene. Microbiol. Ecol. *31*, 305–317.

Shuttleworth, K.L., Sung, J., Kim, E., and Cerniglia, C.E. (2000). Physiological and genetic comparison of two aromatic hydrocarbon-degrading *Sphingomonas* strains. Mol. Cell. *10*, 199–205.

Sikkema, J., de Bont, J.A., and Poolman, B. (1994). Interactions of cyclic hydrocarbons with biological membranes. J. Biol. Chem. *269*, 8022–8028.

Sikkema, J., de Bont, J.A.M., and Poolman, B. (1995). Mechanisms of membrane toxicity of hydrocarbons. Microbiol. Rev. *59*, 201–222.

Singleton, D.R., Ramirez, L.G., and Aitken, M.D. (2009). Characterization of a polycyclic aromatic hydrocarbon degradation gene cluster in a phenanthrene-degrading *Acidovorax* strain. Appl. Environ. Microbiol. *75*, 2613–2620.

Sokolovska, I., Rozenberg, R., Riez, C., Rouxhet, P.G., Agathos, S.N., and Wattiau, P. (2003). Carbon source-induced modifications in the mycolic acid content and cell wall permeability of *Rhodococcus erythropolis* E1. Appl. Environ. Microbiol. *69*, 7019–7027.

Sorokhoh, N.A., Ibrahim, A.S., Ghannoum, M.A., and Radwan, S.S. (1993). High temperature hydrocarbon degradation by *Bacillus stearothermophilus* from oil-polluted Kuwaiti desert. Appl. Microbiol. Biot. *39*, 123–126.

Spain, J.C., and Nishino, S.F. (1987). Degradation of 1,4-Dichlorobenzene by a *Pseudomonas* sp. Appl. Environ. Microbiol. *53*, 1010–1019.

Stach, J.E.M., and Burns, R.G. (2002). Enrichment versus biofilm culture: a functional and phylogenetic comparison of polycyclic aromatic hydrocarbon-degrading microbial communities. Environ. Microbiol. *4*, 169–182.

Story, S.P., Parker, S.H., Kline, J.D., Tzeng, T.-R.J., Mueller, J.G., and Kline, E.L. (2000). Identification of four structural genes and two putative promoters necessary for utilization of naphthalene, phenanthrene, and fluoranthene by *Sphingomonas paucimobilis* var. EPA505. Gene *260*, 155–169.

Stringfellow, W.T., and Aitken, M.D. (1994). Comparative physiology of phenanthrene degradation by two similar pseudomonads isolated from a creosote-contaminated soil. Can. J. Microbiol. *40*, 432–438.

Stringfellow, W.T., and Aitken, M.D. (1995). Competitive metabolism of naphthalene, methylnaphthalenes, and fluorene by phenanthrene-degrading Pseudomonads. Appl. Environ. Microbiol. *61*, 357–362.

Stroud, J.L., Paton, G.I., and Semple, K.T. (2007). Importance of chemical structure on the development of hydrocarbon catabolism in soil. FEMS Microbiol. Lett. *272*, 120–126.

Tao, X.-Q., Lu, G.-N., Dang, Z., Yang, C., and Yi, X.-Y. (2007). A phenanthrene-degrading strain *Sphingomonas* sp. GY2B isolated from contaminated soils. Process Biochem. *42*, 401–408.

Throne-Holst, M., Markussen, S., Winnberg, A., Ellingsen, T.E., Kotlar, H.K., and Zotchev, S.B. (2006). Utilization of n-alkanes by a newly isolated strain of *Acinetobacter venetianus*: The role of two AlkB-type alkane hydroxylases. Appl. Microbiol. Biotech. *72*, 353–360.

Throne-Holst, M., Wentzel, A., Ellingsen, T.E., Kotlar, H-K., and Zotchev, S.B. (2007). Identification of novel genes involved in long-chain n-alkane degradation by *Acinetobacter* sp. strain DSM 17874. Appl. Environ. Microbiol. 73, 3327–3332.

Trenz, S.P., Engesser, K.H., Fischer, P., and Knackmuss, H.-J. (1994). Degradation of fluorene by *Brevibacterium* sp. strain DPO 1361: a novel C-C bond cleavage mechanism VIA 1,10-dihydro-1,10-dihydroxyfluoren-9-one. J. Bacteriol. *176*, 789–795.

Tsitko, I.V., Zaitsev, G.M., Lobanok, A.G., and Salkinoja-Salonen, M.S. (1999). Effect of aromatic compounds on cellular fatty acid composition of *Rhodococcus opacus*. Appl. Environ. Microbiol. *65*, 853–855.

Turnbull, G.A., Ousley, M., Walker, A., Shaw, E., and Morgan, J.A.W. (2001). Degradation of substituted phenylurea herbicides by *Arthrobacter globiformis* strain D47 and characterization of a plasmid-associated hydrolase gene, puhA. Appl. Environ. Microbiol. *67*, 2270–2275.

Unell, M., Kabelitz, N., Jansson, J.K., and Heipieper, H.J. (2007). Adaptation of the psychrotroph *Arthrobacter chlorophenolicus* A6 to growth temperature and the presence of phenols by changes in the anteiso/iso ratio of branched fatty acids. FEMS Microbiol. Lett. *266*, 138–143.

Uyttebroek, M., Breugelmans, P., Janssen, M., Wattiau, P., Joffe, B., Karlson, U., Ortega-Calvo, J.-J., Bastiaens, L., Ryngaert, A., Hausner, M., *et al.* (2006). Distribution of the *Mycobacterium* community and polycyclic aromatic hydrocarbons (PAHs) among different size fractions of a long-term PAH-contaminated soil. Environ. Microbiol. *8*, 836–847.

Uyttebroek, M., Vermeir, S., Wattiau, P., Ryngaert, A., and Springael, D. (2007). Characterization of cultures enriched from acidic polycyclic aromatic hydrocarbon-contaminated soil for growth on pyrene at low pH. Appl. Environ. Microbiol. *73*, 3159–3164.

Uz, I., Duan, Y.P., and Ogram, A. (2000). Characterization of the naphthalene-degrading bacterium *Rhodococcus opacus* M213. FEMS Microbiol. Lett. *185*, 231–238.

Van Beilen, J.B., Li, Z., Duetz, W.A., Smits, T.H.M., and Witholt, B. (2003). Diversity of alkane hydroxylase systems in the environment. Oil Gas Sci. Technol. Rev. IFP. *58*, 427–440.

Van Beilen, J.B., and Funhoff, E.G. (2007). Alcane hydroxylases involved in microbial alkane degradation. Appl. Microbiol. Biot. *74*, 13–21.

Van Hamme, J.D., Odumeru, J.A., and Ward, O.P. (2000). Community dynamics of a mixed-bacterial culture growing on petroleum hydrocarbons in batch cultures. Can. J. Microbiol. *46*, 441–450.

Van Hamme, J.D., Singh, A., and Ward, O.P. (2003). Recent advances in petroleum microbiology. Microbiol. Mol. Biol. R. *67*, 503–549.

Van Herwijnen, R., Wattiau, P., Bastiaens, L., Daal, L., Jonker, L., Springael, D., Govers, H.A.J., and Parsons, J.R. (2003). Elucidation of the metabolic pathway of fluorene and cometabolic pathways of phenanthrene, fluoranthene, anthracene and dibenzothiophene by *Sphingomonas* sp. LB126. Res. Microbiol. *154*, 199–206.

Venosa, A.D., and Zhu, X. (2003). Biodegradation of crude oil contaminating marine shorelines and fresh water wetlands. Spill Sci. Technol. B. *8*, 163–178.

Viamajala, S., Peyton, B.M., Richards, L.A., and Petersen, J.N. (2007). Solubilization, solution equilibria, and biodegradation of PAH's under thermophilic conditions. Chemosphere *66*, 1094–1106.

Vogel, M.T. (1996). Bioaugmentation as a soil bioremediation approach. Curr. Opin. Biotech. *7*, 311–316.

Von der Weid, I., Marques, J.M., Cunha, C.D., Lippi, R.K., Dos Santos, S.C.C., Rosado, A.S., Lins, U., and Seldin, L. (2007). Identification and biodegradation potential of a novel strain of *Dietzia cinnamea* isolated from a petroleum-contaminated tropical soil. System. Appl. Microbiol. *30*, 331–339.

Wakeham, S.G., Schaffner, C., and Giger, W. (1980). Polycyclic aromatic hydrocarbons in recent lake sediments-I. Compounds having anthropogenetic origins. Geochim. Cosmochim. Ac. *44*, 403–413.

Walter, U., Beyer, M., Klein, J., and Rehm, H.-J. (1991). Degradation of pyrene by *Rhodococcus* sp. UW1. Appl. Microbiol. Biot. *34*, 671–676.

Weber, F.J., and de Bont, J.A.M. (1996). Adaptation mechanisms of microorganisms to the toxic effects of organic solvents on membranes. Biochim. Biophys. Acta. *1286*, 225–245.

Weir, S.C., Dupuis, S.P., Providenti, M.A., Lee, H., and Trevors, J.T. (1995). Nutrient-enhanced survival of and phenanthrene mineralization by alginate-encapsulated and free *Pseudomonas* sp. UG14Lr cells in creosote-contaminated soil slurries. Appl. Microbiol. Biot. *43*, 946–951.

Weissenfels, W.D., Beyer, M., and Klein, J. (1990). Degradation of phenanthrene, fluorene, and fluoranthene by pure bacterial cultures. Appl. Microbiol. Biot. *32*, 479–484.

Wentzel, A., Ellingsen, T.E., Kotlar, H.-K., Zotchev, S.B., and Throne-Holst, M. (2007). Bacterial metabolism of long-chain n-alkanes. Appl. Microbiol. Biot. *76*, 1209–1221.

White, K.L. (1986). An overview of immunotoxicology and carcinogenic polycyclic aromatic hydrocarbons. Environ. Carcinog. Rev. *C4*, 163–202.

Whitman, B.E., Lueking, D.R., and Mihelcic, J.R. (1998). Naphthalene uptake by a *Pseudomonas fluorescens* isolate. Can. J. Microbiol. *44*, 1086–1093.

Whyte, L.G., Bourbonnière, L., and Greer, C.W. (1997). Biodegradation of petroleum hydrocarbons by psychrotrophic *Pseudomonas* strains possessing both alkane (alk) and naphthalene (nah) catabolic pathways. Appl. Environ. Microbiol. *63*, 3719–3723.

Whyte, L.G., Hawari, J., Zhou, E., Bourbonnière, L., Inniss, W.E., and Greer, C.W. (1998). Biodegradation of Variable-chain-length alkanes at low temperatures by a psychrotrophic *Rhodococcus* sp. Appl. Environ. Microbiol. *64*, 2578–2584.

Whyte, L.G., Smits, T.H.M., Labbe, D., Witholt, B., Greer, C.W., and van Beilen, J.B. (2002). Gene Cloning and characterization of multiple alkane hydroxylase systems in *Rhodococcus* strains Q15 and NRRL B-16531. Appl. Environ. Biot. *68*, 5933–5942.

Wick, L.Y., Colangelo, T., and Harms, H. (2001). Kinetics of mass-transfer limited bacterial growth on solid PAHs. Environ. Sci. Technol. *35*, 354–361.

Wick, L.Y., Ruiz de Munain, A., Springael, D., and Harms, H. (2002a). Responses of *Mycobacterium* sp. 501T to the low bioavailability of solid anthracene. Appl. Microbiol. Biot. *58*, 378–385.

Wick, L.Y., Wattiau, P., and Harms, H. (2002b). Influence of the growth substrate on the mycolic acid profiles of Mycobacteria. Environ. Microbiol. *4*, 612–616.

Wick, L.Y., Pasche, N., Bernasconi, S.M., Pelz, O., and Harms, H. (2003a). Characterization of multiple-substrate utilization by anthracene-degrading *Mycobacterium frederiksbergense* LB501T. Appl. Environ. Microbiol. *69*, 6133–6142.

Wick, L.Y., Pelz, O., Bernasconi, S.M., Andersen, N., and Harms, H. (2003b). Influence of the growth substrate on ester-linked phospho- and glycolipid fatty acids of pah-degrading *Mycobacterium* sp. LB501T. Environ. Microbiol. *5*, 672–680.

Wiesel, I., Wübker, S.M., and Rehm, H.J. (1993). Degradation of polycyclic aromatic hydrocarbons by an immobilized mixed bacterial culture. Appl. Microbiol. Biot. *39*, 110–116.

Willison, J.C. (2004). Isolation and Characterization of a novel sphingomonad capable of growth with chrysene as sole carbon and energy source. FEMS Microbiol. Lett. *241*, 143–150.

Willumsen, P., Karlson, U., Stackebrandt, E., and Kroppenstedt, R.M. (2001). *Mycobacterium frederiksbergense* sp. nov., a novel polycyclic aromatic hydrocarbon-degrading *Mycobacterium* species. Int. J. Syst. Evol. Micr. *51*, 1715–1722.

Wongsa, P., Tanaka, M., Ueno, A., Hasanuzzaman, M., Yumoto, I., and Okuyama, H. (2004). Isolation and characterization of novel strains of *Pseudomonas aeruginosa* and *Serratia marcescens* possessing high efficiency to degrade gasoline, kerosene, diesel oil, and lubricating oil. Cur. Microbiol. *49*, 415–422.

Xia, Y., Min, H., Rao, G., Lv, Z.-M., Liu, J., Ye, Y.-F., and Duan, X.-J. (2005). Isolation and characterisation of phenanthrene-degrading *Sphingomonas paucimobilis* strain ZX4. Biodegradation *16*, 393–402.

Yamazoe, A., Yagi, O., and Oyaizu, H. (2004). Degradation of Polycyclic aromatic hydrocarbons by a newly isolated dibenzofuran-utilizing *Janibacter* sp. strain YY-1. Appl. Microbiol. Biot. *65*, 211–218.

Ye, D., Siddiqi, M.A., Maccubbin, A.E., Kumar, S., and Sikka, H.C. (1996). Degradation of polynuclear aromatic hydrocarbons by *Sphingomonas paucimobilis*. Environ. Sci. Technol. *30*, 136–142.

Yuste, L., Corbella, M.E., Turiégano, M.J., Karlson, U., Puyet, A., and Rojo, F. (2000). Characterization of bacterial strains able to grow on high molecular weight mass residues from crude oil processing. FEMS Microbiol. Ecol. *32*, 69–75.

Zeinali, M., Vossoughi, M., and Ardestani, S.K. (2008). Degradation of phenanthrene and anthracene by *Nocardia otitidiscavarum* strain TSH1, a moderately thermophilic bacterium. J. Appl. Microbiol. *105*, 398–406.

Zeinali, M., Vossoughiz, M., and Ardestani, S.K. (2007). Characterisation of a moderate thermophilic Nocardia species able to grow on polycyclic aromatic hydrocarbons. Lett. Appl. Microbiol. *45*, 622–628.

Zhang, G.-Y., Ling, J.-Y., Sun, H.-B., Luo, J., Fan, Y.-Y., and Cui, Z.-J. (2009). Isolation and characterization of newly isolated polycyclic aromatic hydrocarbons-degrading *Janibacter anophelis* strain JY11. J. Hazard. Mater. *172*, 580–586.

Zhang, H., Kallimanis, A., Koukkou, A.I., and Drainas, C. (2004). Isolation and characterizaton of novel bacteria degrading polycyclic aromatic hydrocarbons from polluted Greek soils. Appl. Microbiol. Biot. *65*, 124–131.

Zhao, B., Wang, H., Mao, X., and Li, R. (2009). Biodegradation of phenanthrene by a halophilic bacterial consortium under aerobic conditions. Curr. Microbiol. *58*, 205–210.

Zhao, H.-P., Wang, L., Ren, J.-R., Li, Z., Li, M., and Gao, H.-W. (2008). Isolation and characterization of phenanthrene-degrading strains *Sphingomonas* sp. ZP1 and *Tistrella* sp. ZP5. J. Hazard. Mater. *152*, 1293–1300.

Zhao, H.-P., Wu, Q.-S., Wang, L., Zhao, X.-T., and Gao, H.-W. (2009) Degradation of phenanthrene by bacterial strain isolated from soil in oil refinery fields in Shanghai China. J. Hazard. Mater. *164*, 863–869.

Zhou, N.-Y., Fuenmayor, S.L., and Williams, P.A. (2001). nag Genes of *Ralstonia* (formerly *Pseudomonas*) sp. strain U2 encoding enzymes for gentisate catabolism. J. Bacteriol. *183*, 700–708.

Bioremediation of PAH-contaminated Sites: From Pathways to Bioreactors

6

Maria Pouli and Spiros N. Agathos

Abstract

Polycyclic aromatic hydrocarbons (PAHs) are widespread pollutants found in industrial sites linked with petroleum, gas production or other activities involving incomplete combustion of organic matter. These compounds constitute a priority for treatment of contaminated soils and sediments owing to their toxicity and carcinogenicity. Microbial degradation represents an important mechanism of PAH removal and can be used for the treatment of contaminated sites by *in situ* or *ex situ* bioremediation. This technology is taking advantage of a few established peripheral catabolic pathways converting different PAHs into a limited number of central intermediates. Peripheral pathways are typically initiated by dioxygenases, oxidizing PAHs into dihydroxylated intermediates, which, in turn, are catabolized further and degraded via central pathways into TCA cycle metabolites. However, both the hydrophobic characteristics of most PAHs as well as the physicochemical properties of soils diminish the bioavailability of these pollutants and thus limit the degradation capacity of naturally occurring microorganisms for bioremediation of contaminated sites. The success of biotreatment interventions can be enhanced by measures promoting bioavailability (e.g. addition of surfactants or solvents) and/or boosting microbial activity (e.g. by biostimulation or bioaugmentation). The same strategies can be further optimized by implementing appropriately designed bioreactors for faster and more complete removal of recalcitrant PAHs from contaminated soils and sediments.

Introduction

PAHs constitute a widespread class of pollutants released into the environment both by natural processes and anthropogenic activities. They occur in massive quantities from the incomplete combustion of any organic material. PAHs assemble aromatic rings in several arrangements and they comprise exclusively carbon (C) and hydrogen (H) atoms (Fig. 6.1), though substitutions by nitrogen (N) and oxygen commonly occur (heterocyclic compounds). Such chemical structures are characterized by physicochemical properties which turn these substances to highly hydrophobic, hardly bioavailable and persistent pollutants. For these reasons which will be explained further below, these compounds are recalcitrant pollutants with a high impact both on the environment and on human health.

Origin and fate in the environment

Polycyclic compounds are mostly synthetic substances though they may occur as a result of natural phenomena. Their presence in the environment can be a result of pyrolytic and petrogenic processes or even diagenesis. Pyrolysis refers to the incomplete combustion of organic matter including volcanic eruptions, forest fires, burning of coal, wood or crude oil, vehicle exhausts, tobacco smoke, cooking etc. During pyrolysis, the origin of the organic matter and the temperature range affects the distribution and composition of polyaromatic molecules which, in

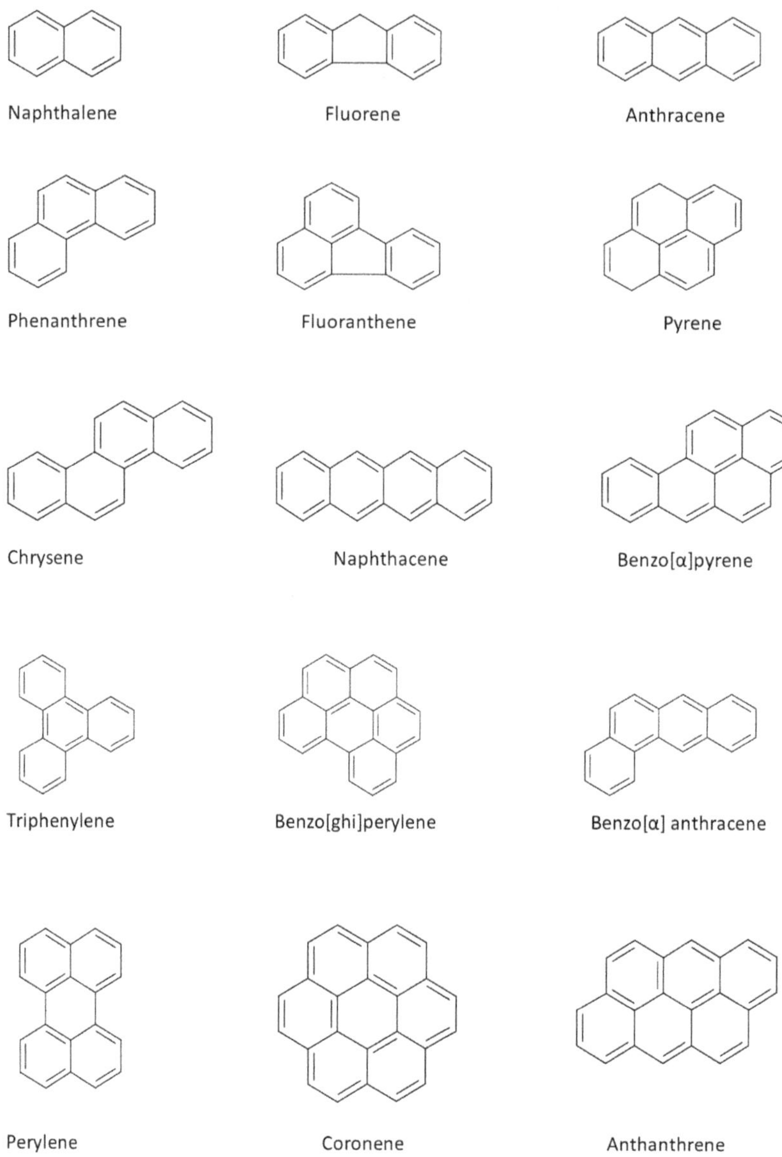

Figure 6.1 Structures of different polycyclic aromatic hydrocarbons.

general, occur as unsubstituted structures while the higher molecular weight (MW) compounds predominate after pyrolytic events (Bouchez-Naitali *et al.*, 2005). Power plants, incineration units, waste treatment processes, accidental spillage and in general the anthropogenic activity in industrialized and highly urbanized areas, are the main contributors to the large amounts of PAHs released in the environment.

Some possible ways through which PAHs enter the environment include volatilization, photo-oxidation, bioaccumulation, leaching, chemical oxidation, biological degradation and adsorption on soil particles. In general, PAHs have low vapour pressures and thus, they do not tend to volatilize apart from the lighter ones that are characterized by low MW (Cerniglia, 1992). The temperature is the driving force influencing the volatilization process and once the polyaromatics enter the gas phase, they can get associated with airborne particulate matter and travel over long distances before being deposited on remote soil or aqueous surfaces. Abiotic transformation may also occur through either direct photolysis by sunlight or formation of hydroxyl radicals which subsequently can augment the polarity, reactivity and toxicity of PAHs (Atkinson and Arey, 1994). Hydrophobic chemical substances can enter the lipophilic membranes of several organisms and bioaccumulate. Several reports have demonstrated data about the distribution, accumulation and different levels of metabolism of three- and four-ring polyaromatics in biological tissues of several species associated with water and soil sediments, such as earth (Ma *et al.*, 1995) and marine worms (Rust *et al.*, 2004), plants (Parrish *et al.*, 2006) and trees (Howsam *et al.*, 2000), mussels (D'Adamo *et al.*, 1997), fish (Baumard *et al.*, 1998) etc. This group of chemicals is considered to be hazardous for human health. Many reports have illustrated the toxic, mutagenic and carcinogenic properties of several PAHs and derivatives (Rubin, 2001), the initiation of carcinogenesis as the liver's enzymatic system cytochrome P450 transforms polyaromatic molecules to reactive intermediates that bind to DNA (Melendez-Colon *et al.*, 1999), the loss of fertility (De Celis *et al.*, 2000), immunosuppressive properties (Laupeze *et al.*, 2002), inhibition of the differentiation of human blood monocytes (van Grevenynghe *et al.*, 2003) etc. For these reasons, the US Environmental Protection Agency has characterized many PAHs as priority pollutants (http://www.epa.gov/) and the European Union has issued a series of guidelines and directives imposing PAH detection limits in food, air and water (Lerda, 2009). The ways that these pollutants can be biochemically available or trapped in soil will be discussed in more detail below.

Aromaticity and physicochemical properties of PAHs

Biochemical persistence of polycyclic systems is directly related to their aromatic character. Aromaticity and, more precisely, the term 'aromatic' was originally conceived to describe components endowed with a distinctive odour. This usage was progressively abandoned with Faraday's discovery of benzene (1825) and Kekulé's cyclic formula for the very same compound (1865); the latter was the first one to use 'aromatic' and 'aromaticity' to describe structural characteristics of benzene-like composites. Today, aromaticity is an essential concept in chemistry and a multidimensional phenomenon described by geometric, energetic and magnetic criteria accounting for the stability, reactivity, molecular structure, and electrochemical properties of many organic compounds. A simplified definition of aromaticity would be that aromatic molecules comprise planar rings of atoms covalently joined with bonds of two different kinds: a ring consists of six carbon atoms, each joined to at least two other carbon atoms by one single and one double bond with a cyclic electron delocalization (Fig. 6.2). The resulting regular-hexagonal structure is characteristic of the aromatic molecules which have $4n+2$ (Hückel's rule) delocalized electrons. The bonds in the aromatic ring are equally shared, equivalent and intermediate between single and double bond lengths, the exact structure of the molecule is somewhere in-between the existing resonance forms and the delocalized electrons contour a continuous cloud (overlapping orbitals) leading to an extra thermodynamic and chemical stability, called resonance energy. These main characteristics are the conventional ones to picture the aromaticity and they point out why the particular electron arrangements are less reactive than ordinary double bonds and attribute to such molecules a unique stability and, thus, chemical and physical properties which reflect energetic stabilization referred to as aromaticity (Balaban, 1980; Krygowski, 1993; Krygowski and Cyranski, 1996; von Schleyer and Jiao, 1996; Krygowski *et al.*, 2000).

The structures being built up by fused benzenoid rings are strongly stabilized and tend to have higher aromaticity and this feature influences the reactivity of these compounds. There are several criteria and more than a dozen indices to measure the aromatic nature of cyclic conjugated hydrocarbons. Though the values of all those indices (mostly structure- and magnetic-based) are not always in full accordance when being compared (Balaban and Randić, 2005), the trends validate that the aromaticity is correlated with the number of carbon atoms and the linearity of the fused cycles into the multibenzenoid forms. Consequently, the more the aromatic rings and

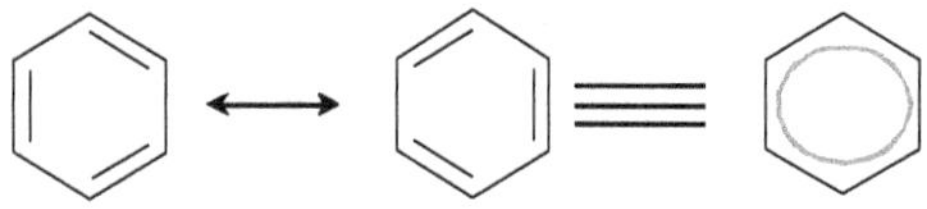

Figure 6.2 Schematic representation of equivalent hexagon structures of the benzene ring which are more accurately represented by the circle (delocalized electrons) within a hexagon diagram.

the higher the molecular weight of polyaromatic structures, the higher their aromaticity and thus, the less reactive these compounds may become. The aromaticity is also related with the assemblage of the rings into the structure: molecules with a linear arrangement tend to be less soluble and resistant to biochemical attack than angular or pericondensed fused ring formations.

The fact that non-polar covalent bonding as explained above occurs in benzenoid hydrocarbons attributes them a series of physical properties having a significant environmental impact. Several chemometric studies on persistent organic pollutants including PAHs (Govers *et al.*, 1984; Brunner *et al.*, 1990; Dunnivant *et al.*, 1992; Todeschini *et al.*, 1995; Ferreira, 2001; Alves de Lima Ribeiro and Ferreira, 2003) have focused on the quantitative structure–activity and structure–property relationships as they have stated the correlation between molecular descriptors (size, shape, symmetry, number of atoms and bonds, intermolecular interactions) with features such as the boiling point (bp; volatility), melting point (mp; solubility predictor), aqueous solubility, vapour pressure (log pL; dispersion in the environment), toxicity, octanol–water partition coefficient (log *P*; hydrophilicity), partitioning descriptors based on the log *P* such as coefficients between sediment/water, gas/water and water/air, etc. In biochemical terms, these analyses confirm the acknowledged polyaromatic properties often stated in the literature: the higher the molecular weight and the number of the fused rings of a PAH molecule, the higher the hydrophobicity and toxicity and the longer its environmental persistence.

This principle can be easily understood by looking at Table 6.1, which illustrates some basic physicochemical properties of several PAHs including the contaminants in many impacted sites. The aqueous solubility of these xenobiotics is a feature of capital importance for environmental processes; it can affect sorption/desorption in soils and, along with the soil properties and the microbial activity, it controls the bioaccessibility and bioavailability of a chemical and, hence, the possible routes of transformation.

Polyaromatics and soil interactions

Once an organic chemical gets in touch with the soil, this molecule may keep its physical state, dissolve in the bulk solution, get sorbed to the soil matrix or evaporate in the vapour phase (Harms and Bosma, 1997). As all hydrophobic organic pollutants (HOPs) including PAHs have low aqueous solubility and low tendencies for volatilization, they avoid the liquid and gaseous phase and they are either free in the bulk water or get mostly associated with the soil matrix where they get sorbed to soil particles. The best-case scenario for elimination of PAHs, sorbed or as solutes, is their microbial utilization either with direct contact of microbial cells or with extracellular enzymes, as it will be discussed below. However, in nature a series of limitation steps occurs, directly linked to the soil–contaminant interactions, which depend not only on the hydrophobic nature of the contaminant but also on the nature of the subsurface system itself.

It is commonly accepted that soil organic matter (SOM) is the predominant sorbent of HOPs and regulates their mobility, even to trace concentrations such as 0.1% w/w. This is mainly attributed to the hydrophobic effect (Xing, 1997; Pignatello, 1998; Weber *et al.*, 1998; Weber *et al.*, 2001; Huang *et al.*, 2003; Pan *et al.*, 2008): while the non-polar compounds are expelled from the water phase, the SOM harbours them as it is the organophilic domain of the subsurface system. The origin, age, source and history of organic matter affects its composition and structure and as several fractions with different physicochemical properties are present, the affinity for non-polar organic compounds varies accordingly (Luthy *et al.*, 1997). SOM is the total soil organic carbon comprising litter, microbial biomass, plant residues (polysaccharides, lignin, proteins, etc.) in various phases of decay (Cuypers *et al.*, 2002).

Table 6.1 Basic properties of some polycyclic aromatic pollutants. Adapted from Nollet (2007)

Pollutant	Molecular weight	Number of fused benzene (and total) rings	Linear formula	Aqueous solubility (mg/l)[1]	Octanol–water partition coefficient (log *P*)
Naphthalene	128	2	$C_{10}H_8$	31.00000	3.30
Fluorene	166	2 (3)	$C_{13}H_{10}$	1.89000	4.18
Anthracene	178	3	$C_{14}H_{10}$	0.43400	4.45
Phenanthrene	178	3	$C_{14}H_{10}$	1.15000	4.46
Fluoranthene	202	3 (4)	$C_{14}H_{10}$	0.26000	5.16
Pyrene	202	4	$C_{16}H_{10}$	0.13500	4.88
Benzo[a]anthracene	228	4	$C_{18}H_{12}$	0.00940	5.76
Chrysene	228	4	$C_{18}H_{12}$	0.00200	5.81
Triphenylene	228	4	$C_{18}H_{12}$	0.00043	5.63
Benzo[a]pyrene	252	5	$C_{20}H_{12}$	0.00162	6.13
Perylene	252	5	$C_{20}H_{12}$	0.00240	6.21
Benzo[ghi]perylene	276	6	$C_{22}H_{12}$	0.00026	6.63
Coronene	300	7	$C_{24}H_{12}$	0.00014	7.36

[1]At 25°C.

It consists mostly of humic substances such as humic acids (HA), fulvic acids (FA) and humin (Pignatello, 1998). HA are insoluble at acidic pH but soluble at higher values, heavier than FA, have a higher hydrogen and carbon content and lower oxygen content than FA, and are more hydrophobic than the latter ones (Gaffney *et al.*, 1996; Ehlers and Loibner, 2006). FA contain more phenolic, carboxylic and ketonic groups, they are more soluble and labile, with less complicated structures than HA, presenting a weaker aromatic character because of their higher H/C mole ratio (Huang *et al.*, 2003; Loffredo and Senesi, 2007). Regarding humin, information is less abundant than HA and FA. It appears to be the most aged and mature fraction with the most complex molecular weight distribution, highly hydrophobic and condensed, and can be further separated to lipid-bound HA and insoluble residues which are strongly associated with inorganic minerals (Pignatello, 1998). The above characteristics are important features of the total organic fraction contributing enormously to the HOPs' sorption to soil (Zhu *et al.*, 2004; Ehlers and Loibner, 2006; Pan *et al.*, 2008; Chefetz and Xing, 2009): higher MW distribution affects the condensation of SOM, condensed SOM is less polar and aliphatic with more aromatic carbon and aromatic moieties, augmenting the aromaticity of the sorbent and rich-aromatic SOM have higher binding PAH affinity as the interactions between the non-polar molecules and SOM indicate. Though there is a controversy about the contributing role of the sorbent's aromaticity and aliphaticity (Chen *et al.*, 2005; Pan *et al.*, 2008; Chefetz and Xing, 2009), there is no doubt that SOM's variable composition reflects different sorbing properties and reactivities, and that the effect of its physical conformation should not be neglected.

Over the past decades, the organic matter was considered to be a partitioning medium with homogeneous sorption behaviour but several phenomena (non-linear sorption, competition between PAHs during sorption and displacement, and hysteresis or irreversible sorption) argued against this assertion. The trend gaining place amongst scientists in order to explain its physical state is the 'polymer model' of SOM, accounting for a soft, rubbery, amorphous domain and a hard, condensed, more heterogeneous 'glassy' one with rigid hydrophobic micropores, first proposed by (Weber *et al.*, 1992; Xing and Pignatello, 1997). The amorphous domain is monolayer, exhibiting a rapid diffusion (in comparison with the condensed), with no competitive effects and no hysteresis as sorption of organic contaminants happens in a linear way (absorption through partition) (Luthy *et al.*, 1997; Chorover and Brusseau, 2008; Pan *et al.*, 2008). The glassy part presents a slow diffusion towards the soft one and also within its own domain where site specific hole-filling occurs in the rigid nanovoids of the condensed matter; the HOPs are partially immobile in these rigid micropores and the hole-filling is associated with irreversible sorption, competition and non-linear sorption of HOPs (Luthy *et al.*, 1997; Pignatello, 1998; Pignatello, 2009). A schematic representation of the sorption/desorption phenomena of the different domains of SOM is given in Fig. 6.3.

Soil can also contain some forms of hard carbon that are not always classified as SOM such as carbon black (soot), kerogen, coal, coke, charcoal, etc., which augment the heterogeneity of soil and differentiate more its sorption variability. Such materials are considered to be of anthropogenic origin (pyrolysis and combustion residues), hydrophobic, very microporous, with high affinity for organic pollutants. Polyaromatics are strongly retained to this fraction in comparison with the amorphous organic part and their desorption may last from decades to centuries (Johnsen and Karlson, 2007), as the high-surface-area of these carbonaceous materials is an important factor of very slow diffusion, nonlinearity and desorption hysteresis for PAHs which are not occluded (Luthy *et al.*, 1997; Koelmans *et al.*, 2006; Pignatello, 2009).

The intra-organic matter diffusion of HOPs is not the only barrier for the mobility of contaminants. Organic macromolecules and mineral grains cement together and form particles, and particles may combine together and form aggregates, principally through SOM adsorption on the clay surface (size ~2 μm) via hydrophobic interactions, H-bonding, water or cation bridging and metal exchange, etc. (Mirsal 2008b) The surface coating is more favourable on alumino-silicate clays which have a large surface area and interlayer reactions are multiplied. During physisorption (hydrophobic effect, electrostatic adsorption, van der Waals interactions, H-bonding), the HOP desorption becomes very unfavourable as the breaking of multiple different bonds should take place (Chorover and Brusseau, 2008). During

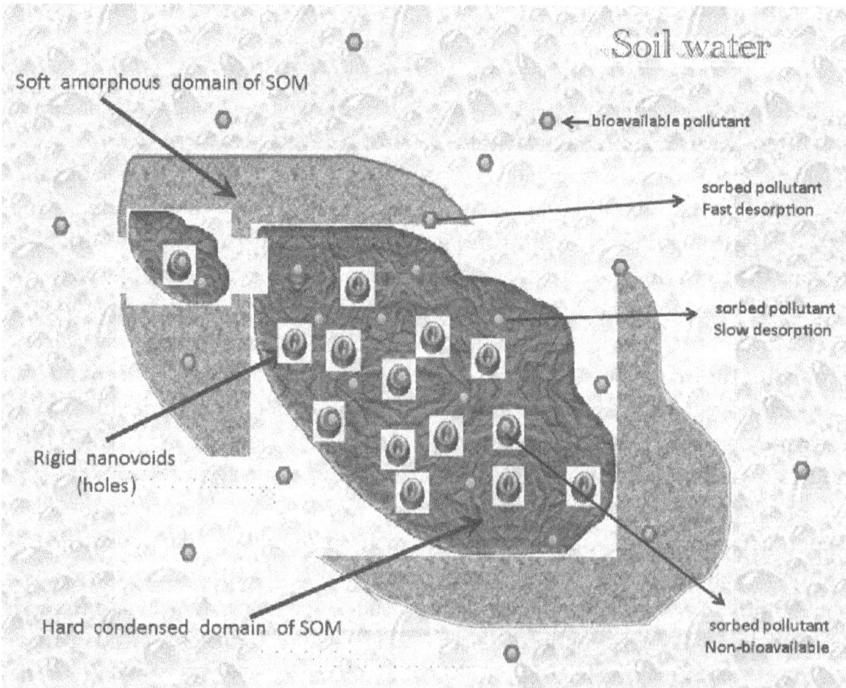

Figure 6.3 Schematic representation of the sorption/desorption of a hydrophobic organic pollutant on the different domains of the soil organic matter (SOM).

chemisorption (participation of organic molecules in the bond-breaking and bond-making surface reactions; Ehlers and Loibner, 2006), cations (i.e. iron, manganese, aluminium) are exchanged by organic ones through ligand substitution. The resulting organoclay complexes are characterized by reinforced stability, a diminished hydrophilicity and subsequent sorption of solute hydrophobic pollutants from the bulk water (Yariv, 2002). These processes inhibit the desorption and transport of HOPs towards bacteria accessible sites since the contaminants' intraparticle diffusion is unable to overcome the binding mechanisms on the internal interlayer surfaces of clays (nanometer scale), the intrapore diffusion in and out of meso- and micro-pores within the grain/particles and the external surfaces of mineral particles/aggregates/precipitates, as well as that within macropores (Pignatello, 1998). During the adsorption phenomena, the intrapore diffusion is responsible for the observed hysteresis of desorption (Pignatello, 2009). Moreover, the pore formation and structure alterations in the SOM glassy domain is a continuous and dynamic process (Pignatello and Xing, 1995) and pollutants are often trapped owing to non-adsorptive phenomena into the pore space as a result of caking, straining or precipitation (Mirsal, 2008b). If we take these facts into consideration, it is understood why soil porosity is considered to be another barrier that sorbed pollutants should overcome in order to enter the bulk liquid and become bioaccessible and why characteristics, such as relative density of pores in the soil, particle/aggregate shape and grain size distribution, have attracted the attention of soil scientists towards the concepts of HOPs sorption/desorption and diffusion. However, a strong HOP affinity to SOM does not necessarily have only negative consequences on PAH mobility. Though there may be a negative correlation between mobility and sorption affinity of organic matter, the humic substances should be also regarded as contaminant carriers (Gaffney *et al.*, 1996): the existence of aromatic and aliphatic moieties facilitates the binding of both hydrophilic and hydrophobic compounds and through colloid properties, they can promote the transport of pollutants when the humics' concentration, the pH and the ionic strength of the medium are favourable. It has been found that humic and fulvic materials dissolved in aqueous solution have a tendency to form micelle-like structures, with hydrophilic shells facing the aqueous phase and hydophobic cores in the centre of the aggregates (Gaffney *et al.*, 1996).

Apart from the nature of a hydrophobic pollutant such as a PAH molecule, the intra-SOM matrix diffusion and pore diffusion are considered to be the rate-limiting steps in the PAHs' transport in soil. For the reasons explained above, they are responsible for the non-linear desorption and hysteresis, accounting for the macroscopic observations of reduced bioavailability (pollutant sequestration) and the recalcitrant fraction of residues bound to the soil matrix. This phenomenon can become more problematic with time as the chemical remains in the subsurface system and becomes even more recalcitrant (ageing).

Finally, the conformation of the soil particles is very important for the spatial distribution of xenobiotics and microbes. Soil bacteria are characterized by a low mobility and are transported mainly by the flowing water. In a water-filled porous medium, several interactions occur and bacteria are present in an attached state to soil particles, mainly onto the SOM layer or patches of particles, and to the air–water interface (Schäfer *et al.*, 1998). The habitat for soil bacteria (average cell size 0.2–0.8 μm) that is protected from predators is considered to be around 2 μm pores, representing less than 50% of the total pore volume and in some cases, 30% of the soil fraction can be around 0.2 μm (Harms and Bosma, 1997). Inevitably, a considerable part of the subsurface system is unreachable to soil bacteria and, consequently, hydrophobic pollutants should gain this particular space where physical contact with microbes can arise. In all cases, these constraints are another factor contributing enormously to PAH mobility and availability in the bulk liquid.

Degradation by microbes

The major route of degradation of aromatic hydrophobic pollutants takes place under aerobic conditions mostly by living organisms. Algae, fungi and bacteria are mainly responsible for these processes. Many species are known to attack polyaromatics in order to yield carbon and energy for maintenance and growth purposes or through the beneficial effect of cometabolism, where non-utilizable organic substances get oxidized at the expense of compounds supporting microbial growth (Horvath, 1972). The best-known bacteria with PAH-degrading capacities include genera such as *Rhodococcus, Sphingomonas, Mycobacterium, Alcaligenes, Nocardioides, Gordonia, Arthrobacter, Pseudomonas, Burkholderia,* etc. and some articles have compiled in detail the currently known degraders and their substrate versatility (Cerniglia, 1992; Juhasz and Naidu, 2000; Kanaly and Harayama, 2000; Balkwill *et al.*, 2006). Bacterial degradation takes place mainly under aerobic conditions as it will be discussed below in some detail. However, destruction of PAHs can occur under anaerobic conditions too. Hydrocarbon-contaminated sites become often anoxic and PAH degradation occurs with nitrate and sulphate as terminal electron acceptors. Many nitrate-, iron- or sulphate-reducing bacteria (i.e. *Acidovorax, Variovorax, Dechlorovorax, Azoarcus, Thauera*) have been isolated under anaerobic conditions with diverse electron acceptors and have shown an advanced ability to degrade mono and polyaromatic hydrocarbons (Coates *et al.*, 1997; Chakraborty *et al.*, 2005; Zhang and Bennett, 2005).

Fungi are also known to have transforming properties towards polyaromatics. Ligninolytic species produce extracellular enzymes with low enough substrate specificity as to decompose the lignin in the soil matrix, an abundant biopolymer with aromatic structure (Kirk and Farrell, 1987). Enzymes such as lignin peroxidases, manganese-dependent peroxidases, phenoloxidases (laccases, tyrosinases), have a non-specific action towards several pollutants with aromatic structure including chlorobiphenyls, pesticides, nitroaromatic compounds, PAHs, synthetic dyes, phenolic compounds, etc. (Field *et al.*, 1992; Canet *et al.*, 1999; Van Aken and Agathos, 2001; Torres *et al.*, 2003; Wesenberg *et al.*, 2003; Rodríguez Couto and Toca Herrera, 2006; Asgher *et al.*, 2008; Mougin *et al.*, 2009). White-rot fungi such as *Phanerochaete* spp., *Bjerkandera* spp., *Pleurotus* spp., *Coriolus* spp. producing these enzymes are often cited in the above reports for their wide and efficient polyaromatic-degrading activity. Moreover, fungal transformation of PAHs may occur intracellularly. Non-ligninolytic species can contribute to PAH destruction via an oxidation process highly regio- and stereo-selective, mediated by cytochrome P450 monooxygenase which produces epoxides; from this point on, epoxides can be rearranged into phenols or get further converted by epoxide hydroxylases (Sutherland, 1992). Though fungal transformation does not necessarily accomplish the total elimination of the pollutants as often as competent bacterial strains, it serves a very important role as a detoxification step by rendering contaminants less toxic. Additionally, fungal transformation products are more water-soluble and, thus, more bioavailable as their transfer to the metabolically active bacterial populations is facilitated in the bulk solution. Apart from the pollutants' transport, another significant role of fungi in soil remediation is represented by the fungal mycelia networks themselves, which serve as 'highways' and mediate bacterial mobilization towards pockets of pollutants (Wick *et al.*, 2010).

Several works have also demonstrated the role of phototrophs such as algae in the elimination of polyaromatics. Cyanobacteria, green algae and diatoms have been shown to remove several compounds with aromatic structures and species such as *Oscillatoria, Agmenellum, Chlamydomonas, Chlorella, Scenedesmus,* and *Selenastrum* illustrate the positive effect of algal transformation towards mono- and polycyclic compounds (Cerniglia, 1992; Semple *et al.*, 1999; Haritash and Kaushik, 2009). The utilization of PAHs may occur in order to derive energy and carbon, though such studies have shown low elimination rates and only partial removal of hydrocarbons in the presence of algal heterotrophic cells. However, the most positive aspect of PAH biotransformation by algae seems to be the synergistic effects in algal-bacterial consortia where the algal presence enhances the degradation potential of the microbiota present. This collaboration is beneficial for both communities as algae produce molecular oxygen, used as terminal electron acceptor for aerobic biotransformation of the organic matter and bacteria in turn release carbon dioxide during the mineralization process, useful for the photosynthetic cycle of algal species.

Ring-hydroxylating dioxygenases of aerobic bacteria

The major route of degradation of aromatic hydrophobic pollutants takes place under aerobic conditions and involves some consecutive steps leading to simple forms of intermediates readily integrated into central metabolism. Bacteria are the main protagonists through the aerobic metabolism of PAHs. During the bacterial transformation of aromatic structures, the initial step occurs via the activation and incorporation of both atoms of molecular oxygen into the aromatic nucleus by a multicomponent enzyme system, the aromatic ring-hydroxylating dioxygenases (RHD). This reaction destabilizes the resistant aromatic structure while leading to the formation of *cis*-dihydrodiols and prepares the persistent compound for further fission. Re-aromatization through a *cis*-diol dehydrogenase is the second step which yields dihydroxylated intermediates which under the consecutive actions of ring-cleaving enzymes get channelled to the central catabolism, such as the tricarboxylic acid cycle (TCA), as illustrated in Fig. 6.4. A thorough discussion of the current state of knowledge on RHD is to be found in Chapter 7.

These enzymes can also act as monooxygenases when they catalyse the insertion of one single atom of oxygen in the aromatic structure. One route of the fluorene catabolic pathway is such an example, as this compound gets commonly oxidized by several microbes with the incorporation of one oxygen atom into the C9 position (Fig. 6.5).

RHDs are multicomponent enzymatic systems consisting of an electron transport chain and a catalytic terminal oxygenase. Their typical arrangement is a three-component complex comprising a ferredoxin reductase, a ferredoxin and the terminal oxygenase's heterologue subunits, α (large) and β (small) (Gibson and Parales, 2000). The electron carrier components transport electrons from NAD(P)H to the terminal oxygenase meaning that the co presence of at least two proteins is required while a source of reductant is constantly needed. A single two-electron transfer from reduced nucleotides NAD(P)H to FAD takes place in the ferredoxin reductase and initiates the reaction. As the NADH can transfer simultaneously both of its electrons, the reductase is required to mediate this transition. Thus the reductase contains flavin, a coenzyme that accepts two electrons and transfers them. The fully reduced form of the FAD provides electrons (one at a time) to the ferredoxin (Butler and Mason, 1997). The ferredoxins are one-electron acceptors, transferring electrons from the reductase to the active site of the terminal oxygenase in order to oxidize the aromatic structure by the incorporation of molecular oxygen. This active site is located at the α subunit which is the catalytic component of the terminal dioxygenase. This subunit contains two regions: the Rieske centre (an iron-sulphur cluster) and the iron binding site. The iron-sulphur cluster serves for the electron transfer (by the reduction and oxidation of iron) to the non-heme iron centre and it is coordinated by two histidines, two cysteines and the iron centre (Parales and Resnick, 2006). The iron centre, an evolved Fe (II) complex is the active site for dioxygen chemistry by means of binding and activating the dioxygen for its final insertion into an aromatic ring (Butler and Mason, 1997). The α subunit plays the major role in controlling the substrate specificity while the β subunit may be involved in the substrate recognition or its role is probably structural related to the stabilization of the whole enzymatic complex (Kauppi *et al*, 1998). All parts are involved in the addition of the dioxygen molecule and necessary for the function of the enzymatic system. A schematic representation of the reaction catalysed by a typical three-component RHD is given in Fig. 6.6.

Though the three-component arrangement is typical for the RHD enzymes, two-component clusters exist also comprising the ferredoxin reductase and the terminal oxygenase while lacking the ferredoxin (Parales and Resnick, 2006). Indeed, based on the electron transfer components present in the structure, a first classification system, known as the Batie system, was proposed to classify the different RHD systems. Three classes of enzymes were established in this scheme: the two-component enzymes (reductase and terminal oxygenase: Class I) and the three-component ones (reductase, ferredoxin and oxygenase; Classes II and III) with further subdivisions. As the

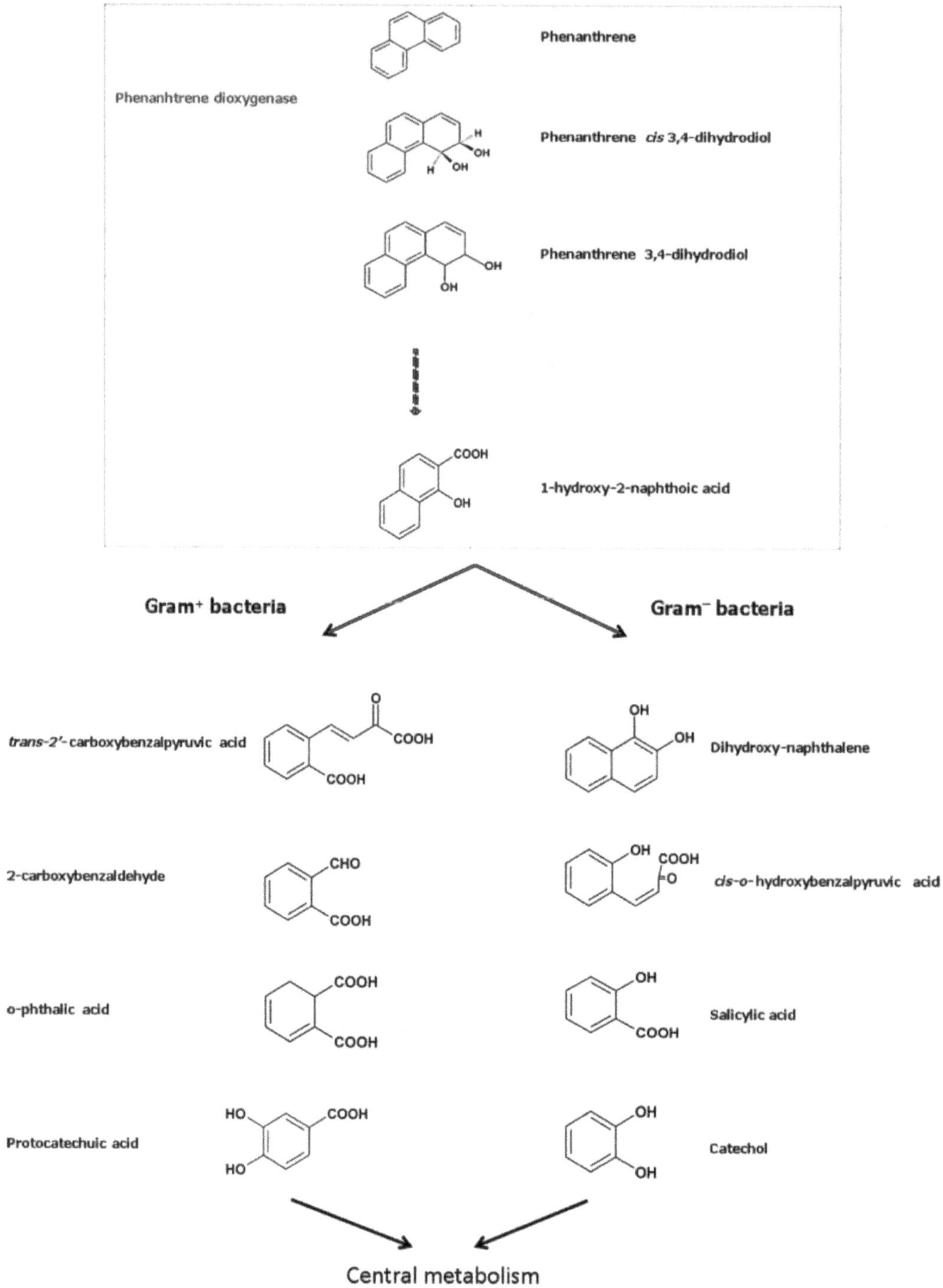

Figure 6.4 The most common bacterial degradation pathway of phenanthrene: the transformation is initiated by a ring hydroxylating dioxygenase at the positions C3 and C4 and the action of consecutive enzymes leads to the formation of the key-metabolite 1-hydroxy-2-naphthoic acid (HNA). From this point, Gram-positive bacteria oxidize this compound through o-phthalic and protocatechuic acids while Gram-negative bacteria proceed through salicylic acid and catechol. Both routes incorporate metabolites to central metabolism.

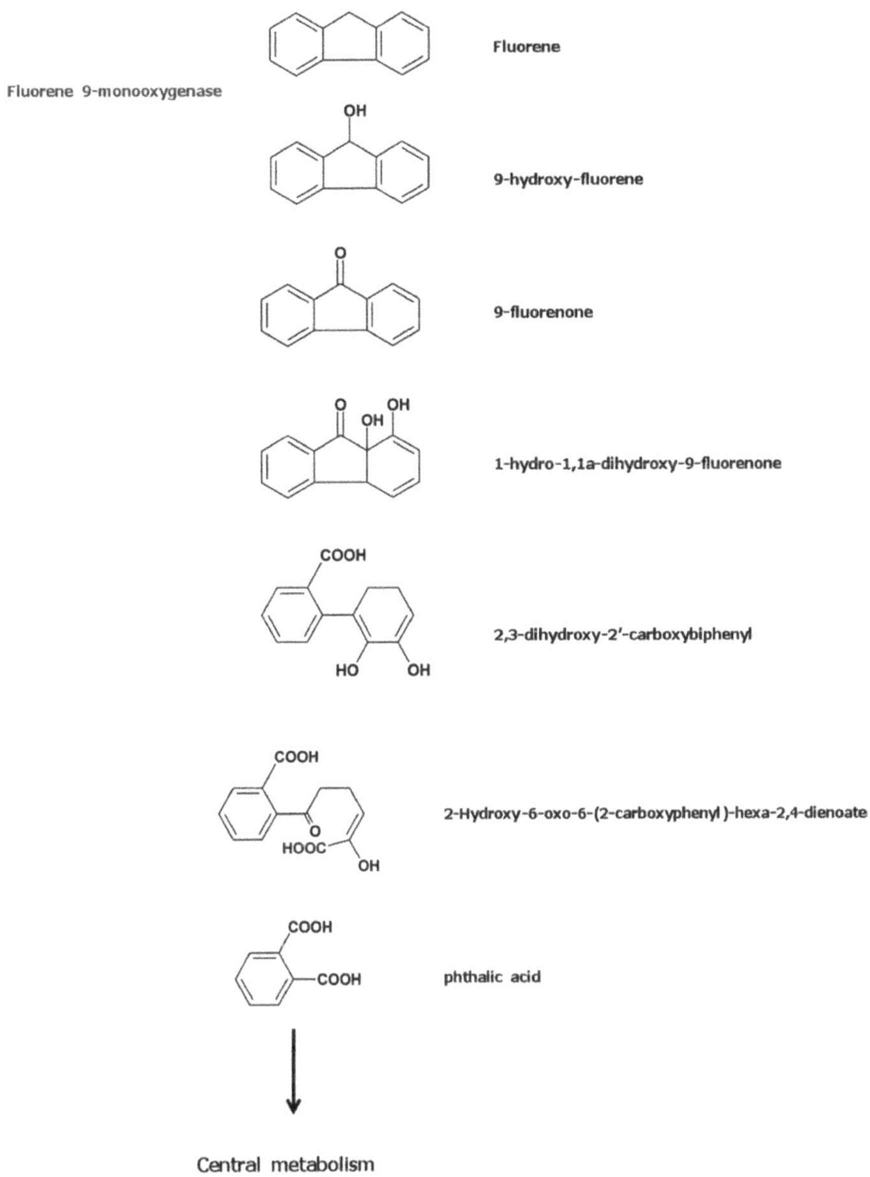

Figure 6.5 The most common bacterial degradation pathway of fluorene proceeds through oxidation at the C9 position by a ring hydroxylating monooxygenase. Under the action of consecutive enzymes, the formation of common metabolites such as phthalic acid takes place before entering central metabolism.

scientific knowledge advanced continuously and new hydroxylating dioxygenases were discovered, this organization failed to classify all the newly discovered enzymatic systems. Afterwards, the Werlen system was proposed (Werlen *et al.*, 1996) which is based on the clustering of the catalytic oxygenase components and this scheme classified the RHDs into four families (naphthalene/PAH, toluene/benzene, biphenyl and toluate/benzoate).

Nowadays, the hydroxylating dioxygenases are classified based on the amino acid sequence comparisons of the catalytic oxygenase α subunits (Gibson and Parales, 2000; Nam *et al.*, 2001a). According to (Gibson and Parales, 2000), four classes are mainly distinguished (Fig. 6.7). The

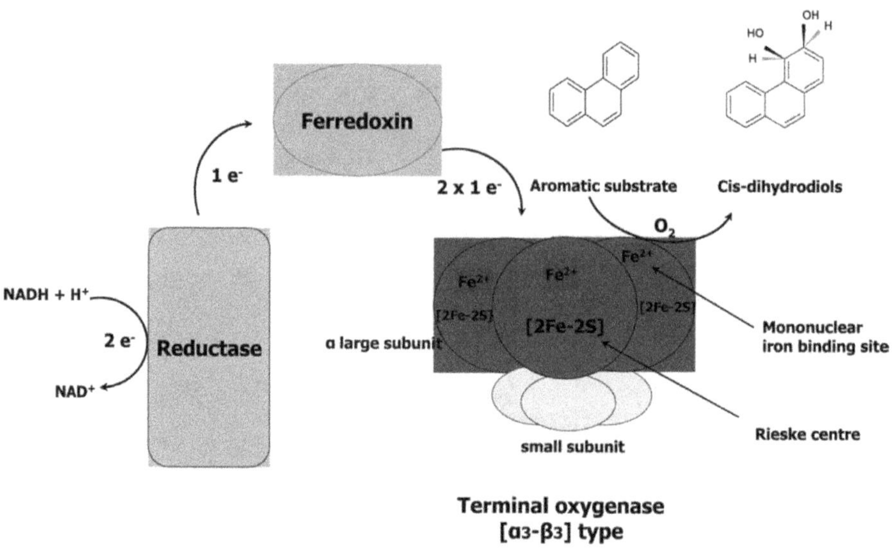

Figure 6.6 Schematic representation of the oxidation of an aromatic compound catalysed by a three-component ring hydroxylating dioxygenase. The NADH electrons are transferred through the reductase and the ferredoxin to the terminal oxygenase in order to effect the addition of two atoms of oxygen to the aromatic substrate.

naphthalene family comprises Gram-negative bacterial enzymes for the degradation of phenanthrene and naphthalene. The benzoate family is a diverse group consisting of enzymes oxidizing aromatic acids. The phthalate class is the most diverse group of enzymes containing both monooxygenases and dioxygenases and presents the largest variability in terms of amino acid sequence and primary substrate. The enzymes of this family have a configuration containing only large subunits in the terminal oxygenase. Most members of the phthalate family lack the β subunits and have only the reductase component in the electron transfer chain. Finally, the toluene/biphenyl class contains enzymes from both Gram-negative and Gram-positive microbes responsible for the transformation of toluene, benzene, chlorobenzenes and biphenyl.

RHD encoding genes in PAH-degrading Gram-negative bacteria

Genetic analysis of PAH degradation by aerobic bacteria was initiated and mostly focused on the degradation of naphthalene by *Pseudomonas* species. A landmark in this effort was the study of *Pseudomonas putida* G7 when its naphthalene catabolic genes (*nah*) were found located on the archetype plasmid NAH7 (Yen and Gunsalus, 1982). Two operons, physically and functionally distinct, were reported: the *nah* operon encoding all the enzymes of the upper catabolic pathway responsible for the conversion of naphthalene to salicylate, and the *sal* operon encoding the enzymes of the lower catabolic pathway where the conversion of salicylate takes places in order to yield acetyl CoA and pyruvate. The two clusters are separated from a region where a regulatory gene (*nahR*) is located, whose product is required for the expression of both operons while the salicylate acts as an inducer (Yen and Gunsalus, 1985). Later on, studies on several *Pseudomonas* strains mainly demonstrated high sequence similarity, gene organization and arrangement, and common biochemical data of the naphthalene catabolic genes. The NAH plasmid pWW60-1 from *P. putida* NCIB9816 (Yang *et al.*, 1994) and the chromosomally encoded genes of *P. stutzeri* AN10 (Bosch *et al.*, 1999; Bosch *et al.*, 2000) were such examples: separate catabolic operons harboured all the necessary catabolic genes, a first operon (*nahAaAbAcAdBFCED*) coding for the

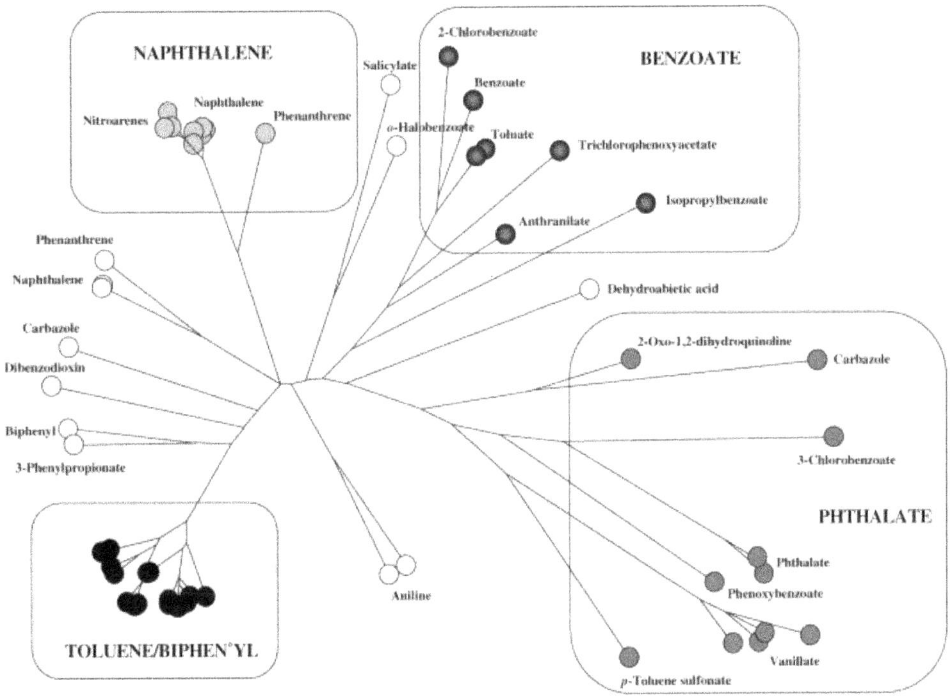

Figure 6.7 Phylogenetic tree of Rieske non-heme iron dioxygenases [adapted from Gibson and Parales (2000) and Wackett (2002)] based on sequence relatedness of the α subunits of the dioxygenase iron–sulfur protein components components. A more detailed version of such a tree is given in Chapter 7, Fig. 7.1.

naphthalene upper degradation pathway and a second one (*nahGTHINLOMKJ*) coding for the lower degradation pathway while both operons were regulated under the action of the NahR protein. High degree of similarity of the upper catabolic pathway (>90% amino acid homology and conserved gene arrangement) with the *nah* genes was found also in several *Pseudomonas* from different locations utilizing diverse aromatic substrates. The so-called *nah*-like genes include *ndo* (Kurkela *et al.*, 1988), *pah* (Kiyohara *et al.*, 1994; Takizawa *et al.*, 1994), and *dox* (Denome *et al.*, 1993).

Several sets of genes isofunctional with the classical *nah*-like sequences have been reported. Some of them show a higher homology with the *nah* ones while others are distantly related. Studies have demonstrated the existence of the *nag* genes of *Ralstonia* sp. strain U2 (a former *Pseudomonas*), all of them organized in a single operon under the regulatory control of a NagR protein (Fuenmayor *et al.*, 1998; Zhou *et al.*, 2001), as well as the *nag* genes of a *Polaromonas* strain which are homologous to those of *Ralstonia* sp. U2, but they are organized in two clusters amongst some other differences (such as two additional genes and two regulators) (Jeon *et al.*, 2006). Though *nah* and *nag* genes display several differences in the gene order and arrangement, they present also striking similarities, a fact which suggests a common ancestry (i.e. separate modules of genes for the conversion of salicylate to central metabolites in both types). A divergent set of genes comprises the *phn* genes, first described in a *Burkholderia* sp., which encode an identical degradation pathway for naphthalene and phenanthrene with the *nah*-like sequences (via salicylate); the *phn* locus however is characterized by many variations in both gene order and sequence similarities in comparison with the *nah*-like genes while the electron carrier components are absent (Laurie and Lloyd-Jones, 1999). Another interesting case was some *Comamonas* strains isolated along with several Pseudomonads from the same environmental sample (Goyal and Zylstra, 1996). Though these bacteria were able to grow at least in the presence of naphthalene or phenanthrene, none of them show any hybridization to the *nah*-like gene probes. None of the *Pseudomonas* spp. did hybridize with the isolated catabolic genes for phenanthrene degradation (*phd*) of one of these strains when used as a probe while another *Comamonas* strain did not show any homology either (Goyal and Zylstra, 1997). Based on these results, the authors suggested that there was one more class of genes responsible for naphthalene/phenanthrene degradation in this sample, distinct from the *nah*-like and *phd* genes.

The members of the *Sphingomonas* genus constitute an original case. The degradative pathways for the transformation of polyaromatic pollutants present nothing unusual in comparison to the described ones above but the transcriptional units demonstrate a complex arrangement. The genes intervening individually in the degradation pathway of a given compound are scattered through several gene clusters. Thus, distinct and highly conserved transcriptional units in terms of organization and sequence harbour genes encoding enzymes responsible for the degradation of monoaromatics, polyaromatics and biphenyls. Genes coding for proteins with consecutive action in a degradative pathway can be distant, on different operons and collocated with other catabolic enzymes. This patchwork arrangement of highly conserved regions in *Sphingomonas* was fully reviewed by Pinyakong and co-workers (Pinyakong *et al.*, 2003a) and it was demonstrated in several degrading species of this genus (Zylstra and Kim, 1997; Romine *et al.*, 1999; Pinyakong *et al.*, 2003b; Schuler *et al.*, 2009). To summarize, the catabolic genes of several microbes (normally clustered and regulated together in operons with distinct roles in the upper and lower degradative pathways) and *Sphingomonas* (mosaic arrangement) code for the very same elimination route for PAHs even if there are some differences in the order and arrangement.

Though the peripheral enzymes responsible for the initial attack of several polyaromatic compounds may be products of conserved gene clusters of higher or lower homology (in terms of order, arrangement or nucleotide sequences), the most common route for Gram-negative bacteria are subsequent biochemical steps leading to central metabolism through the formation of 1-hydroxy-2-naphthoic acid (HNA) and salicylate. However, exceptions concerning the formation of phthalate as metabolite have been recorded (Kiyohara *et al.*, 1982). From this point, normally the aerobic conversion of PAHs proceeds through acetyl CoA and pyruvate while a less common pathway converts the aromatic compounds to fumarate and pyruvate via salicylate and gentisate.

RHD genes in PAH-degrading Gram-positive bacteria

Concerning the Gram-positive PAH-degrading bacteria, the enzymatic steps from the initial attack of PAHs lead to their full transformation via 1-hydroxy-2-naphthoic acid, phthalate and protocatechuate. RHDs from these strains responsible for the transformation of polyaromatic pollutants have been found to be distantly related to the corresponding proteins encoded by Gram-negative species. Particular work on genes encoding RHDs has been done with degraders belonging to the genera *Nocardioides*, *Rhodococcus* and *Mycobacterium*. In the *Nocardioides* sp. strain RP7, the genes coding for the enzymes responsible for the conversion of phenanthrene to phthalate (*phd*) were identified (Iwabuchi and Harayama, 1998; Saito *et al.*, 1999, 2000). The *Rhodococci* have demonstrated a wide metabolic versatility towards several recalcitrant pollutants including PAHs and they are reported quite often to harbour the catabolic genes in large linear plasmids (Larkin *et al.*, 2005). In general, these bacteria contain multiple gene copies of several oxygenases and the cytochrome P450 monooxygenases are very common in these cocci (Larkin *et al.*, 2010). Research on several *Rhodococcus* stains has demonstrated the presence of *nid* (Treadway *et al.*, 1999) and *nar* genes (Larkin *et al.*, 1999) responsible for the conversion of naphthalene. The former case, the *Rhodococcus* sp. I24 was found to harbour both a naphthalene inducible monooxygenase and dioxygenase as well as a toluene dioxygenase set of genes (Priefert *et al.*, 2004). P450 monooxygenase from a *Rhodococcus rubber* showed hydroxylation activity towards naphthalene, fluorene, acenaphthene amongst other aromatic compounds (Liu *et al.*, 2006). Genetic analysis of *R. opacus* strain

SAO101 which was isolated in the presence of dibenzo-*p*-dioxin demonstrated the presence of *nar* genes; the encoded products when cloned and expressed in *E. coli* transformants gave dihydrodiols from compounds such as dibenzofuran, naphthalene and phenanthrene (Kimura *et al.*, 2006).

The extended investigation of *Mycobacterium* sp. strain PYR-1, the first bacterium reported to degrade the HMW pyrene (Heitkamp and Cerniglia, 1988), also demonstrated that its degrading genes responsible for the conversion of phenanthrene and pyrene were quite different from other known genes from Gram-negative bacteria with the same metabolic activity. The cloning, sequencing and functional expression in *E. coli* of its RHD-encoding genes (Khan *et al.*, 2001) showed a novel enzymatic system consisting of the Nid proteins. The *nid* genes were found to be chromosomally located with a different order from the *nid* genes reported in the *Rhodococci* and did not hybridize with the classical *nah* genes. As expected, this strain harbours the necessary genes for the common route of degradation through HNA and phthalate (Stingley *et al.*, 2004). Several studies demonstrated that the *nid* genes are widely distributed amongst several distant mycobacterial species with different substrate oxidizing capacities (Brezna *et al*, 2003). A later study showed the existence of one more independent enzymatic system in this strain favouring mostly the conversion of HMW PAHs (Kim *et al.*, 2006). The existence of multiple initial dioxygenases encoded by different gene clusters in the same bacterium, with preferential activities towards one polyaromatic compound or another was also established for other *Mycobacterium* species (Krivobok *et al.*, 2003; Sho *et al.*, 2004). The presence of multiple RHDs in Mycobacteria may explain the broad substrate specificity with which these strains are equipped and may justify their frequent presence in PAH contaminated soils.

To conclude, all these studies reporting catabolic genes and clusters encoding all the necessary steps for the conversion of the polyaromatic pollutants (either by Gram-negative or Gram-positive aerobic microbes) has confirmed that the peripheral pathways may be numerous for the initial attack. The destabilization of the PAHs is set off by different enzymes with relaxed substrate specificity showing some degree of preferential oxidation at certain C positions, depending on the microorganism or the compound. The biochemical data show that the degradation pathways after these first conversion steps are, however, common and shared between several PAHs, as the only intermediates entering the central microbial metabolism are limited to protocatechuate and catechols.

Abiotic remediation of PAH polluted soils

Given that most of the xenobiotics are of anthropogenic origin and got produced in bulk quantities through the industrialization of the last decades, the accumulation of PAHs in the environment became soon a major problem. The necessity to remediate contaminated soils became evident with the prerequisite of diminishing the risks for human health given that they fit the characteristics (bulk production, long-lived, lipophilic, and biologically active) for bioaccumulation and biomagnification (Grey, 2002; Fent, 2004). For these reasons, decontamination of aromatic-polluted soils has become essential for reducing environmental and health risks.

The clean-up of a polluted soil can be accomplished *in situ* (direct application without removing and destroying or destructuring the soil) and *ex situ* (excavation and subsequent transport and treatment of the soil). For *in situ* treatments, limitations can occur concerning the efficient delivery of oxygen, moisture or nutrients to the subsurface, although the costs are generally lower than *ex situ* processes. The latter are costly, especially for big sites and extended areas while they can cause severe land destruction and may be applicable if the access and transportation can be guaranteed. On the other hand, the off-site processes tend to be more effective as the pollutants are delocalized and can be treated without further environmental risks in specialized installations under fully controlled conditions.

In-site and off-site remediation procedures include both abiotic and biotic practices. In general, the most widely applied restoration technologies comprise physicochemical methods, thermal treatments and the biological capacity for degradation of recalcitrant xenobiotics while a lot of interest is focused on integrated, combined technologies (Atlas and Philp, 2005). The foremost objective while remediating the soil with abiotic treatments is to reach the desorption of soil-bound hydrophobic PAHs from the solid matrix in order to proceed with further actions and achieve outcomes, such as complete or significant (bio)destruction of the pollutants, extraction and removal of the contaminants for further treatment/ disposal and their stabilization and detoxification (Scullion, 2006). Though abiotic techniques for the transformation or destruction of PAHs is not the main subject of this chapter, the most applicable means and methods to treat polluted soil with polyaromatics are discussed briefly below.

Physical and thermal treatments of PAH polluted soils involve the separation, immobilization, removal or destruction of these hazardous substances through physical means, i.e. incineration (Acharya and Ives, 1994), vitrification (Daley, 1989), electroremediation (Reddy and Saichek, 2004; Maturi and Reddy, 2006), thermal desorption (Gan *et al.*, 2009), microwave heating (Appleton *et al.*, 2005), photolysis (Lee, 2003) etc. Regardless of the advantages of these methods (fast treatment time, non-exhaustive site characterization, treatment of a variety of contaminants, etc.), most of the applied physical technologies cannot deal with the destruction but with the removal of pollutants before a chemical and/ or biological treatment process. For xenobiotics with the properties of polyaromatics of HMW (low aqueous solubility, non-mobile, strong affinity for particular soil fractions), these methods seem to be less promising or they promote the first step for further processing (Scullion, 2006). Some thermal treatments such as incineration and vitrification may be fast, applicable to organic pollutants, volume reducing with high destruction and removal efficiencies (Daley, 1989; Hamby, 1996). They are, however, energy intensive (>1000°C), contributing to environmental emissions and health-endangering (inadequate breakdown of targeted pollutants or generation of hazardous byproducts) while they can destroy the soil properties to a great extent (Seeker and Koshland, 1990; Hamby, 1996). A rough classification of the commonly utilized remediating methods is illustrated in Table 6.2.

Chemical technologies involve destruction, transformation, transport or immobilization of pollutants while the applied chemicals agents promote extraction, solubilization and availability of hazardous substances (Rivas, 2006; Mirsal, 2008a). Chemical-bound remedial techniques commonly include the use of agents in order to oxidize the soil-sorbed contaminants. Along with the fact that they present a moderate cost, they are simple to operate and have advanced oxidative dynamics, and, moreover, they have been shown to be highly effective for the reduction or elimination of both the toxicity and the concentration of polyaromatic pollutants within short clean-up times (Hamby, 1996). Numerous field studies and lab-scale experiments have confirmed their potential on the enhanced elimination of PAHs by using oxidants such as hydrogen peroxide- and Fenton's reagent-based techniques (Nam *et al.*, 2001b; Bogan and Trbovic, 2003; Goi and Trapido, 2004; Flotron *et al.*, 2005; Rivas, 2006; Ferrarese *et al.*, 2008), ozone (Masten and Davies, 1997), potassium permanganate (Gates-Anderson *et al.*, 2001; Brown *et al.*, 2003; Ferrarese *et al.*, 2008) etc. The most common limitations occurring during these oxidation treatments are the short lives of the hydroxyl radicals, the reactivity of the reagents with the non-PAH soil constituents and the soil characteristics themselves that may decrease the overall efficiency of degradation (depth, texture, porosity, moisture, pH, organic matter, humic and fulvic acids).

Another class of treatments involves the extraction of the contaminants though it is not a destructive but a transport step. Organic solvents in mixtures or in combination with water can mediate PAH desorption from the soil matrix. The solvents used are normally mild (ethanol, hexane, butanol), the pollutants get desorbed into the liquid phase from which they can be further treated while the solvents are collected and reused. PAH recoveries depend in some way on the nature of the solvent, the amount of pollutant and the sample size while the solid matrix itself can affect greatly the extraction time. In some cases, the solvents may be toxic to the soil fauna and flora and can provoke alteration of the soil organic

Table 6.2 A rough classification of the main abiotic techniques which are commonly used to remediate polycyclic aromatic hydrocarbon (PAH)-contaminated soils. Some major advantages and disadvantages of these treatments are illustrated

Treatments	Most applied	Advantages	Disadvantages
Thermal	Incineration Vitrification Microwave heating Desorption	Fast, reduces toxicity and volume, total destruction, reduces toxicity, mobility and volume, simultaneous treatment of pollutants, not destructive for soil, high selectivity if needed, reduces treatment time, increases mobilization	High capital and operating costs, air-pollution control, energy intensive (>1000°c), expensive, residues to treat, non-destructive method, coupling with other techniques, defective for *in situ* operations, byproducts formation
Physical	Soil washing	Cheap	Effective only for lighter PAHs
Chemical	Oxidation (ozone, hydrogen peroxide, potassium permanganate)	Fast, high efficiency (>90%), simple operation, moderate costs	Environmentally unfriendly, soil- and pollutant-dependent, toxic, exhaustive site/soil characterization need
	Solvent extraction (ASE, PSE, mild)	Short operational times, efficient, selective separation, moderate costs (mild)	Solvent recovery – regeneration, high implementation costs (PSE), affects soil organic matter, dependent on soil characteristics
	Supercritical water extraction	Cheap, green technology, selective	Need for additives
	Supercritical CO_2 extraction	Low cost, non-toxic, stable reactivity	Long extraction times, non-selective for co-contaminants, dependent on soil characteristics
	Transport-facilitating agents (cyclodextrins, surfactants, vegetable oils)	Non-toxic, biodegradable, low cost	Oil residues in soil, dependent on type and quantity of both pollutant and surfactant, age and quantity of contamination

matter and, thus, affect the availability properties of the contaminants. The traditional solvent washing of the soil has been improved through modern techniques such as accelerated solvent extraction (ASE) and pressurized solvent extraction (PSE). When the former is used, solvents are applied at elevated temperature and pressure while the PSE is carried out under supercritical conditions. The main advantages of these extraction techniques in comparison with the conventional methods are the faster extraction times achieved, the possibility for selective separation of substances, the low-volume consumption of solvents and, of course, great PAH recoveries as many studies illustrate (Saim *et al.*, 1997; Berset *et al.*, 1999; Khodadoust *et al.*, 2000; Flotron *et al.*, 2005; Richter *et al.*, 2006; Wang *et al.*, 2007). However, high temperature and pressure may require high costs and may be soil-exhaustive while under extreme conditions of pollution, the choice of the appropriate solvent may be a problem.

Nowadays, more environmentally friendly extraction techniques have gained interest. A characteristic example is the use of water as an extraction medium, though for years its potential use was neglected. Water is cheap, easily available, non-flammable, and non-toxic. Moreover, as water temperature and pressure can be modified above the critical values, this liquid behaves similarly to a dense gas and can act as an organic solvent, therefore some classes of pollutants may be selectively targeted. Water-mediated extractions in supercritical conditions were shown to yield satisfactory results for polyaromatics amongst other polluting substances (Dadkhah and Akgerman, 2002; Kronholm *et al.*, 2002a,b; Anitescu and Tavlarides, 2006; Rivas *et al.*, 2009a). Amongst the supercritical-based fluid extraction technologies, the utilization of carbon dioxide has also gained interest, with promising results (Becnel and Dooley, 1998; Berg *et al.*, 1999; Anitescu and Tavlarides, 2006; Rivas *et al.*,

2009b). It is a cost-effective method, non-toxic, non-exhaustive for the soil and does not alter the soil organic matter (SOM). Supercritical conditions are usually achieved at relatively low levels of temperature and pressure and the reactivity of CO_2 remains quite stable *versus* time. Possible limitations can be the maintenance of a steady CO_2 flow and the reactivity with several soil constituents and/or co-contaminants while the soil characteristics may impose long extraction times for *in situ* applications.

Another large field of application is the use of transport-facilitating agents in order to enhance the extractability, mobility and (bio)availability of PAHs. Experimental data have demonstrated the positive effects of facilitated extraction of polyaromatics from the solid or aqueous phase by vegetable oils (Bogan *et al.*, 2003; Pannu *et al.*, 2004; Gong *et al.*, 2005), cyclodextrins (Lindsey *et al.*, 2003; Maturi and Reddy, 2006; Petitgirard *et al.*, 2009) and surfactants of industrial or biological origin (Fava *et al.*, 2004). These transport agents are non-hazardous and biodegradable and they might constitute economic solutions. Though results on PAH solubilization are encouraging, their efficacy may not be as expected for weathered soils, contaminated with heavy multicomponent hydrocarbon pollution. An undesirable effect of the utilization of vegetable oils is the possibility of oil residues remaining in the soil and, hence, further treatment steps may be required, while for heavily contaminated soil oil saturation may occur (Pannu *et al.*, 2004). Surfactants may have some constraints too. As for a successful utilization, the major part of the initially added amount should be recoverable while for *in situ* applications, surfactant penetration can be challenged from the point of view of soil characteristics (decreased porosity through clogging) and its escaping from the site of interest is quite possible (Hamby, 1996; Mulligan *et al.*, 2001).

Biotic remediation of PAH polluted soils

Biological remediation of soils characterized by polyaromatic contamination can be carried out by a set of biological systems and natural processes in order to clean the polluted medium. The most widely utilized techniques for the elimination of PAHs are related to phytoremediation, application of oxidative enzymes, bioaugmentation and biostimulation. However, it must be said that most of these are still regarded as experimental.

Phytoremediation is an emerging technology that uses plants and trees to clean up or sequester *in situ* organic and inorganic pollutants from the environment. In this type of soil remediation, the main player is the root system which excretes several compounds. These may include PAH-degrading enzymes or substances that increase the bioavailability of pollutants and thus they promote microbial proliferation and enhance microbial activity near the rhizosphere (Bert *et al.*, 2009; Trapp and Karlson, 2001). Phytoremediation is a process considered to be economic, effective, natural and aesthetic, as it is a non-intrusive technology (Sinha *et al.*, 2007). However, phytoremediation has many limitations as plants are not suitable for highly polluted areas (toxicity effects and limited metabolism), tree growth demands long periods for appropriate management of a contaminated site and transformation of pollutants is mostly rhizosphere-limited (Susarla *et al.*, 2002). In addition, these applications are closely dependent on the soil itself: factors such as oxygen, water, nutrients, pH, soil texture, salinity etc. can have major effects upon the vitality and function of roots and associated microflora (Cunningham *et al.*, 1995). In general, phytoremediation is considered to be a slow, long-term remedial technique.

Another potentially useful means of treating sites polluted with polyaromatics is the utilization of enzymes, mostly of fungal origin. Bacteria are the more abundant organisms in soil but fungi can form more voluminous biomass and, in addition, they seem to be more apt to bringing about complete mineralization of persistent organic pollutants such as PAHs. The various enzymes excreted by several fungi and targeting the oxidation of polyaromatics, such as oxidases, peroxidases, laccases and others, gives them some competitive advantages to overcome drawbacks in comparison with degrading bacteria. As cited by Mougin *et al.* (2009), enzymes can be active and more stable in the face of a wide range of physicochemical changes in the soil matrix (pH, temperature, salinity); changes in pollutant concentration may be fatal for microbial cells but enzyme sensitivity is typically not affected to the same extent, enzyme addition into soils is often more feasible than the introduction of whole microorganisms and an enzymatic amendment does not necessarily need the laborious control of several parameters or have the environmental impact that microbes can cause. Moreover, enzymes are not susceptible to predators. The disadvantages that such *in situ* applications may present are the potential loss of activity because of natural degradation of these proteinaceous catalysts, ineffective interactions between substrate and catalyst, or even absorption, immobilization and entrapment of the enzymes in the matrix with important mostly negative consequences on their mobility and activity (Gianfreda and Rao, 2004). Though immobilization is often shown to be beneficial in lab-scale experiments, the interactions of enzymes with clay minerals and SOM do not automatically guarantee the same degree of effectiveness for real-life, *in situ* applications (Wallenstein and Weintraub, 2008). In any case, the properties of enzymes can be better exploited with the continuous advances in biomolecular engineering in order to produce optimized biocatalysts, with high reactivity, enhanced specificity and stability, at a higher scale and at a lower cost without having to face the regulatory constraints that genetically engineered microorganisms (GEMs) may present.

The concept of GEMs has emerged as a solution to the limitations that the utilization of natural degrading microorganisms has revealed in bioremediation approaches based on either bioaugmentation or biostimulation strategies.

Bioaugmentation consists of introducing competent microbes (single species or consortia) into the contaminated soils in order to destroy several recalcitrant pollutants while biostimulation aims to stir up the intrinsic capacity of the natural microbiota by providing the necessary nutrients and localized propitious conditions (e.g. redox potential, pH). Both techniques are meant to enhance natural attenuation in which decontamination takes place without human intervention but only with monitoring of the typically long, natural clean–up process. Several studies have demonstrated the positive effects of bioaugmentation (Trzesicka-Mlynarz and Ward, 1996; Kastner *et al.*, 1998; Hwang and Cutright, 2002; Ruberto *et al.*, 2003; Potin *et al.*, 2004; Jacques *et al.*, 2008; Teng *et al.*, 2010) and biostimulation approaches (Guerin, 1999, 2000; Liebeg and Cutright, 1999; Yu *et al.*, 2005; Silva *et al.*, 2009). For instance, laboratory-scale bioaugmentation using a new *Paracoccus* sp. isolate with the capacity to use fluoranthene, pyrene and benzo[a]pyrene as sole carbon and energy source for growth was found promising for the removal of high-MW PAHs from weathered contaminated soil (Teng *et al.*, 2010). The main advantages and drawbacks of bioaugmentation, including successful but also unsuccessful examples of both techniques, have been reviewed in several articles (Hamby, 1996; Vogel, 1996; van Veen *et al.*, 1997; El Fantroussi and Agathos, 2005; Samanta *et al.*, 2002; Hamdi *et al.*, 2007). Advantages comprise mostly the low cost (installation and operational), non-intensive (low demand for energy, low consumption of chemicals, additives, etc.), environmentally friendly character (natural process, non-disrupting, non-exhaustive) of the bioremediation context. It should also be noted that, contrary to the majority of conventional abiotic treatments, most bioremediation techniques result in full or partial transformation/destruction and not in translocation of the problem. However, there are some factors that may limit their success. In the case of bioaugmentation, the main limitation reflects the frequent inability of competent strains with well-established degrading capacities at the lab-scale to have a similar performance once they get introduced into the soil matrix. No guarantee can be given a priori whether a microbe can survive in the soil jungle or if it will have a long-lasting viability while maintaining high density and catabolic activity of its cells (Verstraete and Top, 1999). If a bacterial candidate is to survive and not to get neutralized by natural predators with a subsequent decline of the initially added biomass, it will have to fit in the ecological niche, compete with the indigenous microorganisms for the available carbon and space in order to sustainably colonize the soil. The autochthonous population is considered to be more stable and adapted to the conditions of its own niche (not counting the possibility of being more 'familiar' with soil pollutants) and thus competition or even antagonistic phenomena for the available resources (C, N, P, oxygen, water, etc.) may arise (van Veen *et al.*, 1997). Apart from biotic factors, abiotic factors may be directly or indirectly the causes for a possible decline in population size and activity of the introduced microbial inoculants. Soil characteristics (porosity, texture, organic matter, humidity, oxygen level, pH, temperature, etc.) have been reported to affect the mobility and colonization of the introduced population in soil (Vogel, 1996; van Veen *et al.*, 1997; Kastner *et al.*, 1998). In addition, failure can be recorded for other reasons such as natural existing substances which inhibit the growth or the catabolic activity of the added organism, utilization of organic substrates other than the polyaromatic contaminant(s), inhibition phenomena, the concentration of the contaminant itself which if it is too low may not support growth, the oligotrophic conditions that in general predominate in the soil matrix, the availability and the age of the pollutants, toxic byproducts or end products, etc. (Shuttleworth and Cerniglia, 1995; Verstraete and Top, 1999; Bamforth and Singleton, 2005; Johnsen *et al.*, 2005). To conclude, many biotic and abiotic factors can hamper an exogenously added microorganism or consortium while the microbes themselves are constantly in a dynamic state of physicochemical response to environmental changes.

Concerning the biostimulation-based remedial approaches, once the indigenous population becomes metabolically active towards the organic pollutants, it finds itself in its own environmental niche. No habitat disturbance is about to occur, the basic survival attributes in the soil microenvironment are already present, the range and variability of the environmental factors exist in the 'memory' of the microbial community and these are advantages to exploit in the decontamination concept by ameliorating some issues (nutrients, aeration, pH correction, pollutants solubilization). Of course, the acclimatization period after a contamination occurs is a decisive factor based on some of these parameters (Leahy and Colwell, 1990). As stated in (Boopathy, 2000), the molecular structure (in terms of possible substrate) and pollutant concentration, the biomass concentration and induction of enzymatic activity, the progressive enrichment and establishment of the hydrocarbon-oxidizing microorganisms, the potential toxicity effect and bioavailability of the contaminants, can affect the adaptation stage of the natural microflora and, ultimately, the outcome of the remediation. Even if the environmental conditions are favourable and the appropriate degrading population is established, PAH contaminated soils and especially the weathered ones are rarely found to be completely cleaned up; this problem is linked to the inaccessibility and unavailability of the polyaromatic compounds and their very slow rate of release which makes them only barely available to microbes in a reasonably short time (Alexander, 2000; Johnsen *et al.*, 2005). Mass transfer of hydrocarbons to cells is an important determinant for bioremediation rate and extent and contributes to the long treatment periods that characterize the context of bioremediation.

In order to avoid the limitations of biotic and abiotic origin occurring during the application of bioremedial approaches, the idea of GEMs has emerged as it aims to create new, effective metabolic pathways without the presence of toxic by- or end-products, to expand the substrate range of microbes, develop their performance, increase their genetic stability related to their catabolic activity, augment the bioavailability of pollutants (e.g. by surfactant production), etc. (Timmis and Pieper, 1999; Samanta *et al.*, 2002). Bacterial adaptation to antibiotic resistance has shown the role of horizontal gene transfer in soil populations and the distribution of catabolic gene clusters involved in the catabolism of xenobiotics provides strong evidence for such transfer (de la Cruz and Davies, 2000). Some studies have demonstrated the potential role of GEMs and genetic mobile elements as contributing to enhanced biodegradation (Garbisu and Alkorta, 1999; Sayler and Ripp, 2000; Top *et al.*, 2002). However, there are many biosafety considerations and questions to be elucidated before a controlled release of GEMs into the environment. Such a release has been proposed for applying a plasmid-mediated bioaugmentation or an *in situ* monitoring of the decontamination process as guided by a biomarked, genetically engineered population.

To conclude, the main problems of the abiotic techniques are their elevated implementation and operational cost, the labour and time involved and the toxicity and impact on humans and the environment, along with the fact that they do not necessarily destroy the contaminants but 'ship' the problem elsewhere without fully resolving it. On the other hand, bioremediation demands long restoration times while the survival and metabolic activity of microbes is easily hampered by numerous and not always adequately controllable factors. Nowadays, the most commonly applied treatments for remediating soils involve the combination of abiotic means with biological degradation.

Ex situ bioremediation of PAH polluted soils and sediments

In many cases, especially when a matrix such as soil or sediment has been heavily polluted with PAHs, *ex situ* bioremediation may be indicated. This implies removal of the contaminated material to a site where an appropriately designed and engineered treatment system can be installed. Typical *ex situ* approaches for PAH treatment include simpler technologies like landfarming and more engineering-intensive setups like bioreactors.

In landfarming the contaminated soil or, more typically, dredged sediment is applied on a dedicated soil surface to allow the indigenous microorganisms to degrade the contaminants aerobically following a treatment involving ploughing on a large exposed surface area (Harmsen, 2004; Atlas and Philp, 2005). This exposure to air enables aerobic microorganisms to biodegrade the contaminants. As discussed earlier in the context of soil, PAHs are also strongly adsorbed by the organic matter in the sediment, hence biodegradation of these molecules in landfarmed sediment is typically a slow process and often high residual concentrations remain for a long time. The biodegradation process is influenced by (a) the desorption and diffusion of the PAHs within soil aggregates towards locations with adequate presence of metabolically active microorganisms, (b) the relative abundance of such microorganisms (potentially increased by biostimulation) and (c) the effectiveness of the aeration during this treatment (Harmsen, 2004). Extensive long-term observation of the behaviour of PAHs of various MWs in landfarmed sediments in the Netherlands has shown that PAHs can be distinguished into fast, slow and very slowly degradable fractions, whose full degradation requires, respectively, time lengths of the order of one year, seven years and 30 or more years (Harmsen, 2004). The concept of bioavailability and its quantification are useful in predicting the extent of biodegradation by modelling (Harmsen, 2004, 2007). Engineering interventions in order to optimize the degradation process (e.g. dewatering, temperature, aeration and biostimulation by nutrient addition) results in speeding up the biodegradation of only the rapidly degradable fraction – in other words, optimization of landfarming is not generally effective (Harmsen, 2004).

Ex situ bioremediation of PAHs in soils or sediments in slurry systems (ranging from simple flasks to more complex bioreactors) appears more promising for contaminant removal and destruction, especially with the help of adapted microbial strains, as illustrated in the case of high naphthalene concentrations (Piskonen *et al.*, 2005; Filonov *et al.*, 2006). More generally, the use of bioreactors offers the possibility of monitoring the system's performance and maintaining optimal operating conditions which ensure the microorganisms' biodegradative activity and, ultimately, shorter remediation times compared with other technologies (Banerjee *et al.*, 1995; Woo and Park, 1999). Since the biodegradation of soil- or sediment-borne PAHs and especially of those of higher MW is highly restricted because of diminished bioavailability (Johnsen *et al.*, 2005; Harmsen, 2007), this limitation can be overcome through the bioreactor's homogenization/mixing system that enhances mass transfer. Moreover, bioreactor technology is fully compatible with specific techniques mentioned previously in the context of *in situ* remediation, that aim at increasing microbial activity and overcoming bioavailability barriers, such as biostimulation, bioaugmentation, the addition of biosurfactants and the incorporation of bulking materials (Atlas and Philp, 2005).

The contribution of a liquid phase, rendering the contaminated soil or sediment into a slurry has been studied in the generic system called slurry bioreactor (SB) which is applicable not only to PAHs but also to many other hydrophobic and recalcitrant pollutants like PCBs, nitroaromatics, halogenated hydrocarbons and other substances found in non-aqueous phase liquids (NAPLs). Process fundamentals and applications of SB together with emerging trends have been recently reviewed (Robles-Gonzalez *et al.*, 2008). The versatility of SB is evident from the potential to optimize not only purely engineering parameters (e.g. mixing and aeration) but also microbial activity through the selection and monitoring of the biodegrader strains(s) (Collina *et al.*, 2005).

The good homogenization capacity of a basic bioreactor design, the rotating drum, has been known for many years and has been used advantageously, for instance, in the biological treatment of anthracene-contaminated soil (Banerjee *et al.*, 1995). Engineering parameters including the degree and mode of mixing, the particle size distribution and the water content in the soil slurry proved decisive in the performance of a rotating drum bioreactor treating PAH-contaminated soil (Woo and Park, 1999). Evolutions of this design include baffled roller bioreactors (rotating cylindrical vessels equipped with baffles) and bead-mill bioreactors (rotating cylindrical vessels partially filled with glass beads). A laboratory-scale baffled roller bioreactor for naphthalene and phenol treatment from a NAPL has validated remarkable increases in mass transfer from the NAPL to the aqueous phase and subsequent high-rate degradation of naphthalene by *Pseudomonas putida* at pollutant concentrations reaching 3 g/l (Nikakhtari *et al.*, 2008). A baffled roller bioreactor and a bead-mill bioreactor were compared for the bioremediation of soil and water contaminated with naphthalene and its derivatives (Yu *et al*, 2006). Although the mass transfer in the bead-mill bioreactor was more than 2-fold higher than that of the baffled bioreactor, the rate of PAH biodegradation was only marginally higher. Therefore the baffled bioreactor design seems preferable, given also its higher effective working volume and its simpler design (Yu *et al.*, 2006). Extensions of this work have confirmed the strong impact of bead loading, bead size and rotation speed on the mass transfer coefficient (Cazakoff *et al.*, 2008) and the group has attempted to obtain insights into the effects of these factors upon scale-up (Wang *et al.*, 2008; Nikakhtari *et al.*, 2010).

The quest for enhancing bioavailability to optimize PAH biodegradation has made use of the concept of the two-phase partitioning bioreactor (TPPB). This concept is based on the use of a biocompatible, water-immiscible organic solvent in which high concentrations of recalcitrant hydrophobic substrates such as PAHs are dissolved. The PAHs partition into the cell-containing aqueous phase at rates determined by the metabolic activity of the cells. After the initial proof of the concept in model systems (Janikowski *et al.*, 2002; MacLeod and Daugulis, 2005) it is important to assess the practical implementation of TPPBs for the degradation of soil- or sediment-borne PAHs on a realistic scale.

In conclusion, bioreactor technology gives the possibility of manipulating a number of environmental parameters that can lead to intensified and fast treatment of PAH-contaminated soils and sediments, as it allows interfacing with several targeted techniques: nutrient and redox modulation (biostimulation), introduction of selected inocula (bioaugmentation), enhancement of pollutant bioavailability (introduction of 'mobilizing' and surface-active agents), etc. In addition, bioreactor technology is progressively interfacing with molecular microbiology tools such as fingerprinting and other culture-independent techniques (e.g. PCR coupled to DGGE) that allow a much more rational monitoring and control of the dynamics of pollutant biodegradation.

References

Acharya, P., and Ives, P. (1994). Incineration at Bayou Bounfouca remediation project. Waste Manag. *14*, 13–26.

Alexander, M. (2000). Aging, bioavailability, and overestimation of risk from environmental pollutants. Environ. Sci. Technol. 34, 4259–4265.

Alves de Lima Ribeiro, F., and Ferreira, M.M.C. (2003). QSPR models of boiling point, octanol-water partition coefficient and retention time index of polycyclic aromatic hydrocarbons. J. Mol. Struct. Theor. Chem. *663*, 109–126.

Anitescu, G., and Tavlarides, L.L. (2006). Supercritical extraction of contaminants from soils and sediments. J. Supercrit. Fluids *38*, 167–180.

Appleton, T.J, Colder, R.I., Kingman, S.W., Lowndes, I.S., and Read, A.G. (2005). Microwave technology for energy-efficient processing of waste. Appl. Energy *81*, 85–113.

Asgher, M., Bhatti, H., Ashraf, M., and Legge, R. (2008). Recent developments in biodegradation of industrial pollutants by white rot fungi and their enzyme system. Biodegradation *19*, 771–783.

Atkinson, R., and Arey, J. (1994). Atmospheric chemistry of gas-phase polycyclic aromatic hydrocarbons: formation of atmospheric mutagens. Environ. Health Perspect. *102*, 117–126.

Atlas, R., and Philp, J. (2005) Bioremediation – Applied Microbial Solutions for Real-World Environmental Cleanup. (ASM Press, American Society for Microbiology: Washington, DC).

Balaban, A.T. (1980). Is aromaticity outmoded? Pure Appl. Chem. *52*, 1409–1429

Balaban, A.T., and Randić, M. (2005). Partitioning of π -electrons in rings of polycyclic conjugated hydrocarbons: Part 6. Comparison with other methods for estimating the local aromaticity of rings in polycyclic benzenoids. J. Math. Chem. 37, 443–453.

Balkwill, D., Fredrickson, J., and Romine, M. (2006). *Sphingomonas* and related genera. In The Prokaryotes, Dworkin, M., Falkow, S., Rosenberg, E., Schleifer, K.-H., Stackebrandt, E., eds. (Spinger: New York).

Bamforth, S.M., and Singleton, I. (2005). Bioremediation of polycyclic aromatic hydrocarbons: current knowledge and future directions. J. Chem. Tech. Biotechnol. *80*, 723–736.

Banerjee, D.K., Fedorak, P.M., Hashimoto, A., Masliyah, J.H., Pickard, M.A., and Gray, M.R. (1995). Monitoring the biological treatment of anthracene-contaminated soil in a rotating-drum bioreactor. Appl. Microbiol. Biotechnol. *43*, 521–528.

Baumard, P., Budzinski, H., Garrigues, P., Sorbe, J.C., Burgeot, T., and Bellocq, J. (1998). Concentrations of PAHs (polycyclic aromatic hydrocarbons) in various marine organisms in relation to those in sediments and to trophic level. Mar. Pollut. Bull. *36*, 951–960.

Becnel, J.M., and Dooley, K.M. (1998). Supercritical fluid extraction of polycyclic aromatic hydrocarbon mixtures from contaminated soils. Ind. Engineer. Chem. Res. 37, 584–594.

Berg, B.E., Lund, H.S., Kringstad, A., and Kvernheim, A.L. (1999). Routine analysis of hydrocarbons, PCB and PAH in marine sediments using supercritical CO2 extraction. Chemosphere *38*, 587–599.

Berset, J.D., Ejem, M., Holzer, R., and Lischer, P. (1999). Comparison of different drying, extraction and detection techniques for the determination of priority polycyclic aromatic hydrocarbons in background contaminated soil samples. Anal. Chim. Acta *383*, 263–275.

Bert, V., Seuntjens, P., Dejonghe, W., Lacherez, S., Thuy, H., and Vandecasteele, B. (2009). Phytoremediation as a management option for contaminated sediments in tidal marshes, flood control areas and dredged sediment landfill sites. Environ. Sci. Pollut. Res. *16*, 745–764.

Bogan, B.W., and Trbovic, V. (2003). Effect of sequestration on PAH degradability with Fenton's reagent: roles of total organic carbon, humin, and soil porosity. J. Hazard. Mater. *100*, 285–300.

Bogan, B.W., Trbovic, V., and Paterek, J.R. (2003). Inclusion of vegetable oils in Fenton's chemistry for remediation of PAH-contaminated soils. Chemosphere *50*, 15–21.

Boopathy, R. (2000). Factors limiting bioremediation technologies. Bioresour. Technol. *74*, 63–67.

Bosch, R., García-Valdés, E., and Moore, E.R.B. (1999). Genetic characterization and evolutionary implications of a chromosomally encoded naphthalene-degradation upper pathway from *Pseudomonas stutzeri* AN10. Gene *236*, 149–157.

Bosch, R., García-Valdés, E., and Moore, E.R.B. (2000). Complete nucleotide sequence and evolutionary significance of a chromosomally encoded naphthalene-degradation lower pathway from *Pseudomonas stutzeri* AN10. Gene *245*, 65–74.

Bouchez-Naitali, M., Blanchet, D., Haeseler, F., and Vandecasteele, J.P. (2008). Biodegradation of polycyclic aromatic hydrocarbons (PAHs). In Petroleum Microbiology, Vandecasteele, J.P., ed. (Editions Technip: Paris, France), pp. 342–412.

Brezna, B., Khan, A.A., and Cerniglia, C.E. (2003). Molecular characterization of dioxygenases from polycyclic aromatic hydrocarbon-degrading *Mycobacterium* spp. FEMS Microbiol. Lett. *223*, 177–183.

Brown, G.S,, Barton, L.L., and Thomson, B.M. (2003). Permanganate oxidation of sorbed polycyclic aromatic hydrocarbons. Waste Manag. *23*, 737–740.

Brunner, S., Hornung, E., Santl, H., Wolff, E., Piringer, O.G., Altschuh, J., and Brueggemann, R. (1990). Henry's law constants for polychlorinated biphenyls: experimental determination and structure–property relationships. Environ. Sci. Technol. *24*, 1751–1754.

Butler, C.S., and Mason, J. (1997). Structure-function analysis of the bacterial aromatic ring hydroxylating dioxygenases. In Advances in Microbial Physiology, Poole, R.K., ed. (Academic Press, London, UK), pp. 47–84.

Canet, R., Lopez-Real, J., and Beck, A.J. (1999). Overview of polycyclic aromatic hydrocarbon biodegradation by white-rot fungi. Land Contam. Reclam. 7, 191–197.

Cazakoff, D., Nemati, M., and Hill, G.A., (2008) Naphthalene mass transfer from a non-aqueous-phase-liquid (NAPL) in rotating baffled and bead mill bioreactors, Sep. Sci. Technol. *43*, 2103–2116.

Cerniglia, C.E. (1992). Biodegradation of polycyclic aromatic hydrocarbons. Biodegradation 3, 351–368.

Chakraborty, R., O'Connor, S.M., Chan, E., and Coates, J.D. (2005). Anaerobic degradation of benzene, toluene, ethylbenzene, and xylene compounds by *Dechloromonas* strain RCB. Appl. Environ. Microbiol. *71*, 8649–8655.

Chefetz, B., and Xing, B. (2009). Relative role of aliphatic and aromatic moieties as sorption domains for organic compounds: a review. Environ. Sci. Technol. *43*, 1680–1688.

Chen, B., Johnson, E.J., Chefetz, B., Zhu, L., and Xing, B. (2005). Sorption of polar and nonpolar aromatic organic contaminants by plant cuticular materials: role of polarity and accessibility. Environ. Sci. Technol. *39*, 6138–6146.

Chorover, J., and Brusseau, M.L. (2008). Kinetics of sorption – desorption. In Kinetics of Water-Rock Interaction, Brantley, S.L.K.J., White, A.F., ed. (Springer: New York), pp. 109–149.

Coates, J.D., Woodward, J., Allen, J., Philp, P., and Lovley, D.R. (1997). Anaerobic degradation of polycyclic aromatic hydrocarbons and alkanes in petroleum-contaminated marine harbor sediments. Appl. Environ. Microbiol. *63*, 3589–3593.

Collina, E., Bestetti, G., Di Gennaro, P., Franzetti, A., Gugliersi, F., Lasagni, M., and Pitea, D. (2005). Naphthalene biodegradation kinetics in an aerobic slurry-phase bioreactor. Environ. Int. *31*, 167–171

de la Cruz, F., and Davies, J. (2000). Horizontal gene transfer and the origin of species: lessons from bacteria. Trends Microbiol. *8*, 128–133.

Cunningham, S.D., Berti, W.R., and Huang, J.W. (1995). Phytoremediation of contaminated soils. Trends Biotechnol. *13*, 393–397.

Cuypers, C., Grotenhuis, T., Nierop, K.G.J., Franco, E.M., de Jager, A., and Rulkens, W. (2002). Amorphous and condensed organic matter domains: the effect of persulfate oxidation on the composition of soil/sediment organic matter. Chemosphere *48*, 919–931.

D'Adamo, R., Pelosi, S., Trotta, P., and Sansone, G. (1997). Bioaccumulation and biomagnification of polycyclic aromatic hydrocarbons in aquatic organisms. Marine Chem. *56*, 45–49.

Dadkhah, A.A., and Akgerman, A. (2002). Hot water extraction with in situ wet oxidation: PAHs removal from soil. J. Hazard. Mater. *93*, 307–320.

Daley, P.S. (1989). Cleaning up sites with on-site process plants. Environ. Sci.Technol. *23*, 912–916.

De Celis, R., Feria-Velasco, A., Gonzalez-Unzaga, M., Torres-Calleja, J., and Pedron-Nuevo, N. (2000). Semen quality of workers occupationally exposed to hydrocarbons. Fertil. Steril. *73*, 221–228.

Denome, S.A., Stanley, D.C., Olson, E.S., and Young, K.D. (1993). Metabolism of dibenzothiophene and naphthalene in *Pseudomonas* strains: complete DNA sequence of an upper naphthalene catabolic pathway. J. Bacteriol. *175*, 6890–6901.

Dunnivant, F.M., Elzerman, A.W., Jurs, P.C., and Hasan, M.N. (1992). Quantitative structure–property relationships for aqueous solubilities and Henry's law constants of polychlorinated biphenyls. Environ. Sci. Technol. *26*, 1567–1573.

Ehlers, G.A.C., and Loibner, A.P. (2006). Linking organic pollutant (bio)availability with geosorbent properties and biomimetic methodology: A review of geosorbent characterisation and (bio)availability prediction. Environ. Pollut. *141*, 494–512.

El Fantroussi, S., and Agathos, S.N. (2005). Is bioaugmentation a feasible strategy for pollutant removal and site remediation? Curr. Opin. Microbiol. *8*, 268–275.

Fava, F., Berselli, S., Conte, P., Piccolo, A., and Marchetti, L. (2004). Effects of humic substances and soya lecithin on the aerobic bioremediation of a soil historically contaminated by polycyclic aromatic hydrocarbons (PAHs). Biotechnol. Bioeng. *88*, 214–223.

Fent, K. (2004). Ecotoxicological effects at contaminated sites. Toxicology *205*, 223–240.

Ferrarese, E., Andreottola, G., and Oprea, I.A. (2008). Remediation of PAH-contaminated sediments by chemical oxidation. J. Hazard. Mater. *152*, 128–139.

Ferreira, M.M.C. (2001). Polycyclic aromatic hydrocarbons: a QSPR study. Chemosphere *44*, 125–146.

Field, J.A., de Jong, E., Feijoo Costa, G., and de Bont, J.A. (1992). Biodegradation of polycyclic aromatic hydrocarbons by new isolates of white rot fungi. Appl. Environ. Microbiol. *58*, 2219–2226.

Filonov, A.E., Puntus, I.F., Karpov, A.V., Kosheleva, I.A., Akhmetov, L.I., Yonge, D.R., Petersen, J.N., and Boronin, A.M. (2006) Assessment of naphthalene biodegradation efficiency of *Pseudomonas* and *Burkholderia* strains tested in soil model systems, J. Chem. Technol. Biotechnol. *81*, 216–224.

Flotron, V., Delteil, C., Padellec, Y., and Camel, V. (2005). Removal of sorbed polycyclic aromatic hydrocarbons from soil, sludge and sediment samples using the Fenton's reagent process. Chemosphere *59*, 1427–1437.

Fuenmayor, S.L., Wild, M., Boyes, A.L., and Williams, P.A. (1998). A gene cluster encoding steps in conversion of naphthalene to gentisate in *Pseudomonas* sp. strain U2. J. Bacteriol. *180*, 2522–2530.

Gaffney, J.S., Marley, N.A and Clark, S.B. (1996). Humic and fulvic acids and organic colloidal materials in the environment. In Humic and Fulvic Acids, American Chemical Society Symposium Series, vol. 651, Gaffney, J.S., Marley, N.A., and Clark, S.B., eds. (ACS: Washington, DC) pp. 2–16.

Gan, S., Lau, E.V., and Ng H.K. (2009). Remediation of soils contaminated with polycyclic aromatic hydrocarbons (PAHs). J. Hazard. Mater. *172*, 532–549.

Garbisu, C., and Alkorta, I. (1999). Utilization of genetically engineered microorganisms (GEMs) for bioremediation. J. Chem. Tech. Biotechnol. *74*, 599–606.

Gates-Anderson, D.D., Siegrist, R.L., and Cline, S.R., (2001). Comparison of potassium permanganate and hydrogen peroxide as chemical oxidants for organically contaminated soils. J. Environ. Eng. *127*, 337–347.

Gianfreda, L., and Rao, M.A. (2004). Potential of extra cellular enzymes in remediation of polluted soils: a review. Enzyme Microb. Technol. *35*, 339–354.

Gibson, D.T., and Parales, R.E. (2000). Aromatic hydrocarbon dioxygenases in environmental biotechnology. Curr. Opin. Biotechnol., *11*, 236–243.

Goi, A., and Trapido, M. (2004). Degradation of polycyclic aromatic hydrocarbons in soil: The fenton reagent versus ozonation. Environ. Technol. *25*, 155–164.

Gong, Z., Alef, K., Wilke, B.M., and Li, P. (2005). Dissolution and removal of PAHs from a contaminated soil using sunflower oil. Chemosphere *58*, 291–298.

Govers, H., Ruepert, C., and Aiking, H. (1984). Quantitative structure–activity relationships for polycyclic aromatic hydrocarbons: Correlation between molecular connectivity, physico-chemical properties, bioconcentration and toxicity in *Daphnia pulex*. Chemosphere *13*, 227–236.

Goyal, A.K., and Zylstra, G.J. (1996). Molecular cloning of novel genes for polycyclic aromatic hydrocarbon degradation from *Comamonas testosteroni* GZ39. Appl. Environ. Microbiol. *62*, 230–236.

Goyal, A.K., and Zylstra, G.J. (1997). Genetics of naphthalene and phenanthrene degradation by *Comamonas testosteroni*. J. Ind. Microbiol. Biotechnol. *19*, 401–407.

Gray JS (2002). Biomagnification in marine systems: the perspective of an ecologist. Mar. Pollut. Bull. *45*, 46–52.

van Grevenynghe, J., Rion, S., Le Ferrec, E., Le Vee, M., Amiot, L., Fauchet, R., and Fardel, O. (2003). Polycyclic aromatic hydrocarbons inhibit differentiation of human monocytes into macrophages. J. Immunol. *170*, 2374–2381.

Guerin, T. (2000). The differential removal of aged polycyclic aromatic hydrocarbons from soil during bioremediation. Environ. Sci. Pollut. Res. *7*, 19–26.

Guerin, T.F. (1999). Bioremediation of phenols and polycyclic aromatic hydrocarbons in creosote contaminated soil using *ex situ* land treatment. J. Hazard. Mater. *65*, 305–315.

Hamby, D.M. (1996). Site remediation techniques supporting environmental restoration activities--a review. Sci. Total Environ. *191*, 203–224.

Hamdi, H., Benzarti, S., Manusadzianas, L., Aoyama, I., and Jedidi, N. (2007). Bioaugmentation and biostimulation effects on PAH dissipation and soil ecotoxicity under controlled conditions. Soil Biol. Biochem. *39*, 1926–1935.

Haritash, A.K., and Kaushik, C.P. (2009). Biodegradation aspects of polycyclic aromatic hydrocarbons (PAHs): A review. J. Hazard. Mater. *169*, 1–15.

Harms, H., and Bosma, T.N.P. (1997). Mass transfer limitation of microbial growth and pollutant degradation. J. Ind. Microbiol. Biotechnol. *18*, 97–105.

Harmsen, J. (2004). Landfarming of Polycyclic Aromatic Hydrocarbons and Mineral Oil Contaminated Sediments, Alterra Scientific Contribution 14. (Wageningen: the Netherlands).

Harmsen, J. (2007). Measuring bioavailability: From a scientific approach to standard methods. J. Environ. Qual. *36*, 1420–1428.

Horvath, R. (1972). Microbial co-metabolism and the degradation of organic compounds in nature. Bacteriol. Rev. *36*, 146–155.

Howsam, M., Jones, K.C., and Ineson, P. (2000). PAHs associated with the leaves of three deciduous tree species. I – Concentrations and profiles. Environ. Pollut. *108*, 413–424.

Huang, W., Peng, P., Yu, Z., and Fu, J. (2003). Effects of organic matter heterogeneity on sorption and desorption of organic contaminants by soils and sediments. Appl. Geochem. *18*, 955–972.

Hwang, S., and Cutright, T.J. (2002). Biodegradability of aged pyrene and phenanthrene in a natural soil. Chemosphere *47*, 891–899.

Iwabuchi, T., and Harayama, S. (1998). Biochemical and molecular characterization of 1-hydroxy-2-naphthoate dioxygenase from *Nocardioides* sp. KP7. J. Biol. Chem. *273*, 8332–8336.

Jacques, R.J.S., Okeke, B.C., Bento, F.M., Teixeira, A.S., Peralba, M.C.R., and Camargo, F.A.O. (2008). Microbial consortium bioaugmentation of a polycyclic aromatic hydrocarbons contaminated soil. Bioresour. Technol. *99*, 2637–2643.

Janikowski, T.B., Velicogna, D., Punt, M., and Daugulis, A.J. (2002). Use of a two-phase partitioning bioreactor for degrading polycyclic armatic hydrocarbons by a *Sphingomonas* sp. Appl. Microbiol. Biotechnol. *59*, 368–376.

Jeon, C.O., Park, M., Ro, H.-S., Park, W., and Madsen, E.L. (2006). The naphthalene catabolic (nag) genes of Polaromonas naphthalenivorans CJ2: evolutionary implications for two gene clusters and novel regulatory control. Appl. Environ. Microbiol. *72*, 1086–1095.

Johnsen, A., and Karlson, U. (2007). Diffuse PAH contamination of surface soils: environmental occurrence, bioavailability, and microbial degradation. Appl. Microbiol. Biotechnol. *76*, 533–543.

Johnsen, A.R., Wick, L.Y., and Harms, H. (2005). Principles of microbial PAH-degradation in soil. Environ. Pollut. *133*, 71–84.

Juhasz, A.L., and Naidu, R. (2000). Bioremediation of high molecular weight polycyclic aromatic hydrocarbons: a review of the microbial degradation of benzo[a] pyrene. Int. Biodeter. Biodegr. *45*, 57–88.

Kanaly, R.A., and Harayama, S. (2000). Biodegradation of high-molecular-weight polycyclic aromatic hydrocarbons by bacteria. J. Bacteriol. *182*, 2059–2067.

Kastner, M., Breuer-Jammali, M., and Mahro, B. (1998). Impact of inoculation protocols, salinity, and pH on the degradation of polycyclic aromatic hydrocarbons (PAHs) and survival of PAH-degrading bacteria introduced into soil. Appl. Environ. Microbiol. *64*, 359–362.

Kauppi, B., Lee, K., Carredano, E., Parales, R.E., Gibson, D.T., Eklund, H., and Ramaswamy, S. (1998). Structure of an aromatic ring-hydroxylating dioxygenase-naphthalene 1,2-dioxygenase. Structure *6*, 571–586.

Khan, A.A., Wang, R.F., Cao, W.W., Doerge, D.R., Wennerstrom, D., and Cerniglia, C.E. (2001). Molecular cloning, nucleotide sequence, and expression of genes encoding a polycyclic aromatic ring dioxygenase from *Mycobacterium* sp. strain PYR-1. Appl. Environ. Microbiol. *67*, 3577–3585.

Khodadoust, A.P., Bagchi, R., Suidan, M.T., Brenner, R.C., and Sellers, N.G. (2000). Removal of PAHs from highly contaminated soils found at prior manufactured gas operations. J. Hazard. Mater. *80*, 159–174.

Kim, S.J., Kweon, O., Freeman, J.P., Jones, R.C., Adjei, M.D., Jhoo, J.W., Edmondson, R.D., and Cerniglia, C.E. (2006). Molecular cloning and expression of genes encoding a novel dioxygenase involved in low- and high-molecular-weight polycyclic aromatic hydrocarbon degradation in *Mycobacterium vanbaalenii* PYR-1. Appl. Environ. Microbiol. *72*, 1045–1054.

Kimura, N., Kitagawa, W., Mori, T., Nakashima, N., Tamura, T., and Kamagata, Y. (2006). Genetic and biochemical characterization of the dioxygenase involved in lateral dioxygenation of dibenzofuran from *Rhodococcus opacus* strain SAO101. Appl. Microbiol. Biotechnol. *73*, 474–484.

Kirk, T.K., and Farrell, R.L. (1987). Enzymatic 'combustion': the microbial degradation of lignin. Annu. Rev. Microbiol. *41*, 465–501.

Kiyohara, H., Nagao, K., Kouno, K., and Yano, K. (1982). Phenanthrene-degrading phenotype of *Alcaligenes faecalis* AFK2. Appl. Environ. Microbiol. *43*, 458–461

Kiyohara, H., Torigoe, S., Kaida, N., Asaki, T., Iida, T., Hayashi, H., and Takizawa, N. (1994). Cloning and characterization of a chromosomal gene cluster, pah, that encodes the upper pathway for phenanthrene and naphthalene utilization by *Pseudomonas putida* OUS82. J. Bacteriol. *176*, 2439–2443.

Koelmans, A.A., Jonker, M.T.O., Cornelissen, G., Bucheli, T.D., Van Noort, P.C.M., and Gustafsson, Ö. (2006). Black carbon: The reverse of its dark side. Chemosphere *63*, 365–377.

Krivobok, S., Kuony, S., Meyer, C., Louwagie, M., Willison, J.C., and Jouanneau, Y. (2003). Identification of pyrene-induced proteins in *Mycobacterium* sp. strain 6PY1: evidence for two ring-hydroxylating dioxygenases. J. Bacteriol. *185*, 3828–3841.

Kronholm, J., Desbands, B., Hartonen, K., and Riekkola, M.L. (2002a). Environmentally friendly laboratory-scale remediation of PAH-contaminated soil by using pressurized hot water extraction coupled with pressurized hot water oxidation. Green Chem. *4*, 213–219.

Kronholm, J., Kalpala, J., Hartonen, K., and Riekkola, M.-L. (2002b). Pressurized hot water extraction coupled with supercritical water oxidation in remediation of sand and soil containing PAHs. J. Supercrit. Fluids *23*, 123–134.

Krygowski, T.M. (1993). Crystallographic studies of inter- and intramolecular interactions reflected in aromatic character of pi-electron systems. J. Chem. Inf. Comput. Sci. *33*, 70–78.

Krygowski, T.M., and Cyranski, M. (1996). Separation of the energetic and geometric contributions to the aromaticity. Part IV. A general model for the [pi]-electron systems. Tetrahedron *52*, 10255–10264.

Krygowski, T.M., Cyranski, M.K., Czarnocki, Z., Haefelinger, G., and Katritzky, A.R. (2000). Aromaticity: a theoretical concept of immense practical importance. Tetrahedron *56*, 1783–1796.

Kurkela, S., Lehvaslaiho, H., Palva, E.T, Teeri and T.H. (1988). Cloning, nucleotide sequence and characterization of genes encoding naphthalene dioxygenase of *Pseudomonas putida* strain NCI9816. Gene *7392*, 355–362.

Larkin, M.J., Allen, C.C., Kulakov, L.A., and Lipscomb, D.A. (1999). Purification and characterization of a novel naphthalene dioxygenase from *Rhodococcus* sp. strain NCIMB12038. J. Bacteriol. *181*, 6200–6204.

Larkin, M.J., Kulakov, L.A., and Allen, C.C.R. (2005). Biodegradation and *Rhodococcus* – masters of catabolic versatility. Curr. Opin. Biotechnol. *16*, 282–290.

Larkin, M.J., Kulakov, L.A., and Allen, C.C.R. (2010). *Rhodococcus*: genetics and functional genomics. In Handbook of Hydrocarbon and Lipid Microbiology, Timmis, K.N., ed. (Springer: New York), pp. 1345–1353.

Laupeze, B., Amiot, L., Sparfel, L., Le Ferrec, E., Fauchet, R., and Fardel, O. (2002). Polycyclic aromatic hydrocarbons affect functional differentiation and maturation of human monocyte-derived dendritic cells. J. Immunol. *168*, 2652–2658.

Laurie, A.D., and Lloyd-Jones, G. (1999). The *phn* genes of *Burkholderia* sp. strain RP007 constitute a divergent gene cluster for polycyclic aromatic hydrocarbon catabolism. J. Bacteriol. *181*, 531–540.

Leahy, J.G., and Colwell, R.R. (1990). Microbial degradation of hydrocarbons in the environment. Microbiol. Mol. Biol. Rev. *54*, 305–315.

Lee, R. (2003). Photo-oxidation and photo-toxicity of crude and refined oils. Spill Sci. Technol. Bull. *8*, 157–162.

Lerda, D. 2009. Polycylic Aromatic Hydrocarbons (PAHs) Factsheet, 2nd edition. (Institute for Reference Materials and Measurements, European Commission: Geel, Belgium).

Liebeg, E.W., and Cutright, T.J. (1999). The investigation of enhanced bioremediation through the addition of macro and micro nutrients in a PAH contaminated soil. Int. Biodeter. Biodegr. *44*, 55–64.

Lindsey, M.E., Xu, G., Lu, J., and Tarr, M.A. (2003). Enhanced Fenton degradation of hydrophobic organics by simultaneous iron and pollutant complexation with cyclodextrins. Sci. Total Environ. *307*, 215–229.

Liu, L., Schmid, R., and Urlacher, V. (2006). Cloning, expression, and characterization of a self-sufficient cytochrome P450 monooxygenase from *Rhodococcus ruber* DSM 44319. App. Microbiol. Biotechnol. *72*, 876–882.

Loffredo, E., and Senesi, N. (2008). The role of natural organic matter (humic substances) on adsorption of pesticides possessing endocrine disruptor activity. In The Fate of Persistent Organic Pollutants in the Environment, Mehmetli, E., and Koumanova, B., eds. (Springer: Dordrecht), pp. 369–383.

Luthy, R.G., Aiken, G.R., Brusseau, M.L., Cunningham, S.D., Gschwend, P.M., Pignatello, J.J., Reinhard, M., Traina, S.J., Weber, W.J., and Westall, J.C. (1997). Sequestration of hydrophobic organic contaminants by geosorbents. Environ. Sci. Technol. *31*, 3341–3347.

Ma, W.C., Immerzeel, J., and Bodt, J. (1995). Earthworm and food interactions on bioaccumulation and disappearance in soil of polycyclic aromatic hydrocarbons: studies on phenanthrene and fluoranthene. Ecotoxicol. Environ. Saf. *32*, 226–232.

MacLeod, C.T., and Daugulis, A.J. (2005). Interfacial effects in a two-phase partitioning bioreactor: degradation of polycyclic aromatic hydrocarbons (PAHs) by a hydrophobic *Mycobacterium*. Proc. Biochem. *40*, 1799–1805.

Masten, S.J., and Davies, S.H.R. (1997). Efficacy of *in situ* ozonation for the remediation of PAH contaminated soils. J. Contam. Hydrol. *28*, 327–335.

Maturi, K., and Reddy, K.R. (2006). Simultaneous removal of organic compounds and heavy metals from soils by electrokinetic remediation with a modified cyclodextrin. Chemosphere *63*, 1022–1031.

Melendez-Colon, V.J., Luch, A., Seidel, A., and Baird, W.M. (1999). Cancer initiation by polycyclic aromatic hydrocarbons results from formation of stable DNA adducts rather than apurinic sites. Carcinogenesis *20*, 1885–1891.

Mirsal, I.A. (2008a). Planning and realisation of soil remediation. In Soil Pollution: Origin, Monitoring and Remediation, Mirsal, I.A., ed. (Springer: New York), pp. 265–281.

Mirsal, I.A. (2008b). Pollution mechanisms and soil-pollutants interaction. In Soil Pollution: Origin, Monitoring and Remediation, Mirsal, I.A., ed. (Springer: New York), pp. 175–197.

Mougin, C., Boukcim, H., and Jolivalt, C. (2009). Soil bioremediation strategies based on the use of fungal enzymes. In Advances in Applied Bioremediation, Singh, A., Ward, O.P., and Kuhad, R.C., eds (Springer: New York), pp. 123–149.

Mulligan, C.N., Yong, R.N., and Gibbs, B.F. (2001). Surfactant-enhanced remediation of contaminated soil: a review. Engin. Geo. *60*, 371–380.

Nam, J.-W., Nojiri, H., Yoshida, T., Habe, H., Yamane, H., and Omori, T. (2001a). New classification system for oxygenase components involved in ring-hydroxylating oxygenations. Biosci. Biotechnol. Biochem. *65*, 254–263.

Nam, K., Rodriguez, W., and Kukor, J.J. (2001b). Enhanced degradation of polycyclic aromatic hydrocarbons by biodegradation combined with a modified Fenton reaction. Chemosphere *45*, 11–20.

Nikakhtari, H., Kumar, P., Nemati, M., and Hill, G.A. (2008). Mass transfer and bioremediation of aromatics from NAPL in a baffled roller bioreactor. Process Saf. Environ. *86*, 252–258.

Nikakhtari, H., Song, W.N., Kumar, P., Nemati, M., and Hill, G.A. (2010). Scale up of diesel oil biodegradation in a baffled roller bioreactor, Chemosphere *79*, 1010–1016.

Nollet, L.M.L., (2007). Handbook of Water Analysis. (CRC Press: Taylor and Francis Group New York).

Pan, B., Ning, P., and Xing, B. (2008). Part IV – sorption of hydrophobic organic contaminants. Environ. Sci. Pollut. Res. *15*, 554–564.

Pannu, J.K., Singh, A., and Ward, O.P. (2004). Vegetable oil as a contaminated soil remediation amendment: application of peanut oil for extraction of polycyclic aromatic hydrocarbons from soil. Process Biochem. *39*, 1211–1216.

Parales, R.E., and Resnick, S.M. (2006). Aromatic ring hydroxylating dioxygenases. In *Pseudomonas,* Vol. 4, Ramos, J.-L., and Levesque, R.C., eds. (Springer: New York), pp. 287–341.

Parrish, Z.D., White, J.C., Isleyen, M., Gent, M.P.N., Iannucci-Berger, W., Eitzer, B.D., Kelsey, J.W., and Mattina, M.I. (2006). Accumulation of weathered polycyclic aromatic hydrocarbons (PAHs) by plant and earthworm species. Chemosphere *64*, 609–618.

Petitgirard, A., Djehiche, M., Persello, J., Fievet, P., and Fatin-Rouge, N. (2009). PAH contaminated soil remediation by reusing an aqueous solution of cyclodextrins. Chemosphere *75*, 714–718.

Pignatello, J.J. (1998). Soil organic matter as a nanoporous sorbent of organic pollutants. Adv. Colloid Interface Sci. 76–77, 445–467.

Pignatello, J.J. (2009). Bioavailability of contaminants in soil. In Advances in Applied Bioremediation, Singh, A., Ward, O.P., and Kuhad, R.C., eds (Springer: New York), pp. 35–71.

Pignatello, J.J., and Xing, B. (1995). Mechanisms of slow sorption of organic chemicals to natural particles. Environ. Sci. Technol. *30*, 1–11.

Pinyakong, O., Habe, H., and Omori, T. (2003a). The unique aromatic catabolic genes in sphingomonads degrading polycyclic aromatic hydrocarbons (PAHs). J. Gen. Appl. Microbiol. *49*, 1–19.

Pinyakong, O., Habe, H., Yoshida, T., Nojiri, H., and Omori, T. (2003b). Identification of three novel salicylate 1-hydroxylases involved in the phenanthrene degradation of *Sphingobium* sp. strain P2. Biochem. Biophys. Res. Commun. *301*, 350–357.

Piskonen, R., Nyyssonen, M., Rajamaki, T., and Itavaara, M. (2005). Monitoring of accelerated naphthalene-biodegradation in a bioaugmented soil slurry. Biodegradation *16*, 127–134.

Potin, O., Rafin, C., and Veignie, E. (2004). Bioremediation of an aged polycyclic aromatic hydrocarbons (PAHs)-contaminated soil by filamentous fungi isolated from the soil. Int. Biodeter. Biodegr. *54*, 45–52.

Priefert, H., O'Brien, X.M., Lessard, P.A., Dexter, A.F., Choi, E.E., Tomic, S., Nagpal, G., Cho, J.J., Agosto, M., Yang, L., *et al.* (2004). Indene bioconversion by a toluene inducible dioxygenase of *Rhodococcus* sp. I24. Appl. Microbiol. Biotechnol. *65*, 168–176.

Reddy, K.R., and Saichek, R.E. (2004). Enhanced electrokinetic removal of phenanthrene from clay soil by periodic electric potential application. J. Environ. Sci. Health A. Tox. Hazard. Subst Environ. Eng. *39*, 1189–1212.

Richter, P., Jiménez, M., Salazar, R., and Maricán, A. (2006). Ultrasound-assisted pressurized solvent extraction for aliphatic and polycyclic aromatic hydrocarbons from soils. J. Chromatogr. A *1132*, 15–20.

Rivas, F.J. (2006). Polycyclic aromatic hydrocarbons sorbed on soils: A short review of chemical oxidation based treatments. J. Hazard. Mater. *138*, 234–251.

Rivas, J., Gimeno, O., de la Calle, R.G., Portela, J.R., and de la Ossa, E.M. (2009a). Remediation of PAH spiked soils: Concentrated H2O2 treatment/continuous hot water extraction-oxidation. J. Hazard. Mater. *168*, 1359–1365.

Rivas, J., Gimeno, O., Mantell, C., Portela, J.R., Martinez de la Ossa, E.J., and de la Calle, R.G. (2009b). Supercritical CO2 extraction of PAHs on spiked soil: Co-solvent effect and solvent regeneration by ozonization. J. Hazard. Mater. *162*, 777–784.

Robles-Gonzalez, I.V., Fava, F., and Poggi-Varaldo, H.M. (2008). A review of slurry bioreactors for bioremediation of soils and sediments. Microb. Cell Fact. *7*, 5

Rodríguez Couto, S., and Toca Herrera, J.L. (2006). Industrial and biotechnological applications of laccases: A review. Biotechnol. Adv. *24*, 500–513.

Romine, M.F., Stillwell, L.C., Wong, K.-K., Thurston, S.J., Sisk, E.C., Sensen, C., Gaasterland, T., Fredrickson, J.K., and Saffer, J.D. (1999). Complete sequence of a 184-kilobase catabolic plasmid from *Sphingomonas aromaticivorans* F199. J. Bacteriol. *181*, 1585–1602.

Ruberto, L., Vazquez, S.C., and Mac Cormack, W.P. (2003). Effectiveness of the natural bacterial flora, biostimulation and bioaugmentation on the bioremediation of a hydrocarbon contaminated Antarctic soil. Int. Biodeter. Biodegr. *52*, 115–125.

Rubin, H. (2001). Synergistic mechanisms in carcinogenesis by polycyclic aromatic hydrocarbons and by tobacco smoke: a bio-historical perspective with updates. Carcinogenesis *22*, 1903–1930.

Rust, A.J., Burgess, R.M., Brownawell, B.J., and McElroy, A.E. (2004). Relationship between metabolism and bioaccumulation of benzo[alpha]pyrene in benthic invertebrates. Environ. Toxicol. Chem. *23*, 2587–2593.

Saim, Na, Dean, J.R., Abdullah, M.P., and Zakaria, Z. (1997). Extraction of polycyclic aromatic hydrocarbons from contaminated soil using Soxhlet extraction, pressurised and atmospheric microwave-assisted extraction, supercritical fluid extraction and accelerated solvent extraction. J. Chromatogr. *791*, 361–366.

Saito, A,. Iwabuchi, T., and Harayama, S. (1999). Characterization of genes for enzymes involved in the phenanthrene degradation in *Nocardioides* sp. KP7. Chemosphere *38*, 1331–1337.

Saito, A., Iwabuchi, T., and Harayama, S. (2000). A novel phenanthrene dioxygenase from *Nocardioides* sp. strain KP7: expression in *Escherichia coli*. J. Bacteriol. *182*, 2134–2141.

Samanta, S.K., Singh, O.V., and Jain, R.K. (2002). Polycyclic aromatic hydrocarbons: environmental pollution and bioremediation. Trends Biotechnol. *20*, 243–248.

Sayler, G.S., and Ripp, S. (2000). Field applications of genetically engineered microorganisms for bioremediation processes. Curr. Opin. Biotechnol *11*, 286–289.

Schäfer, A., Ustohal, P., Harms, H., Stauffer, F., Dracos, T., and Zehnder, A.J.B. (1998). Transport of bacteria in unsaturated porous media. J. Contam. Hydrol. *33*, 149–169.

von Schleyer, P.R., and Jiao, H. (1996). What is aromaticity? Pure Appl. Chem. *68*, 209–218.

Schuler, L., Jouanneau, Y., Ní Chadhain, S., Meyer, C., Pouli, M., Zylstra, G., Hols, P., and Agathos, S.N. (2009). Characterization of a ring-hydroxylating dioxygenase from phenanthrene-degrading *Sphingomonas* sp. strain LH128 able to oxidize benz[a]anthracene. Appl. Microbiol. Biotechnol. *83*, 465–475.

Scullion, J. (2006). Remediating polluted soils. Naturwissenschaften *93*, 51–65.

Seeker, R.W., and Koshland, C.P. (1990). Combustion byproduct formation: an overview. Combust. Sci. Technol. *47*, 1–6.

Semple, K.T., Cain, R.B., and Schmidt, S. (1999). Biodegradation of aromatic compounds by microalgae. FEMS Microbiol. Lett. *170*, 291–300.

Sho, M., Hamel, C., and Greer, C.W. (2004). Two distinct gene clusters encode pyrene degradation in *Mycobacterium* sp. strain S65. FEMS Microbiol. Ecol. *48*, 209–220.

Shuttleworth, K., and Cerniglia, E. (1995). Environmental aspects of PAH biodegradation. Appl. Biochem. Biotechnol. *54*, 291–302.

Silva, Í.S., Santos, EdCd, Menezes, CRd, Faria, AFd, Franciscon, E., Grossman, M., and Durrant, L.R. (2009). Bioremediation of a polyaromatic hydrocarbon contaminated soil by native soil microbiota and bioaugmentation with isolated microbial consortia. Bioresour. Technol. *100*, 4669–4675.

Sinha, R.K., Herat, S., and Tandon, P.K. (2007). Phytoremediation: role of plants in contaminated site management. In Environmental Bioremediation Technologies, Singh, S., Tripathi, R.D., eds (Springer: Heidelberg, Germany), pp. 315–330.

Stingley, R.L., Khan, A.A., and Cerniglia, C.E. (2004). Molecular characterization of a phenanthrene degradation pathway in *Mycobacterium vanbaalenii* PYR-1. Biochem. Biophys. Res. Commun. *322*, 133–146.

Susarla, S., Medina, V.F., and McCutcheon, S.C. (2002). Phytoremediation: An ecological solution to organic chemical contamination. Ecol. Eng. *18*, 647–658.

Sutherland, J.B. (1992). Detoxification of polycyclic aromatic hydrocarbons by fungi. J. Ind. Microbiol. Biotechnol. *9*, 53–61.

Takizawa, N., Kaida, N., Torigoe, S., Moritani, T., Sawada, T., Satoh, T., and Kiyohara, H. (1994). Identification and characterization of genes encoding polycyclic aromatic hydrocarbon dioxygenase and polycyclic aromatic hydrocarbon dihydrodiol dehydrogenase in *Pseudomonas putida* OUS82. J. Bacteriol. *176*, 2444–2449.

Teng, Y., Luo, Y.M., Sun, M.M., Liu, Z.J., Li, Z.G., and Christie, P. (2010). Effect of bioaugmentation by Paracoccus sp. strain HPD-2 on the soil microbial community and removal of polycyclic aromatic hydrocarbons from an aged contaminated soil. Bioresour. Technol. *101*, 3437–3443.

Timmis, K.N., and Pieper, D.H. (1999). Bacteria designed for bioremediation. Trends Biotechnol. *17*, 200–204.

Todeschini, R., Gramatica, P., Provenzani, R., and Marengo, E. (1995). Weighted holistic invariant molecular descriptors. Part 2. Theory development and applications on modeling physicochemical properties of polyaromatic hydrocarbons. Chemometr. Intell. Lab. Syst. *27*, 221–229.

Top, E.M., Springael, D., and Boon, N. (2002). Catabolic mobile genetic elements and their potential use in bioaugmentation of polluted soils and waters. FEMS Microbiol. Ecol. *42*, 199 – 208.

Torres, E., Bustos-Jaimes, I., and Le Borgne, S. (2003). Potential use of oxidative enzymes for the detoxification of organic pollutants. Appl. Catal. B. Environ. *46*, 1–15.

Trapp, S., and Karlson, U. (2001). Aspects of phytoremediation of organic pollutants. J. Soils Sediments *1*, 37–43.

Treadway, S.L., Yanagimachi, K.S., Lankenau, E., Lessard, P.A., Stephanopoulos, G., and Sinskey, A.J. (1999). Isolation and characterization of indene bioconversion genes from *Rhodococcus* strain I24. Appl. Microbiol. Biotechnol. *51*, 786–793.

Trzesicka-Mlynarz, D., and Ward, O.P. (1996). Degradation of fluoranthene in a soil matrix by indigenous and introduced bacteria. Biotechnol. Lett. *18*, 181–186.

van Veen, J.A., van Overbeek, L.S., and van Elsas, J.D. (1997). Fate and activity of microorganisms introduced into soil. Microbiol. Mol. Biol. Rev. *61*, 121–135.

Verstraete, W., and Top, E.M. (1999). Soil clean-up: lessons to remember. Int. Biodeter. Biodegr. *43*, 147–153.

Vogel, T.M. (1996). Bioaugmentation as a soil bioremediation approach. Curr. Opin. Biotechnol. *7*, 311–316.

Wackett, L.P. (2002). Mechanism and applications of Rieske non-heme iron dioxygenases. Enzyme Micob. Technol. *31*, 577–587.

Wallenstein, M.D., and Weintraub, M.N. (2008). Emerging tools for measuring and modeling the *in situ* activity of soil extracellular enzymes. Soil Biol. Biochem. *40*, 2098–2106.

Wang, W., Meng, B., Lu, X., Liu, Y., and Tao, S. (2007). Extraction of polycyclic aromatic hydrocarbons and organochlorine pesticides from soils: A comparison between Soxhlet extraction, microwave-assisted extraction and accelerated solvent extraction techniques. Anal. Chim. Acta. *602*, 211–222.

Wang, Y., Riess, R., Nemati, M., Hill, G.A., and Headly, J. (2008). Scale-up impacts on mass transfer and bioremediation of suspended naphthalene particles in bead mill bioreactors. Bioresour. Technol. *99*, 8143–8150.

Weber, W.J., McGinley, P.M., and Katz, L.E. (1992). A distributed reactivity model for sorption by soils and sediments. 1. Conceptual basis and equilibrium assessments. Environ. Sci. Technol. *26*, 1955–1962.

Weber, W.J., Huang, W., and Yu, H. (1998). Hysteresis in the sorption and desorption of hydrophobic organic contaminants by soils and sediments: 2. Effects of soil organic matter heterogeneity. J. Contam. Hydrol. *31*, 149–165.

Weber, W.J., LeBoeuf, E.J., Young, T.M., and Huang, W. (2001). Contaminant interactions with geosorbent organic matter: insights drawn from polymer sciences. Water Res. 35, 853–868.

Werlen, C., Kohler, H.-PE., and van der Meer, J.R. (1996). The broad substrate chlorobenzene dioxygenase and cis-chlorobenzene dihydrodiol dehydrogenase of *Pseudomonas* sp. strain P51 are linked evolutionarily to the enzymes for benzene and toluene degradation. J. Biol. Chem. *271*, 4009–4016.

Wesenberg, D., Kyriakides, I., and Agathos, S.N. (2003). White-rot fungi and their enzymes for the treatment of industrial dye effluents. Biotechnol. Adv. *22*, 161–187.

Wick, L.Y., Furuno, S., and Harms, H. (2010). Fungi as transport vectors for contaminants and contaminant-degrading bacteria. In Handbook of Hydrocarbon and Lipid Microbiology, Timmis, K.N., ed. (Springer: New York), pp. 1555–1561.

Woo, S.H., and Park, J.M. (1999). Evaluation of drum bioreactor performance used for decontamination of soil polluted with polycyclic aromatic hydrocarbons. J. Chem. Technol. Biotechnol. *74*, 937–944.

Xing, B. (1997). The effect of the quality of soil organic matter on sorption of naphthalene. Chemosphere *35*, 633–642.

Xing, B., and Pignatello, J.J. (1997). Dual-mode sorption of low-polarity compounds in glassy poly(vinyl chloride) and soil organic matter. Environ. Sci. Technol. *31*, 792–799.

Yang, Y., Chen, R., and Shiaris, M, (1994). Metabolism of naphthalene, fluorene and phenanthrene: preliminary characterization of a cloned gene cluster from *Pseudomonas putida* NCIB9816. J. Bacteriol. *176*, 2158–2164.

Yariv, S. (2002). Introduction to organo–clay complexes and interactions. In Organo–Clay Complexes and Interactions. Yariv, S., and Cross, H., eds (Marcel Dekker: New York), pp. 39–111.

Yen, K.M., and Gunsalus, I.C. (1982). Plasmid gene organization: naphthalene/salicylate oxidation. Proc. Natl. Acad. Sci. U.S.A. *79*, 874–878.

Yen, K.M., and Gunsalus, I.C. (1985). Regulation of naphthalene catabolic genes of plasmid NAH7. J. Bacteriol. *162*, 1008–1013.

Yu, K.S.H., Wong, A.H.Y., Yau, K.W.Y., Wong, Y.S., and Tam, N.F.Y. (2005). Natural attenuation, biostimulation and bioaugmentation on biodegradation of polycyclic aromatic hydrocarbons (PAHs) in mangrove sediments. Mar. Pollut. Bull. *51*, 1071–1077.

Yu, R., Nemati, M., Hill, G.A., and Headly, J. (2006). Mass transfer and bioremediation of naphthalene and methyl naphthalenes in baffled and bead mill bioreactors. Can. J. Chem. Eng. *84*, 349–355.

Zhang, C., and Bennett, G.N. (2005). Biodegradation of xenobiotics by anaerobic bacteria. Appl. Microbiol. Biotechnol. *67*, 600–618.

Zhou, N.-Y., Fuenmayor, S.L., and Williams, P.A. (2001). nag genes of *Ralstonia* (formerly *Pseudomonas*) sp. strain U2 encoding enzymes for gentisate catabolism. J. Bacteriol. *183*, 700–708.

Zhu, D., Hyun, S., Pignatello, J.J., and Lee, L.S. (2004). Evidence for π-π electron donor–acceptor interactions between π-donor aromatic compounds and π-acceptor sites in soil organic matter through pH effects on sorption. Environ. Sci. Technol. *38*, 4361–4368.

Zylstra, G.J., and Kim, E. (1997). Aromatic hydrocarbon degradation by *Sphingomonas yanoikuyae* B1. J. Ind. Microbiol. Biotechnol. *19*, 408–414.

Ring-hydroxylating Dioxygenases Involved in PAH Biodegradation: Structure, Function and Biodiversity

7

Yves Jouanneau, Florence Martin, Serge Krivobok and John C. Willison

Abstract

The first step in the biodegradation of PAHs by aerobic bacteria is catalysed by metalloenzymes known as ring-hydroxylating dioxygenases (RHDs). Because of the hydrophobic nature and chemical resistance of PAHs, their initial attack by RHDs is a difficult reaction, which is critical to the whole degradation process. This chapter gives an overview of the current knowledge on the genetics, structure, catalytic mechanism and diversity of RHDs involved in PAH degradation. In the past decade, the crystal structures of 10 RHDs have been determined, giving insights into the mechanism of substrate recognition and regioselectivity of dioxygenation. The reaction catalysed by the archetypal naphthalene dioxygenase has been investigated in detail, thus providing a better understanding of the RHD catalytic mechanism. Studies on the catabolic genes responsible for PAH degradation in several bacterial taxa have highlighted the great phylogenetic diversity of RHDs. The implementation of culture-independent methods has afforded means to further explore the environmental diversity of PAH-degrading bacteria and RHDs. Recent advances in this field now allow the *in situ* identification of bacteria responsible for pollutant removal. Further biotechnological developments based on microarrays and functional metagenomics should lead to the conception of molecular tools useful for the bioremediation of PAH-contaminated ecosystems.

Introduction

In the biodegradation of numerous aromatic hydrocarbons including polycyclic aromatic hydrocarbons (PAHs) by aerobic microorganisms, oxygenases play a central role. Most common oxygenases are either heme-containing cytochromes P450 that catalyse single hydroxylations, or dioxygenases that incorporate both oxygen atoms of the O_2 molecule into the substrate. Dioxygenases are metalloenzymes that generally contain non-heme iron at the active site and fall into two categories: ring-hydroxylating dioxygenases (RHDs), which catalyse a double hydroxylation on two adjacent carbons of the substrate, and ring-cleaving dioxygenases, which catalyse the opening of the ring of catecholic substrates (Harayama *et al.*, 1992). RHDs are multicomponent enzymes (EC 1.14.12.-) consisting of an oxygenase associated with one or two specific electron carriers. They utilize NAD(P)H as a reductant and transfer electrons to the active site for activation of dioxygen, a prerequisite to substrate hydroxylation. In most cases, the oxygenase component is hexameric ($\alpha_3\beta_3$), with each alpha subunit containing a Rieske-type [2Fe–2S] cluster and a mononuclear iron centre where the dioxygenation reaction takes place. RHDs are widely distributed in various taxa among Gram-positive and Gram-negative bacteria. To date, over 1300 RHDs have been identified on the basis of sequence similarities and have been compiled in the NCBI databases. Current knowledge, based on well-studied enzymes such as naphthalene dioxygenase, indicates that RHDs have relatively broad substrate specificities.

This chapter will focus primarily on RHDs involved in PAH degradation with special emphasis on enzyme systems that have been described in the last decade. RHDs from representative PAH-degrading bacteria isolated so far are compared on the basis of their genetic organization, sequence relatedness, and substrate specificity. Aspects related to the structure and catalytic mechanism of RHDs are only briefly discussed, but reviews on these topics have been published recently (Ferraro *et al.*, 2005; Kovaleva and Lipscomb, 2008). A section is devoted to recent studies exploring the biodiversity of PAH-degrading bacteria and associated RHDs in various terrestrial and aquatic environments using culture-independent methods. Finally, we present a brief outlook on the possible strategies to improve PAH bioremediation through bioengineering of RHDs. For general aspects of PAH degradation or RHD biocatalysis that are not covered in this chapter, the reader is referred to recently published reviews (Habe and Omori, 2003; Parales and Resnick, 2006; Doyle *et al.*, 2008; Peng *et al.*, 2008).

General properties and classification

Ring-hydroxylating dioxygenases, also called Rieske non-heme iron oxygenases are multicomponent enzymes generally consisting of a catalytic component with hydroxylase activity, associated with one or two electron carrier proteins. In most cases, the hydroxylase component is itself composed of two subunits, an alpha subunit of about 50 kDa and a beta subunit of around 20 kDa, which assemble in an $\alpha_3\beta_3$ heterohexamer. However, the hydroxylase component of carbazole dioxygenase contains only alpha subunits (Sato *et al.*, 1997). Each alpha subunit consists of two domains, the N-terminal Rieske domain, which contains a [2Fe–2S] cluster, and the C-terminal catalytic domain, which contains a mononuclear ferrous ion close to the substrate-binding site. The catalytic component needs electrons in order to activate oxygen at each cycle of hydroxylation of the substrate. Two auxiliary proteins, a ferredoxin and a flavin-containing oxidoreductase, usually provide the necessary reductant at the expense of NAD(P)H oxidation. Sometimes, the two proteins are fused in a single polypeptide chain, as is the case for benzoate dioxygenase (Karlsson *et al.*, 2002) and anthranilate dioxygenase (Eby *et al.*, 2001).

RHDs share common structural features, including quaternary structure and conserved residues in the Rieske and catalytic domains of the alpha subunit that are involved in binding of the metal centres (see below). However, they differ considerably in terms of amino acid sequence and with respect to the structure of the associated electron carriers. Almost 20 years ago, Batie *et al.* (1992) proposed a classification of dioxygenases based on the number of associated electron carriers and the nature of their redox centres. Three classes of RHDs were defined, comprising two-component enzyme systems such as the phthalate dioxygenase from *Burkholderia cepacia* (Correll *et al.*, 1992) (class I), three-component enzymes containing an FAD-containing reductase and a [2Fe–2S] ferredoxin, as exemplified by the biphenyl dioxygenase (Furusawa *et al.*, 2004) (class II), and three-component enzymes in which the reductase contains both a FAD and a [2Fe–2S] cluster, as represented by the naphthalene dioxygenase (Kauppi *et al.*, 1998) (class III). Later, new RHDs were identified and characterized, some of which did not fit into this classification system. This was the case for the carbazole dioxygenase from *Pseudomonas* sp. CA10 (Sato *et al.*, 1997) and for three-component enzymes in which the ferredoxin contains a [3Fe–4S] cluster instead of the more common [2Fe–2S] cluster (Martin and Mohn, 1999; Takagi *et al.*, 2005). Nam *et al.* (2001) therefore proposed an alternative classification system based on amino acid sequence similarities between the alpha subunits of hydroxylase components. RHDs were then organized into four families of proteins according to phylogenetic criteria, each family corresponding to types of enzyme with similar substrate specificities. For example, type II enzymes are represented by benzoate/toluate dioxygenases, type III enzymes by naphthalene and PAH dioxygenases, and type IV enzymes by biphenyl and toluene dioxygenases. Type I enzymes have in common the presence of a hydroxylase composed of alpha subunits only, and show greater disparity than the other families in terms of amino acid sequences and substrate range. An advantage of this classification system lies in the possibility to classify any new oxygenase sequence derived from a nucleotide sequence, even though the associated electron carriers are unknown. Such a situation is encountered in bacterial genomes where catabolic genes are dispersed, and especially when genes encoding hydroxylase components do not lie close to the genes encoding electron carriers. This is frequently the case in PAH-degrading bacteria such as Mycobacteria and Sphingomonadaceae (see below). Recently, a more sophisticated classification system has been proposed, which takes into account the sequences of associated electron carriers in the phylogenetic analysis of RHDs (Kweon *et al.*, 2008). In addition to the four subclasses of RHDs described by Nam *et al.*, a new group was defined comprising enzymes found in Gram-positive PAH-degrading bacteria (Saito *et al.*, 2000; Krivobok *et al.*, 2003; Kim *et al.*, 2006). A distinctive feature of these Type V RHDs is that their associated ferredoxin contains a [3Fe–4S] cluster, in contrast to type III PAH dioxygenases from Gram-negative bacteria, which use [2Fe–2S] ferredoxins. A phylogenetic tree based on alignments of α subunit sequences of representative RHDs is shown in Fig. 7.1.

Genetics and diversity of RHDs involved in PAH degradation

Genetics and diversity of RHDs in Gram-negative bacteria

Genes responsible for PAH degradation were first studied in *Pseudomonas* strains isolated on naphthalene as carbon source. Naphthalene degradation genes are generally clustered in two closely linked operons carried on a specific plasmid, as initially reported for the NAH7 plasmid of *P. putida* G7 (Yen and Gunsalus, 1982; Yen and Serdar, 1988). The upper operon contains genes coding for the naphthalene dioxygenase (*nahAaAbAcAd*) as well as genes specifying the enzymes catalysing six additional oxidation steps to produce salicylate (*nahABFCQED*). The lower operon also called the *sal* operon, (*nahGTHINLOMKJ*) encodes the enzymes required to convert salicylate to Krebs cycle intermediates. Between the two operons and oriented in the opposite direction lies the *nahR* gene encoding a LysR-type transcriptional regulator, which controls the expression of both operons (Schell, 1985; Tropel and van der Meer, 2004). The NahR regulator activates transcription by binding to specific sequences in the promoter regions of the two *nah* operon and within its own promoter (autoregulation), in a manner dependent on the presence of salicylate. The organization of the naphthalene degradation genes and their nucleotide sequence was found to be very similar in other *Pseudomonas* strains, including NCIB 9816-4 (Simon *et al.*, 1993) and *P. stutzeri* (Bosch *et al.*, 1999; Bosch *et al.*, 2000).

In *Comamonas testosteroni* strains isolated on phenanthrene, the genes encoding the initial dioxygenase did not hybridize with the *nahA* probes, indicating that their sequences diverge significantly from that of counterparts found in *Pseudomonas* strains (Goyal and Zylstra, 1996). The organization of the phenanthrene catabolic genes (*phd*) was similar to that observed in *Pseudomonas*, except that a gene encoding a dihydrodiol dehydrogenase (*phdB*) was inserted between the genes encoding phenanthrene dioxygenase (Goyal and Zylstra, 1997) (Fig. 7.2). A remarkably similar organization was found in the gene cluster responsible for phenanthrene degradation in *Alcaligenes faecalis* AKF2 (Kiyohara *et al.*, 1982). However, the latter strain degrades phenanthrene via the o-phthalic acid pathway found in Gram-positive bacteria, whereas *C. testosteroni* strains utilize the salicylate pathway, like Pseudomonads.

A recently isolated phenanthrene degrader identified as an *Acidovorax* strain (NA3) was shown to contain a catabolic cluster highly similar to that found in strain AKF2 (Singleton *et al.*, 2009). The alpha subunit of strain NA3 phenanthrene dioxygenase exhibited the highest amino acid sequence identity with counterparts from AFK2 and *Burkholderia* strains. On the basis of transcriptional analysis by real-time qPCR, the phenanthrene degradation genes were found to be highly expressed in cells exposed to phenanthrene or naphthalene, but were not expressed in the presence of other PAHs.

The phenanthrene degrader *Burkholderia* sp. strain RP007 contains a cluster of genes similar in sequence to those found in strain AKF2 but showing an arrangement different from that described in the two strains above (Laurie and Lloyd-Jones, 1999) (Fig. 7.2). The catabolic cluster *phnRSFECDAcAdB* contains nine genes including the

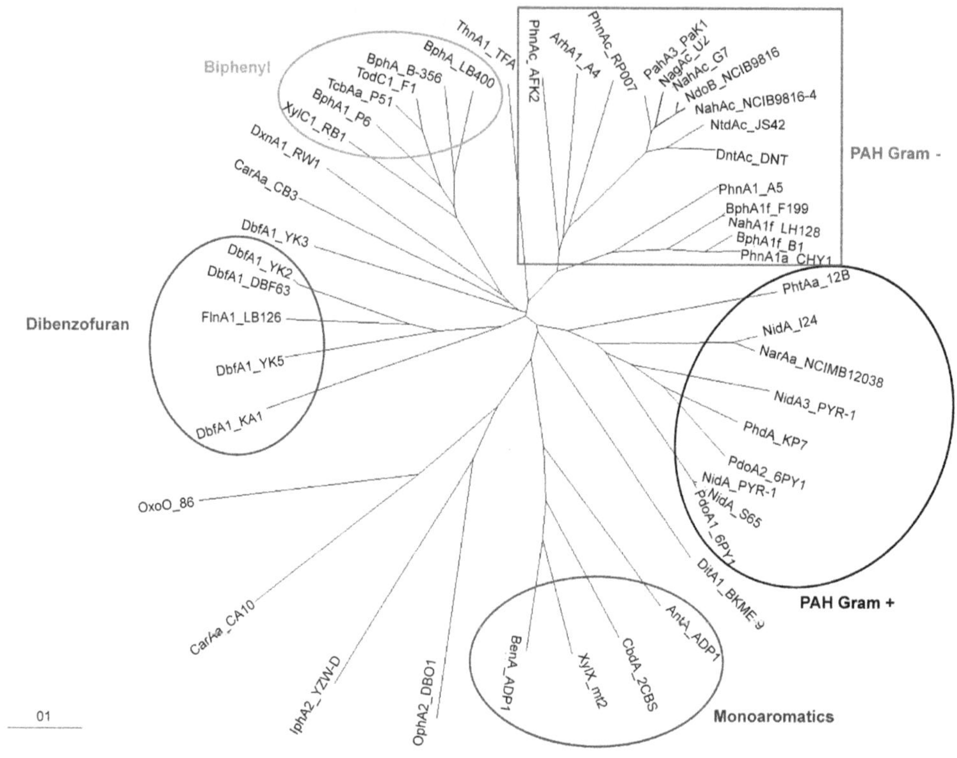

Figure 7.1 Phylogenetic tree of alpha subunits of selected RHDs. Sequence alignment and the generation of a tree by neighbour-joining analysis were performed using Clustal X software. The scale bar represents substitutions per amino acid. Sequences were grouped in subfamilies according to substrate specificity. Alpha subunits were from the following dioxygenase/organism systems: PhnAc_AFK2, *Alcalingenes faecalis* AFK2 (AB024945); ArhA1_A4, *Sphingomonas* sp. strain A4 (BAD34447); PhnAc_RP007, *Burkholderia* sp. RP007 (AF061751); PahA3_PaK1, *Pseudomonas aeruginosa* PaK1 (D84146); NagAc_U2, *Ralstonia* sp. U2 (AF036940); NahAc_G7, *Pseudomonas putida* G7 (YP_534822); NdoB_NCIB9816, *Pseudomonas putida* NCIB9816 (M23914); NahAc_NCIB9816–4, *Pseudomonas putida* NCIB9816–4 (NP_863072); NtdAc_JS42, *Pseudomonas* sp. JS42 (U49504); DntAc_DNT, *Burkholderia* sp. DNT (U62430); PhnA1_A5, *Cycloclasticus* sp. strain A5 (AB102786); BphA1f_F199, *Sphingomonas aromaticivorans* F119 (AF079317); PhnA1f_LH128, *Sphingomonas* sp. strain LH128 (EU024112); BphA1f_B1, *Sphingomonas yanoikuyae* B1 (ABM91740); PhnA1a_CHY1, *Sphingomonas* sp. strain CHY-1 (AJ633551); PhtAa_12B, *Arthrobacter keyseri* 12B (AF331043); NidA_I24, *Rhodococcus* sp. I24 (AF121905); NarAa_NCIMB12038, *Rodococcus* sp. NCIMB12038 (AF082663); NidA3_PYR-1, *Mycobacterium vanbaalenii* PYR-1 (AAY85176); PhdA_KP7, *Nocardioides* sp. KP7 (AB017794); PdoA2_6PY1, *Mycobacterium* sp. strain 6PY1 (CAD38643); NidA_PYR-1, *Mycobacterium vanbaaleni* PYR-1 (AF249301); NidA_S65, *Mycobacterium* sp. S65 (AF 546904); PdoA1_6PY1, *Mycobacterium* sp. strain 6PY1 (CAD38647); DitA1_BKME-9, *Pseudomonas abietaniphila* BKME-9 (AF119621); AntA_ADP1, *Acinetobacter calcoaceticus* ADP1 (AF071556); CbdA_2CBS, *Burkholderia cepacia* 2CBS (X79076); XylX_mt2, *Pseudomonas putida* mt2 (M64747); BenA_ADP1, *Acinetobacter calcoaceticus* ADP1 (AF009224); OphA2_DBO1, *Burkholderia cepacia* DBO1 (AF095748); IphA2_YZW-D, *Comamonas testosteroni* YZW-D (AY923836); CarAa_CA10, *Pseudomonas resinovorans* CA10 (AB088420); OxoO_86, *Pseudomonas putidas* 86 (Y12655); DbfA1_KA1, *Sphingomonas* sp. KA1 (BAF03470); DbfA1_YK5, *PaeniBacillus* sp. strain YK5 (BAF 53401); FlnA1_LB126, *Sphingomonas* sp. strain LB126 (EU024110); DbfA1_DBF63, *Terrabacter* sp. strain DBF63 (BAC75993); DbfA1_YK2, *Rhodococcus* sp. strain YK2 (BAC00802); DbfA1_YK3, *Terrabacter* sp. strain YK3 (BAC06602); CarAa_CB3, *Sphingomonas* CB3 (AF 060489); DxnA1_RW1, *Sphingomonas* sp. RW1 (X72850); XylC1_RB1, *Cycloclasticus oligotrophus* RB1 (U51165); BphA1_P6, *Rhodococcus globerulus* P6 (X80041); TcbAa_P51, *Pseudomonas* sp. P51 (U15298); TodC1_F1, *Pseudomonas putida* F1 (J04996); BphA_B-356, -356, *Comamonas testosteroni* B-356 (U47637); BphA_LB400, *Burkholderia xenovorans* LB400 (M86348); ThnA1_TFA, *Sphingopyxis macrogoltabida* TFA (AF157565).

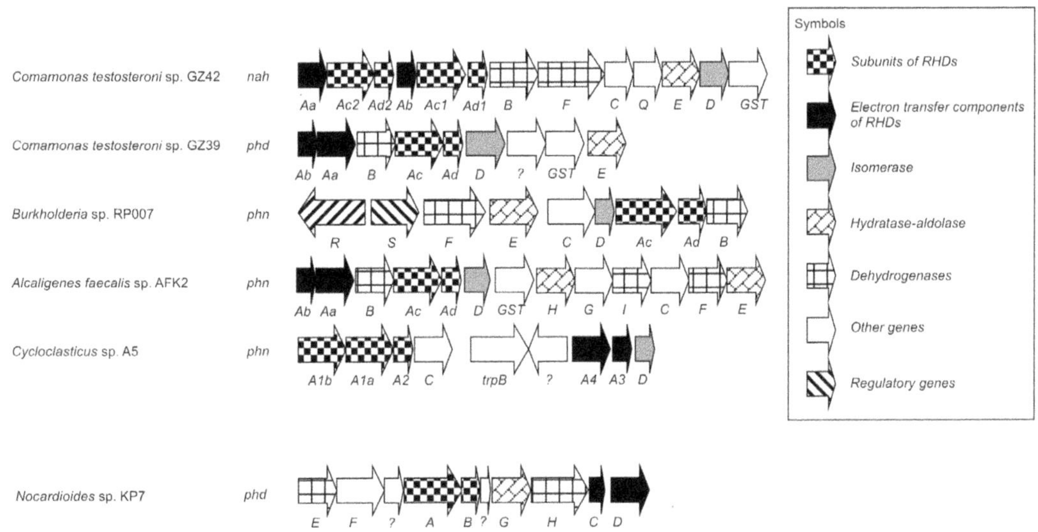

Figure 7.2 Genetic organization of polycyclic aromatic hydrocarbon (PAH) catabolic genes in representative Proteobacteria. The relevant gene cluster from the phenanthrene-degrading Gram-positive bacterium *Nocardioides* sp. KP7 is also shown for comparison. The figure was adapted from Schuler (2008) with permission.

phnAcAd coding for the hydroxylase component of phenanthrene dioxygenase but lacks the genes specifying associated electron carriers. Sequence analysis suggested that PhnR and PhnS are transcriptional regulators, the latter being of the LysR-type, similar to NahR.

In *Ralstonia* sp. strain U2, isolated on naphthalene, the catabolic genes also form a tight cluster *nagAaGHAbAcAdBFCQED* in which two genes, *nagGH,* appear to be inserted between the *nagAa-d* genes encoding the naphthalene dioxygenase (Fuenmayor *et al.*, 1998). In contrast to the previously discussed PAH degraders, strain U2 does not convert salicylate into catechol but into gentisate thanks to a salicylate 5-dioxygenase coded by the *nagGH* genes (Zhou *et al.*, 2001). The *nag* genes are positively regulated via NagR, a regulator very similar to NahR, which is also activated by salicylate (Jones *et al.*, 2003).

Another naphthalene degrader, called *Polaromonas naphthalenivorans* CJ2, also exhibited a degradation pathway involving gentisate as an intermediate. The cluster of *nag* genes of CJ2 is very similar in organization and sequences to that found in *Ralstonia* U2 (Jeon *et al.*, 2006). These genes are controlled, at the transcriptional level, by the regulator NagR1, which functions in a salicylate-dependent manner as mentioned above. Contrary to strain U2, strain CJ2 lacks the *nagY, nagM* and *nagN* genes, one of which encodes a chemotaxis protein. Moreover, strain CJ2 contains another set of genes responsible for the complete degradation of gentisate, which is regulated by NagR2 another type of regulator that is not activated by salicylate nor by naphthalene.

Bacteria belonging to the genus Cycloclasticus are thought to play a major role in the degradation of PAHs in marine environments (Kasai *et al.*, 2002). In *Cycloclasticus* strain A5, which is able to grow on naphthalene, methylnaphthalenes and phenanthrene, the *phnA1A2* genes encoding the dioxygenase and the *phnA3A4* encoding a ferredoxin and a reductase are part of a 10.5-kb gene cluster (Kasai *et al.*, 2003) (Fig. 7.2). The dioxygenase of strain A5 showed moderate sequence homology with those from other PAH degraders, the closely related enzymes being dioxygenases from Sphingomonads (60% and 52% identities at best for the α and β subunits of the hydroxylase, respectively).

Sphingomonads

The genus *Sphingomonas* initially designated a group of bacteria known for their ability to degrade a wide range of organic pollutants, including dibenzo-p-dioxin and dibenzofuran (Armengaud *et al.*, 1998), pesticides (Kumari *et al.*, 2002) and PAHs (Leys *et al.*, 2004). This taxonomic group has been split into new genera called *Sphingobium, Novosphingobium, Sphingopyxis* in 2001 (Takeuchi *et al.*, 2001). Genetic analyses of PAH-degrading strains showed that catabolic genes had a unique organization where genes involved in polyaromatic degradation were interspersed with xyl genes responsible for monoaromatic degradation (Fig. 7.3) (Kim and Zylstra, 1999; Romine *et al.*, 1999; Pinyakong *et al.*, 2003a). In *Novosphingobium aromaticivorans* F199, 79 genes thought to be associated with the metabolism or transport of aromatic hydrocarbons are carried on a 184-kb plasmid called pNL1 (Romine *et al.*, 1999). As many as six gene pairs possibly encoding the α and β subunits of distinct hydroxylases were identified on pNL1, but their specific function was unknown (Fig. 7.3). In *Sphingomonas* CHY-1, a strain able to grow on chrysene (Willison, 2004), the pair of genes encoding the initial PAH dioxygenase has been identified and genetic evidence showed that this dioxygenase was the only enzyme required to catalyse the primary oxidation of PAHs (Demaneche *et al.*, 2004). The cluster of catabolic genes in strain F199 and in related strains, contain a single copy each of a gene encoding a ferredoxin and a gene encoding a reductase, located more than 20 kb apart. This observation suggested that the multiple terminal hydroxylases coded by the catabolic cluster shared the same electron carriers. Consistent with this idea, it was found that the initial PAH dioxygenase from strain CHY-1, called PhnI, and another hydroxylase coded by *phnA1bA2b* (PhnII), utilized the same electron carriers (Demaneche *et al.*, 2004; Jouanneau *et al.*, 2007; Fig. 7.3).

The PhnII enzyme, which shares many structural features with classical three-component dioxygenases, has been characterized as a salicylate 1-hydroxylase that converts salicylate and several substituted derivates of salicylate to catecholic products (Jouanneau *et al.*, 2007). This enzyme substitutes for the monomeric salicylate hydroxylases found in Pseudomonads although the two types of enzymes are phylogenetically unrelated. The three-component enzyme from *Sphingomonas* CHY-1 also differs from the monomeric type in that it is inactive on 1-hydroxy-2-naphthoate, and therefore does not participate in the degradation pathway of 3-ring PAHs such as phenanthrene or anthracene. In *Sphingobium* P2, it was found that three of the multiple hydroxylases encoded by the catabolic cluster were salicylate hydroxylases analogous to PhnII (Pinyakong *et al.*, 2003b).

Salicylate 5-hydroxylase from *Ralstonia* U2 and anthranilate dioxygenase from *Burkholderia cepacia* DBO1 (Chang *et al.*, 2003) are phylogentically related to three component salicylate hydroxylases with which they form a distinct subgroup among RHDs (Fig. 7.1).

Sphingomonas macrogolitabida TFA has been studied for its ability to utilize tetralin (1,2,3,4-tetrahydronaphthalene) as a carbon source. The *thn* genes responsible for this degradation are clustered in two separate and divergently transcribed operons (Moreno-Ruiz *et al.*, 2003). The *thnA1A2* genes encoding the catalytic component of the initial dioxygenase are located in the main operon, whereas the *thnA3A4* genes coding for the associated electron carriers are part of the other operon. The oxygenase component of this enzyme appeared to be distantly related to PAH dioxygenases from Gram-negative bacteria (Fig. 7.1).

The dioxin dioxygenase from *Sphingomonas* RW1, which catalyses an angular attack on dibenzo-*p*-dioxin and dibenzofuran, is encoded by genes dispersed in the genome (Armengaud *et al.*, 1998). In the main locus, the *dxnA1A2* genes encoding the hydroxylase component are associated with *dxnB*, which encodes a ring-cleavage dioxygenase that catalyses the next degradation step. The genes encoding the associated ferredoxin and reductase components are located in two distant and separate loci. The deduced amino acid sequence of the terminal dioxygenase showed weak homology with counterparts from other bacteria, the closest relatives being biphenyl dioxygenases (Armengaud *et al.*, 1998).

Genetics and diversity of RHDs in Gram-positive bacteria

Gram-positive bacteria able to degrade PAHs fall into two categories, those which can attack fluorene, dibenzofuran and similar substrates by

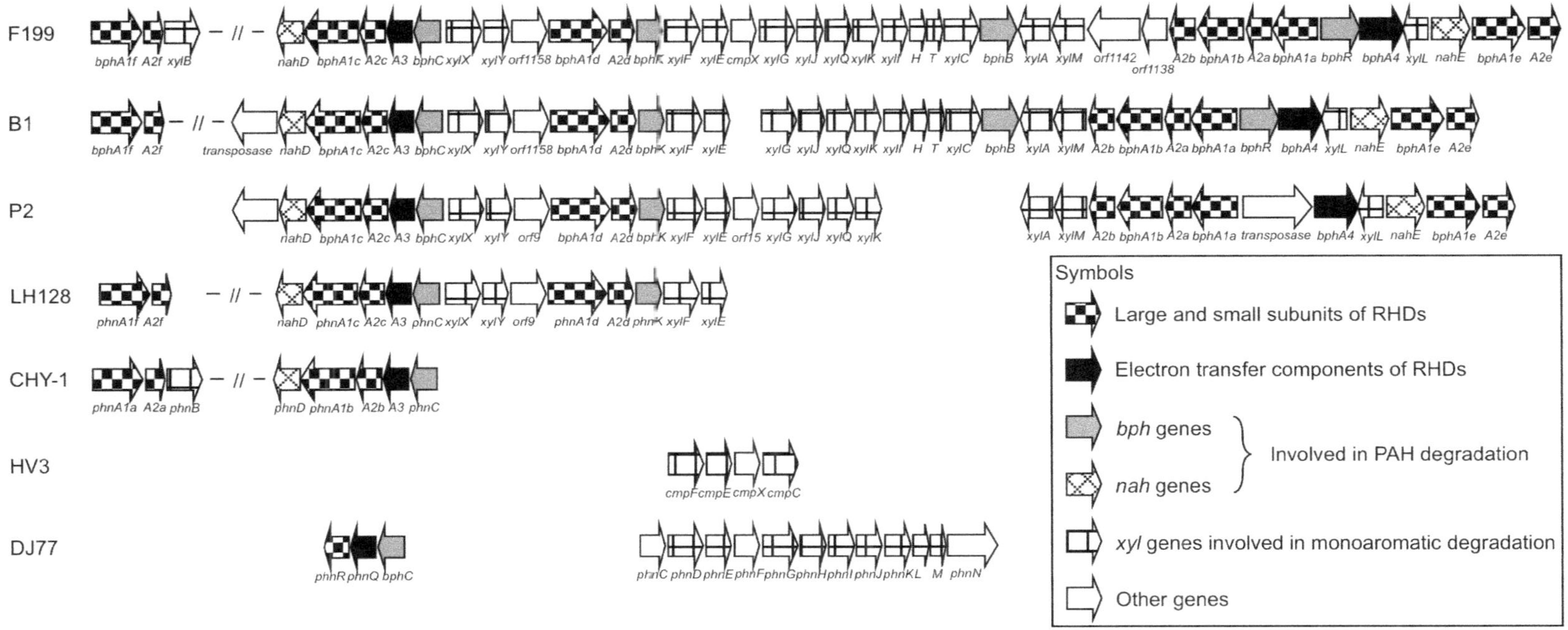

Figure 7.3 Genetic organization of polycyclic aromatic hydrocarbon (PAH), biphenyl and xylene catabolic genes in Sphingomonads. Gene clusters are from *Novosphingobium aromaticivorans* F199, *Sphingobium yanoikuyae* B1, *Sphingobium* sp. P2, *Sphingomonas* sp. LH128, *Sphingomonas* sp. CHY-1, *Sphingobium agrestis* HV3, and *Sphingomonas chungbukensis* DJ77. The figure was adapted from Schuler (2008) with permission.

angular dioxygenation, including *Terrabacter* and *Rhodococcus* strains, and those which can utilize phenanthrene and 4-ring PAHs, including *Nocardioides* and *Mycobacterium* strains.

In the first category, *Terrabacter* sp. strain DBF63 is one of the best-studied dibenzofuran and fluorene degraders. The initial attack on these compounds is catalysed by an angular dioxygenase, the catalytic component of which is encoded by *dbfA1A2* (Habe *et al.*, 2003; Habe *et al.*, 2004). These genes are part of a large cluster including the *flnBdbfA1A2flnEflnD1D2flnC* operon and genes involved in phthalate degradation. A gene (*dbfA3*) was later identified downstream from *dbfA1A2*, which encodes a [3Fe–4S] ferredoxin able to serve as electron donor to the hydroxylase component of the dioxygenase (Takagi *et al.*, 2005). The initial dioxygenase catalyses two types of reaction on fluorene, first a single hydroxylation on the carbon in the C-9 position to yield 9-fluorenol, then an angular dihydroxylation to give 1,1a-dihydroxy-1-hydro-9-fluorenone. This compound is then converted by FlnB to 2′-carboxy-2,3-dihydroxybiphenyl, which is further transformed to phthalate in two steps catalysed by FlnD1D2 and FlnC. Genes homologous to dbfA1A2 were identified in other Gram-positive dibenzofuran degraders, including *Terrabacter* sp. strain YK3 (Iida *et al.*, 2002), various *Rhodococcus* strains (Aly *et al.*, 2008), but also in two *Sphingomonas* strains able to metabolize carbazole (Shintani *et al.*, 2007) and fluorene (Schuler *et al.*, 2008). In *Terrabacter* YK3 and *Rhodococcus* YK2, the genes responsible for dibenzofuran degradation, including the *dfdA1A2A3A4* coding for the angular dioxygenase, are carried on a plasmid (Iida *et al.*, 2002). Recently, the dfd genes were found to be regulated by a new type of regulator (DfdR) that activates transcription in response to the presence of dibenzofuran or a few other hydrophobic hydrocarbons (Iida *et al.*, 2009b). In *Rhodococcus* HA01, two inducible angular dioxygenases encoded by *dfdA1A2A3A4* and *DbfA1A2*, have been described (Aly *et al.*, 2008). The first set of genes was highly similar to that found in *Terrabacter* YK3, whereas the second set was closely related to that coding for dibenzofuran dioxygenase from strain DBF63. The two enzymes, which appeared to be simultaneously expressed in strain HA01, exhibited complementary activities towards chlorinated substrates, with DfdA1A2 performing the angular dioxygenation of 3-chlorodibenzofuran and DbfA1A2 transforming 2-chlorodibenzofuran.

Two *Sphingomonas* strains were found to degrade fluorene through a degradation pathway similar to that described in *Terrabacter*, suggesting an initial attack by angular dioxygenation. In *Sphingomonas* sp. LB126, the operon responsible for fluorene degradation exhibited a gene organization similar to that found in strain DBF63, but the gene products were only distantly related, as exemplified by the α and β subunits of the dioxygenase, which showed 63% and 52% sequence identities with DbfA1 and DbfA2, respectively (Schuler *et al.*, 2008). The oxygenase from strain LB126 converted fluorene to 9-fluorenol and subsequently to the angular dihydrodiol product like the DBF63 enzyme. It also transformed dibenzofuran and attacked other 2- and 3-ring PAHs, fluorene and phenanthrene being the preferred substrates (Schuler *et al.*, 2008).

The marine strain *Nocardioides* sp. KP7 was one of the first PAH-degrading Gram-positive bacterium to be genetically characterized (Saito *et al.*, 1999, 2000). The genes responsible for the oxidative breakdown of phenanthrene to phthalate were found to be different in terms of organization and sequence from what had been previously reported in Gram-negative bacteria (Fig. 7.2). The main cluster *phdEFABGHCD* contained the genes coding for the α and β subunits of phenanthrene dioxygenase (*phdAB*), preceded by genes encoding a dihydrodiol dehydrogenase (*phdE*) and an extradiol dioxygenase (*phdF*). The four genes located downstream specified an aldolase-hydratase (*phdG*), an aldehyde dehydrogenase (*phdH*), a ferredoxin (*phdC*) and a reductase (*phdD*), the latter two gene products being the electron carriers specifically associated with phenanthrene dioxygenase. Altogether, the enzymes coded by this cluster catalysed five oxidation steps to convert phenanthrene to 1-hydroxy 2-naphthoate. The latter intermediate was further metabolized to phthalate by three enzymes coded by the *phdIJK* gene cluster located about 6-kb apart (Saito *et al.*, 2000). The dioxygenase from strain KP7 was the first member of a new subclass of enzymes essentially found in *Mycobacterium* and *Rhodococcus* strains, forming a distinctive phylogenetic group (Fig. 7.1) and featuring a [3Fe–4S] ferredoxin as typical electron carrier.

Several members of the genus *Mycobacterium* have been studied for their ability to degrade high molecular weight PAHs, including pyrene and fluoranthene (Heitkamp *et al.*, 1988; Khan *et al.*, 2001; Vila *et al.*, 2001; Krivobok *et al.*, 2003; Schneider *et al.*, 1996; Sho *et al.*, 2004; Stingley *et al.*, 2004). The genes encoding what appeared to be the hydroxylase component of a pyrene dioxygenase were first described in *M. vanbaalenii* PYR-1, and designated *nidBA* (*nid* for *n*aphthalene *i*n*d*uced; Khan *et al.*, 2001). Analysis of protein profiles by 2D gel electrophoresis indicated that six proteins were induced by pyrene and phenanthrene, one of which was the alpha subunit of the dioxygenase (NidA). In another pyrene degrader, called *Mycobacterium* 6PY1, a thorough proteomic analysis of pyrene- and phenanthrene-induced polypeptides identified 23 proteins, four of which were subunits of two dioxygenase terminal components named Pdo1 and Pdo2 (Krivobok *et al.*, 2003). The other induced proteins showed significant similarities with enzymes previously implicated in phenanthrene metabolism in *Nocardioides* KP7. The genes encoding Pdo1 and Pdo2 were located in two separate loci, and no genes encoding possibly associated electron carriers were found in either locus. Nevertheless, the two dioxygenases could be successfully expressed in *E. coli* and their activity was enhanced by co-expressing the PhdC and PhdD electron carriers from strain KP7. Amino acid sequence comparison indicated that Pdo1 was highly similar to the pyrene dioxygenase from strain PYR-1, whereas Pdo2 was related to the phenanthrene dioxygenase from strain KP7 (Krivobok *et al.*, 2003). Other *Mycobacterium* strains able to degrade pyrene or fluoranthene were found to express RHDs homologous to Pdo1 or Pdo2 in a regulated manner (Sho *et al.*, 2004; Liang *et al.*, 2006; Pagnout *et al.*, 2007). More recently, Kim *et al.* assigned a gene product to nearly all steps of the pyrene degradation pathway in strain PYR-1 based on a detailed analysis involving exhaustive genomic and proteomic data (Kim *et al.*, 2007). However, some of the proposed gene functions remain hypothetical and need to be assessed experimentally. The PAH catabolic genes are part of a large cluster extending over 150 kb and encoding multiple terminal components of oxygenases (Kim *et al.*, 2008). Although many of these oxygenases have still unknown functions, at least four of them have been assigned a catalytic role, including a phenanthrene dioxygenase homologous to Pdo2 from 6PY1, a phthalate dioxygenase encoded by *phtAaAb*, another PAH dioxygenase encoded by *nidA3B3*, and the pyrene dioxygenase discussed above (*nidBA*). The NidA3B3 terminal oxygenase showed a preference for fluoranthene (Kim *et al.*, 2006) and was expressed to a relatively high level in *M. vanbaalenii* PYR-1 when grown on this PAH as substrate (Kweon *et al.*, 2007). The entire cluster contains only two electron carrier genes closely linked to *phtAaAb*, *phtAc* encoding a [3Fe–4S] ferredoxin and *phtAd* encoding a reductase, suggesting that the multiple hydroxylases specified by the catabolic cluster share the same electron carriers, a situation similar to that encountered in Sphingomonads. Interestingly, the genomes of four other PAH-degrading *Mycobacterium* strains, including *M. gilvum* strain PYR-GCK (NC009338), *Mycobacterium* sp. KMS (NC008705), MCS (NC008146) and JLS (NC009077) appeared to contain a catabolic cluster very similar in gene sequence and organization to that found in strain PYR-1, except that a large portion of the cluster was duplicated on the chromosome (strain PYR-GCK) or on plasmids (strains MCS and KMS).

The naphthalene degrader *Rhodococcus* sp. strain NCIMB12038 contains a RHD whose hydroxylase component showed little sequence similarities with naphthalene dioxygenases from Pseudomonads but significant homology with the Gram-positive RHDs described above (Larkin *et al.*, 1999).

The catalytic mechanism of RHDs: the example of naphthalene dioxygenase from *Pseudomonas* sp. NCBI 9816-4

Chemically speaking, the initial oxygenation of the aromatic rings is the most difficult catalytic step in the aerobic degradation of PAHs. Bacterial RHDs catalyse the incorporation of the two atoms of molecular oxygen into one of the highly stable aromatic ring, thus allowing further breakdown of the substrate molecule by subsequent dehydrogenation and ring cleavage. The mechanism of the reaction catalysed by RHDs has been studied in detail with the naphthalene dioxygenase from *P. putida* as model enzyme, taking advantage of the crystal structure of the terminal oxygenase component, which was determined in 1998 (Kauppi *et al.*, 1998). The dihydroxylation of naphthalene occurs at the enzyme active site, which is a ferrous ion bound to the alpha subunit through the side chains of two His and one Asp residues. The reaction involves reduction and cleavage of molecular oxygen, and insertion of the oxygen atoms on the C-1 and C-2 carbon atoms of the substrate to yield *cis-(1R,2S)*-dihydroxy-1,2-dihydronaphthalene. Although the reductase and ferredoxin components are required for continuous catalytic turnover of the enzyme, it has been demonstrated that the hydroxylase component alone can catalyse the dioxygenation of naphthalene in single turnover experiments (Wolfe *et al.*, 2001). As isolated, the oxygenase component is unreactive with oxygen. However, upon stoichiometric reduction of the Rieske cluster, naphthalene and oxygen react with the hydroxylase to form the dihydrodiol product. Stopped-flow analyses and spectroscopic measurements indicated that product formation was concomitant with the oxidation of the Rieske and mononuclear iron centres. Moreover, it was found that the oxygenase component could convert naphthalene to dihydrodiol in the presence of hydrogen peroxide, a reaction that stopped after one turnover with the product trapped in the active site (Wolfe and Lipscomb, 2003). This reaction, known as the peroxide shunt, suggested that the normal catalytic cycle of the dioxygenase involves a peroxo intermediate, and that the release of the product requires reduction of the mononuclear iron centre. Altogether, the results obtained with naphthalene dioxygenase led Wolfe and Lipscomb (2003) to propose a mechanism for the reaction cycle of RHDs. In this scheme, the oxygenase component is reduced first, then binds sequentially the substrate and O_2 to form an Fe(III)-hydroperoxo intermediate. The exact nature of the reactive species generating the dihydrodiol product is still unknown, but it seems likely that the release of the reaction product requires another reduction step to regenerate the ferrous state of the mononuclear centre. Using phthalate dioxygenase as a model, other authors found that this scheme did not fit their data because single turnover reactions yielded one half the expected amount of product and ended with the active site in the ferrous state (Tarasev and Ballou, 2005). It has been proposed that the catalytic cycle of phthalate dioxygenase required the participation of the reductase component to deliver one electron equivalent per cycle.

The occurrence of a hydroperoxo intermediate in the reaction cycle has been substantiated by the discovery that oxygen binds side-on at the active site of naphthalene dioxygenase and by the structural description of a ternary complex of the oxygenase component with O_2 and indole (Karlsson *et al.*, 2003). The particular binding mode of oxygen on the active site iron and the rigorous positioning of the substrate would afford a mechanism whereby both atoms of O_2 react in concert with the substrate to form the dihydrodiol. However, it is still uncertain whether it is an Fe(III)-hydroperoxo or an Fe(V)-oxo-hydroxo reactive species that eventually attacks the substrate to yield the dihydrodiol product (Kovaleva and Lipscomb, 2008).

Comparison of available structures of RHD catalytic components

The catalytic component of the naphthalene dioxygenase from *Pseudomonas* sp. strain NCIB9816-4 (NDO) was the first enzyme to be structurally characterized (Kauppi *et al.*, 1998). Since then, nine crystal structures from other dioxygenase systems have been reported (Ferraro *et al.*, 2005), including three enzymes able to attack PAHs, namely the oxygenases from *Rhodococcus* NCIMB12038 (Gakhar *et al.*, 2005), *Sphingobium yanoikuyae* B1 (Ferraro *et al.*, 2007) and *Sphingomonas* CHY-1 (Jakoncic *et al.*, 2007a; Jakoncic *et al.*, 2007b). All enzymes exhibit an $\alpha_3\beta_3$ quaternary structure, except for the carbazole oxygenase from *P. resinovorans* CA10 (Nojiri *et al.*, 2005)and the oxoquinoline 8-monooxygenase from *P. putida* strain 86 (Martins *et al.*, 2005), which are homotrimers consisting of alpha subunits only. The following discussion will be focused on enzymes implicated in PAH degradation.

All $\alpha_3\beta_3$ oxygenases exhibit a mushroom-like overall structure similar to that reported for NDO, with the trimer of alpha subunits sitting on a base consisting of the beta subunits. The alpha/beta pairs of subunits have almost identical folds and are arranged around a threefold axis of symmetry. The alpha subunits are composed of two domains, the N-terminal Rieske domain that contains the [2Fe–2S] cluster and the catalytic domain containing the substrate binding site and the mononuclear iron centre. The fold of the Rieske domain is similar to that found in the cytochrome bc1 complex of the respiratory chain. The cluster is bound to the protein through four conserved ligand residues, two cysteines and two histidines. In the catalytic domain, the iron atom is coordinated by two histidines and one aspartate (as a bidentate) and one water molecule. The last position of coordination (axial) is occupied by the side chain oxygen atom of a conserved asparagine residue but the O–Fe distance is >3.75 Å. The distance between the Rieske centre and the mononuclear iron is about 44 Å within the same alpha subunit. The Rieske centre is, however, much closer to the iron centre of the neighbouring subunit (12 Å), indicating that electron transfer occurs via the metal centres of two adjacent alpha subunits. In addition, these two metal centres are connected through a hydrogen bond network involving a conserved aspartate residue (Asp205 in NDO). This residue binds a histidine ligand of the iron centre (His208) and a histidine ligand of the Rieske cluster (His104), thus forming a plausible bridge for electron transfer (Kauppi *et al.*, 1998). The essential role of the Asp205 residue was demonstrated in experiments where its replacement through site-directed mutagenesis resulted in a total loss of enzyme activity (Parales *et al.*, 1999). Its role in electron transfer between the metal centres is, however, controversial. A study using anthranilate dioxygenase and variants carrying substitutions of the conserved Asp residue concluded that this residue rather served to maintain the protonation state and reduction potential of the Rieske cluster (Beharry *et al.*, 2003).

The substrate-binding site appears as a hydrophobic pocket open to the solvent on one side and ending at the iron centre on the other. Seventeen residues lining and shaping the substrate binding site have been identified in NDO (Carredano *et al.*, 2000), three of which are strictly conserved, including the two histidine ligands of the ferrous iron (His208 and His213 in NDO) and the Asp205 residue discussed above. Other semi-conserved residues include Asn201, which is replaced by a glutamine in the biphenyl dioxygenase from *Rhodococcus* RHA1, and Phe202, Gly251 and Phe358, which are replaced by other residues only in trimeric (α3) oxygenases. Differences observed in the active site residues among dioxygenases might explain in part differences in substrate specificity (Ferraro *et al.*, 2005). Indeed, interactions between substrates and active site residues are believed to control the orientation of the substrate relative to the ferrous centre and eventually the regio- and enantio-selectivity of the catalytic reaction.

Although the nitrobenzene dioxygenase (NBDO) acts chiefly on monoaromatic substrates, its structure is remarkably similar to that of NDO (Friemann *et al.*, 2005). Only five differences were observed in the residues lining the substrate-binding site. The crystal structure of NBDO in complex with nitrobenzene showed that nitrobenzene is maintained in the correct position for catalysis through hydrogen bonding between the nitro group of the substrate and a specific Asn residue (Asn258). Replacement of this residue by a valine (the analogous residue in NDO) changed the regio-selectivity of the reaction with 2-nitrotoluene, yielding 2-nitro benzyl alcohols rather than catechol (Ferraro *et al.*, 2005).

The oxygenase from strain NCIMB12038 is structurally similar to NDO, although the two enzymes share only 30% sequence identity. Moreover, indole was found to bind to the active site of the enzyme in the same position as that described for the NDO–indole complex (Gakhar *et al.*, 2005), although a few differences were observed in the residues lining the active site.

The crystal structure of the broad-substrate-range dioxygenase from *Sphingomonas* CHY-1 (Jakoncic *et al.*, 2007b) and that of the highly similar biphenyl dioxygenase from *Sphingobium yanoikuyae* B1 (Ferraro *et al.*, 2007) were published almost at the same time. The following description concerns the former enzyme.

The RHD from strain CHY-1 is unique in that it can oxidize a very large panel of PAHs ranging from naphthalene to the five-ring benzo[a] pyrene (Jouanneau *et al.*, 2006). The catalytic component, called PhnI, exhibits a folding pattern very similar to that of NDO, the α carbon chains of the two proteins being superimposable within a 1.2 Å root-mean-square deviation. However, PhnI features the largest substrate-binding pocket ever reported for a RHD (Fig. 7.4). This 12-Å-long hydrophobic cavity extends from the iron centre on one side to the solvent on the other. The entrance of the cavity is delimited by two flexible loops, which might control the access of the substrates to the active site. Compared with NDO, two leucine residues on loop L1 (Leu223 and Leu226) are moved towards the solvent, thus elongating the catalytic pocket by at least 2 Å (Jakoncic *et al.*, 2007a). Conserved residues line the part of the cavity proximal to the iron centre, whereas Phe350, Leu356 and Phe404 shape the central and distal parts. Molecular modelling indicated that the substrate binding cavity was large enough to accommodate the 5-ring substrate benzo[a]pyrene, with minor changes in the positions of lining residues (Fig. 7.4). The predicted position of this substrate suggested a hydroxylation of carbons C-9 and C-10 (Jakoncic *et al.*, 2007a), which is consistent with the position of the hydroxyls on the experimentally obtained dihydrodiol (Jouanneau *et al.*, 2006). Further modelling experiments gave plausible positions for other substrates, such as phenanthrene end benz(a)anthracene, in the substrate binding pocket, which were also in good agreement with the regioselectivity of the hydroxylation reactions. This suggested that the large size and the particular topology of the substrate-binding site of PhnI were the main determinants of its broad substrate specificity.

Carbazole dioxygenase is a singular RHD in that it catalyses angular hydroxylation and because its catalytic component is made up of alpha subunits only. The crystal structure of the trimeric protein exhibits a doughnut-like shape in which interactions between specific loops of adjacent alpha subunits stabilize the trimer (Nojiri *et al.*, 2005). The fold of the alpha subunit is roughly similar to that in the $\alpha_3\beta_3$ NDO structure, although the two proteins share only 17% sequence identity. The substrate-binding site differs from that of other RHDs in that the distal

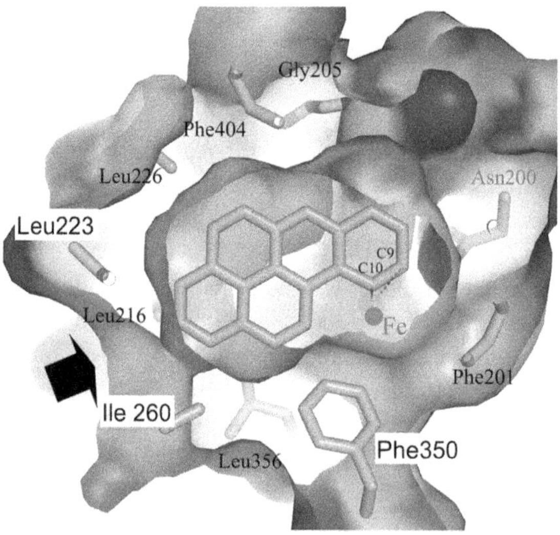

Figure 7.4 Representation of the substrate-binding pocket of the RHD from *Sphingomonas* CHY-1 and modelling of a benzo[a]pyrene molecule bound at the active site. The entrance into the pocket is indicated by the broad arrow. Residues that are thought to be important for regioselectivity (Phe 350) or substrate specificity (Leu 223 and Ile 260) are highlighted. The figure was reproduced from Jakoncic *et al.* (2007a) with permission.

part of the hydrophobic pocket is lined by rather polar residues. Docking simulation of carbazole in the substrate-binding site indicated that, in the predicted position, the carbon C-1 and C-9a of carbazole were the carbon atoms closest to the iron centre, consistent with an angular dioxygenation reaction. Carbazole dioxygenase is the only RHD system for which the structure of a binary complex between the oxygenase and the ferredoxin components has been solved (Ashikawa *et al.*, 2006). In the complex, three ferredoxin molecules were bound at the intersubunit boundary of the trimeric oxygenase, each protein–protein interaction involving a combination of electrostatic, hydrophobic and hydrogen bonds between specific residues of the two partners. Protein interactions and conformational changed brought the Rieske clusters of each partner to a distance of around 12 Å suitable for electron transfer. The ferredoxin binding site has been described as a depression on the surface of the oxygenase trimer at the interface of adjacent alpha subunits. A similar depression has been observed between the alpha and beta subunits of NDO, suggesting that it might be the binding site for the ferredoxin in this hexameric oxygenase (Ashikawa *et al.*, 2006). Analysis of a ternary complex between the oxygenase and ferredoxin components and carbazole showed that upon binding of the substrate, conformational changes resulted in the closure of a lid over it, as if it were trapped in the substrate binding pocket.

Substrate specificity of RHDs with respect to PAHs

RHDs are generally endowed with relaxed substrate specificity, thus allowing bacteria to initiate the degradation of a wide range of aromatic hydrocarbons. In addition, they are remarkable catalysts in that they can perform several types of oxidation reaction, including *cis*-dihydroxylation, monooxygenation, desaturation and sulfoxidation. The impressive range of oxidative reactions catalysed by naphthalene dioxygenase has been reviewed by Resnick and Gibson (1996). Many of the reactions catalysed by RHDs, especially the generation of various *cis*-dihydroarenediols, might be exploited in the synthesis of products useful in chemistry and medicine (Hudlicky *et al.*, 1999).

Despite the fact that RHDs can attack a wide range of substrates, most of them are only active on a limited number of the 16 priority PAHs, essentially those comprising two or three rings. Some RHDs, like carbazole dioxygenase, specialize in the angular *cis*-dihydroxylation of tricyclic substrates containing one heterocycle. At the other extreme, a few RHDs able to oxidize 4- and 5-ring PAHs have been described, mainly in strains belonging to the Sphingomonads or the *Mycobacterium* genus, and quite different substrate specificities have been observed. Here, the discussion has been focused on the catalytic abilities of the latter category of RHDs as well as on angular dioxygenases.

Angular dioxygenases

The carbazole dioxygenase from *Pseudomonas resinovorans* CA10 can perform angular dioxygenation of several carbazole-like compounds, including dibenzofuran, dibenzo-*p*-dioxin, as well as regular *cis*-dihydroxylation on a variety of two- and three-ring substrates such as naphthalene and biphenyl. It can also transform dibenzothiophene to the corresponding sulfoxide and fluorene to 9-fluorenol (Nojiri *et al.*, 1999). Studies on variants of the CA10 RHD obtained by site-directed mutagenesis indicated that three residues lining the substrate-binding pocket were important for controlling angular *cis*-dihydroxylation (Uchimura *et al.*, 2008). The dibenzofuran dioxygenase from *Terrabacter* DBF63 showed similar substrate specificity except that it could not transform carbazole (Kasuga *et al.*, 2001). The RHD from the fluorene degrader *Sphingomonas* LB126 exhibited a broader substrate range since it proved capable of angular dioxygenation on fluorene and on the three tricyclic substrates mentioned above. In addition, the LB126 RHD could transform phenanthrene and anthracene but generated unusual dihydrodiol products, i.e. *cis*-9,10-dihydro-9,10-dihydroxy-phenanthrene for the former and a compound tentatively identified as 2,3-dihydro-2,3-dihydroxyanthracene for the latter (Schuler *et al.*, 2008). These compounds are considered byproducts that cannot be further degraded through the main catabolic pathway of

PAH degraders. They might arise from incorrect positioning of the substrates in the enzyme active site.

Regular dioxygenases

Naphthalene dioxygenase from *P. putida* sp. strain 9816–4 converts indole to indigo (Ensley *et al.*, 1983) and dihydroxylates biphenyl and PAHs up to three rings with high regio- and enantioselectivity (Resnick *et al.*, 1996; Parales *et al.*, 2000). Studies of variants obtained by site-directed mutagenesis indicated that the enzyme could tolerate single amino acid substitutions near the active site with only minor changes to its specificity (Parales *et al.*, 2000). However, replacement of Phe352 by smaller amino acids resulted in variants with altered regioselectivity. While the wild-type enzyme converted biphenyl to the *cis*-2,3-dihydrodiol with minor amounts of the *cis*-3,4-dihydrodiol, all the variants tested produced a large excess of the 3,4-isomer. Likewise, phenanthrene was mainly hydroxylated in the 3,4 positions by the wild-type enzyme, whereas mutant enzymes produced more of the 1,2-dihydrodiol. The enantioselectivity of the reaction was also altered (Parales *et al.*, 2000). Interestingly, replacement of the phenylalanine residue equivalent to NDO Phe352 by smaller residues in other RHDs, including biphenyl dioxygenase (Suenaga *et al.*, 2002) and tetrachlorobenzene dioxygenase (Pollmann *et al.*, 2003), resulted in significant alterations of oxygenase specificity. The important role of the Phe352 residue in NDO has been studied from a structural point of view by elucidating the 3D structure of a F352V variant, either alone or as a complex with substrates (Ferraro *et al.*, 2006). This variant exhibited a broader substrate-binding pocket, allowing phenanthrene to bind in a different position than in the wild-type enzyme in which the C1 and C2 carbons are close to the mononuclear iron centre. These results are consistent with previous enzyme assays showing that this variant mainly generated *cis*-1,2-dihydroxy-1,2-dihydrophenanthrene rather than the 3,4-dihydrodiol. They also provide a structural basis for the notion that the orientation of the substrate binding at the active site is the primary determinant of the regio- and stereoselectivity of RHD-catalysed oxygenation reactions.

Besides NDO from strain NCIB 9816-4, only a few RHDs have been studied for their ability to utilize PAHs as substrates. In addition, most published data have been obtained with E. coli strains producing recombinant RHDs using various expression systems, thus precluding meaningful comparisons of activities towards PAHs. The RHD from *Sphingomonas* CHY-1 exhibits one of the broadest substrate ranges ever reported, since it is able to attack nine of the 16 priority PAHs. Also, it is one of the rare cases where the catalytic activity towards each PAH was measured using the purified enzyme, thus allowing the kinetic parameters and the stoichiometry of the dioxygenase reaction to be determined (Jouanneau *et al.*, 2006). The dioxygenase activity was maximal with naphthalene and declined as a function of the substrate ring number, to reach 2% of the maximum with benzo[a]pyrene. With naphthalene as substrate, the CHY-1 dioxygenase exhibited an apparent k_{cat} of 1.82 ± 0.03/s and a specificity constant of 2.0 ± 0.3 μM–1/s. These values are comparable to those reported for biphenyl dioxygenase (k_{cat} = 7.0 ± 0.2/s, specificity constant of 1.2 μM/1/s; Imbeault *et al.*, 2000) and 2-nitrotoluene dioxygenase (k_{cat} = 2.4/; Parales *et al.*, 2005) and indicate that RHDs are intrinsically slow enzymes. Most PAHs were converted to a single dihydrodiol, except for fluorene, fluoranthene and benz[a]anthracene, which yielded up to three diols (Jouanneau *et al.*, 2006). The four-ring benz[a]anthracene was transformed to three dihydrodiol isomers bearing hydroxyls in positions 1,2-, 8,9- and 10,11-. The 1,2- and 10,11-dihydrodiols were found to be substrates of the enzyme and were converted in a subsequent reaction to a *bis-cis*-dihydrodiol hydroxylated on carbon positions 1,2,10 and 11. It was observed that the coupling between substrate oxidation and O_2 (and NADH) utilization varied depending on PAHs and decreased with substrate size. Uncoupling was associated with a release of hydrogen peroxide, which might be toxic for the cells if such a side reaction also occurs *in vivo*. Hence, the unproductive NADH oxidation that was associated mainly to 4- and 5-ring PAH oxidation might in part explain the slow metabolism and poor yields observed during bacterial growth on such PAHs.

The RHD from *Sphingobium yanoikuyae* B1 exhibits 91–93% amino acid sequence identity with its counterpart from strain CHY-1, and yet displays significant differences in substrate specificity (Chadhain *et al.*, 2007). The purified B1 RHD apparently utilizes biphenyl as its preferred substrate and was therefore referred to as a biphenyl dioxygenase (Yu *et al.*, 2007), although it falls into the subgroup of PAH RHDs (from Gram-negative bacteria) based on phylogenetic analysis (Fig. 7.1) and on its substrate specificity. Like the CHY-1 RHD, the B1 enzyme was also shown to dihydroxylate naphthalene, phenanthrene and benzo[a]pyrene, the latter two substrates yielding two dihydrodiol isomers instead of one (Yu *et al.*, 2007). Previous studies using a mutant strain derived from strain B1 indicated that the same RHD was capable of converting chrysene (Boyd *et al.*, 1997) and benz[a]anthracene (Gibson *et al.*, 1975), the latter substrate giving three dihydrodiols just like the CHY-1 RHD.

The phenanthrene degrader Sphinogomonas LH128 produces an RHD exhibiting 78% (α subunit) and 63% (β subunit) identities with the CHY-1 enzyme (Schuler *et al.*, 2009). Its substrate range was found to be similar, with a marked preference for naphthalene. Nevertheless, the LH128 RHD was less competent in the dioxygenation of large PAHs than the CHY-1 enzyme, as it failed to transform benzo[a]pyrene.

The differences in substrate specificity between the three related enzymes above may be tentatively interpreted in structural terms, based on the crystallographic data known for two of them. The substrate-binding pocket is almost identical in all three enzymes as judged from the conservation of the residues lining this cavity. Two conservative changes are however observed in the B1 enzyme (Ile replaces Val202 and Leu replaces Ile260). Much more variation can be noticed in the sequences of the two loops located at the entrance of the active site, which are thought to control substrate access (Jakoncic *et al.*, 2007b). Differences in substrate specificity might also be due to changes of resides more distant from the active site.

This notion is supported by studies regarding biphenyl dioxygenases in which three specific regions of the alpha subunit have been shown to determine the selectivity of enzymes towards PCBs (Furukawa *et al.*, 2004).

The RHD from *Cycloclasticus* A5, which is distantly related to the PAH dioxygenases from Sphingomonads, exhibits a markedly different selectivity. Based on biotransformation assays using a recombinant *E. coli* strain overexpressing this RHD, it was found that the enzyme was able to oxidize biphenyl, naphthalene, phenanthrene and methylnaphthalenes, but it failed to transform anthracene and larger PAHs (Kasai *et al.*, 2003).

The RHD from the acenaphthene degrader *Sphingomonas* sp. strain A4 was found to oxidize acenaphthylene, naphthalene, phenanthrene, anthracene and fluoranthene, although strain A4 could not grow on the latter four hydrocarbons (Pinyakong *et al.*, 2004). Unlike RHDs from other Sphingomonadaceae, the A4 enzyme is related, although distantly, to RHDs from Betaproteobacteria, the phenanthrene dioxygenase from *Burkholderia* RP007 being the closest homologue (56% identity).

Gram-positive bacteria belonging to the *Mycobacterium* genus were found to degrade a range of PAHs up to five rings in a process probably involving multiple RHDs since several such enzymes were expressed in PAH-grown cells (see section 'Genetics and diversity of RHDs in Gram-postive bacteria'). In *M. vanbaalenii* PYR-1, two RHDs encoded by *nidBA* and *nidA3B3* have been shown to dihydroxylate the 4-ring PAHs pyrene and fluoranthene, respectively (Khan *et al.*, 2001; Kim *et al.*, 2006). Based on biotransformation assays, NidA3B3 was also found to attack several two- to four-ring PAHs including naphthalene, phenanthrene, anthracene and pyrene but preferentially transformed fluoranthene, consistent with the finding that this enzyme was maximally induced in cells grown with that PAH as sole substrate. In *Mycobacterium* 6PY1, two RHDs with markedly different PAH specificities have been studied: the Pdo1enzyme preferentially attacked pyrene but also converted phenanthrene to two dihydrodiols bearing hydroxyls in position 3,4-, and 9,10-, whereas the Pdo2 enzyme only dihydroxylated two- and three-ring PAHs, phenanthrene being the best substrate (Krivobok *et al.*, 2003). All data reported so far on the PAH selectivity of *Mycobacterium* RHDs were from biotransformation assays using *E. coli* recombinant systems in which dioxygenases were generally poorly expressed. In addition, the associated electron carriers being unknown, oxygenase components were expressed alone or in combination with heterologous electron carriers, thus resulting in non-optimal and probably underestimated enzyme activities. Recently, the Pdo1 and Pdo2 oxygenase components from strain 6PY1 were obtained in highly active soluble forms in *E. coli* using non-commercial expression systems (Krivobok, Meyer and Jouanneau, unpublished work). Moreover, two genes designated *phtAc* and *phtAd* encoding a [3Fe–4S] ferredoxin and a reductase, respectively, were cloned from the *Mycobacterium* 6PY1 genome and successfully co-expressed with the oxygenase components. The co-expression of these electron carriers resulted in a 8-fold (Pdo1) or 2-fold (Pdo2) enhancement of the enzyme activity in biotransformation assays (unpublished results). Pdo1 appeared as a unique RHD in that it could utilize three- and some four-ring PAHs as substrates, including phenanthrene, anthracene, fluoranthene and pyrene but was unable to hydroxylate smaller substrates like naphthalene and biphenyl. Confronted with the problem of poor expression of RHDs from Gram-positive actinomycetes in *E. coli*, Pagnout *et al.* (2007) utilized *Mycobacterium smegmatis* as host for overproducing a RHD from the PAH-degrader *Mycobacterium* SNP11. In this expression system, the SNP11 RHD, which is very similar to Pdo2 (>98% sequence identity), efficiently oxidized phenanthrene and anthracene but also fluorene and fluoranthene. More recently, an expression system involving a *Rhodococcus* strain as host for studying recombinant RHDs has been developed. Interestingly, three RHDs expressed in this system exhibited different regio-specificities and relative preference towards dibenzofurans and chlorinated dioxins compared with the enzymes produced in *E. coli* (Iida *et al.*, 2009). Hopefully, convenient expression systems for overproduction of actinobacterial RHDs in highly active form will emerge, thus allowing purification and further characterization of these enzymes, which show PAH selectivities quite different from those of their Gram-negative counterparts.

Exploring the natural biodiversity of RHD involved in PAH degradation

Early studies on the biodiversity of PAH-degrading bacteria relied on microbial culture and selection on defined growth media as a means to isolate strains with the desired traits. This approach appeared inappropriate when it was realized that a large majority of the bacterial species present in the environment (>90%) were not culturable under standard laboratory growth conditions. Various culture-independent methods were then developed to study microbial diversity, most of which were based on DNA extraction from environmental samples followed by PCR-amplification of selected molecular markers i.e. genes (or gene regions) encoding the 16S ribosomal RNA subunit [see Kirk *et al.* (2004) for review]. Bacterial species identification and classification based on sequence analysis of 16S rRNA genes recovered from the environment is now the most commonly used method to probe biodiversity, although some pitfalls and biases due to PCR must be taken into consideration (Wintzingerode *et al.*, 1997). In the last decade, studies targeting a specific population of bacteria in complex environment such as soils have emerged, owing to the implementation of new techniques involving *in situ* labelling of a population of interest with a naturally rare stable isotope and subsequent isolation and analysis of the labelled DNA. This approach known as stable isotope probing (SIP) has been used recently to identify PAH degraders in soils or sediments (Jeon *et al.*, 2003; Singleton *et al.*, 2005). The diversity of RHDs in soils has been assessed directly by PCR amplification and sequence analysis of relevant gene portions recovered from environmental samples (Yeates *et al.*, 2000). Here, the contribution of recent studies to our understanding of the biodiversity of PAH-degraders and associated RHDs is discussed.

Relative abundance and diversity of known PAH degraders in the environment

Since Sphingomonadaceae are very often found in PAH-polluted sites, their diversity has been investigated by PCR-based analysis of 16S rRNA genes using family specific primers (Leys *et al.*,

2004). Sequence analysis revealed that most 16S rRNA genes were significantly different from those of known isolates, suggesting that species different from those studied so far are present in PAH-contaminated soils. A similar study was conducted to examine the community composition of fast-growing *Mycobacterium* in various contaminated soils (Leys *et al.*, 2005). Species identified included known PAH degraders such as *M. austroafricanum* and *M. frederiksbergense* but also *M. tusciae*, a species not previously reported to degrade PAHs.

In marine environment, classical isolation of PAH degraders from contaminated sediment in the Gulf of Mexico led to 23 strains which all belonged to the *Cycloclasticus* genus (Geiselbrecht *et al.*, 1998). Several culture-independent studies conducted all around the world showed that bacterial communities involved in PAH degradation in coastal and estuarine sediments were actually dominated by *Cycloclasticus* in all marine environments tested (Kasai *et al.*, 2002; McKew *et al.*, 2007; Zhou *et al.*, 2009; Niepceron *et al.*, 2010). Members belonging to this genus were also found to be most abundant in sediment of the deep ocean (Cui *et al.*, 2008). On the other hand, in a coastal environment heavily impacted by oil spill, members of the classes Alphaproteobacteria (*Sphingomonadaceae*) and Actinobacteria (*Mycobacterium, Rhodococcus*) were found to prevail (Alonso-Gutierrez *et al.*, 2009).

Direct analysis of the RHD biodiversity through PCR-based approaches

In recent studies, RHD biodiversity in PAH-contaminated marine environments was investigated by PCR-based analysis of a portion of the dioxygenase alpha subunit gene. *Cycloclasticus*-like RHD genes appeared to be well represented in coastal sediments from Patagonia (Lozada *et al.*, 2008), consistent with the dominance of this genus in such habitats. Seven other RHD gene types were detected, five of which were distantly related to RHD sequences from known bacteria and two of them clustered with *nahAc*-like (*Pseudomonas*) and *phnAc*-like (*A. faecalis* AFK2) genes. On this other hand, RHDs similar to those found in Pseudomonads (*nahAc*) and Betaproteobacteria (*nagAc, pahAc*) were mostly detected in marine microbial mats subjected to an oil contamination (Bordenave *et al.*, 2008). In mangrove sediments, RHD genes similar to those found in Actinobacteria and in Sphingomonads were detected by PCR in bacterial isolates recovered from these habitats (Zhou *et al.*, 2006).

Differences in the observed biodiversity of RHDs in PAH-polluted sites might reflect the natural variability of microbial communities in geographically distant sites. On the other hand, differences might also be due to the PCR-based methods used for amplification of the RHD genes. The genes encoding the α subunits of RHDs are highly variable in sequence (see Fig. 7.1) and various sets of degenerate primers have been used to amplify different gene regions (Ni Chadhain *et al.*, 2006; Bordenave *et al.*, 2008; Lozada *et al.*, 2008). None of the proposed primer sets has the potential to amplify all types of RHD genes and biases in the PCR-based approaches are therefore to be expected. Nevertheless, one can use primer sets specific for a defined group of RHDs, for instance those found in Gram-positive PAH degraders. Analysis of subantartic marine sediments using a PCR-based approach with primers targeting Gram-positive RHDs led to the identification of 14 groups of genes, most of which were related to dioxygenases genes found in *Mycobacterium, Rhodococcus, Nocardioides* and *Terrabacter* (Marcos *et al.*, 2009). As a means to detect a wide range of RHD genes without compromising specificity, Ni Chadain *et al.* (2006) designed generic primers to amplify a small PCR product encoding a relatively conserved region of the α subunit around the Rieske-cluster binding site. Although the primers were successfully tested against DNA from various isolated bacteria, PCR products obtained from environmental samples were essentially related to RHDs from Gram-negative bacteria. No product could be attributed to Actinobacteria RHD, although this bacterial group represented a significant part of the microbial diversity as determined from 16S rRNA genotyping. Another drawback of this approach was the limited information gained from the sequences of the short PCR products representing only about 6% of the α subunit gene length. Yet, this study was one of the first to combine the analysis of rRNA profiles and RHD profiles, providing useful information for assessing the bioremediation potential of a contaminated soil. In another study, a similar PCR-based approach was used to examine the RHD diversity of a coal tar contaminated aquifer in relation with the naphthalene degradation potential (Yagi and Madsen, 2009). Results indicated a broad diversity dominated by nagAc-type genes, the relative abundance of which fluctuated substantially over a 9-month period as determined by quantitative PCR. Good agreement was observed between in field gene expression as monitored by nagAc specific mRNA and the naphthalene degradation potential in the aquifer.

As a means to determine the occurrence or relative abundance of specific RHD genes in a given soil or sediment, methods based on quantitative PCR have been developed using specific primers (Baldwin *et al.*, 2003; Park and Crowley, 2006; DeBruyn *et al.*, 2007; Cebron *et al.*, 2008). Owing to the sequence divergence between RHDs from Gram-positive and Gram-negative bacteria, Cebron *et al.* (2008) designed one set of primers for each RHD category. The primer sets were used for quantification of RHD genes in soil by real-time qPCR. Data could be normalized relative to the 16S rRNA gene copy number, thus allowing an estimation of the PAH degrader population of up to 1% of the whole bacterial community. RHD gene enumeration by qPCR was used as a means to monitor biodegradation activity during treatment of a BTEX contaminated site. Fluctuations in the RHD levels detected were correlated with dynamic changes of the microbial population as a function of oxygen availability (Nebe *et al.*, 2009).

The presence of RHDs genes as revealed by PCR amplification in a given environmental sample does not prove that the corresponding enzymes are expressed *in situ*. In order to identify RHDs that are expressed *in situ*, one approach consists of analysing specific transcripts by RT-PCR using bulk RNA extracted from environmental samples. For example, transcripts of *nahAc*-like genes corresponding to 28 gene sequences potentially coding for naphthalene dioxygenases were detected in groundwater from a coal tar contaminated site (Wilson *et al.*, 1999). More recently, real-time PCR of RHD transcripts and microarrays were employed to monitor changes in the diversity and degradation potential of high Arctic soil bacteria upon hydrocarbon pollution (Yergeau *et al.*, 2009).

Identification of PAH-degraders *in situ* by stable isotope probing

Even though DNA or RNA-based molecular methods to investigate the microbial biodiversity of polluted sites have made significant progresses especially during the last decade, it is still a challenge to identify those bacterial species that are most active at removing pollutants *in situ*. In order to meet this challenge, one of the best experimental approach employed so far, called stable isotope probing (SIP), was first described 10 years ago (Radajewski *et al.*, 2000). The method involves incorporation of a rare stable isotope, most often ^{13}C or ^{15}N, into the biological material of a subpopulation of microorganisms that can metabolize the organic compounds of interest, here the PAHs. For this purpose, a fully labelled organic compound is incubated with the microbial population of an environmental sample to allow selective incorporation of the isotope in targeted bacteria. Subsequent isolation of bulk DNA or RNA from the environmental sample, followed by separation and recovery of labelled nucleic acids by density centrifugation on cesium salt gradient, allows analysis of labelled biomarkers (i.e. rRNA gene sequences) and identification of relevant target bacteria (Dumont and Murrell, 2005; Kreuzer-Martin and Jarman, 2007; Neufeld *et al.*, 2007). Using a field-based SIP strategy to track naphthalene degraders on a coal tar contaminated site, Jeon *et al.* found out that bacteria most active on the dicyclic hydrocarbon belonged to the class of betaproteobacteria (Jeon *et al.*, 2003). The dominant naphthalene degrader was isolated and identified as a previously undescribed *Polaromonas* strain, host of a RHD related to that found in *Comamonas* strains. This study brought clear evidence that the bacteria having a major role in pollutant removal *in situ* had nothing to do with the great majority of naphthalene degraders studied so far, mainly Pseudomonads derived from classical microbiological selection and cultivation methods.

In a SIP study targeting naphthalene- and phenanthrene-degrading bacteria in a bioreactor treating PAH-contaminated soil, Singleton *et al.* (2005) observed that the bacterial groups detected differed depending on the PAH-substrate used as probe. Bacteria selected on naphthalene primarily belonged to the *Pseudomonas* and *Ralstonia* genera, whereas those selected on phenanthrene were dominated by *Acidovorax*. The same authors examined the diversity of pyrene-degrading bacteria from the microbial community of a similar bioreactor (Singleton *et al.*, 2006). The primary species identified based on 16S rRNA genotyping were uncultivated members of the beta and gamma classes of proteobacteria. In a later study also using ^{13}C-pyrene as a probe, relevant degraders were tracked in creosote-polluted soil under static conditions or under agitation (Jones *et al.*, 2008). Although pyrene mineralization was stimulated by agitation, the dominant bacterial degraders identified based on 16S rRNA sequences under either conditions were very similar. The rRNA gene sequences obtained were closely related to those of uncultured Gammaproteobacteria, and were also remarkably similar to those previously found as dominant in a slurry bioreactor (Singleton *et al.*, 2006). None of the sequences matched that of Actinobacteria rRNA, although members of this class, especially fast-growing *Mycobacteria*, are the most studied, cultivated group of pyrene-degrading bacteria.

SIP techniques eliminate the need for growth outside the natural microenvironment, thereby reducing bias associated with growth in artificial media and permitting the exploration of the vast majority of uncultured bacteria present in natural or contaminated environments. It is also a suitable approach for studying changes in the composition of the subpopulation of degraders as a function of contaminant concentration or availability, although such investigations are still rare. In this respect, we have recently initiated an analysis of the biodiversity of PAH degraders in the soil of a wetland station collecting road runoff. Using ^{13}C-phenanthrene as a probe, we observed that the population of PAH degraders changed dramatically as a function of phenanthrene availability (Martin and Jouanneau, unpublished). Possible limitations to the SIP approach are the uncertainties linked to carbon cross-feeding; that is the incorporation of stable isotope into the biomass of trophically related bacteria, through uptake of labelled metabolites or scavenging of labelled biomass. DeRito *et al.* (2005) showed that ^{13}C-labelled biomass from primary phenol-degraders could be used by other soil bacteria. However, cross-feeding can be avoided by reducing the incubation of microbial populations with the labelled substrate to a time shorter than that required for its complete mineralization, which can be monitored by $^{13}CO_2$ respirometry (Padmanabhan *et al.*, 2003).

Beyond the advantage of straightforward rRNA-based identification of bacteria responsible for pollutant degradation, the selective labelling of DNA *in situ* offers the possibility of PCR based analysis of any functional gene involved in the degradation. In this respect, the diversity of RHDs involved in PAH degradation might be explored by PCR amplification of relevant portion of the genes.

Strategies for improving PAH bioremediation based on current knowledge on RHDs

From the current knowledge of their catalytic properties, RHDs appear as rather slow enzymes with limited activity towards large polycyclic hydrocarbons. Most studied enzymes cannot oxidize PAHs composed of more than three aromatic rings, and even RHDs displaying a broader substrate specificity, such as the *Sphingomonas* CHY-1 RHD, are poorly efficient in the oxidation of four- and five-ring substrates (Jouanneau *et al.*, 2006). Given that RHDs catalyse a critical step in PAH degradation, which can limit substrate processing through the downstream catabolic pathway, strategies based on genetic engineering to improve enzyme activity or expand substrate specificity might be implemented to enhance bioremediation, especially with respect to recalcitrant four- and five-ring PAHs. So far, bioengineering studies were essentially focused on biphenyl dioxygenases in attempts to improve the biodegradation of polychlorobiphenyls (PCBs) (Furukawa, 2000; Barriault *et al.*, 2002; Furukawa *et al.*, 2004). Such studies aimed at improving the substrate specificity and catalytic efficiency of biphenyl dioxygenases towards PCB are reviewed in detail in Chapter 8 of this book. Generally, the most successful approach employed to improve the catalytic activity of dioxygenases relied on directed molecular evolution involving DNA shuffling of the gene of interest according to Stemmer (1994). In the case of RHDs, only the gene encoding the α subunit of the catalytic component has been the target of DNA shuffling, since this subunit contains the active site and the substrate-binding pocket, which determines specificity. In an interesting development of this technique, Vezina *et al.* amplified partial sequences of the bphA gene encoding the dioxygenase alpha subunit from soil DNA and used them in shuffling experiments to improve the activity of biphenyl dioxygenase towards 2,2′-substituted PCB, thus taking advantage of the catalytic biodiversity available in soil for engineering new enzymes (Vezina *et al.*, 2007).

The naphthalene dioxygenase from *Ralstonia* strain U2 was engineered to oxidize nitroaromatic compounds (Keenan *et al.*, 2005). Saturation mutagenesis of the residue F350 in the alpha subunit of the enzyme led to the isolation of a variant with acquired dioxygenase activity towards dinitrotoluenes. Additional mutagenesis through DNA shuffling identified two other residues that influenced the regiospecificity of the enzyme and allowed the generation of variant with three substitutions, which exhibited enhanced activity towards dinitrotoluenes. As another option to expand the range of substrates oxidized by a given bacterium, it has been proposed to co-express in a single cell catalytic components of dioxygenases sharing the same electron carriers. An *E. coli* strain carrying two such recombinant oxygenases could simultaneously oxidize naphthalene and dinitrotoluene (Keenan and Wood, 2006).

In techniques such as DNA shuffling, large mutant libraries are generated, and the selection of enzymes with the desired activity requires the implementation of efficient screening strategies (Tee and Schwaneberg, 2007). For biphenyl dioxygenases, screening is based on the detection of a yellow product generated from biphenyl or PCBs by two enzymes that are co-expressed with the RHD in recombinant cells, which convert the biphenyl dihydrodiol to a coloured cleavage product. A similar strategy might not be applicable for PAH dioxygenases, because the cleavage product obtained from the first ring cleavage of PAHs is not easy to detect. Hence, the selection of RHDs with higher activity towards PAHs, especially four- and five-ring hydrocarbons, through DNA shuffling would require another high throughput assay for measuring the activity of recombinant RHDs. In this respect, Joern *et al.* (2001) have developed a solid-phase screen for the detection of dioxygenase activity in recombinant clones. In the assay, the dihydrodiol produced by the dioxygenase reaction is indirectly measured after chemical or enzymatic conversion to an oxidation product that reacts with Gibbs reagent to form a blue complex. After screening 9000 clones using this procedure, a variant of toluene dioxygenase was selected with an activity twice as high as that of the wild-type enzyme (Joern *et al.*, 2001). In practice, we found that the method is laborious, not very sensitive and inadequate for the detection of highly unstable dihydroxylated PAH such as 1,2-dihydroxyanthracene. As an alternative, we recently developed an assay based on the direct fluorescence detection of the dihydrodiols formed, which has been adapted to the microplate format. Although the assay is limited to PAH yielding dihydrodiols with high enough intrinsic fluorescence (anthracene, chrysene, pyrene, benzo[a]pyrene), preliminary results suggest that it is suitable for the screening of RHDs with activity towards four- and five-ring PAHs (Jouanneau, unpublished).

Conclusions and perspectives

Over the last decade, considerable progress in our understanding of RHD catalytic function has been achieved thanks to detailed structural analyses of representative oxygenase components through X-ray crystallography. The regio- and stereo-selectivity of RHDs can be easily explained from studies on enzyme–substrate complexes showing the critical role of the substrate-binding pocket in the positioning of the substrate relative to the active-site iron atom. The size and topology of the substrate-binding site are certainly important factors in controlling enzyme specificity, but the structural determinants responsible for substrate recognition are still largely unknown. This is an important issue in the context of PAH biodegradation, because the most genotoxic four- and five-ring hydrocarbons are poor substrates for most RHDs. Taking advantage of the accumulated structural information on RHDs, one can try to improve RHD catalytic efficiency or enlarge substrate range towards recalcitrant PAHs by employing bioengineering strategies based on rational design or directed molecular evolution. Such strategies, especially those involving DNA shuffling, have been successfully implemented to improve the substrate range of biphenyl dioxygenases towards PCBs (Furukawa, 2000; Furukawa *et al.*, 2004). Hence, provided that efficient screening systems are developed to select for enzymes with desired characters, significant improvements in the performances of PAH oxidizing RHDs may be expected from future bioengineering studies, especially those based on molecular directed evolution.

In recent years, new tools have emerged to explore the biodiversity of highly complex microbial populations in their environmental habitats. In particular, it becomes possible to identify bacteria that are most active in the degradation of pollutants *in situ* without the need to cultivate and isolate them. In the case of PAH polluted sites, the use of approaches such as stable isotope probing led to the discovery that the dominant PAH degraders were different from the bacterial species that had previously been isolated and studied under standard laboratory conditions. This observation is consistent with the general assumption that >90% of bacterial species present in the environment are unknown, uncultured microorganisms. Consequently, knowledge derived from PAH degraders studied so far need to be revisited by focusing on species that play a major role in pollutant removal *in situ*. In this respect, it will be necessary to develop novel cultivation conditions and isolation procedures in order to study these bacterial species and eventually to identify their PAH catabolic enzymes. Further developments of culture-independent methods might combine stable isotope labelling with *in situ* hybridization for the analysis of single cells (Huang *et al.*, 2009), or with metagenomics to couple sequence information from a specific subpopulation with metabolic function (Leigh *et al.*, 2007; Kalyuzhnaya *et al.*, 2008).

PCR-based methods were also developed to analyse directly the biodiversity of RHDs in environmental samples. Since the genes encoding RHDs display great sequence variability, and no region is sufficiently conserved to design universal primers, many studies relying on a single set of degenerate primers introduce biases resulting in a limited sequence sampling of the biodiversity. This problem can be overcome by using multiple sets of primers to amplify different families of RHDs. For instance, two sets of primers were designed to amplify separately RHDs from Gram-positive and Gram-negative PAH degraders (Cebron *et al.*, 2008). Yet, these PCR-based methods do not provide unambiguous evidence that the detected RHD genes are all related to the metabolic function of interest and expressed under the studied conditions. Again, labelling methods such DNA-based or RNA-based stable isotope-probing have the potential to specifically target pollutant degraders and their associated RHDs. For example, bacteria and functional genes associated with biphenyl degradation were identified in a PCB-polluted site using DNA-SIP coupled to analyses including gene detection by DNA microarray and RHD-specific PCR amplification (Leigh *et al.*, 2007). Similar approaches might be used to target RHDs and other functional genes involved in PAH degradation in polluted sites. The resulting genetic information might be useful for the development of quantitative analysis of specific catabolic genes such as real time qPCR, and for the conception of specific microarrays designed to detect relevant genes and their transcripts *in situ*. The development of such molecular tools to detect, quantify and monitor bacterial populations involved in PAH degradation, as well as their specific catabolic enzymes, is likely to be of benefit to bioremediation technologies in the near future.

Acknowledgements

We thank L. Schuler for providing Fig. 7.2 and 7.3. This work was supported by a PhD fellowship from the Région Rhône-Alpes to F. Martin, and by grants from the CNRS and the University of Grenoble.

References

Alonso-Gutierrez, J., Figueras, A., Albaiges, J., Jimenez, N., Vinas, M., Solanas, A.M., and Novoa, B. (2009). Bacterial communities from shoreline environments (Costa da Morte, Northwestern Spain) affected by the Prestige oil spill. Appl. Environ. Microbiol. *75*, 3407–3418.

Aly, H.A.H., Huu, N.B., Wray, V., Junca, H., and Pieper, D.H. (2008). Two angular dioxygenases contribute to the metabolic versatility of dibenzofuran-degrading *Rhodococcus* sp. strain HA01. Appl. Environ. Microbiol. *74*, 3812–3822.

Armengaud, J., Happe, B., and Timmis, K.N. (1998). Genetic analysis of dioxin dioxygenase of *Sphingomonas* sp. Strain RW1: catabolic genes dispersed on the genome. J. Bacteriol. *180*, 3954–3966.

Ashikawa, Y., Fujimoto, Z., Noguchi, H., Habe, H., Omori, T., Yamane, H., and Nojiri, H. (2006). Electron transfer complex formation between oxygenase and ferredoxin components in Rieske nonheme iron oxygenase system. Structure *14*, 1779–1789.

Baldwin, B.R., Nakatsu, C.H., and Nies, L. (2003). Detection and enumeration of aromatic oxygenase genes by multiplex and real-time PCR. Appl. Environ. Microbiol. *69*, 3350–3358.

Barriault, D., Plante, M.M., and Sylvestre, M. (2002). Family shuffling of a targeted bphA region to engineer biphenyl dioxygenase. J. Bacteriol. *184*, 3794–3800.

Batie, C.J., Ballou, D.P., and Correll, C.C. (1992). Phthalate dioxygenase reductase and related flavin-iron–sulfur containing electron transferases. In Chemistry and Biochemistry of Flavoenzymes, Müller, F., ed. (CRC Press: Boca Raton, FL), pp. 543–556.

Beharry, Z.M., Eby, D.M., Coulter, E.D., Viswanathan, R., Neidle, E.L., Phillips, R.S., and Kurtz, D.M. (2003). Histidine ligand protonation and redox potential in the Rieske dioxygenases: Role of a conserved aspartate in anthranilate 1,2-dioxygenase. Biochemistry *42*, 13625–13636.

Bordenave, S., Goni-urriza, M., Vilette, C., Blanchard, S., Caumette, P., and Duran, R. (2008). Diversity of ring-hydroxylating dioxygenases in pristine and oil contaminated microbial mats at genomic and transcriptomic levels. Environ. Microbiol. *10*, 3201–3211.

Bosch, R., Garcia-Valdes, E., and Moore, E.R. (1999). Genetic characterization and evolutionary implications of a chromosomally encoded naphthalene-degradation upper pathway from *Pseudomonas stutzeri* AN10. Gene *236*, 149–157.

Bosch, R., Garcia-Valdes, E., and Moore, E.R. (2000). Complete nucleotide sequence and evolutionary significance of a chromosomally encoded naphthalene-degradation lower pathway from *Pseudomonas stutzeri* AN10. Gene *245*, 65–74.

Boyd, D.R., Sharma, N.D., Agarwal, R., Resnick, S.M., Schocken, M.J., Gibson, D.T., Sayer, J.M., Yagi, H., and Jerina, D.M. (1997). Bacterial dioxygenase-catalysed dihydroxylation and chemical resolution routes to enantiopure cis-dihydrodiols of chrysene. J. Chem. Soc. Perkin Trans. *1*, 1715–1723.

Carredano, E., Karlsson, A., Kauppi, B., Choudhury, D., Parales, R.E., Parales, J.V., Lee, K., Gibson, D.T., Eklund, H., and Ramaswamy, S. (2000). Substrate binding site of naphthalene 1,2-dioxygenase: functional implications of indole binding. J. Mol. Biol. *296*, 701–712.

Cebron, A., Norini, M.P., Beguiristain, T., and Leyval, C. (2008). Real-time PCR quantification of PAH-ring hydroxylating dioxygenase (PAH-RHD alpha) genes from Gram-positive and Gram-negative bacteria in soil and sediment samples. J. Microbiol. Meth. *73*, 148–159.

Chadhain, S.M.N., Moritz, E.M., Kim, E., and Zylstra, G.J. (2007). Identification, cloning, and characterization of a multicomponent biphenyl dioxygenase from *Sphingobium yanoikuyae* B1. J. Ind. Microbiol. Biotechnol. *34*, 605–613.

Chang, H.-K., Mohseni, P., and Zylstra, G.J. (2003). Characterization and regulation of the genes for a novel anthranilate 1,2-dioxygenase from *Burkholderia cepacia* DBO1. J. Bacteriol. *185*, 5871–5881.

Correll, C.C., Batie, C.J., Ballou, D.P., and Ludwig, M.L. (1992). Phthalate dioxygenase reductase: a modular structure for electron transfer from pyridine nucleotides to [2Fe–2S]. Science *258*, 1604–1610.

Cui, Z.S., Lai, Q.L., Dong, C.M., and Shao, Z.Z. (2008). Biodiversity of polycyclic aromatic hydrocarbon-degrading bacteria from deep sea sediments of the Middle Atlantic Ridge. Environ. Microbiol. *10*, 2138–2149.

DeBruyn, J.M., Chewning, C.S., and Sayler, G.S. (2007). Comparative quantitative prevalence of Mycobacteria and functionally abundant *nidA*, *nahAc*, and *nagAc* dioxygenase genes in coal tar contaminated sediments. Environ. Sci. Technol. *41*, 5426–5432.

Demaneche, S., Meyer, C., Micoud, J., Louwagie, M., Willison, J.C., and Jouanneau, Y. (2004). Identification and functional analysis of two aromatic ring-hydroxylating dioxygenases from a *Sphingomonas* strain degrading various polycyclic aromatic hydrocarbons. Appl. Environ. Microbiol. *70*, 6714–6725.

DeRito, C.M., Pumphrey, G.M., and Madsen, E.L. (2005). Use of field-based stable isotope probing to identify adapted populations and track carbon flow through a phenol-degrading soil microbial community. Appl. Environ. Microbiol. *71*, 7858–7865.

Doyle, E., Muckian, L., Hickey, A.M., and Clipson, N. (2008). Microbial PAH degradation. Adv. Appl. Microbiol. *65*, 27–66.

Dumont, M.G., and Murrell, J.C. (2005). Stable isotope probing – linking microbial identity to function. Nature Rev. Microbiol. 3, 499–504.

Eby, D.M., Beharry, Z.M., Coulter, E.D., Kurtz, D.M., Jr., and Neidle, E.L. (2001). Characterization and evolution of anthranilate 1,2-dioxygenase from *Acinetobacter* sp. strain ADP1. J. Bacteriol. *183*, 109–118.

Ensley, B.D., Ratzkin, B.J., Osslund, T.D., Simon, M.J., Wackett, L.P., and Gibson, D.T. (1983). Expression of naphthalene oxidation genes in *Escherichia coli* results in the biosynthesis of indigo. Science 222, 167–169.

Ferraro, D.J., Gakhar, L., and Ramaswamy, S. (2005). Rieske business: structure-function of Rieske non-heme oxygenases. Biochem. Biophys. Res. Commun. *338*, 175–190.

Ferraro, D.J., Okerlund, A.L., Mowers, J.C., and Ramaswamy, S. (2006). Structural basis for regioselectivity and stereoselectivity of product formation by naphthalene 1,2-dioxygenase. J. Bacteriol. *188*, 6986–6994.

Ferraro, D.J., Brown, E.N., Yu, C.L., Parales, R.E., Gibson, D.T., and Ramaswamy, S. (2007). Structural investigations of the ferredoxin and terminal oxygenase components of the biphenyl 2,3-dioxygenase from *Sphingobium yanoikuyae* BI. BMC Struct. Biol. 7.

Friemann, R., Ivkovic-Jensen, M.M., Lessner, D.J., Yu, C.L., Gibson, D.T., Parales, R.E., Eklund, H., and Ramaswamy, S. (2005). Structural insight into the dioxygenation of nitroarene compounds: the crystal structure of nitrobenzene dioxygenase. J. Mol. Biol. *348*, 1139–1151.

Fuenmayor, S.L., Wild, M., Boyes, A.L., and Williams, P.A. (1998). A gene cluster encoding steps in conversion of naphthalene to gentisate in *Pseudomonas* sp. strain U2. J. Bacteriol. *180*, 2522–2530.

Furukawa, K. (2000). Engineering dioxygenases for efficient degradation of environmental pollutants. Cur. Opin. Biotechnol. *11*, 244–249.

Furukawa, K., Suenaga, H., and Goto, M. (2004). Biphenyl dioxygenases: Functional versatilities and directed evolution. J. Bacteriol. *186*, 5189–5196.

Furusawa, Y., Nagarajan, V., Tanokura, M., Masai, E., Fukuda, M., and Senda, T. (2004). Crystal structure of the terminal oxygenase component of biphenyl dioxygenase derived from *Rhodococcus* sp. strain RHA1. J. Mol. Biol. *342*, 1041–1052.

Gakhar, L., Malik, Z.A., Allen, C.C. R., Lipscomb, D.A., Larkin, M.J., and Ramaswamy, S. (2005). Structure and increased thermostability of *Rhodococcus* sp. naphthalene 1,2-dioxygenase. J. Bacteriol. *187*, 7222–7231.

Geiselbrecht, A.D., Hedlund, B.P., Tichi, M.A., and Staley, J.T. (1998). Isolation of marine polycyclic aromatic hydrocarbon (PAH)-degrading *Cycloclasticus* strains from the Gulf of Mexico and comparison of their PAH degradation ability with that of Puget Sound *Cycloclasticus* strains. Appl. Environ. Microbiol. *64*, 4703–4710.

Gibson, D.T., Mahadevan, V., Jerina, D.M., Yogi, H., and Yeh, H.J. (1975). Oxidation of the carcinogens benzo[a]pyrene and benzo[a]anthracene to dihydrodiols by a bacterium. Science *189*, 295–297.

Goyal, A.K., and Zylstra, G.J. (1996). Molecular cloning of novel genes for polycyclic aromatic hydrocarbon degradation from *Comamonas testosteroni* GZ39. Appl. Environ. Microbiol. *62*, 230–236.

Goyal, A.K., and Zylstra, G.J. (1997). Genetics of naphthalene and phenanthrene degradation by *Comamonas testosteroni*. J. Ind. Microbiol. Biotechnol. *19*, 401–407.

Habe, H., and Omori, T. (2003). Genetics of polycyclic aromatic hydrocarbon metabolism in diverse aerobic bacteria. Biosci. Biotech. Biochem. *67*, 225–243.

Habe, H., Miyakoshi, M., Chung, J., Kasuga, K., Yoshida, T., Nojiri, H., and Omori, T. (2003). Phthalate catabolic gene cluster is linked to the angular dioxygenase gene in *Terrabacter* sp. strain DBF63. Appl. Microbiol. Biotechnol. *61*, 44–54.

Habe, H., Chung, J.S., Kato, H., Ayabe, Y., Kasuga, K., Yoshida, T., Nojiri, H., Yamane, H., and Omori, T. (2004). Characterization of the upper pathway genes for fluorene metabolism in *Terrabacter* sp. strain DBF63. J. Bacteriol. *186*, 5938–5944.

Harayama, S., Kok, M., and Neidle, E.L. (1992). Functional and evolutionary relationships among diverse oxygenases. Annu. Rev. Microbiol. *46*, 565–601.

Heitkamp, M.A., Freeman, J.P., Miller, D.W., and Cerniglia, C.E. (1988). Pyrene degradation by a *Mycobacterium* sp.: identification of ring oxidation and ring fission products. Appl. Environ. Microbiol. *54*, 2556–2565.

Huang, W.E., Ferguson, A., Singer, A.C., Lawson, K., Thompson, I.P., Kalin, R.M., Larkin, M.J., Bailey, M.J., and Whiteley, A.S. (2009). Resolving genetic functions within microbial populations: *In situ* analyses using rRNA and mRNA stable isotope probing coupled with single-cell Raman-fluorescence *in situ* hybridization. Appl. Environ. Microbiol. *75*, 234–241.

Hudlicky, T., Gonzalez, D., and Gibson, D.T. (1999). Enzymatic dihydroxylation of aromatics in enantioselective synthesis: expanfing asymetric methodology. Aldrichimica Acta *32*, 35–62.

Iida, T., Mukouzaka, Y., Nakamura, K., and Kudo, T. (2002). Plasmid-borne genes code for an angular dioxygenase involved in dibenzofuran degradation by *Terrabacter* sp. strain YK3. Appl. Environ. Microbiol. *68*, 3716–3723.

Iida, T., Moteki, Y., Nakamura, K., Taguchi, K., Otagiri, M., Asanuma, M., Dohmae, N., Usami, R., and Kudo, T. (2009a). Functional expression of three Rieske non-heme iron oxygenases derived from Actinomycetes in *Rhodococcus* species for investigation of their degradation capabilities of dibenzofuran and chlorinated dioxins. Biosci. Biotech. Biochem. *73*, 822–827.

Iida, T., Waki, T., Nakamura, K., Mukouzaka, Y., and Kudo, T. (2009b). The GAF-like-domain-containing transcriptional regulator DfdR Is a sensor protein for dibenzofuran and several hydrophobic aromatic compounds. J. Bacteriol. *191*, 123–134.

Imbeault, N.Y., Powlowski, J.B., Colbert, C.L., Bolin, J.T., and Eltis, L.D. (2000). Steady-state kinetic characterization and crystallization of a polychlorinated biphenyl-transforming dioxygenase. J. Biol. Chem. *275*, 12430–12437.

Jakoncic, J., Jouanneau, Y., Meyer, C., and Stojanoff, V. (2007a). The catalytic pocket of the ring-hydroxylating dioxygenase from *Sphingomonas* CHY-1. Biochem. Biophys. Res. Commun. *352*, 861–866.

Jakoncic, J., Jouanneau, Y., Meyer, C., and Stojanoff, V. (2007b). The crystal structure of the ring-hydroxylating dioxygenase from *Sphingomonas* CHY-1. Febs J. *274*, 2470–2481.

Jeon, C.O., Park, M., Ro, H.S., Park, W., and Madsen, E.L. (2006). The naphthalene catabolic (*nag*) genes of *Polaromonas naphthalenivorans* CJ2: Evolutionary implications for two gene clusters and novel regulatory control. Appl. Environ. Microbiol. 72, 1086–1095.

Jeon, C.O., Park, W., Padmanabhan, P., DeRito, C., Snape, J.R., and Madsen, E.L. (2003). Discovery of a bacterium, with distinctive dioxygenase, that is responsible for *in situ* biodegradation in contaminated sediment. Proc. Natl. Acad. Sci. U.S.A. *100*, 13591–13596.

Joern, J.M., Sakamoto, T., Arisawa, A.A., and Arnold, F.H. (2001). A versatile high throughput screen for dioxygenase activity using solid-phase digital imaging. J. Biomol. Scr. *6*, 219–223.

Jones, M.D., Singleton, D.R., Carstensen, D.P., Powell, S.N., Swanson, J.S., Pfaender, F.K., and Aitken, M.D. (2008). Effect of incubation conditions on the enrichment of pyrene-degrading bacteria identified by stable-isotope probing in an aged, PAH-contaminated soil. Microbial Ecol. *56*, 341–349.

Jones, R.M., Britt-Compton, B., and Williams, P.A. (2003). The naphthalene catabolic (nag) genes of *Ralstonia* sp. strain U2 are an operon that is regulated by NagR, a LysR-type transcriptional regulator. J. Bacteriol. *185*, 5847–5853.

Jouanneau, Y., Meyer, C., Jakoncic, J., Stojanoff, V., and Gaillard, J. (2006). Characterization of a naphthalene dioxygenase endowed with an exceptionally broad substrate specificity toward polycyclic aromatic hydrocarbons. Biochemistry *45*, 12380–12391.

Jouanneau, Y., Micoud, J., and Meyer, C. (2007). Purification and characterization of a three-component salicylate 1-hydroxylase from *Sphingomonas* sp. strain CHY-1. Appl. Environ. Microbiol. *73*, 7515–7521.

Kalyuzhnaya, M.G., Lapidus, A., Ivanova, N., Copeland, A.C., McHardy, A.C., Szeto, E., Salamov, A., Grigoriev, I.V., Suciu, D., Levine, S.R., *et al.* (2008). High-resolution metagenomics targets specific functional types in complex microbial communities. Nature Biotechnol. *26*, 1029–1034.

Karlsson, A., Beharry, Z.M., Eby, D.M., Coulter, E.D., Neidle, E.L., Kurtz, D.M., Eklund, H., and Ramaswamy, S. (2002). X-ray crystal structure of benzoate 1,2-dioxygenase reductase from *Acinetobacter* sp. strain ADP1. J. Mol. Biol. *318*, 261–272.

Karlsson, A., Parales, J.V., Parales, R.E., Gibson, D.T., Eklund, H., and Ramaswamy, S. (2003). Crystal structure of naphthalene dioxygenase: side-on binding of dioxygen to iron. Science *299*, 1039–1042.

Kasai, Y., Kishira, H., and Harayama, S. (2002). Bacteria belonging to the genus *Cycloclasticus* play a primary role in the degradation of aromatic hydrocarbons released in a marine environment. Appl. Environ. Microbiol. *68*, 5625–5633.

Kasai, Y., Shindo, K., Harayama, S., and Misawa, N. (2003). Molecular characterization and substrate preference of a polycyclic aromatic hydrocarbon dioxygenase from *Cycloclasticus* sp. strain A5. Appl. Environ. Microbiol. *69*, 6688–6697.

Kasuga, K., Habe, H., Chung, J.S., Yoshida, T., Nojiri, H., Yamane, H., and Omori, T. (2001). Isolation and characterization of the genes encoding a novel oxygenase component of angular dioxygenase from the Gram-positive dibenzofuran-degrader *Terrabacter* sp. strain DBF63. Biochem. Biophys. Res. Commun. *283*, 195–204.

Kauppi, B., Lee, K., Carredano, E., Parales, R.E., Gibson, D.T., Eklund, H., and Ramaswamy, S. (1998). Structure of an aromatic-ring-hydroxylating dioxygenase-naphthalene 1,2- dioxygenase. Structure *6*, 571–586.

Keenan, B.G., Leungsakul, T., Smets, B.F., Mori, M.A., Henderson, D.E., and Wood, T.K. (2005). Protein engineering of the archetypal nitroarene dioxygenase of *Ralstonia* sp. strain U2 for activity on aminonitrotoluenes and dinitrotoluenes through alpha-subunit residues leucine 225, phenylalanine 350, and glycine 407. J. Bacteriol. *187*, 3302–3310.

Keenan, B.G., and Wood, T.K. (2006). Orthric Rieske dioxygenases for degrading mixtures of 2,4-dinitrotoluene/naphthalene and 2-amino-4,6-dinitrotoluene/4-amino-2,6-dinitrotoluene. Appl. Microbiol. Biotechnol. *73*, 827–838.

Khan, A.A., Wang, R.F., Cao, W.W., Doerge, D.R., Wennerstrom, D., and Cerniglia, C.E. (2001). Molecular cloning, nucleotide sequence, and expression of genes encoding a polcyclic aromatic ring dioxygenase from *Mycobacterium* sp. strain PYR-1. Appl. Environ. Microbiol. *67*, 3577–3585.

Kim, E., and Zylstra, G.J. (1999). Functional analysis of genes involved in biphenyl, naphthalene, phenanthrene, and m-xylene degradation by *Sphingomonas yanoikuyae* B1. J. Ind. Microbiol. Biotechnol. *23*, 294–302.

Kim, S.J., Kweon, O.G., Freeman, J.P., Jones, R.C., Adjei, M.D., Jhoo, J.W., Edmondson, R.D., and Cerniglia, C.E. (2006). Molecular cloning and expression of genes encoding a novel dioxygenase Involved in low- and high-molecular-weight polycyclic aromatic hydrocarbon degradation in *Mycobacterium vanbaalenii* PYR-1. Appl. Environ. Microbiol. *72*, 1045–1054.

Kim, S.J., Kweon, O., Jones, R.C., Edmondson, R.D., and Cerniglia, C.E. (2008). Genomic analysis of polycyclic aromatic hydrocarbon degradation in *Mycobacterium vanbaalenii* PYR-1. Biodegradation *19*, 859–881.

Kirk, J.L., Beaudette, L.A., Hart, M., Moutoglis, P., Klironomos, J.N., Lee, H., and Trevors, J.T. (2004). Methods of studying soil microbial diversity. J. Microbiol. Meth. *58*, 169–188.

Kiyohara, H., Nagao, K., Kouno, K., and Yano, K. (1982). Phenanthrene-degrading phenotype of *Alcaligenes faecalis* AFK2. Appl. Environ. Microbiol. *43*, 458–461.

Kovaleva, E.G., and Lipscomb, J.D. (2008). Versatility of biological non-heme Fe(II) centers in oxygen activation reactions. Nature Chem. Biol. *4*, 186–193.

Kreuzer-Martin, H.W., and Jarman, K.H. (2007). Stable isotope ratios and forensic analysis of microorganisms. Appl. Environ. Microbiol. *73*, 3896–3908.

Krivobok, S., Kuony, S., Meyer, C., Louwagie, M., Willison, J.C., and Jouanneau, Y. (2003). Identification of pyrene-induced proteins in *Mycobacterium* sp. 6PY1: Evidence for two ring-hydroxylating dioxygenases. J. Bacteriol. *185*, 3828–3841.

Kumari, R., Subudhi, S., Suar, M., Dhingra, G., Raina, V., Dogra, C., Lal, S., van der Meer, J.R., Holliger, C., and Lal, R. (2002). Cloning and characterization of lin genes responsible for the degradation of hexachlorocyclohexane isomers by *Sphingomonas paucimobilis* strain B90. Appl. Environ. Microbiol. *68*, 6021–6028.

Kweon, O., Kim, S.J., Jones, R.C., Freeman, J.P., Adjei, M.D., Edmondson, R.D., and Cerniglia, C.E. (2007). A polyomic approach to elucidate the fluoranthene-degradative pathway in *Mycobacterium vanbaalenii* PYR-1. J. Bacteriol. *189*, 4635–4647.

Kweon, O., Kim, S.-J., Baek, S., Chae, J.-C., Adjei, M., Baek, D.-H., Kim, Y.-C., and Cerniglia, C.E. (2008). A new classification system for bacterial Rieske non-heme iron aromatic ring-hydroxylating oxygenases. BMC Biochemistry 9, 11.

Larkin, M.J., Allen, C.C., Kulakov, L.A., and Lipscomb, D.A. (1999). Purification and characterization of a novel naphthalene dioxygenase from *Rhodococcus* sp. strain NCIMB12038. J. Bacteriol. *181*, 6200–6204.

Laurie, A.D., and Lloyd-Jones, G. (1999). The *phn* genes of *Burkholderia* sp. strain RP007 constitute a divergent gene cluster for polycyclic aromatic hydrocarbon catabolism. J. Bacteriol. *181*, 531–540.

Leigh, M.B., Pellizari, V.H., Uhlik, O., Sutka, R., Rodrigues, J., Ostrom, N.E., Zhou, J.H., and Tiedje, J.M. (2007). Biphenyl-utilizing bacteria and their functional genes in a pine root zone contaminated with polychlorinated biphenyls (PCBs). Isme J. *1*, 134–148.

Leys, N.M., Ryngaert, A., Bastiaens, L., Verstraete, W., Top, E.M., and Springael, D. (2004). Occurrence and phylogenetic diversity of *Sphingomonas* strains in soils contaminated with polycyclic aromatic hydrocarbons. Appl. Environ. Microbiol. *70*, 1944–1955.

Leys, N.M., Ryngaert, A., Bastiaens, L., Wattiau, P., Top, E.M., Verstraete, W., and Springael, D. (2005). Occurrence and community composition of fast-growing *Mycobacterium* in soils contaminated with polycyclic aromatic hydrocarbons. FEMS Microbiol. Ecol. *51*, 375–388.

Liang, Y., Gardner, D.R., Miller, C.D., Chen, D., Anderson, A.J., Weimer, B.C., and Sims, R.C. (2006). Study of biochemical pathways and enzymes involved in pyrene degradation by *Mycobacterium* sp. strain KMS. Appl. Environ. Microbiol. 72, 7821–7828.

Lozada, M., Mercadal, J.P.R., Guerrero, L.D., Di Marzio, W.D., Ferrero, M.A., and Dionisi, H.M. (2008). Novel aromatic ring-hydroxylating dioxygenase genes from coastal marine sediments of Patagonia. BMC Microbiol. *8*.

Marcos, M.S., Lozada, M., and Dionisi, H.M. (2009). Aromatic hydrocarbon degradation genes from chronically polluted Subantarctic marine sediments. Lett. Appl. Microbiol. *49*, 602–608.

Martin, V.J., and Mohn, W.W. (1999). A novel aromatic-ring-hydroxylating dioxygenase from the diterpenoid-degrading bacterium *Pseudomonas abietaniphila* BKME-9. J. Bacteriol. *181*, 2675–2682.

Martins, B.M., Svetlitchnaia, T., and Dobbek, H. (2005). 2-oxoquinoline 8-monooxygenase oxygenase component: Active site modulation by Rieske-[2Fe–2S] center oxidation/reduction. Structure *13*, 817–824.

McKew, B.A., Coulon, F., Osborn, A.M., Timmis, K.N., and McGenity, T.J. (2007). Determining the identity and roles of oil-metabolizing marine bacteria from the Thames estuary, UK. Environ. Microbiol. *9*, 165–176.

Moreno-Ruiz, E., Hernaez, M.J., Martinez-Perez, O., and Santero, E. (2003). Identification and functional characterization of *Sphingomonas macrogolitabida* strain TFA genes involved in the first two steps of the tetralin catabolic pathway. J. Bacteriol. *185*, 2026–2030.

Nam, J.W., Nojiri, H., Yoshida, T., Habe, H., Yamane, H., and Omori, T. (2001). New classification system for oxygenase components involved in ring- hydroxylating oxygenations. Biosci. Biotechnol. Biochem. *65*, 254–263.

Nebe, J., Baldwin, B.R., Kassab, R.L., Nies, L., and Nakatsu, C.H. (2009). Quantification of aromatic oxygenase genes to evaluate enhanced bioremediation by oxygen releasing materials at a gasoline-contaminated site. Environ. Sci. Technol. *43*, 2029–2034.

Neufeld, J.D., Vohra, J., Dumont, M.G., Lueders, T., Manefield, M., Friedrich, M.W., and Murrell, J.C. (2007). DNA stable-isotope probing. Nature Protocols *2*, 860–866.

Ni Chadhain, S.M., Norman, R.S., Pesce, K.V., Kukor, J.J., and Zystra, G.J. (2006). Microbial dioxygenase gene population shifts during polycyclic aromatic hydrocarbon biodegradation. Appl. Environ. Microbiol. 72, 4078–4087.

Niepceron, M., Portet-Koltalo, F., Merlin, C., Motelay-Massei, A., Barray, S., and Bodilis, J. (2010). Both *Cycloclasticus* spp. and *Pseudomonas* spp. as PAH-degrading bacteria in the Seine estuary (France). FEMS Microbiol. Ecol. *71*, 137–147.

Nojiri, H., Nam, J.W., Kosaka, M., Morii, K.I., Takemura, T., Furihata, K., Yamane, H., and Omori, T. (1999). Diverse oxygenations catalyzed by carbazole 1,9a-dioxygenase from *Pseudomonas* sp. strain CA10. J. Bacteriol. *181*, 3105–3113.

Nojiri, H., Ashikawa, Y., Noguchi, H., Nam, J.W., Urata, M., Fujimoto, Z., Uchimura, H., Terada, T., Nakamura, S., Shimizu, K., *et al.* (2005). Structure of the terminal oxygenase component of angular dioxygenase, carbazole 1,9a-dioxygenase. J. Mol. Biol. *351*, 355–370.

Padmanabhan, P., Padmanabhan, S., DeRito, C., Gray, A., Gannon, D., Snape, J.R., Tsai, C.S., Park, W., Jeon, C., and Madsen, E.L. (2003). Respiration of 13C-labeled substrates added to soil in the field and subsequent 16S rRNA gene analysis of 13C-labeled soil DNA. Appl. Environ. Microbiol. *69*, 1614–1622.

Pagnout, C., Frache, G., Poupin, P., Maunit, B., Muller, J.F., and Ferard, J.F. (2007). Isolation and characterization of a gene cluster involved in PAH degradation in *Mycobacterium* sp. strain SNP11: Expression in *Mycobacterium smegmatis* mc 155. Res. Microbiol. *158*, 175–186.

Parales, R.E., Huang, R., Yu, C.L., Parales, J.V., Lee, F.K., Lessner, D.J., Ivkovic-Jensen, M.M., Liu, W., Friemann, R., Ramaswamy, S., *et al.* (2005). Purification, characterization, and crystallization of the components of the nitrobenzene and 2-nitrotoluene dioxygenase enzyme systems. Appl. Environ. Microbiol. *71*, 3806–3814.

Parales, R.E., Parales, J.V., and Gibson, D.T. (1999). Aspartate 205 in the catalytic domain of naphthalene dioxygenase is essential for activity. J. Bacteriol. *181*, 1831–1837.

Parales, R.E., Lee, K., Resnick, S.M., Jiang, H.Y., Lessner, D.J., and Gibson, D.T. (2000). Substrate specificity of naphthalene dioxygenase: Effect of specific amino acids at the active site of the enzyme. J. Bacteriol. *182*, 1641–1649.

Parales, R.E., and Resnick, S.M. (2006). Aromatic ring hydroxylating dioxygenases. In *Pseudomonas*, Ramos, J.L., and Levesque, R.C., eds. (Springer: New York), pp. 287–340.

Park, J.W., and Crowley, D.E. (2006). Dynamic changes in nahAc gene copy numbers during degradation of naphthalene in PAH-contaminated soils. Appl. Environ. Microbiol. 72, 1322–1329.

Peng, R.H., Xiong, A.S., Xue, Y., Fu, X.Y., Gao, F., Zhao, W., Tian, Y.S., and Yao, Q.H. (2008). Microbial biodegradation of polyaromatic hydrocarbons. FEMS Microbiol. Rev. 32, 927–955.

Pinyakong, O., Habe, H., and Omori, T. (2003a). The unique aromatic catabolic genes in sphingomonads degrading polycyclic aromatic hydrocarbons (PAHs). J. Gen. Appl. Microbiol. 49, 1–19.

Pinyakong, O., Habe, H., Yoshida, T., Nojiri, H., and Omori, T. (2003b). Identification of three novel salicylate 1-hydroxylases involved in the phenanthrene degradation of *Sphingobium* sp. strain P2. Biochem. Biophys. Res. Commun. *301*, 350–357.

Pinyakong, O., Habe, H., Kouzuma, A., Nojiri, H., Yamane, H., and Omori, T. (2004). Isolation and characterization of genes encoding polycyclic aromatic hydrocarbon dioxygenase from acenaphthene and acenaphthylene degrading *Sphingomonas* sp. strain A4. Fems Microbiol. Lett. *238*, 297–305.

Pollmann, K., Wray, V., Hecht, H.J., and Pieper, D.H. (2003). Rational engineering of the regioselectivity of TecA tetrachlorobenzene dioxygenase for the transformation of chlorinated toluenes. Microbiology *149*, 903–913.

Radajewski, S., Ineson, P., Parekh, N.R., and Murrell, J.C. (2000). Stable-isotope probing as a tool in microbial ecology. Nature *403*, 646–649.

Resnick, S.M., and Gibson, D.T. (1996). Regio- and stereospecific oxidation of fluorene, dibenzofuran, and dibenzothiophene by naphthalene dioxygenase from *Pseudomonas* sp. strain NCIB 9816-4. Appl. Environ. Microbiol. *62*, 4073–4080.

Resnick, S.M., Lee, K., and Gibson, D.T. (1996). Diverse reactions catalyzed by naphthalene dioxygenase from *Pseudomonas* sp. strain NCIB 9816. J. Ind. Microbiol. Biotechnol. *17*, 438–457.

Romine, M.F., Stillwell, L.C., Wong, K.K., Thurston, S.J., Sisk, E.C., Sensen, C., Gaasterland, T., Fredrickson, J.K., and Saffer, J.D. (1999). Complete sequence of a 184-kilobase catabolic plasmid from *Sphingomonas aromaticivorans* F199. J. Bacteriol. *181*, 1585–1602.

Saito, A., Iwabuchi, T., and Harayama, S. (1999). Characterization of genes for enzymes involved in the phenanthrene degradation in *Nocardioides* sp. KP7. Chemosphere *38*, 1331–1337.

Saito, A., Iwabuchi, T., and Harayama, S. (2000). A novel phenanthrene dioxygenase from *Nocardioides* sp. strain KP7: Expression in *Escherichia coli*. J. Bacteriol. *182*, 2134–2141.

Sato, S.I., Nam, J.W., Kasuga, K., Nojiri, H., Yamane, H., and Omori, T. (1997). Identification and characterization of genes encoding carbazole 1,9a-dioxygenase in *Pseudomonas* sp. strain CA10. J. Bacteriol. *179*, 4850–4858.

Schell, M.A. (1985). Transcriptional control of the nah and sal hydrocarbon-degradation operons by the *nahR* gene product. Gene *36*, 301–309.

Schneider, J., Grosser, R., Jayasimhulu, K., Xue, W., and Warshawsky, D. (1996). Degradation of pyrene, benz[a]anthracene, and benzo[a]pyrene by *Mycobacterium* sp. strain RJGII-135, isolated from a former coal gasification site. Appl. Environ. Microbiol. *62*, 13–19.

Schuler, L. (2008) Identification of genes involved in the degradation of polycyclic aromatic hydrocarbons by two *Sphingomonas spp*. PhD thesis (Catholic University of Louvain: Louvain, Belgium).

Schuler, L., Ni Chadhain, S.M., Jouanneau, Y., Meyer, C., Zylstra, G.J., Hols, P., and Agathos, S.N. (2008). Characterization of a novel angular dioxygenase from fluorene-degrading Spingomonas sp. strain LB126. Appl. Environ. Microbiol. *74*, 1050–1057.

Schuler, L., Jouanneau, Y., Ni Chadhain, S.M., Meyer, C., Pouli, M., Zylstra, G.J., Hols, P., and Agathos, S.N. (2009). Characterization of a ring-hydroxylating dioxygenase from phenanthrene-degrading *Sphingomonas* sp. strain LH128 able to oxidize benz[a]anthracene. Appl. Microbiol. Biotechnol. *83*, 465–475.

Shintani, M., Urata, M., Inoue, K., Eto, K., Habe, H., Omori, T., Yamane, H., and Nojiri, H. (2007). The *Sphingomonas* plasmid pCAR3 is involved in complete mineralization of carbazole. J. Bacteriol. *189*, 2007–2020.

Sho, M., Hamel, C., and Greer, C.W. (2004). Two distinct gene clusters encode pyrene degradation in *Mycobacterium* sp. strain S65. FEMS Microbiol. Ecol. *48*, 209–220.

Simon, M.J., Osslund, T.D., Saunders, R., Ensley, B.D., Suggs, S., Harcourt, A., Suen, W.C., Cruden, D.L., Gibson, D.T., and Zylstra, G.J. (1993). Sequences of genes encoding naphthalene dioxygenase in *Pseudomonas putida* strains G7 and NCIB 9816-4. Gene *127*, 31–37.

Singleton, D.R., Powell, S.N., Sangaiah, R., Gold, A., Ball, L.M., and Aitken, M.D. (2005). Stable-isotope probing of bacteria capable of degrading salicylate, naphthalene, or phenanthrene in a Bioreactor treating contaminated soil. Appl. Environ. Microbiol. *71*, 1202–1209.

Singleton, D.R., Sangaiah, R., Gold, A., Ball, L.M., and Aitken, M.D. (2006). Identification and quantification of uncultivated Proteobacteria associated with pyrene degradation in a bioreactor treating PAH-contaminated soil. Environ. Microbiol. *8*, 1736–1745.

Singleton, D.R., Ramirez, L.G., and Aitken, M.D. (2009). Characterization of a polycyclic aromatic hydrocarbon degradation gene cluster in a phenanthrene-degrading *Acidovorax* strain. Appl. Environ. Microbiol. *75*, 2613–2620.

Stemmer, W.P. (1994). DNA shuffling by random fragmentation and reassembly: *in vitro* recombination for molecular evolution. Proc. Natl. Acad. Sci. U.S.A. *91*, 10747–10751.

Stingley, R.L., Khan, A.A., and Cerniglia, C.E. (2004). Molecular characterization of a phenanthrene degradation pathway in *Mycobacterium vanbaalenii* PYR-1. Biochem. Biophys. Res. Commun. *322*, 133–146.

Suenaga, H., Watanabe, T., Sato, M., Ngadiman, and Furukawa, K. (2002). Alteration of regiospecificity in biphenyl dioxygenase by active-site engineering. J. Bacteriol. *184*, 3682–3688.

Takagi, T., Habe, H., Yoshida, T., Yamane, H., Omori, T., and Nojiri, H. (2005). Characterization of [3Fe–4S] ferredoxin DbfA3, which functions in the angular dioxygenase system of sp. strain DBF63. Appl. Microbiol. Biotechnol. *68*, 336–345.

Takeuchi, M., Hamana, K., and Hiraishi, A. (2001). Proposal of the genus *Sphingomonas* sensu stricto and three new genera, *Sphingobium*, *Novosphingobium* and *Sphingopyxis*, on the basis of phylogenetic and chemotaxonomic analyses. Int. J. Syst. Evol. Microbiol. *51*, 1405–1417.

Tarasev, M., and Ballou, D.P. (2005). Chemistry of the catalytic conversion of phthalate into its cis-dihydrodiol during the reaction of oxygen with the reduced form of phthalate dioxygenase. Biochemistry *44*, 6197–6207.

Tee, K.L., and Schwaneberg, U. (2007). Directed evolution of oxygenases: Screening systems, success stories and challenges. Combin. Chem. High Throughput Scr. *10*, 197–217.

Tropel, D., and van der Meer, J.R. (2004). Bacterial transcriptional regulators for degradation pathways of aromatic compounds. Microbiol. Mol. Biol. Rev. *68*, 474–500.

Uchimura, H., Horisaki, T., Umeda, T., Noguchi, H., Usami, Y., Li, L., Terada, T., Nakamura, S., Shimizu, K., Takemura, T., *et al.* (2008). Alteration of the substrate specificity of the angular dioxygenase carbazole 1,9a-dioxygenase. Biosci. Biotech. Biochem. *72*, 3237–3248.

Vezina, J., Barriault, D., and Sylvestre, M. (2007). Family shuffling of soil DNA to change the regiospecificity of *Burkholderia xenovorans* LB400 biphenyl dioxygenase. J. Bacteriol. *189*, 779–788.

Vila, J., Lopez, Z., Sabate, J., Minguillon, C., Solanas, A. M., and Grifoll, M. (2001). Identification of a novel metabolite in the degradation of pyrene by *Mycobacterium* sp. strain AP1: Actions of the isolate on two- and three-ring polycyclic aromatic hydrocarbons. Appl. Environ. Microbiol. *67*, 5497–5505.

Willison, J.C. (2004). Isolation and characterization of a novel sphingomonad capable of growth with chrysene as sole carbon and energy source. FEMS Microbiol. Lett. *241*, 143–150.

Wilson, M.S., Bakermans, C., and Madsen, E.L. (1999). *In situ*, real-time catabolic gene expression: Extraction and characterization of naphthalene dioxygenase mRNA transcripts from groundwater. Appl. Environ. Microbiol. *65*, 80–87.

Wintzingerode, F., Göbel, U.B., and Stackebrandt, E. (1997). Determinatiion of microbial diversity in environmental samples: pitfalls of PCR-based rRNA analysis. FEMS Microbiol. Rev. *21*, 213–229.

Wolfe, M.D., and Lipscomb, J.D. (2003). Hydrogen peroxide-coupled cis-diol formation catalyzed by naphthalene 1,2-dioxygenase. J. Biol. Chem. *278*, 829–835.

Wolfe, M.D., Parales, J.V., Gibson, D.T., and Lipscomb, J.D. (2001). Single turnover chemistry and regulation of O-2 activation by the oxygenase component of naphthalene 1,2-dioxygenase. J. Biol. Chem. *276*, 1945–1953.

Yagi, J.M., and Madsen, E.L. (2009). Diversity, abundance, and consistency of microbial oxygenase expression and biodegradation in a shallow contaminated aquifer. Appl. Environ. Microbiol. 75, 6478–6487.

Yeates, C., Holmes, A.J., and Gillings, M.R. (2000). Novel forms of ring-hydroxylating dioxygenases are widespread in pristine and contaminated sails. Environ. Microbiol. 2, 644–653.

Yen, K.M., and Gunsalus, I.C. (1982). Plasmid gene organization: naphthalene/salicylate oxidation. Proc. Natl. Acad. Sci. U.S.A. *79*, 874–878.

Yen, K.M., and Serdar, C.M. (1988). Genetics of naphthalene catabolism in Pseudomonads. CRC Crit. Rev. Microbiol. *15*, 247–268.

Yergeau, E., Arbour, M., Brousseau, R., Juck, D., Lawrence, J.R., Masson, L., Whyte, L.G., and Greer, C.W. (2009). Microarray and Real-Time PCR analyses of the responses of high-arctic soil bacteria to hydrocarbon pollution and bioremediation treatments. Appl. Environ. Microbiol. *75*, 6258–6267.

Yu, C.L., Liu, W., Ferraro, D.J., Brown, E.N., Parales, J.V., Ramaswamy, S., Zylstra, G.J., Gibson, D.T., and Parales, R.E. (2007). Purification, characterization, and crystallization of the components of a biphenyl dioxygenase system from *Sphingobium yanoikuyae* B1. J. Ind. Microbiol. Biotechnol. *34*, 311–324.

Zhou, H.W., Guo, C.L., Wong, Y.S., and Tam, N.F.Y. (2006). Genetic diversity of dioxygenase genes in polycyclic aromatic hydrocarbon-degrading bacteria isolated from mangrove sediments. FEMS Microbiol. Lett. *262*, 148–157.

Zhou, H.W., Wong, A.H.Y., Yu, R.M.K., Park, Y.D., Wong, Y.S., and Tam, N.F.Y. (2009). Polycyclic Aromatic Hydrocarbon-Induced Structural Shift of Bacterial Communities in Mangrove Sediment. Microbial Ecol. *58*, 153–160.

Zhou, N.Y., Fuenmayor, S.L., and Williams, P.A. (2001). *nag* genes of *Ralstonia* (formerly *Pseudomonas*) sp. strain U2 encoding enzymes for gentisate catabolism. J. Bacteriol. *183*, 700–708.

Engineering Microbial Enzymes and Plants to Promote PCB Degradation in Soil: Current State of Knowledge

8

Michel Sylvestre and Jean-Patrick Toussaint

Abstract

The fate of polychlorinated biphenyls (PCBs) in soil and sediments is driven by a combination of interacting processes including several known biological processes. Under anaerobic conditions some bacteria use organohalides (including PCBs) as terminal acceptors. This process is responsible for the depletion of highly chlorinated congeners. Under aerobic conditions, PCBs are oxidized and mineralized by fungi through various pathways involving ligninolytic enzymes and monooxygenases and by bacteria through an initial dioxygenation reaction. Furthermore, several investigations have brought evidence that the rhizosphere provides a remarkable ecological niche to enhance the PCB degradation process by rhizobacteria. In this review, we will briefly summarize our current knowledge regarding the four above-mentioned biological processes involved in PCB degradation. Currently, the biochemistry of the anaerobic PCB-degrading process is still poorly understood. In the case of fungal enzymes, it is not yet clear which of the ligninolytic or monooxygenase systems prevails in PCB degradation. However, the bacterial oxidative enzymes have been investigated extensively. Furthermore, recent studies suggest that designing processes based on plant–microbe association are very promising avenues to remediate PCB-contaminated sites. In this review emphasis will be placed on the current state of knowledge regarding the strategies that are proposed to engineer PCB-degrading bacterial oxidative enzymes and PCB-degrading plant-microbe systems to degrade PCBs.

Introduction

Many hydrophobic organic compounds including polychlorinated biphenyls (PCBs) are priority soil contaminants because of their toxicity and tendency to persist in the environment. Prototrophic soil bacteria are major contributors to the biological processes involved in the degradation of these pollutants is soil. The impressive nutritional versatility of soil bacteria is largely attributed to the capacity of their degradative enzymes to catalyse reactions involving organic compounds other than their natural substrate. The sequential enzymatic reactions involved in the mineralization of organic compounds are organized into pathways that are used to generate the carbon supply and the energy required for growth. However, most of the enzymatic reactions that involve substrates analogues are fortuitous and do not provide the carbon and energy required by the cells to drive their metabolism. In this way, many man-made pollutants such as PCBs are co-metabolically transformed in the environment (Sylvestre, 1995). The ability of enzymes to undergo a relaxation of their specificities without loss of function and to exhibit catalytic activity towards a range of structurally distinct substrates are critical to expend metabolic versatility. However, *in situ* bacterial-based bioremediation processes to destroy persistent organic pollutants (POPs) such as PCBs are limited by the failure of existing microbial catabolic enzymes to mineralize them.

Bioavailability is another critical factor influencing the processes involved in the biological degradation of POPs (Blackburn and Hafker, 1993; Estrella *et al.*, 1993; Maier, 2000), mainly because many of the POPs are hydrophobic and have a tendency to adsorb to soil organic matter and sediment. Surface active agents such as rhamnolipids can facilitate, to some extent, the degradation of PCBs (Barriault and Sylvestre, 1993; Lajoie *et al.*, 1994; Billingsley *et al.*, 1999; Wu *et al.*, 2000; Fava and DiGioia, 2001; Fava and Ciccotosto, 2002; Fava and Piccolo, 2002) as well as polycyclic aromatic hydrocarbons (PAHs) (Aronstein and Alexander, 1993; Tiehm, 1994; Burd and Ward, 1996; Jimenez and Bartha, 1996; Volkering *et al.*, 1997; Boonchan *et al.*, 1998; Wang *et al.*, 1998; Willumsen *et al.*, 1998; Zheng and Obbard, 2001). Unfortunately, surface active agents alone are not sufficient to allow complete remediation of these pollutants because of the lack of efficient bacterial catabolic enzymes to oxidize them. Thus, the degradation of PCBs and other POPs will require the engineering of new bacteria carrying modified more efficient degradative enzymes (Shannon and Unterman, 1993; Timmis and Pieper, 1999; Furukawa, 2000).

The fate of contaminants in the environment is controlled by a combination of interacting physical, chemical and biological processes. Physical processes are responsible for the movement of contaminants through the soil and subsurface away from their source. Chemical and biological processes determine the extent to which compounds will be transformed. As seen above, microbial cells are major contributors to the biological degradation processes as they carry enzymes to degrade POPs and they also produce surface active compounds to increase their bioavailability. However, microbial cells are not the sole drivers of the biodegradation processes as plants contribute significantly to them. Plants have the capacity to withstand relatively high concentrations of organic contaminants. Therefore, they can be used to uptake and degrade organic compounds (phytotransformation), to uptake and concentrate organics (phytoextraction or rhizofiltration), to remove contaminants from the subsurface and volatilize them (phytovolatilization), to sequester contaminants in soil (phytostabilization), but a major contribution of plants to biodegradation processes is their ability to enhance microbial degradation in the rhizosphere (rhizoremediation). The rhizosphere is the zone of soil surrounding plant roots where the biology and chemistry of the soil are influenced by the roots. Plants release root exudates providing nutrients to microbial populations in the rhizosphere (Villacieros *et al.*, 2003). Evidences show that some bacteria are genetically endowed with the machinery to colonize plant roots. They are thus advantageously competitive compared with other soil bacteria (Zhuang *et al.*, 2007). In return, rhizobacteria produce plant growth stimulators or they suppress pathogens through competition and antibiotics production (Zhuang *et al.*, 2007). Many *Pseudomonas* and *Burkholderia* species count among the major plant growth-promoting rhizobacteria (PGPR). Importantly, these bacterial genera are also among the most versatile bacteria involved in the degradation of POPs, and thus are likely candidates to degrade POPs pumped up in the rhizosphere by plant roots. Furthermore, plants exude chemicals such as phenylpropanoids and terpenes that not only act as signals to stimulate the root colonization process by PGPR (Villacieros *et al.*, 2003), but also induce catabolic pathways involved in POP degradation (Donnelly *et al.*, 1994; Fletcher and Hegde, 1995; Hernandez *et al.*, 1997; Leigh *et al.*, 2002; Singer *et al.*, 2003b). Narasimhan *et al.* (2003) showed that PCB removal by *Pseudomonas putida* PML2, a phenylpropanoid-utilizing and PCB-degrading rhizobacteria, was significantly lower in the rhizosphere of an *Arabidopsis thaliana* mutant exuding less flavonoids than in the rhizosphere of the wild-type plant.

In this review we will summarize the latest developments regarding the microbial catabolic pathways involved in PCB degradation and how they can be modified to enhance their catalytic activities towards PCBs and we will discuss some of the novel strategies that have been suggested to improve the rhizoremediation processes.

Chemistry and use of PCBs

PCBs are synthetic industrial oils that were manufactured from 1929 until their use began to be restricted in the 1970s and their production banned in most countries in the late 1970s. Because of their excellent chemical and thermal stabilities they were used on a large scale as plasticizers, flame retardants, lubricants, fluid heat exchangers, compression fluid. The extensive application of the PCBs resulted in their wide distribution in the environment. Produced by direct chlorination of biphenyl, PCBs have been marketed as mixtures containing a fraction (60–80) of the 209 possible congeners, particularly congeners containing three to six chlorine atoms. The congeners are designated by describing the position of the chlorine atoms on the biphenyl ring or, more simply, by the IUPAC numbering system. The congeners differ in their physical properties according to the number and the position of chlorine atoms. The high-chlorinated biphenyls are less water-soluble and less volatile than the low-chlorinated ones. Importantly, the degree and position of chlorine substitution influence their structural feature. Thus, the non-*ortho* substituted congeners have a tendency to take a co-planar configuration miming coplanar compounds such as dioxins (Safe, 1990) (Fig. 8.1). This fact has major consequences on their interaction with bioactive compounds and especially with enzymes since enzyme–substrate interactions are highly specific. This explains why, as we will see below, only some of the 209 PCB congeners can be degraded by the bacterial biphenyl catabolic pathways.

PCB degradation pathways

There are three known pathways by which microorganisms influence the fate of PCBs in the environment. Under anaerobic conditions some anaerobic bacteria can use organohalides (including PCBs) as terminal acceptors. Under aerobic conditions, PCBs are oxidized by fungi through various pathways involving ligninolytic enzymes and monooxygenases, and by bacteria through an initial dioxygenation reaction. Altogether, several pathways are involved in PCB degradation in natural environment. In this section we will briefly summarize the current knowledge for these pathways and in the case of the bacterial oxidative pathway we will describe in more details each of the enzymatic step involved and how these enzymes can be engineered to enhance PCB degradation.

Anaerobic pathway

Anaerobic degradation of PCBs was first described in the mid-1980s (Brown *et al.*, 1987). Several investigations have shown that this process is responsible for the depletion of highly chlorinated congeners of water sediments (Wiegel and Wu, 2000) resulting in major altered distribution of PCB congener patterns in these anaerobic environments. In this process, the *meta* and *para* positions are the preferred positions for dechlorination resulting in an accumulation of the *ortho*-substituted congeners in the environment (Maltseva *et al.*, 1999). Several examples of co-metabolic reductive dehalogenation catalysed by fortuitous reactions involving enzymes and co-factors involved in other anaerobic processes have

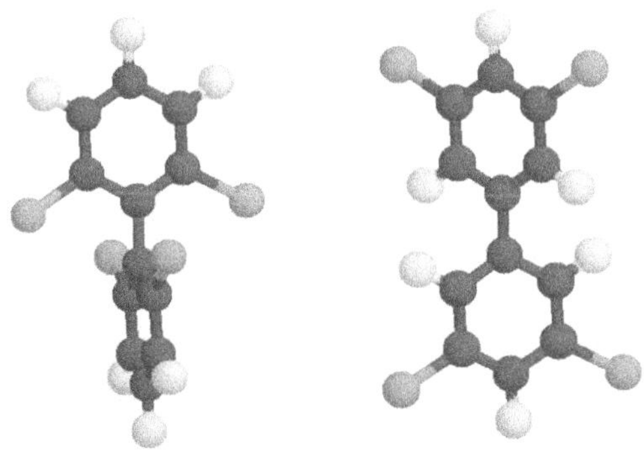

Figure 8.1 Three-dimensional (3D) structure of 2,2′,6,6′- and 3,3′,4,4′-tetrachlorobiphenyl, which are two congeners exhibiting very distinct 3D structure. (White balls, hydrogen atoms; grey balls, chlorine atoms).

been demonstrated (Holliger *et al.*, 2003; Smidt and de Vos, 2004). The microorganisms involved in this process are however, rather difficult to obtain in pure or axenic cultures, which makes it difficult to determine the biochemical pathways involved and the environmental parameters affecting the process. Other bacteria able to use organochlorine compounds as electron acceptor in a process called dehalorespiration have recently been described. Dehalorespiration has now been demonstrated in several bacterial genera, including *Desulfitobacterium, Dehalococcoides, Dehalospirillum, Dehalobacter, Desulfomonile, Desulfovibrio* and several others (Smidt and de Vos, 2004). However, up until now, only two among these genera have been associated to PCB dehalogenation. *Desulfitobacterium dehalogenans,* which is capable of *ortho*-dehalogenation of a variety of chloroaromatic chemicals can also dehalogenate hydroxylated PCBs (Utkin *et al.*, 1994). More recently, several pure bacterial isolates belonging to the genus *Dehalococcoides* were found to dehalogenate PCB congeners (Fennell *et al.*, 2004; May *et al.*, 2008; Adrian *et al.*, 2009). Notably, *Dehalococcoides* sp. strain DBDB1, a bacteria isolated for its ability to grow using chlorinated benzenes as respiratory electron acceptors, was found to effect *meta* and *para* dehalogenation reaction on a number of chlorobiphenyl congeners (Adrian *et al.*, 2009). This is the first example where a single bacterial isolate was demonstrated to be able to repeat the complex PCB dehalogenation patterns produced by environmental anaerobic consortia.

Therefore, bacteria of the genus *Dehalococcoides* are currently considered to be widely major contributors to the dehalogenation of PCBs in the environment. They are slow-growing strictly anaerobic tiny disk-shaped bacteria. Their taxonomic status is still unclear (Bedard, 2007), but based on 16S rRNA phylogeny, *Dehalococcoides* fall within Chloroflexi and their cell wall composition is closer to that of the Archae than to Gram-negative or positive bacteria. A notable feature is the presence of multiple reductive dehalogenase genes homologues that are presumed to be responsible for the dehalogenation reactions these organisms can perform. The genome of three members of *Dehalococcoides* has now been established and several reductive dehalogenases have now been partially purified and characterized, including PceA, an enzyme able to dehalogenate tetrachloroethene and also dichlorophenol. We should expect more progress on the characterization of the biology and biochemistry of these intriguing organisms in the near future.

Fungal pathway

In soils, PCBs are mineralized by the ubiquitous white-rot fungi, which comprise fungi capable of degrading lignin extensively. Lignin is an extremely recalcitrant polymeric component of wood, which is mineralized essentially only by the white-rot fungi and some bacteria belonging to the *Actinomycetes*. Three extracellular enzymes are involved in lignin degradation, two heme-containing peroxidases, lignin peroxidase (LiP) and manganese peroxidase (MnP) and a copper-containing phenoloxidase (laccase) (Mester and Tien, 2000). In addition, some of the white-rot fungi produce a large number of the P450 type cytochrome monooxygenases which are presumed to be involved in further degradation of peroxidase-depolymerized lignin (Doddapaneni and Yadav, 2005). In these fungi, lignin-degrading enzymes are produced under nutrient-limited conditions, whereas P450 monooxygenases are produced under nutrient-rich conditions. Of particular interest is the fungus *Phanerocheate chrysosporium,* which is a ligninolytic organism possessing both LiP and MnP systems, and contains over 148 predicted genes encoding P450 enzymes, representing almost 1% of the entire genome (Martinez *et al.*, 2004). This indicates the importance of the latter enzyme in the fungal metabolism. Several other non-ligninolytic fungi also found in soils, such as *Cunninghamella elegans,* are endowed of tremendous catabolic possibilities which are attributed to heme-containing enzymes belonging to the group of cytochrome P450 (Cerniglia, 1997).

Fungi possessing these systems have the ability to degrade a wide range of organic, complex compounds, including many POPs. Thus several investigations have shown that fungi can metabolize PAHs, PCBs, DDT and explosives such as trinitrotoluene (Pointing, 2001). In the mid-1980s, Eaton (1985) reported that nitrogen-limited cultures of *P. chrysosporium* mineralized

7% of uniformly labelled ^{14}C Aroclor 1254 (a commercial PCB mixture) within 22 days. Since then, many investigations have confirmed the ability of fungal oxidative enzymes to degrade a large number of xenobiotics, including PCBs (Pointing, 2001).

While the role of LiPs and MnPs in the biotransformation of many xenobiotics under nutrient-limited conditions has been established more than 20 years ago (Mester and Tien, 2000), the exact role played by P450 monooxygenases is still not clear. Several investigations are reporting PCB degradations under nitrogen-limiting conditions (Pointing, 2001) but in other investigations PCB removal was independent of ligninolytic enzyme production as degradation was observed under both nutrient-rich and -poor media (Yadav *et al.*, 1995). Moreover, studies based on ligninolytic negative mutants or on P450 inhibitors have provided indirect evidence that P450 enzymes are involved in the metabolism of several xenobiotics, including phenanthrene and chlorophenoxyacetate (Sutherland *et al.*, 1991; Yadav and Reddy, 1992). Elucidation of the role of P450 enzymes under nutrient-limited conditions has received less attention because of overwhelming past interest in peroxidases. However it seems likely that both fungal ligninolytic and monooxygenase oxidative systems are involved in the degradation and mineralization of many POPs, including PCBs.

Unfortunately, efforts to isolate and purify components of the P450 systems responsible for the degradation of xenobiotics have been met with limited success in all fungal species (van den Brink *et al.*, 1998). However, Wang *et al.* (2000) recently cloned and sequenced a P450 gene from *C. elegans* in *Escherichia coli*. Furthermore, Yadav *et al.* (2003) cloned two P450 genes CYP63A1 and CYP63A2 from *P. chrysosporium* in *E. coli*. Unfortunately, in all cases the proteins were not expressed as active enzymes in *E. coli*. It is, however, noteworthy that Yadav and Loper (2000) cloned the sole cytochrome P450 reductase gene from *P. chrysosporium* in *E. coli*. Strategies such as cloning of truncated proteins as was used for the expression of human P450 in *E. coli* (Guengerich, 2005) will most probably be required to express the fungal enzyme in bacteria for characterization.

Altogether data accumulated from the literature indicate that investigating individual P450 genes from fungi with the aim of elucidating their role in the metabolism of xenobiotics will be a fastidious task. A strategy using expression profiling should be a much better way of reaching this goal.

Bacterial biphenyl catabolic pathway

In aerobic bacteria PCBs are degraded by the biphenyl catabolic pathway enzymes (Abramowicz, 1990) (Fig. 8.2). Four enzymatic steps are required to transform biphenyl/chlorobiphenyls into corresponding benzoate/chlorobenzoates. Chlorobenzoates are then degraded through other pathways (Furukawa, 2000). The identification of PCB metabolites (Ahmad *et al.*, 1991; Ahmed and Focht, 1973; Bedard *et al.*, 1986; Furukawa *et al.*, 1978; Furukawa *et al.*, 1979; Massé *et al.*, 1989; Massé *et al.*, 1984) as well as the identification of the genes involved (Ahmad *et al.*, 1990; Commandeur *et al.*, 1997; Erickson and Mondello, 1992; Fukuda *et al.*, 1994; Kikuchi *et al.*, 1994; Sylvestre *et al.*, 1996b; Taira *et al.*, 1992) have allowed the elucidation of the various enzymatic steps required to convert chlorobiphenyls into chlorobenzoates. The first step of this pathway is catalysed by the biphenyl 2,3-dioxygenase (BPDO) (Haddock and Gibson, 1995; Haddock *et al.*, 1995; Hurtubise *et al.*, 1995; Hurtubise *et al.*, 1996; Nadim *et al.*, 1987). The enzyme introduces one molecule of oxygen on two vicinal *ortho-meta* carbons of the aromatic ring (Fig. 8.2). The metabolite, *cis*-2,3-dihydro-2,3-dihydroxybiphenyl is then re-aromatized by the 2,3-dihydro-2,3-dihydroxybiphenyl-2,3-dehydrogenase (Barriault *et al.*, 1999; Hulsmeyer *et al.*, 1998; Khan *et al.*, 1996). The catechol metabolite is then cleaved by the 2,3-dihydroxybiphenyl-1,2-dioxygenase to generate 2-hydroxy-6-oxo-6-phenylhexa-2,4-dienoic acid (HOPDA) (Eltis *et al.*, 1993; Han *et al.*, 1995; Khan *et al.*, 1996; Taira *et al.*, 1988), which is then hydrolysed to produce benzoic and pentanoic acids (Hatta *et al.*, 1998; Kim *et al.*, 1996; Lau *et al.*, 1996; Mukerjee-Dhar *et al.*, 1998; Seah *et al.*, 2000, 2001; Seeger *et al.*, 1995). The clear role of other genes associated to the *bph* operon in PCB degradation has not yet been elucidated. This is the case of *bphK* a gene that is presumed

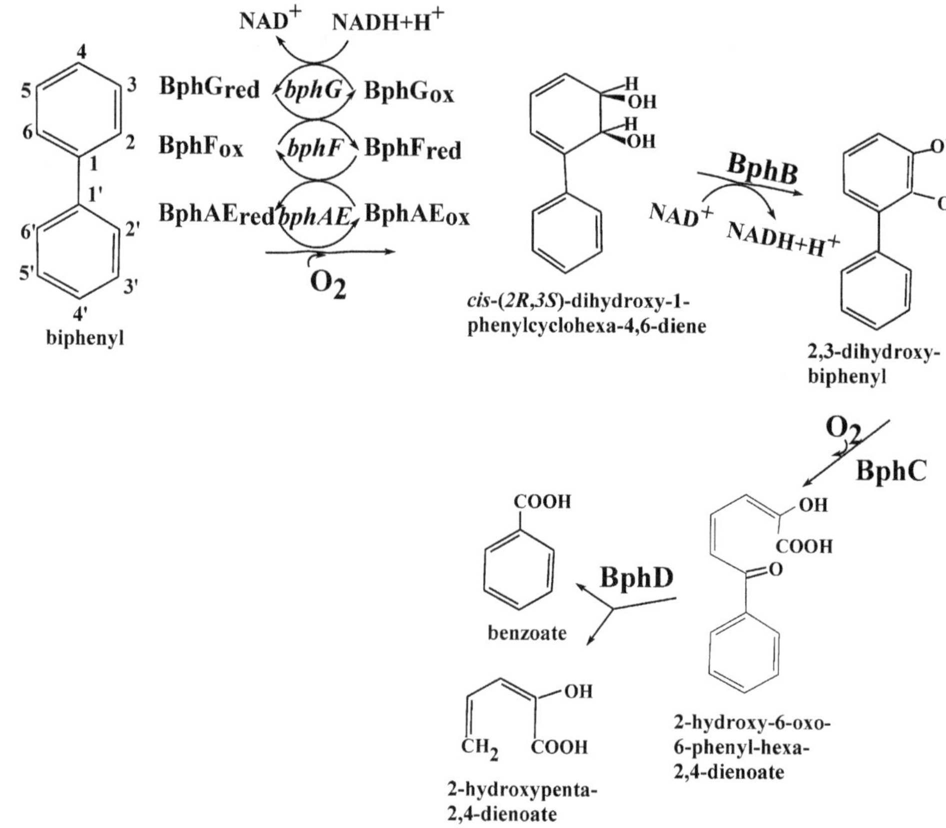

Figure 8.2 Bacterial biphenyl catabolic pathway of *B. xenovorans* LB400. BphAE, BphF and BphG are respectively the oxygenase component, the ferredoxine and the ferredoxin reductase components of the 2,3-biphenyl dioxygenase; BphB is the 2,3-dihydro-2,3-dihydroxybiphenyl dehydrogenase; BphC is the 2,3-dihydroxybiphenyl 1,2-dioxygenase; BphD is the 2-hydroxy-6-oxo-6-phenylhexa-2,4-dienoate hydrolase.

to be involved in PCB dehalogenation (Brennan *et al.*, 2009; Tocheva *et al.*, 2006).

A proper understanding of the biochemical processes by which PCBs are degraded is the key to the successful engineering of efficient PCB-degrading bacteria. In the following paragraphs we will clarify the role played by each of the enzymes of the biphenyl catabolic pathway in PCB degradation and we will summarize the current knowledge of biochemical and biocatalytic properties of these enzymes.

Biphenyl dioxygenase

BPDO introduces one molecule of oxygen on two vicinal ortho–meta carbons of the aromatic ring (Fig. 8.2). Many investigations have addresses the mechanisms by which BPDO is modified to interact with new substrate analogues. BPDO is of particular interest not only for its potential application as biocatalyst to oxygenate persistent pollutants such as PCBs (Furukawa *et al.*, 2004; Sylvestre, 2004), chlorodibenzofurans (Seeger *et al.*, 2001; L'Abbée *et al.*, 2005; Mohammadi and Sylvestre, 2005), but it can also serve as catalyst to manufacture fine chemicals (Resnick and Gibson, 1996; Seeger *et al.*, 2003; Shindo *et al.*, 2005; Kagami *et al.*, 2008). BPDO comprises three components: the iron–sulfur oxygenase (hereinafter will be referred to as BphAE) is an heterohexamer comprising three α (Mr = 51,000) and three β (Mr = 22,000) subunits, the ferredoxin (BphF, Mr = 12,000) and the ferredoxin reductase (BphG, Mr = 43,000). The encoding genes for *Burkholderia xenovorans* LB400 (Erickson and Mondello, 1992), which is the best PCB-degrader of natural origin, are bphA (BphAE α subunit), bphE (BphAE β subunit), bphF (BphF) and bphG (BphG). BphAE interacts directly with the substrate to catalyse a stereospecific 2,3-dihydroxylation of biphenyl (Resnick and Gibson, 1996) (Fig. 8.2).

Engineering the biphenyl dioxygenase to enhance its activity towards chlorobiphenyls and other biphenyl analogues will require an understanding of the mechanisms by which the amino acids of the enzyme catalytic pocket interact with the substrate to bind and orient it. Based on many investigations done in the 1990s, the C-terminal portion of the dioxygenase α subunit was found to carry the major determinants of substrate specificity (Erickson and Mondello, 1993; Suyama *et al.*, 1996; Kimura *et al.*, 1997; Mondello *et al.*, 1997; Kumamaru *et al.*, 1998; Suenaga *et al.*, 1999, 2001, 2002; Barriault *et al.*, 2002; Zielinski *et al.*, 2002, 2003, 2006; Barriault and Sylvestre, 2004). While a wide range of investigations have addressed the role of the BPDO large subunit, only scarce reports have addressed the role of the small subunit. The clear role of the β subunit is still unresolved, but reports have shown an involvement in substrate specificity (Hirose *et al.*, 1994; Hurtubise *et al.*, 1998). Similarly, the role of the two other enzyme components, the ferredoxin reductase and the ferredoxin, on enzyme specificities is not clear. Although the ferredoxin from *B. xenovorans* LB400 and the ferredoxin reductase from *Pseudomonas* sp. KKS102 were crystallized and their structure analysed (Senda *et al.*, 2000; Couture *et al.*, 2001), we do not know yet if their association with the oxygenase component has any allosteric impact on the conformation of the $\alpha_3\beta_3$ oxygenase component. However, recent investigations involving carbazole 1,9a-dioxygenases provided evidence of conformational changes occurring of the oxygenase component to suitably align its Rieske [2Fe–2S] cluster with that of the ferredoxin component (Ashikawa *et al.*, 2006; Inoue *et al.*, 2009). The effect of these conformational changes on overall activity an specificity has yet to be evaluated. In the case of BPDOs, no investigation has yet reported the crystal structure of the oxygenase–ferredoxin complexes.

Beside region III amino acids, which is a stretch of seven amino acids comprising Thr^{335}Phe^{336}Asn^{337}Asn^{338}Ile^{339}Arg^{340}Ile^{341} in LB400 BphA (Mondello *et al.*, 1997; Barriault *et al.*, 2002; Barriault and Sylvestre, 2004; Mohammadi and Sylvestre, 2005), residues Asn^{377}Phe^{378} (Mondello *et al.*, 1997; Suenaga *et al.*, 1999), Phe^{384} (Suenaga *et al.*, 2002) and Gln^{255} Ile^{258} and Ala^{268} (Suenaga *et al.*, 2001; Barriault *et al.*, 2002) of LB400 BphA were also found to influence the catalytic properties of the dioxygenase.

Available information shows that the two rings of the biphenyl molecule are not in a parallel position, but are skewed inside the catalytic pocket of *Rhodococcus jostii* RHA1 and *Pandorae pnomenusa* dioxygenases (Furusawa *et al.*, 2004; Gomez-Gil *et al.*, 2007). This information is interesting since the *meta-para* substituted congeners, the most toxic ones, are known to adopt a coplanar conformation. Dibenzofuran can be regarded as a biphenyl molecule in which both rings are held in a coplanar configuration through the ether bridge that connects two *ortho*-carbons (Fig. 8.3). This molecule is poorly oxygenated by LB400 BphAE (Seeger *et al.*, 2001; L'Abbée *et al.*, 2005), however, changing only two amino acid residues, Thr335Phe336 of *B. xenovorans* LB400 BphA by Ala335Met336 generated variant *p4* BphAE that, compared with LB400 BphAE, exhibited an expended substrate range including persistent PCBs such as 2,6-dichlorobiphenyl (Barriault and Sylvestre, 2004) and co-planar analogues of biphenyl such as dibenzofuran (Mohammadi and Sylvestre, 2005). Subsequently, changing Asn338Ile341Leu409 of *p4* BphA to Gln338Val341Phe409 generated RR41 BphA that oxygenates dibenzofuran more efficiently than *p4* BphA (Mohammadi and Sylvestre, 2005). This suggests that region III residues play a crucial role in positioning the non-reactive biphenyl ring inside the catalytic cavity and altering this region appears a promising avenue to enhance the activity towards co-planar analogues of biphenyl. However, structural analysis of *R. jostii* RHA1 BphA1A2 (Furusawa *et al.*, 2004) and comparison with other aryl hydroxylating dioxygenases, has confirmed that Phe378 and Phe384 are near the substrate inside the catalytic pocket and thus probably interact with it to determine the specificity and regiospecificity. In a recent report, Ferraro *et al.* (2006) clearly showed the involvement of Phe352 (corresponding to Phe378 of LB400 BphA) of the naphthalene dioxygenase

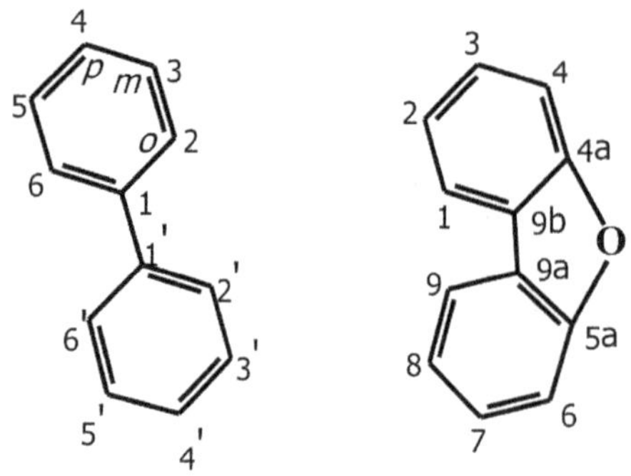

Figure 8.3 Comparison between the structure of biphenyl and dibenzofuran.

large subunit in substrate orientation. This residue alters the space available for substrate orientation inside the catalytic pocket. Furthermore, other residues that according to a model based on the structure of the naphthalene dioxygenase and which were not in direct contact with the substrate were also found to influence the enzyme specificity and regiospecificity (Zielinski *et al.*, 2003; Vézina *et al.*, 2007). Nevertheless, whether they are at a distance that can permit a direct interaction with the substrate, or whether they are farther from the substrates, the precise mechanism by which the substrate is bound and oriented inside the catalytic pocket is yet unknown. However, X-ray structural analysis of several BPDOs has now been reported. The structure of *R. jostii* RHA1 BphA1A2 (Furusawa *et al.*, 2004) was the first to be reported. More recently, it was shown that *P. pnomenusa* B-356 was able to oxygenate congeners such as 2,6-dichlorobiphenyl and 2,4,4′-trichlorobiphenyl that are poorly degraded by *B. xenovorans* LB400. Structural analysis of the 2,6-dichlorobiphenyl-complexed B356 BphAE and comparison to a modelled structure of BphAELB400 provided evidence that the replacement of Phe336 of LB400 BphAE by Ile334 of B-356 BphAE creates more space around C-4 of the distal biphenyl ring facilitating the binding of para-substituted congeners (Gomez-Gil *et al.*, 2007). However, the reason why B-356 BphAE can oxygenate 2,6-dichlorobiphenyl better than LB400 BphAE is yet unsolved. The structure of *Sphingobium yanoikuyae* B1 BphAE has also been reported (Ferraro *et al.*, 2007), however, this enzyme is structurally more similar to the naphthalene dioxygenase of *Pseudomonas* sp. NCIB 9816-4 (Kauppi *et al.*, 1998) than to other biphenyl known dioxygenase. Since *Pseudomonas* sp. NCIB 9816-4 can transform several chlorobiphenyls (Barriault and Sylvestre, 1999) it is likely that some of the naturally occurring naphthalene dioxygenases might have evolved to expend or even modify their range of substrate to replace naphthalene with biphenyl ring.

Altogether, the information currently available on the structural features involved in substrate specificities and regiospecificities of aryl-hydroxylating dioxygenases are still limited and they all converge to support the conclusion that in practice, random rather than rational approaches are more suitable to engineer new better performing enzyme. Approaches have been employed altering the C-terminal portion randomly or at targeted sites and exploiting the natural diversity of *bphA* (Camara *et al.*, 2007) through strategies of artificial evolution. A study combining both approaches, by shuffling soil DNA encoding segment of BphA has generated BphA variants with novel regioselectivities (Vézina *et al.*, 2007). The role of several residues of BPDOs large subunit on substrate specificity has also been addressed by comparing the primary sequences of various BphAs deduced from PCR amplicons obtained from cultured bacteria as well as soil DNA (Witzig *et al.*, 2006; Vézina *et al.*, 2008). However, the precise role of each amino acids residue lining the catalytic cavity or those more distant that influence the conformation of key residues involved in substrate positioning and catalytic activities is still unclear. Crystal structure analyses of wild type and evolved enzymes complexed to chlorinated substrate will be required to help understanding the mechanisms by which these enzymes evolve to expend their substrate specificities. Until then, directed evolution strategies based on random mutagenesis and gene shuffling remain the best strategy to evolve the oxygenases.

2,3-Dihydro-2,3-dihydroxybiphenyl dehydrogenase

The second step in the sequence of reactions of the biphenyl catabolic pathway is catalysed by a dehydrogenase to generate a dihydroxy metabolite

(Fig. 8.2). The enzyme is a flavin containing homotetramer (Sylvestre *et al.*, 1996a; Hulsmeyer *et al.*, 1998).

The enzyme has been crystallized and its structure has been determined (Hulsmeyer *et al.*, 1998). Three residues, Ser^{142}, Tyr^{155}, and Lys^{159} that were postulated to be involved in catalytic activity of *B. xenovorans* LB400 BphB are conserved in *P. pnomenusa* B-356 BphB (Vedadi et *al.*, 2000). It was suggested that these residues function as a 'catalytic triad' like other members of the short-chain alcohol dehydrogenase/reductase (SDR) family (Vedadi *et al.*, 2000).

Substitution of each triad residue was examined in B-356 BphB (Vedadi *et al.*, 2000). These three residues were shown to be critical for enzyme activity. In addition, B-356 BphB showed a strong preference for NAD(+) over NADP(+), with a 260-fold higher specificity constant (k(cat)/K(m)) (Vedadi *et al.*, 2000). It was suggested that inefficient use of NADP(+) by BphB might partly be due to the presence of an aspartate residue at position 36.

Data from experimentation done with His-tagged-purified B-356 BphB and LB400 BphB preparations have shown that this enzyme can catalyse the dehydrogenation of a very broad range of substrates, including the 3,4-dihydroxylated metabolite produced from the oxygenation of 2,2′,5,5′-tetrachlorobiphenyl (Barriault *et al.*, 1998). Thus, BphB does not appear to represent a major blockage in the PCB degradation pathway.

2,3-Dihydroxybiphenyl 1,2-dioxygenase

The third reaction of the pathway is catalysed by the 2,3-dihydroxybiphenyl 1,2-dioxygenase (2,3-DHDB) which is an extradiol type catecholic ring cleavage dioxygenase (Fig. 8.2). The enzyme catalyses the addition of two atoms of oxygen to the catechol ring of the substrate resulting in its cleavage. It converts 2,3-dihydroxybiphenyl to 2-hydroxy-6-oxo-6-phenylhexa-2,4-dienoic acid (HOPDA) (Furukawa and Arimura, 1987; Eltis *et al.*, 1993). Extradiol type dioxygenases typically contain one non-heme iron (Fe(II)) in their active site (Cerdan *et al.*, 1995; Han *et al.*, 1995).

The enzyme being very sensitive to oxygen (Eltis *et al.*, 1993), anaerobically purified BphC were crystalized from *Burkholderia* sp. LB400 (Han *et al.*, 1995) and from *Pseudomonas* sp. KKS102 (Uragami *et al.*, 2001) and their structure has been determined. The enzyme is a homo-octamer. Each monomer contains two domains. Within each domain are two copies of a substructure comprising four β strands and one α helix in sequential order βαβββ. Thus the monomer contains four copies of this unit within two domains, suggesting genetic duplication at two points in the evolutionary process of the enzyme (Han *et al.*, 1995).

There are three critical features about this enzyme that influence its catalytic properties towards PCBs. These are the high sensitivity to 3-chlorocatechol, the substrate inhibition caused by *ortho*-chlorinated dihydroxybiphenyl metabolites and their inability to cleave *meta-para* hydroxylated dihydroxybiphenyl metabolites.

Data obtained with *P. pnomenusa* B-356 (Sondossi *et al.*, 1992), with *Pseudomonas stutzeri* (Vrana *et al.*, 1996) and with a PCB-degrading consortia (Guilbeault *et al.*, 1994) have provided evidence that chlorocatechol inhibition has a high impact on the degradation of congeners producing 3-chlorocatechol. Thus, when 3-chlorobiphenyl is converted to 3-chlorobenzoate, the biphenyl-associated benzoate dioxygenase can convert 3-chlorobenzoate to 3-chlorocatechol in several PCB degrading strains. The efficiency of conversion is low (Sondossi *et al.*, 1992). However, because of the high sensitivity of the 2,3-DHDB to 3-chlorocatechol (Hein *et al.*, 1998), small amounts of this metabolite is needed to severely interfere with the catalytic conversion of chlorobiphenyls to chlorobenzoates.

The mechanism involved in 3-chlorocatechol inhibition has been elucidated by Vaillancourt *et al.* (2002) who showed that 3-chlorocatechol inhibits the enzyme by oxidation of the active site Fe(II) to Fe(III). The 2,3-DHDB-catalysed cleavage of 3-chlorocatechol yielded predominantly 2-pyrone-6-carboxilic acid and 2-hydroxymuconic acid, consistent with transient formation of acyl chloride. However, the enzyme was not covalently modified by this acyl chloride (Vaillancourt *et al.*, 2002).

Nevertheless, it is to be expected that catecholic inhibition of BphC will not restrict PCB degradation in engineered bacteria that can degrade chlorobenzoates efficiently. Thus, hybrid bacteria carrying efficient chlorobiphenyl and chlorocatechol degradation pathways were able to grow efficiently on chlorobiphenyls (Mokross *et al.*, 1990; Havel and Reineke, 1991, 1993; Adams *et al.*, 1992; Hrywna *et al.*, 1999). Furthermore, catechol 2,3-dioxygenases able to cleave efficiently 3-chlorocatechol have been described (Mars *et al.*, 1999) and the domain of the enzyme that is responsible for suicide inhibition has been identified (Mars *et al.*, 1999). It is therefore likely that engineering new enzymes and/or new PCB degrading bacteria can overcome the metabolic blockages resulting from chlorocatechol inhibition.

The second feature of the 2,3-DHBD that has a high impact on PCB degradation is its sensitivity to *ortho*-chlorinated dihydroxylated substrates. Dai *et al.* (2002) have shown that 2′,6′-dichloro-2,3-dihydroxybiphenyl has high affinity with the dioxygenase and it binds it, acting as suicide inhibitors, preventing productive reaction with substrate that are normally easily cleaved by the enzyme. Structural analysis has shown that the 2′-chlorine substituent binds in contact with conserved side chains that are believed to define the oxygen-binding site. Furthermore, the 2′-chlorine has the potential to partially inhibit oxygen binding to the binary complex by affecting the orientation of the oxygen relative to the catalytic iron (Dai *et al.*, 2002). However, a study comparing the catalytic properties of *B. xenovorans* 2,3-DHDB with those of two 2,3-DHDB isozymes produced by *Rhodococcus globerulus* P6 showed that in spite of the fact that all DHDB shared the same inability to cleave 2,6′-dichloro-2,3-dihydroxybiphenyl, each enzyme exhibits quite distinct catalytic properties towards other dihydroxylated metabolites of PCBs (Vaillancourt *et al.*, 2003; Fortin *et al.*, 2005a). For example, some of the worst substrates for one enzyme were the best for the other. This means that the 2,3-DHDB substrate specificities vary greatly depending on the enzyme structure and can most probably be changed by enzyme engineering.

The third feature of the 2,3-DHDB that impacts on PCB degradation is its inability to cleave the *meta-para* hydroxylated metabolites (Eltis *et al.*, 1993; Hein *et al.*, 1998; Taira *et al.*, 1988). Complete mineralization of PCB congeners whose degradation is initiated by 3,4-dihydroxylation thus requires the presence of enzymes that are able to transform these compounds. Fortunately, the homologous enzyme 1,2-dihydroxynaphthalene dioxygenase (Dox G) of *Pseudomonas* sp. C18 (Denome *et al.*, 1993) was found to catalyse the ring cleavage of these *meta-para*-hydroxylated metabolites more efficiently than BphC (Barriault *et al.*, 1998). Altogether, considering that the structure of BphC is known and that homologous extradiol dioxygenases possess complementary properties to that of BphC, the odds for the development of new more efficient engineered enzyme by *in vitro* directed evolution or by rational design are very promising. In this respect, variant DoxG exhibiting higher ability to cleave 3,4-dihydroxybiphenyl metabolites were obtained by directed evolution (Fortin *et al.*, 2005b), showing that the engineering of extradiol dioxygenases exhibiting the ability to further degrade metabolites produced from doubly *ortho-meta*-substituted congeners such as 2,2′,5,5′-tetrachlorobiphenyl is feasible.

2-Hydroxy-6-oxo-6-phenylhexa-2,4-dienoate hydrolase (HOPDA hydrolase)

The last step of the *bph* pathway is catalysed by a hydrolase. The enzyme has been crystalized from *R. jostii* RHA1 (Nandhagopal *et al.*, 2001). It is an octameric enzyme with a 422-point group symmetry (Nandhagopal *et al.*, 2001). The subunit can be divided into core and lid domains. The active site of the enzyme is situated in the substrate-binding pocket, which is located between the two domains (Nandhagopal *et al.*, 2001). The substrate-binding pocket can be divided into hydrophobic and hydrophilic regions.

Regarding the features of interest to engineer more efficient PCB degrading enzymes, HOPDAs bearing chlorine atoms on the phenyl ring are in general good substrates for the HOPDA hydrolase isolated from strain LB400 (Seah *et al.*, 2000). However 3- and 4-chloro-HOPDA are poor substrates and competitive inhibitors of the enzyme (Seah *et al.*, 2000, 2001). An important consequence of this feature is the fact that several chlorobiphenyls that bear chlorine on both rings will be converted to the corresponding chloro-HOPDA as dead-end metabolites or to chloroacetophenones that are produced spontaneously from chloro-HOPDA (Seah *et al.*, 2000). This is the case for instance for 4,4'CB. Seah *et al.* (2001) have compared the catalytic properties of homologous *B. xenovorans* LB400 and R. globerulus P6 HOPDA hydrolases. Fortunately, they found significant variations. Thus unlike LB400 BphD, P6 BphD is more specific for 9- and 10-chloro-HOPDA than for HOPDA. Furthermore, unlike LB400 BphD, 4-chloro-HOPDA competitively inhibits P6 BphD more efficiently than 3-chloro-HOPDA. Therefore, there are chances that comparative investigations help design new-engineered BphD with features that allow a more efficient degradation of congeners that bear chlorine atoms on both rings.

Exploiting plants and bacteria to phytoremediate PCBs

Phytoremediation can be defined as a process by which plants interact with contaminants in the soil resulting in the uptake, degradation, or stabilization of the latter. This promising biotechnology has therefore a lot of potential in alleviating contaminated soils from PCBs. It is generally accepted that plants can metabolize xenobiotics, including PCBs, through a three-phase process, also dubbed the 'green liver model' (Sandermann, 1994), in which a soil contaminant will be 'activated' by hydroxylation, then conjugated with a plant molecule (e.g. sugar) before being sequestrated into a vacuole or in the cell wall (Van Aken *et al.*, 2010). However, it has been demonstrated that the first phase of the degradation of PCBs in plants may result in the formation and accumulation of hydroxylated PCBs, which are more toxic than the parent compounds (Rezek *et al.*, 2008). Moreover, other drawbacks can limit the efficiency of phytoremediation. Firstly, not all plant species can tolerate high levels of contaminants (Glick, 2010). Even though some plants can hyper-accumulate contaminants, they quite often do not produce enough biomass to be sufficiently efficient at removing the harmful compounds, or simply have a slow growth rate (Zhuang *et al.*, 2007; Dimkpa *et al.*, 2009). Secondly, because plants are autotrophic organisms, they do not have the capacity to completely transform xenobiotics (e.g. PCBs), mainly because of a lack of catabolic enzymes required to do so, which results in slow and/or partial transformation of such compounds (Eapen *et al.*, 2007; Van Aken, 2008). Furthermore, the efficiency of a plant to metabolize PCBs is highly dependent on both the plant species and PCB congeners (Rezek *et al.*, 2008; Sylvestre *et al.*, 2009). Therefore, further research is needed in order to successfully exploit the full potential of phytoremediation as an efficient and safe technology to use.

Alternatively, harnessing the potential of plants and their root-associated microorganisms in order to facilitate the remediation of PCBs in contaminated soils, a process also referred to as rhizoremediation (Kuiper *et al.*, 2004; Mackova *et al.*, 2009), has shown a lot of potential in the past decades (Glick, 2010). Plants release a diverse array of compounds (or exudates) in the rhizosphere, partly as a result of their secondary metabolism, which then directly or indirectly affect microorganisms (Lambers *et al.*, 2009). Root exudates (most of which being organic compounds) may act as signal molecules in the rhizosphere (Bais *et al.*, 2004; Shaw *et al.*, 2006; Steinkellner *et al.*, 2007), but can also be used as nutrients by microorganisms (Lambers *et al.*, 2009), thus promoting their growth in the vicinity of the roots. Because some of the organic compounds found in the root exudates (e.g. flavonoids, terpenes) share chemical similarity with contaminants such as PCBs, it has been suggested that they can act as co-metabolites or pathway inducers to stimulate rhizobacterial degradation of xenobiotics (Donnelly *et al.*, 1994; Fletcher and Hegde, 1995; Singer *et al.*, 2003b). For instance, the growth of *Pseudomonas* spp., a major group of rhizobacteria capable of degrading PCBs that are commonly found in the rhizosphere (Kuiper *et al.*, 2004), has been shown to be enhanced when the bacteria grew in the vicinity of certain plant species or on purified compounds found in root exudates (Donnelly *et al.*, 1994; Leigh *et al.*, 2002, 2006). Hence, several authors have demonstrated in the past decades that the plant–rhizobacteria association can greatly influence the degradation of PCBs in contaminated soils (Fletcher and Hegde, 1995; Focht, 1995; Gilbert and Crowley, 1997; Hernandez *et al.*, 1997; Leigh *et al.*, 2002, 2006; Narasimhan *et al.*, 2003; Singer *et al.*, 2003a,b; Mackova *et al.*, 2007).

Although the perspective of using rhizoremediation as a tool to remove persistent contaminants from the soil is quite appealing, only a few bacterial strains have the capacity of degrading highly chlorinated PCBs (e.g. *B. xenovorans* LB400). Most of these bacteria are not endowed with the genetic background to colonize plants roots and, in addition, they cannot fully degrade complex PCBs. Therefore the use of transgenic plants and/or bacteria may offer yet another potential avenue with regards to the degradation of POPs in contaminated soils (Barac *et al.*, 2004; Macek *et al.*, 2008; Sylvestre *et al.*, 2009; Van Aken *et al.*, 2010). By introducing one or several genes from organisms that can degrade xenobiotic pollutants to candidate plants or bacteria, it is now possible to improve the capacity of plants to catabolize persistent organic pollutants such as PCBs (Barac *et al.*, 2004; Eapen *et al.*, 2007). Several hosts and donors species have now been tested for the production of transgenic plants for various xenobiotic pollutant degradation as reviewed summarily by Eapen *et al.* (2007) and by Van Aken *et al.* (2010).

Since accumulation of toxic metabolites produced by plants P450 systems and release of *trans*-diols may hamper efficient PCB removal by combined plant-rhizobacteria metabolism, it has been suggested that transgenic plants producing bacterial PCB-degrading enzymes can overcome these difficulties and be advantageously used for PCB-rhizoremediation processes (Mohammadi *et al.*, 2007; Macek *et al.*, 2008). Replacing P450 monooxygenase by a Rieske-type aryl-hydroxylating dioxygenase such as BPDO requires minimally the simultaneous expression of three genes *bphA, bphE, bphF*. The reductase component encoded in bacteria by *bphG* can be replaced by plant ferredoxin reductases (Hurtubise *et al.*, 1995), but all three components (the oxygenase, the ferredoxin and the ferredoxin reductase) are required for full activity (Hurtubise *et al.*, 1996). Analyses of tobacco plants transiently expressing *B. xenovorans* LB400 genes encoding the BPDO components or transformed with them (Mohammadi *et al.*, 2007) have shown that each components can be produced individually as active protein in plants. Furthermore, active BphAE and BphG were co-purified from *Nicotiana benthamiana* leaves agroinfiltrated with pGreen-*bphA+bphE* + pGreen-*bphG*. However, none of the transgenic plants co-transformed with pGreen-*bphA+bphE+bphG* + pGreen-*bphF* (Mohammadi *et al.*, 2007) produced the three enzyme components simultaneously. In addition the frequency of viable plants among the transformants was low. This suggests that simultaneous expression of all four BPDO genes in transgenic tobacco is hampered by genetic or physiological reasons.

Among the possible physiological reasons, BPDO might interfere with plant cell development. BPDOs can oxidize various homocyclic and heterocyclic arenes (Misawa *et al.*, 2002, 2005), including plant-derived flavonoids (Seeger *et al.*, 2003). As a consequence, BPDO might affect plant metabolism by altering the structure of small signal molecules that are involved in early steps of plant development, although no direct evidence is available. This hypothesis needs confirmation since several human P450s that are likely to oxidize plant small chemicals and thus interfere with their metabolism were successfully cloned into plants. For example, human P450s CYP1A1, CYP2B6, CYP2E1, CYP2C9 that are part of a small set of CYPs accounting for 90–95% of xenobiotic metabolism in liver (Guengerich, 2005) were successfully cloned in plants (Eapen *et al.*, 2007; Doty, 2008; Kawahigashi *et al.*, 2008) to increase tolerance to herbicides

Unlike BPDO, 2,3-DHBD consists of a single homo-octamer component. Therefore, a single gene is required to produce active enzyme. Macek *et al.* (2008) have recently discussed an approach to overcome the inability of plants to cleave dihydroxybiphenyls by cloning 2,3-DHBD into plants. Gene bphC from *P. pnomenusa* B-356 was successfully cloned in *Nicotiana tabacum* (Francova *et al.*, 2003; Novakova *et al.*, 2009) and gene *dbfB* encoding 2,3,2′-trihydroxybiphenyl 2,3-dioxygenase from *Terrabacter* sp. DBF63 dibenzofuran pathway was cloned in *Arabidopsis* (Uchida *et al.*, 2005). Interestingly, plants expressing 2,3-DHBD were more resistant to PCBs than non-transgenic ones (Macek *et al.*, 2005; Novakova *et al.*, 2009). This feature might be attributed to the fact that 2,3-DHBD can remove 2,3-dihydroxy-chlorobiphenyls derived from PCBs (Mackova *et al.*, 2007) which are potentially toxic to plants. Although toxicity of 2,3-dihydroxy-chlorobiphenyls to plants has never been examined directly, Camara *et al.* (2004) have shown they are toxic to bacterial cells, Novakova *et al.* (2009) have shown they are toxic to tobacco plants and Lovecka *et al.* (2004) have shown that monohydroxylated PCB metabolites are toxic to plants depending on number and position of the chlorine atoms. Furthermore, the toxicity of catechol to plants has been clearly demonstrated by Liao *et al.* (2006). These are among the many observations supporting the use of transgenic plants producing 2,3-DHBD for rhizoremediation of PCB-contaminated sites since these plants are likely to be more resistant to the PCB metabolites produced by plants and their associated rhizobacteria.

Conclusion

PCBs are complex mixtures of biphenyl analogues where the structural properties of each individual congener vary significantly depending on the number and position of chlorine atoms on the biphenyl skeleton. This variability impacts considerably on the biodegradability of the mixture. Several biological systems have evolved to recycle xenobiotics in the biosphere, and these systems are available to remove persistent organic pollutants such as PCB. Some anaerobic bacteria, especially of the genus *Dehalococcoides* can dehalogenate highly chlorinated congeners, principally onto *meta* and *para* carbons to reduce the chlorine charge of the molecule. It has been suggested that this system be exploited in a dual anaerobic-aerobic process to clean-up PCB contaminated site (Abramowicz, 1990). However, although our knowledge about anaerobic dehalogenations has progressed considerably over the last decade, further investigations are required to fully understand the process, including the enzyme reactions that are involved. Fungi are able to degrade many xenobiotics and to mineralize them. However, the mineralization process is slow because it is based on the use of many enzymes (including ligninolytic peroxidases and cytochrome P450 monooxygenases) that act non-specifically on PCBs and their metabolites. This system is more useful for the removal of the toxic charge caused by PCBs by transforming them into hydroxylated metabolites that are then coupled to humic acids. The biphenyl catabolic pathway of aerobic bacteria has been studied extensively for its ability to transform several PCB congeners. Results obtained in recent decades show the feasibility of engineering enzymes and bacteria (especially the biphenyl dioxygenase and the 2,3-dihydroxybiphenyl 1,2-dioxygenase) for the degradation of the most persistent congeners which include the *meta-para*-substituted congeners, which are co-planar, and the doubly *ortho*-substituted, which are produced from anaerobic degradation of highly chlorinated PCBs. Finally, plants contribute actively to the restoration of PCB-contaminated sites. Although they can accumulate and metabolize PCB, their metabolism is slow and can generate more toxic metabolites. However, plants can act in association with rhizosphere bacteria to efficiently degrade PCB. Plants can be transformed with PCB degrading genes to initiate PCB degradation and to detoxify some of the bacterial metabolites of PCBs. They also produce chemicals such as terpenes and flavonoids that can stimulate the bacterial degradation of PCBs. In the near future, it is expected that genetically engineered plants and rhizobacteria will be available to develop an efficient PCB-removal process.

References

Abramowicz, D.A. (1990). Aerobic and anaerobic biodegradation of PCBs: a review. Crit. Rev. Biotechnol. *10*, 241–251.

Adams, R.H., Huang, C.M., Higson, F.K., Brenner, V., and Focht, D.D. (1992). Construction of a 3-chlorobiphenyl-utilizing recombinant from an intergeneric mating. Appl. Environ. Microbiol. *58*, 647–654.

Adrian, L., Dudkova, V., Demnerova, K., and Bedard, D.L. (2009). '*Dehalococcoides*' sp. strain CBDB1 extensively dechlorinates the commercial polychlorinated biphenyl mixture aroclor 1260. Appl. Environ. Microbiol. *75*, 4516–4524.

Ahmad, D., Massé, R., and Sylvestre, M. (1990). Cloning and expression of genes involved in 4-chlorobiphenyl transformation by *Pseudomonas testosteroni*: homology to polychlorobiphenyl-degrading genes in other bacteria. Gene *86*, 53–61.

Ahmad, D., Sylvestre, M., Sondossi, M., and Massé, R. (1991). Bioconversion of 2-hydroxy-6-oxo-6-(4′-chlorophenyl)hexa-2,4-dienoic acid, the *meta*-cleavage product of 4-chlorobiphenyl. J. Gen. Microbiol. *137*, 1375–1385.

Ahmed, M., and Focht, D.D. (1973). Degradation of polychlorinated biphenyls by two species of *Achromobacter*. Can. J. Microbiol. *19*, 47–52.

Aronstein, B.N., and Alexander, M. (1993). Effect of a non-ionic surfactant added to the soil surface on the biodegradation of aromatic hydrocarbons within the soil. Appl. Microbiol. Biotechnol. *39*, 386–390.

Ashikawa, Y., Fujimoto, Z., Noguchi, H., Habe, H., Omori, T., Yamane, H., and Nojiri, H. (2006). Electron transfer complex formation between oxygenase and ferredoxin components in Rieske nonheme iron oxygenase system. Structure *14*, 1779–1789.

Bais, H.P., Park, S.W., Weir, T.L., Callaway, R.M., and Vivanco, J.M. (2004). How plants communicate using the underground information superhighway. Trends Plant. Sci. *9*, 26–32.

Barac, T., Taghavi, S., Borremans, B., Provoost, A., Oeyen, L., Colpaert, J.V., Vangronsveld, J., and van der Lelie, D. (2004). Engineered endophytic bacteria improve phytoremediation of water-soluble, volatile, organic pollutants. Nat. Biotechnol. *22*, 583–588.

Barriault, D., and Sylvestre, M. (1993). Factors affecting PCB degradation by an implanted bacterial strain in soil microcosms. Can. J. Microbiol. *39*, 594–602.

Barriault, D., and Sylvestre, M. (1999). Functionality of biphenyl 2,3-dioxygenase components in naphthalene 1,2-dioxygenase. Appl. Microbiol. Biotechnol. *51*, 592–597.

Barriault, D., and Sylvestre, M. (2004). Evolution of the biphenyl dioxygenase BphA from *Burkholderia xenovorans* LB400 by random mutagenesis of multiple sites in region III. J. Biol. Chem. *279*, 47480–47488.

Barriault, D., Durand, J., Maaroufi, H., Eltis, L.D., and Sylvestre, M. (1998). Degradation of polychlorinated biphenyl metabolites by naphthalene-catabolizing enzymes. Appl. Environ. Microbiol *64*, 4637–4642.

Barriault, D., Vedadi, M., Powlowski, J., and Sylvestre, M. (1999). cis-2,3-dihydro-2,3-dihydroxybiphenyl dehydrogenase and cis-1,2-dihydro-1,2-dihydroxynaphathalene dehydrogenase catalyze dehydrogenation of the same range of substrates. Biochem. Biophys. Res. Commun. *260*, 181–187.

Barriault, D., Plante, M.M., and Sylvestre, M. (2002). Family shuffling of a targeted bphA region to engineer biphenyl dioxygenase. J. Bacteriol. *184*, 3794–3800.

Bedard, D.L. (2007). *Dehalococcoides*: Nature's anaerobic superbug. In Proceedings of the 4th Symposium on Biosorption and Bioremediation, Mackova, M., Macek, T., Demnerova, K., Pazlar, V., and Novakova, M., eds. (Institute of Chemical Technology Prague: Prague, Czech Republic), pp. 7–10.

Bedard, D.L., Unterman, R., Bopp, L.H., Brennan, M.J., Haberl, M.L., and Johnson, C. (1986). Rapid assay for screening and characterizing microorganisms for the ability to degrade polychlorinated biphenyls. Appl. Environ. Microbiol. *51*, 761–768.

Billingsley, K.A., Backus, S.M., and Ward, O.P. (1999). Effect of surfactant solubilization on biodegradation of polychlorinated bipbenyl congeners by *Pseudomonas* LB400. Appl. Microbiol. Biotechnol. *52*, 255–260.

Blackburn, J.W., and Hafker, W.R. (1993). The impact of biochemistry, bioavailability and bioactivity on the selection of bioremediation techniques. Trends Biotechnol. *11*, 328–333.

Boonchan, S., Britz, M.L., and Stanley, G.A. (1998). Surfactant-enhanced biodegradation of high molecular weight polycyclic aromatic hydrocarbons by *Stenotrophomonas maltophilia*. Biotechnol. Bioeng. *59*, 482–494.

Brennan, E., McGuinness, M., and Dowling, D.N. (2009). Bioinformatic analysis and *in vitro* site-directed mutagenesis of conserved amino acids in BphK(LB400), a specific bacterial glutathione transferase. Int. Biodeter. Biodegr. *63*, 928–932.

van den Brink, H.J.M., van Gorcom, R.F.M., van den Hondel, C.A.M.J.J., and Punt, P.J. (1998). Cytochrome P450 enzyme systems in fungi. Fungal Gen. Biol. *23*, 1–17.

Brown, J.F.J., Bedard, D.L., Brennan, M.J., Carnahan, J.C., Feng, H., and Wagner, R.E. (1987). Polychlorinated biphenyl dechlorination in aquatic sediments. Science *236*, 709–712.

Burd, G., and Ward, O.P. (1996). Bacterial degradation of polycyclic aromatic hydrocarbons on agar plates: The role of biosurfactants. Biotechnol. Techniques *10*, 371–374.

Camara, B., Herrera, C., Gonzalez, M., Couve, E., Hofer, B., and Seeger, M. (2004). From PCBs to highly toxic metabolites by the biphenyl pathway. Environ. Microbiol. *6*, 842–850.

Camara, B., Seeger, M., Gonzalez, M., StandfussGabisch, C., Kahl, S., and Hofer, B. (2007). Generation by a widely applicable approach of a hybrid dioxygenase showing improved oxidation of polychlorobiphenyls. Appl. Environ. Microbiol. 73, 2682–2689.

Cerdan, P., Rekik, M., and Harayama, S. (1995). Substrate specificity differences between two catechol 2,3-dioxygenases encoded by the TOL and NAH plasmids from *Pseudomonas putida*. Eur. J. Biochem. *229*, 113–118.

Cerniglia, C.E. (1997). Fungal metabolism of polycyclic aromatic hydrocarbons: Past, present and future applications in bioremediation. J. Ind. Microbiol. Biot. *19*, 324–333.

Commandeur, L.C.M., May, R.J., Mokross, H., Bedard, D.L., Reineke, W., Govers, H.A.J., and Parsons, J.R. (1997). Aerobic degradation of polychlorinated biphenyls by Alcaligenes sp. JB1: Metabolites and enzymes. Biodegradation 7, 435–443.

Couture, M.M.J., Colbert, C.L., Babini, E., Rosell, F.I., Mauk, A.G., Bolin, J.T., and Eltis, L.D. (2001). Characterization of BphF, a Rieske-type ferredoxin with a low reduction potential. Biochemistry *40*, 84–92.

Dai, S., Vaillancourt, F.H., Maaroufi, H., Drouin, N.M., Neau, D.B., Snieckus, V., Bolin, J.T., and Eltis, L.D.

(2002). Identification and analysis of a bottleneck in PCB biodegradation. Nat. Struct. Biol. *9*, 934–939.

Denome, S.A., Stanley, D.C., Olson, E.S., and Young, K.D. (1993). Metabolism of dibenzothiophene and naphthalene in *Pseudomonas* strains: complete DNA sequence of an upper naphthalene catabolic pathway. J. Bacteriol. *175*, 6890–6901.

Dimkpa, C., Weinand, T., and Asch, F. (2009). Plant–rhizobacteria interactions alleviate abiotic stress conditions. Plant Cell Environ. *32*, 1682–1694.

Doddapaneni, H., and Yadav, J.S. (2005). Microarray-based global differential expression profiling of P450 monooxygenases and regulatory proteins for signal transduction pathways in the white rot fungus *Phanerochaete chrysosporium*. Mol. Genet. Genomics *274*, 454–466.

Donnelly, P.K., Hegde, R.S., and Fletcher, J.S. (1994). Growth of PCB-degrading bacteria on compounds from photosynthetic plants. Chemosphere *28*, 981–988.

Doty, S.L. (2008). Enhancing phytoremediation through the use of transgenics and endophytes. New Phytol. *179*, 318–333.

Eapen, S., Singh, S., and D'Souza, S.F. (2007). Advances in development of transgenic plants for remediation of xenobiotic pollutants. Biotechnol. Adv. *25*, 442–451.

Eaton, D.C. (1985). Mineralization of polychlorinated biphenyls by *Phanerochaete chrysosporium* a ligninolytic fungus. Enzyme Micob. Technol. *7*, 194–196.

Eltis, L.D., Hofmann, B., Hecht, H.J., Lunsdorf, H., and Timmis, K.N. (1993). Purification and crystallization of 2,3-dihydroxybiphenyl 1,2-dioxygenase. J. Biol. Chem. *268*, 2727–2732.

Erickson, B.D., and Mondello, F.J. (1992). Nucleotide sequencing and transcriptional mapping of the genes encoding biphenyl dioxygenase, a multicomponent polychlorinated-biphenyl-degrading enzyme in *Pseudomonas* strain LB400. J. Bacteriol. *174*, 2903–2912.

Erickson, B.D., and Mondello, F.J. (1993). Enhanced biodegradation of polychlorinated biphenyls after site-directed mutagenesis of a biphenyl dioxygenase gene. Appl. Environ. Microbiol. *59*, 3858–3862.

Estrella, M.R., Brusseau, M.L., Maier, R.S., Pepper, I.L., Wierenga, P.J., and Miller, R.M. (1993). Biodegradation, sorption, and transport of 2,4-dichlorophenoxyacetic acid in saturated and unsaturated soils. Appl. Environ. Microbiol. *59*, 4266–4273.

Fava, F., and Ciccotosto, V.F. (2002). Effects of randomly methylated-beta-cyclodextrins (RAMEB) on the bioavailability and aerobic biodegradation of polychlorinated biphenyls in three pristine soils spiked with a transformer oil. Appl. Microbiol. Biotechnol. *58*, 393–399.

Fava, F., and DiGioia, D. (2001). Soya lecithin effects on the aerobic biodegradation of polychlorinated biphenyls in an artificially contaminated soil. Biotechnol. Bioeng. *72*, 177–184.

Fava, F., and Piccolo, A. (2002). Effects of humic substances on the bioavailability and aerobic biodegradation of polychlorinated biphenyls in a model soil. Biotechnol. Bioeng. *77*, 204–211.

Fennell, D.E., Nijenhuis, I., Wilson, S.F., Zinder, S.H., and Haggblom, M.M. (2004). *Dehalococcoides ethenogenes* strain 195 reductively dechlorinates diverse chlorinated aromatic pollutants. Environ. Sci. Technol. *38*, 2075–2081.

Ferraro, D.J., Okerlund, A.L., Mowers, J.C., and Ramaswamy, S. (2006). Structural basis for regioselectivity and stereoselectivity of product formation by naphthalene 1,2-dioxygenase. J. Bacteriol. *188*, 6986–6994.

Ferraro, D.J., Brown, E.N., Yu, C.L., Parales, R.E., Gibson, D.T., and Ramaswamy, S. (2007). Structural investigations of the ferredoxin and terminal oxygenase components of the biphenyl 2,3-dioxygenase from *Sphingobium yanoikuyae* B1. BMC Struct. Biol. 7, 10.

Fletcher, J.S., and Hegde, R.S. (1995). Release of phenols by perennial plant-roots and their potential importance in bioremediation. Chemosphere *31*, 3009–3016.

Focht, D.D. (1995). Strategies for the improvement of aerobic metabolism of polychlorinated-biphenyls. Curr. Opin. Biotech. *6*, 341–346.

Fortin, P.D., Lo, A.T.F., Haro, M.A., Kaschabek, S.R., Reineke, W., and Eltis, L.D. (2005a). Evolutionarily divergent extradiol dioxygenases possess higher specificities for polychlorinated biphenyl metabolites. J. Bacteriol. *187*, 415–421.

Fortin, P.D., MacPherson, I., Neau, D.B., Bolin, J.T., and Eltis, L.D. (2005b). Directed evolution of a ring-cleaving dioxygenase for polychlorinated biphenyl degradation. J. Biol. Chem. *280*, 42307–42314.

Francova, K., Sura, M., Macek, T., Szekeres, M., Bancos, S., Demnerova, K., Sylvestre, M., and Mackova, M. (2003). Preparation of plants containing bacterial enzyme for degradation of polychlorinated biphenyls. Fresen. Environ. Bull. *12*, 309–313.

Fukuda, M., Yasukochi, Y., Kikuchi, Y., Nagata, Y., Kimbara, K., Horiuchi, H., Takagi, M., and Yano, K. (1994). Identification of the bphA and bphB genes of *Pseudomonas* sp. strain KKS102 involved in degradation of biphenyl and polychlorinated biphenyls. Biochem. Biophys. Res. Commun. *202*, 850–856.

Furukawa, K. (2000). Biochemical and genetic bases of microbial degradation of polychlorinated biphenyls (PCBs). J. Gen. Appl. Microbiol. *46*, 283–296.

Furukawa, K., and Arimura, N. (1987). Purification and properties of 2,3-dihydroxybiphenyl dioxygenase from polychlorinated biphenyl-degrading *Pseudomonas pseudoAlcaligenes* and *Pseudomonas aeruginosa* carrying the cloned bphC gene. J. Bacteriol. *169*, 924–927.

Furukawa, K., Matsumura, F., and Tonomura, K. (1978). *Alcaligenes* and *Acinetobacter* strains capable of degrading polychlorinated biphenyls. Agric. Biol. Chem. *42*, 543–548.

Furukawa, K., Suenaga, H., and Goto, M. (2004). Biphenyl dioxygenases: functional versatilities and directed evolution. J. Bacteriol. *186*, 5189–5196.

Furukawa, K., Tomizuka, N., and Kamibayashi, A. (1979). Effect of chlorine substitution on the bacterial metabolism of various polychlorinated biphenyls. Appl. Environ. Microbiol. *38*, 301–310.

Furusawa, Y., Nagarajan, V., Tanokura, M., Masai, E., Fukuda, M., and Senda, T. (2004). Crystal structure of the terminal oxygenase component of biphenyl dioxygenase derived from *Rhodococcus* sp. strain RHA1. J. Mol. Biol. *342*, 1041–1052.

Gilbert, E.S., and Crowley, D.E. (1997). Plant compounds that induce polychlorinated biphenyl biodegradation by *Arthrobacter* sp. strain B1B. Appl. Environ. Microbiol. *63*, 1933–1938.

Glick, B.R. (2010). Using soil bacteria to facilitate phytoremediation. Biotechnol. Adv. *28*, 367–374.

Gomez-Gil, L., Kumar, P., Barriault, D., Bolin, J.T., Sylvestre, M., and Eltis, L.D. (2007). Characterization of biphenyl dioxygenase of Pandoraea pnomenusa B-356 as a potent polychlorinated biphenyl-degrading enzyme. J. Bacteriol. *189*, 5705–5715.

Guengerich, P.F. (2005). Human cytochrome P450 enzymes. In Cytochrome P450: Structure, mechanism, and Biochemistry, Ortiz de Montellao, P.R., ed. (Kluwer Academic/Plenum Publishers: New York), pp. 377–530.

Guilbeault, B., Sondossi, M., Ahamad, D., and Sylvestre, M. (1994). Factors affecting the enhancement of PCB degradative ability of soil microbial populations. Int. Biodeter. Biodegr. 33, 73–91.

Haddock, J.D., and Gibson, D.T. (1995). Purification and characterization of the oxygenase component of biphenyl 2,3-dioxygenase from *Pseudomonas* sp. strain LB400. J. Bacteriol. *177*, 5834–5839.

Haddock, J.D., Horton, J.R., and Gibson, D.T. (1995). Dihydroxylation and dechlorination of chlorinated biphenyls by purified biphenyl 2,3-dioxygenase from *Pseudomonas* sp. strain LB400. J. Bacteriol. *177*, 20–26.

Han, S., Eltis, L.D., Timmis, K.N., Muchmore, S.W., and Bolin, J.T. (1995). Crystal structure of the biphenyl-cleaving extradiol dioxygenase from a PCB-degrading pseudomonad. Science *270*, 976–980.

Hatta, T., Shimada, T., Yoshihara, T., Yamada, A., Masai, E., Fukuda, M., and Kiyohara, H. (1998). Meta-fission product hydrolases from a strong PCB degrader *Rhodococcus* sp. RHA1. J. Ferment. Bioeng. *85*, 174–179.

Havel, J., and Reineke, W. (1991). Total degradation of various chlorobiphenyls by cocultures and *in vivo* constructed hybrid pseudomonads. FEMS Microbiol. Lett. *62*, 163–169.

Havel, J., and Reineke, W. (1993). Microbial degradation of chlorinated acetophenones. Appl Environ Microbiol *59*, 2706–2712.

Hein, P., Powlowski, J., Barriault, D., Hurtubise, Y., Ahmad, D., and Sylvestre, M. (1998). Biphenyl-associated *meta*-cleavage dioxygenases from *Comamonas testosteroni* B-356. Can. J. Microbiol. *44*, 42–49.

Hernandez, B.S., Koh, S.C., Chial, M., and Focht, D.D. (1997). Terpene-utilizing isolates and their relevance to enhanced biotransformation of polychlorinated biphenyls in soil. Biodegradation *8*, 153–158.

Hirose, J., Suyama, A., Hayashida, S., and Furukawa, K. (1994). Construction of hybrid biphenyl (*bph*) and toluene (*tod*) genes for functional analysis of aromatic ring dioxygenases. Gene *138*, 27–33.

Holliger, C., Regeard, C., and Diekert, G. (2003). Dehalogenation: Microbial processes and environmental applications. In Dehalogenation by Anaerobic Bacteria, Häggblom, I.B. M., ed. (Kluwer Academic Publisher: New York), pp. 115–157.

Hrywna, Y., Tsoi, T.V., Maltseva, O.V., Quensen, J.F., 3rd, and Tiedje, J.M. (1999). Construction and characterization of two recombinant bacteria that grow on *ortho*- and *para*-substituted chlorobiphenyls. Appl. Environ. Microbiol. *65*, 2163–2169.

Hulsmeyer, M., Hecht, H.J., Niefind, K., Hofer, B., Eltis, L.D., Timmis, K.N., and Schomburg, D. (1998). Crystal structure of cis-biphenyl-2,3-dihydrodiol-2,3-dehydrogenase from a PCB degrader at 2.0 A resolution. Protein Sci. *7*, 1286–1293.

Hurtubise, Y., Barriault, D., Powlowski, J., and Sylvestre, M. (1995). Purification and characterization of the *Comamonas testosteroni* B-356 biphenyl dioxygenase components. J. Bacteriol. *177*, 6610–6618.

Hurtubise, Y., Barriault, D., and Sylvestre, M. (1996). Characterization of active recombinant his-tagged oxygenase component of *Comamonas testosteroni* B-356 biphenyl dioxygenase. J. Biol. Chem. 271. *271*, 8152–8156–8152–8156.

Hurtubise, Y., Barriault, D., and Sylvestre, M. (1998). Involvement of the terminal oxygenase beta subunit in the biphenyl dioxygenase reactivity pattern toward chlorobiphenyls. J. Bacteriol. *180*, 5828–5835.

Inoue, K., Ashikawa, Y., Umeda, T., Abo, M., Katsuki, J., Usami, Y., Noguchi, H., Fujimoto, Z., Terada, T., Yamane, H., *et al.* (2009). Specific interactions between the ferredoxin and terminal oxygenase components of a class IIB Rieske nonheme iron oxygenase, carbazole 1,9a-dioxygenase. J. Mol. Biol. *392*, 436–451.

Jimenez, I.Y., and Bartha, R. (1996). Solvent-augmented mineralization of pyrene by a *Mycobacterium* sp. Appl. Environ. Microb. *62*, 2311–2316.

Kagami, O., Shindo, K., Kyojima, A., Takeda, K., Ikenaga, H., Furukawa, K., and Misawa, N. (2008). Protein engineering on biphenyl dioxygenase for conferring activity to convert 7-hydroxyflavone and 5,7-dihydroxyflavone (chrysin). J. Biosci. Bioeng. *106*, 121–127.

Kauppi, B., Lee, K., Carredano, E., Parales, R.E., Gibson, D.T., Eklund, H., and Ramaswamy, S. (1998). Structure of an aromatic-ring-hydroxylating dioxygenase-naphthalene 1,2- dioxygenase. Structure *6*., 571–586.

Kawahigashi, H., Hirose, S., Ohkawa, H., and Ohkawa, Y. (2008). Transgenic rice plants expressing human p450 genes involved in xenobiotic metabolism for phytoremediation. J. Mol. Microbiol. Biotechnol. *15*, 212–219.

Khan, A.A., Wang, R.F., Nawaz, M.S., Cao, W.W., and Cerniglia, C.E. (1996). Purification of 2,3-dihydroxybiphenyl 1,2-dioxygenase from *Pseudomonas putida* OU83 and characterization of the gene (bphC). Appl. Environ. Microbiol. *62*, 1825–1830.

Kikuchi, Y., Yasukochi, Y., Nagata, Y., Fukuda, M., and Takagi, M. (1994). Nucleotide sequence and functional analysis of the *meta*-cleavage pathway involved in biphenyl and polychlorinated biphenyl degradation in *Pseudomonas* sp. strain KKS102. J. Bacteriol. *176*, 4269–4276.

Kim, E., Kim, Y., and Kim, C.K. (1996). Genetic structures of the genes encoding 2,3-dihydroxybiphenyl 1,2-dioxygenase and 2-hydroxy-6-oxo-6-phenylhexa-2,4-dienoic acid hydrolase from biphenyl- and 4-chlorobiphenyl-degrading *Pseudomonas* sp. strain DJ-12. Appl. Environ. Microbiol. *62*, 262–265.

Kimura, N., Nishi, A., Goto, M., and Furukawa, K. (1997). Functional analyses of a variety of chimeric dioxygenases constructed from two biphenyl dioxygenases that are similar structurally but different functionally. J. Bacteriol. *179*, 3936–3943.

Kuiper, I., Lagendijk, E.L., Bloemberg, G.V., and Lugtenberg, B.J.J. (2004). Rhizoremediation: A beneficial plant–microbe interaction. Mol. Plant Microbe In. *17*, 6–15.

Kumamaru, T., Suenaga, H., Mitsuoka, M., Watanabe, T., and Furukawa, K. (1998). Enhanced degradation of polychlorinated biphenyls by directed evolution of biphenyl dioxygenase. Nat. Biotechnol. *16*, 663–666.

L'Abbée, J.B., Barriault, D., and Sylvestre, M. (2005). Metabolism of dibenzofuran and dibenzo-p-dioxin by the biphenyl dioxygenase of *Burkholderia xenovorans* LB400 and *Comamonas testosteroni* B-356. Appl. Microbiol. Biotechnol. *67*, 506–514.

Lajoie, C.A., Layton, A.C., and Sayler, G.S. (1994). Cometabolic oxidation of polychlorinated biphenyls in soil with a surfactant-based field application vector. Appl. Environ. Microbiol. *60*, 2826–2833.

Lambers, H., Mougel, C., Jaillard, B., and Hinsinger, P. (2009). Plant–microbe-soil interactions in the rhizosphere: an evolutionary perspective. Plant Soil *321*, 83–115.

Lau, P.C.K., Garnon, J., Labbé, D., and Wang, Y. (1996). Location and sequence analysis of a 2-hydroxy-6-oxo-6-phenylhexa-2,4-dienoate hydrolase-encoding gene (bpdF) of the biphenyl/polychlorinated biphenyl degradation pathway in *Rhodococcus* sp. M5. Gene *171*, 53–57.

Leigh, M.B., Fletcher, J.S., Fu, X.O., and Schmitz, F.J. (2002). Root turnover: An important source of microbial substrates in rhizosphere remediation of recalcitrant contaminants. Environ. Sci. Technol. *36*, 1579–1583.

Leigh, M.B., Prouzova, P., Mackova, M., Macek, T., Nagle, D.P., and Fletcher, J.S. (2006). Polychlorinated biphenyl (PCB)-degrading bacteria associated with trees in a PCB-contaminated site. Appl. Environ. Microbiol. 72, 2331–2342.

Liao, Y., Zhou, X., Yu, J., Cao, Y.J., Li, X., and Kuai, B.K. (2006). The key role of chlorocatechol 1,2-dioxygenase in phytoremoval and degradation of catechol by transgenic Arabidopsis. Plant Physiol. *142*, 620–628.

Lovecka, P., Mackova, M., Demnerova, K. (2004). Metabolic products of PCBs in bacteria and plants – comparison of their toxicity and genotoxicity. In Proceedings ESEB 2004, Verstraete, W., ed. (Taylor and Francis Group: London), pp. 299–302.

Macek, T., Sura, M., Pavlikova, D., Francova, K., Scouten, W.H., Szekeres, M., Sylvestre, M., and Mackova, M. (2005). Can tobacco have a potentially beneficial effect to our health? Z Naturforsch [C] *60*, 292–299.

Macek, T., Kotrba, P., Svatos, A., Novakova, M., Demnerova, K., and Mackova, M. (2008). Novel roles for genetically modified plants in environmental protection. Trends Biotechnol. *26*, 146–152.

Mackova, M., Vrchotova, B., Francova, K., Sylvestre, M., Tomaniova, M., Lovecka, P., Demnerova, K., and Macek, T. (2007). Biotransformation of PCBs by plants and bacteria – consequences of plant–microbe interactions. Eur. J. Soil Biol. *43*, 233–241.

Mackova, M., Prouzova, P., Stursa, P., Ryslava, E., Uhlik, O., Beranova, K., Rezek, J., Kurzawova, V., Demnerova, K., and Macek, T. (2009). Phyto/rhizoremediation studies using long-term PCB-contaminated soil. Environ. Sci. Pollut. Res. Int. *16*, 817–829.

Maier, R.M. (2000). Bioavailability and its importance to bioremediation. In Bioremediation, Valdes, J.J., ed. (Academic Press: Netherlands), pp. 59–78.

Maltseva, O.V., Tsoi, T.V., Quensen, J.F., Fukuda, M., and Tiedje, J.M. (1999). Degradation of anaerobic reductive dechlorination products of Aroclor 1242 by four aerobic bacteria. Biodegradation *10*, 363–371.

Mars, A.E., Kingma, J., Kaschabek, S.R., Reineke, W., and Janssen, D.B. (1999). Conversion of 3-chlorocatechol by various catechol 2,3-dioxygenases and sequence analysis of the chlorocatechol dioxygenase region of *Pseudomonas putida* GJ31. J. Bacteriol. *181*, 1309–1318.

Martinez, D., Larrondo, L.F., Putnam, N., Gelpke, M.D., Huang, K., Chapman, J., Helfenbein, K.G., Ramaiya, P., Detter, J.C., Larimer, F., *et al.* (2004). Genome sequence of the lignocellulose degrading fungus *Phanerochaete chrysosporium* strain RP78. Nat. Biotechnol. *22*, 695–700.

Massé, R., Messier, F., Peloquin, L., Ayotte, C., and Sylvestre, M. (1984). Microbial biodegradation of 4-chlorobiphenyl, a model compound of chlorinated biphenyls. Appl. Environ. Microbiol. *47*, 947–951.

Massé, R., Messier, F., Ayotte, C., Lévesque, M.-F., and Sylvestre, M. (1989). A Comprehensive gas chromatographic/Mass spectrometric analysis of 4-chlorobiphenyl bacterial degradation products. Biomed. Environ. Mass *18*, 27–47.

May, H.D., Miller, G.S., Kjellerup, B.V., and Sowers, K.R. (2008). Dehalorespiration with polychlorinated biphenyls by an anaerobic ultraMicrobacterium. Appl. Environ. Microbiol. *74*, 2089–2094.

Mester, T., and Tien, M. (2000). Oxidation mechanism of ligninolytic enzymes involved in the degradation of environmental pollutants. Int. Biodeter. Biodegrad. *46*, 51–59.

Misawa, N., Shindo, K., Takahashi, H., Suenaga, H., Iguchi, K., Okazaki, H., Harayama, S., and Furukawa, K. (2002). Hydroxylation of various molecules including heterocyclic aromatics using recombinant *Escherichia coli* cells expressing modified biphenyl dioxygenase genes. Tetrahedron *58*, 9605–9612.

Misawa, N., Nakamura, R., Kagiyama, Y., Ikenaga, H., Furukawa, K., and Shindo, K. (2005). Synthesis of vicinal diols from various arenes with a heterocyclic, amino or carboxyl group by using recombinant *Escherichia coli* cells expressing evolved biphenyl dioxygenase and dihydrodiol dehydrogenase genes. Tetrahedron *61*, 195–204.

Mohammadi, M., and Sylvestre, M. (2005). Resolving the profile of metabolites generated during oxidation of dibenzofuran and chlorodibenzofurans by the biphenyl catabolic pathway enzymes. Chem. Biol. *12*, 835–846.

Mohammadi, M., Chalavi, V., Novakova-Sura, M., Laliberte, J.F., and Sylvestre, M. (2007). Expression of bacterial biphenyl-chlorobiphenyl dioxygenase genes in tobacco plants. Biotechnol. Bioeng. *97*, 496–505.

Mokross, H., Schmidt, E., and Reineke, W. (1990). Degradation of 3-chlorobiphenyl by *in vivo* constructed hybrid pseudomonads. FEMS Microbiol. Lett. *59*, 179–185.

Mondello, F.J., Turcich, M.P., Lobos, J.H., and Erickson, B.D. (1997). Identification and modification of biphenyl dioxygenase sequences that determine the specificity of polychlorinated biphenyl degradation. Appl. Environ. Microbiol. *63*, 3096–3103.

Mukerjee-Dhar, G., Hatta, T., Shimura, M., and Kimbara, K. (1998). Analysis of changes in congener selectivity during PCB degradation by *Burkholderia* sp. strain TSN101 with increasing concentrations of PCB and characterization of the *bphBCD* genes and gene products. Arch. Microbiol. *169*, 61–70.

Nadim, L.M., Schocken, M.J., Higson, F.K., Gibson, D.T., Bedard, L.D., Bopp, L.H., and Mondello, F.J. (1987). Bacterial oxidation of polychlorinated biphenyls. In Proceedings of the US EPA Thirteenth Annual Research Symposium on Land Disposal, Remedial Action, Incineration, and Treatment of Hazardous Wast, Agency, E.P., ed. (Environmental Protection Agency: Cincinnati, OH), pp. 395–402.

Nandhagopal, N., Yamada, A., Hatta, T., Masai, E., Fukuda, M., Mitsui, Y., and Senda, T. (2001). Crystal structure of 2-hydroxyl-6-oxo-6-phenylhexa-2,4-dienoic acid (HPDA) hydrolase (BphD enzyme) from the *Rhodococcus* sp. strain RHA1 of the PCB degradation pathway. J. Mol. Biol. *309*, 1139–1151.

Narasimhan, K., Basheer, C., Bajic, V.B., and Swarup, S. (2003). Enhancement of plant–microbe interactions using a rhizosphere metabolomics-driven approach and its application in the removal of polychlorinated biphenyls. Plant Physiol. *132*, 146–153.

Novakova, M., Mackova, M., Chrastilova, Z., Viktorova, J., Szekeres, M., Demnerova, K., and Macek, T. (2009). Cloning the bacterial *bphC* gene into *Nicotiana tabacum* to improve the efficiency of PCB phytoremediation. Biotechnol. Bioeng. *102*, 29–37.

Pointing, S.B. (2001). Feasibility of bioremediation by white-rot fungi. Appl. Microbiol. Biotechnol. *57*, 20–33.

Resnick, S.M., and Gibson, D.T. (1996). Oxidation of 6,7-dihydro-5H-benzocycloheptene by bacterial strains expressing naphthalene dioxygenase, biphenyl dioxygenase, and toluene dioxygenase yields homochiral monol or cis-diol enantiomers as major products. Appl. Environ. Microbiol. *62*, 1364–1368.

Rezek, J., Macek, T., Mackova, M., Triska, J., and Ruzickova, K. (2008). Hydroxy-PCBs, methoxy-PCBs and hydroxy-methoxy-PCBs: Metabolites of polychlorinated biphenyls formed *in vitro* by tobacco cells. Environ. Sci. Technol. *42*, 5746–5751.

Safe, S. (1990). Polychlorinated biphenyls (PCBs), dibenzo-p-dioxins (PCDDs), dibenzofurans (PCDFs), and related compounds: environmental and mechanistic considerations which support the development of toxic equivalency factors (TEFs). CRC Crit. Rev. Toxicol. *21*, 51–88.

Sandermann, H. (1994). Higher-plant metabolism of xenobiotics – The green liver concept. Pharmacogenetics *4*, 225–241.

Seah, S.Y., Labbe, G., Nerdinger, S., Johnson, M.R., Snieckus, V., and Eltis, L.D. (2000). Identification of a serine hydrolase as a key determinant in the microbial degradation of polychlorinated biphenyls. J. Biol. Chem. *275*, 15701–15708.

Seah, S.Y.K., Labbe, G., Kaschabek, S.R., Reifenrath, F., Reineke, W., and Eltis, L.D. (2001). Comparative specificities of two evolutionarily divergent hydrolases involved in microbial degradation of polychlorinated biphenyls. J. Bacteriol. *183*, 1511–1516.

Seeger, M., Timmis, K.N., and Hofer, B. (1995). Conversion of chlorobiphenyls into phenylhexadienoates and benzoates by the enzymes of the upper pathway for polychlorobiphenyl degradation encoded by the bph locus of *Pseudomonas* sp. strain LB400. Appl. Environ. Microbiol. *61*, 2654–2658.

Seeger, M., Camara, B., and Hofer, B. (2001). Dehalogenation, denitration, dehydroxylation, and angular attack on substituted biphenyls and related compounds by a biphenyl dioxygenase. J. Bacteriol. *183*, 3548–3555.

Seeger, M., Gonzalez, M., Camara, B., Munoz, L., Ponce, E., Mejias, L., Mascayano, C., Vasquez, Y., and Sepulveda-Boza, S. (2003). Biotransformation of natural and synthetic isoflavonoids by two recombinant microbial enzymes. Appl. Environ. Microbiol. *69*, 5045–5050.

Senda, T., Yamada, T., Sakurai, N., Kubota, M., Nishizaki, T., Masai, E., Fukuda, M., and Mitsuidagger, Y. (2000). Crystal structure of NADH-dependent ferredoxin reductase component in biphenyl dioxygenase. J. Mol. Biol. *304*, 397–410.

Shannon, M.J.R., and Unterman, R. (1993). Evaluating bioremediation: distinguishing fact from fiction. Annu. Rev. Microbiol. *47*, 715–738.

Shaw, L.J., Morris, P., and Hooker, J.E. (2006). Perception and modification of plant flavonoids signals by rhizosphere microorganisms. Environ. Microbiol. *8*, 1867–1880.

Shindo, K., Nakamura, R., Osawa, A., Kagami, O., Kanoh, K., Furukawa, K., and Misawa, N. (2005). Biocatalytic synthesis of monocyclic arene-dihydrodiols and -diols by *Escherichia coli* cells expressing hybrid toluene/biphenyl dioxygenase and dihydrodiol dehydrogenase genes. J. Mol. Catal. B-Enzym. 35, 134–141.

Singer, A.C., Crowley, D.E., and Thompson, I.P. (2003a). Secondary plant metabolites in phytoremediation and biotransformation. Trends Biotechnol. *21*, 123–130.

Singer, A.C., Smith, D., W.A., J., Hathuc, K., and Crowley, D.E. (2003b). Impact of the plant rhizosphere and augmentation on remediation of polychlorinated biphenyl contaminated soil. Environ. Toxicol. Chem. *22*, 1998–2004.

Smidt, H., and de Vos, W.M. (2004). Anaerobic microbial dehalogenation. Annu. Rev. Microbiol. *58*, 43–73.

Sondossi, M., Sylvestre, M., and Ahmad, D. (1992). Effects of chlorobenzoate transformation on the *Pseudomonas testosteroni* biphenyl and chlorobiphenyl degradation pathway. Appl. Environ. Microbiol. *58*, 485–495.

Steinkellner, S., Lendzemo, V., Langer, I., Schweiger, P., Khaosaad, T., Toussaint, J.P., and Vierheilig, H. (2007). Flavonoids and strigolactones in root exudates as signals in symbiotic and pathogenic plant–fungus interactions. Molecules *12*, 1290–1306.

Suenaga, H., Nishi, A., Watanabe, T., Sakai, M., and Furukawa, K. (1999). Engineering a hybrid pseudomonad to acquire 3,4-dioxygenase activity for polychlorinated biphenyls. J. Biosci. Bioeng. *87*, 430–435.

Suenaga, H., Mitsuoka, M., Ura, Y., Watanabe, T., and Furukawa, K. (2001). Directed evolution of biphenyl dioxygenase: Emergence of enhanced degradation capacity for benzene, toluene, and alkylbenzenes. J. Bacteriol. *183*, 5441–5444.

Suenaga, H., Watanabe, T., Sato, M., Ngadiman, and Furukawa, K. (2002). Alteration of regiospecificity in biphenyl dioxygenase by active-site engineering. J. Bacteriol. *184*, 3682–3688.

Sutherland, J.B., Selby, A.L., Freeman, J.P., Evans, F.E., and Cerniglia, C.E. (1991). Metabolism of phenanthrene by *Phanerochaete chrysosporium*. Appl. Environ. Microbiol. *57*, 3310–3316.

Suyama, A., Iwakiri, R., Kimura, N., Nishi, A., Nakamura, K., and Furukawa, K. (1996). Engineering hybrid pseudomonads capable of utilizing a wide range of aromatic hydrocarbons and of efficient degradation of trichloroethylene. J. Bacteriol. *178*, 4039–4046.

Sylvestre, M. (1995). Biphenyl/chlorobiphenyls catabolic pathway of *Comamonas testosteroni* B-356: Prospect for use in bioremediation. Int. Biodeter. Biodeg. *34*, 189–211.

Sylvestre, M. (2004). Genetically modified organisms to remediate polychlorinated biphenyls. Where do we stand? Int. Biodeter. Biodeg. *54*, 153–162.

Sylvestre, M., Hurtubise, Y., Barriault, D., Bergeron, J., and Ahmad, D. (1996a). Characterization of active recombinant 2,3-dihydro-2,3-dihydroxybiphenyl dehydrogenase from *Comamonas testosteroni* B-356 and sequence of the encoding gene (*bphB*). Appl. Environ. Microbiol. *62*, 2710–2715.

Sylvestre, M., Sirois, M., Hurtubise, Y., Bergeron, J., Ahmad, D., Shareck, F., Barriault, D., Guillemette, I., and Juteau, J.M. (1996b). Sequencing of *Comamonas testosteroni* strain B-356-biphenyl/chlorobiphenyl dioxygenase genes: evolutionary relationships among Gram-negative bacterial biphenyl dioxygenases. Gene *174*, 195–202.

Sylvestre, M., Macek, T., and Mackova, M. (2009). Transgenic plants to improve rhizoremediation of polychlorinated biphenyls (PCBs). Curr. Opin. Biotechnol. *20*, 242–247.

Taira, K., Hayase, N., Arimura, N., Yamashita, S., Miyazaki, T., and Furukawa, K. (1988). Cloning and nucleotide sequence of the 2,3-dihydroxybiphenyl dioxygenase gene from the PCB-degrading strain of *Pseudomonas paucimobilis* Q1. Biochemistry 27, 3990–3996.

Taira, K., Hirose, J., Hayashida, S., and Furukawa, K. (1992). Analysis of bph operon from the polychlorinated biphenyl-degrading strain of *Pseudomonas pseudoAlcaligenes* KF707. J. Biol. Chem. *267*, 4844–4853.

Tiehm, A. (1994). Degradation of polycyclic aromatic hydrocarbons in the presence of synthetic surfactants. Appl. Environ. Microbiol. *60*, 258–263.

Timmis, K.N., and Pieper, D.H. (1999). Bacteria designed for bioremediation. Trends Biotechnol. *17*, 200–204.

Tocheva, E.I., Fortin, P.D., Eltis, L.D., and Murphy, M.E. (2006). Structures of ternary complexes of BphK, a bacterial glutathione S-transferase that reductively dechlorinates polychlorinated biphenyl metabolites. J. Biol. Chem. *281*, 30933–30940.

Uchida, E., Ouchi, T., Suzuki, Y., Yoshida, T., Habe, H., Yamaguchi, I., Omori, T., and Nojiri, H. (2005). Secretion of bacterial xenobiotic-degrading enzymes from transgenic plants by an apoplastic expressional system: An applicability for phytoremediation. Environ. Sci. Technol. 39, 7671–7677.

Uragami, Y., Senda, T., Sugimoto, K., Sato, N., Nagarajan, V., Masai, E., Fukuda, M., and Mitsui, Y. (2001). Crystal structures of substrate free and complex forms of reactivated BphC, an extradiol type ring-cleavage dioxygenase. J. Inorg. Biochem. *83*, 269–279.

Utkin, I., Woese, C., and Wiegel, J. (1994). Isolation and characterization of *Desulfitobacterium dehalogenans* gen. nov., sp. nov., an anaerobic bacterium which reductively dechlorinates chlorophenolic compounds. Int. J. Syst. Bacteriol. *44*, 612–619.

Vaillancourt, F.H., Labbe, G., Drouin, N.M., Fortin, P.D., and Eltis, L.D. (2002). The mechanism-based inactivation of 2,3-dihydroxybiphenyl 1,2-dioxygenase by catecholic substrates. J. Biol. Chem. *277*, 2019–2027.

Vaillancourt, F.H., Haro, M.A., Drouin, N.M., Karim, Z., Maaroufi, H., and Eltis, L.D. (2003). Characterization of extradiol dioxygenases from a polychlorinated biphenyl-degrading strain that possess higher specificities for chlorinated metabolites. J. Bacteriol. *185*, 1253–1260.

Van Aken, B. (2008). Transgenic plants for phytoremediation: helping nature to clean up environmental pollution. Trends Biotechnol. *26*, 225–227.

Van Aken, B., Correa, P.A., and Schnoor, J.L. (2010). Phytoremediation of polychlorinated biphenyls:

New trends and promises. Environ. Sci. Technol. *44*, 2767–2776.

Vedadi, M., Barriault, D., Sylvestre, M., and Powlowski, J. (2000). Active site residues of cis-2,3-dihydro-2,3-dihydroxybiphenyl dehydrogenase from *Comamonas testosteroni* strain B-356. Biochemistry *39*, 5028–5034.

Vézina, J., Barriault, D., and Sylvestre, M. (2007). Family shuffling of soil DNA to change the regiospecificity of *Burkholderia xenovorans* LB400 biphenyl dioxygenase. J. Bacteriol. *189*, 779–788.

Vézina, J., Barriault, D., and Sylvestre, M. (2008). Diversity of the C-terminal portion of the biphenyl dioxygenase large subunit. J. Mol. Microbiol. Biotechnol. *15*, 139–151.

Villacieros, M., Power, B., Sanchez-Contreras, M., Lloret, J., Oruezabal, R.I., Martin, M., Fernandez-Pinas, F., Bonilla, I., Whelan, C., Dowling, D.N., *et al.* (2003). Colonization behaviour of *Pseudomonas fluorescens* and *Sinorhizobium meliloti* in the alfalfa (Medicago Sativa) rhizosphere. Plant Soil *251*, 47–54.

Volkering, F., Breure, A.M., and Rulkens, W.H. (1997). Microbiological aspects of surfactant use for biological soil remediation. Biodegradation *8*, 401–417.

Vrana, B., Dercova, K., Balaz, S., and Sevcikova, A. (1996). Effect of chlorobenzoates on the degradation of polychlorinated biphenyls (PCB) by *Pseudomonas stutzeri*. World J. Microb. Biot. *12*, 323–326.

Wang, J.M., Marlowe, E.M., MillerMaier, R.M., and Brusseau, M.L. (1998). Cyclodextrin-enhanced biodegradation of phenanthrene. Environ. Sci. Technol. *32*, 1907–1912.

Wang, R.F., Cao, W.W., Khan, A.A., and Cerniglia, C.E. (2000). Cloning, sequencing, and expression in *Escherichia coli* of a cytochrome P450 gene from *Cunninghamella elegans*. FEMS Microbiol. Lett. *188*, 55–61.

Wiegel, J., and Wu, Q. (2000). Microbial reductive dehalogenation of polychlorinated biphenyls. FEMS Microbiol. Ecol. *32*, 1–15.

Willumsen, P.A., Karlson, U., and Pritchard, P.H. (1998). Response of fluoranthene-degrading bacteria to surfactants. Appl. Microbiol. Biotechnol. *50*, 475–483.

Witzig, R., Junca, H., Hecht, H.J., and Pieper, D.H. (2006). Assessment of toluene/biphenyl dioxygenase gene diversity in benzene-polluted soils: links between benzene biodegradation and genes similar to those encoding isopropylbenzene dioxygenases. Appl. Environ. Microbiol. *72*, 3504–3514.

Wu, Q., Sowers, K.R., and May, H.D. (2000). Establishment of a polychlorinated biphenyl-dechlorinating microbial consortium, specific for doubly flanked chlorines, in a defined, sediment-free medium. Appl. Environ. Microbiol. 66. *66*, 49–5349–5353.

Yadav, J.S., and Reddy, C.A. (1992). Non-involvement of lignin peroxidases and manganese peroxidases in 2,4,5-trichlorophenoxyacetic acid degradation by *Phanerochaete chrysosporium*. Biotechnol. Lett. *14*, 1089–1092.

Yadav, J.S., and Loper, J.C. (2000). Cytochrome P450 oxidoreductase gene and its differentially terminated cDNAs from the white rot fungus *Phanerochaete chrysosporium*. Curr. Genet. *37*, 65–73.

Yadav, J.S., Quensen, J.F., Tiedje, J.M., and Reddy, C.A. (1995). Degradation of polychlorinated biphenyl mixtures (Aroclors 1242, 1254, and 1260) by the white rot fungus *Phanerochaete chrysosporium* as evidenced by congener-specific analysis. Appl. Environ. Microbiol. *61*, 2560–2565.

Yadav, J.S., Soellner, M.B., Loper, J.C., and Mishra, P.K. (2003). Tandem cytochrome P450 monooxygenase genes and splice variants in the white rot fungus *Phanerochaete chrysosporium*: cloning, sequence analysis, and regulation of differential expression. Fungal Genet. Biol. *38*, 10–21.

Zheng, Z.M., and Obbard, J.P. (2001). Effect of non-ionic surfactants on elimination of polycyclic aromatic hydrocarbons (PAHs) in soil-slurry by *Phanerochaete chrysosporium*. J. Chem. Technol. Biotechnol. *76*, 423–429.

Zhuang, X.L., Chen, J., Shim, H., and Bai, Z.H. (2007). New advances in plant growth-promoting rhizobacteria for bioremediation. Environ. Int. *33*, 406–413.

Zielinski, M., Backhaus, S., and Hofer, B. (2002). The principal determinants for the structure of the substrate-binding pocket are located within a central core of a biphenyldioxygenase alpha subunit. Microbiology *148*, 2439–2448.

Zielinski, M., Kahl, S., Hecht, H.J., and Hofer, B. (2003). Pinpointing biphenyl dioxygenase residues that are crucial for substrate interaction. J. Bacteriol. *185*, 6976–6980.

Zielinski, M., Kahl, S., Standfuss-Gabisch, C., Camara, B., Seeger, M., and Hofer, B. (2006). Generation of novel-substrate-accepting biphenyl dioxygenases through segmental random mutagenesis and identification of residues involved in enzyme specificity. Appl. Environ. Microbiol. 72, 2191–2199.

Catabolic Plasmids and Mobile Genetic Elements Involved in the Degradation of Non-metal Xenobiotic Compounds

9

Amalia S. Afendra, Maria Parapouli and Constantin Drainas

Abstract

During the last century, xenobiotic pollutants harmful to environment and health were dramatically increased as a consequence of human activities, such as petroleum industries, agro-industries, household or commercial use. In the polluted areas a large diversity of bacteria with the ability to use these compounds as carbon and/or nitrogen source were developed and proved to be useful for bioremediation applications. Their biodegradation properties depend on genes, which were modified, recombined and improved accordingly over the years helping bacteria adapt to the new harsh xenobiotic environments. Over time, catabolic genes were evolutionary grouped in clusters, established through genetic rearrangements in transmissible regions of mobile genetic elements, and spread by horizontal gene transfer among bacteria of different genera or taxa coexisting in polluted areas. This chapter focuses on plasmids and other mobile genetic elements which carry genes or gene clusters coding for catabolic enzymes involved in the degradation of a number of industrially important xenobiotic pollutants. These include chlorinated and polychlorinated compounds, phthalates, sulfur compounds and some major groups of pesticides. The origin and evolution of these catabolic pathways to different genera is also reviewed.

Introduction

Virgin natural environments untouched from human menace are becoming scarcer in our days. From the time of industrial revolution, and in particular ever since petrol as a fossil fuel became the major source of energy in the developing societies, the concentration of non-metal pollutants in the environment increased exponentially. Ocean and fresh waters, sediments, coasts, banks, soils, especially those in close vicinity to urban and suburban areas, are overloaded with xenobiotics – substances unfamiliar to degradation pathways of higher organisms including humans. In spite of this unfamiliarity, in most cases the lipophilic nature of many of non-metal pollutants renders them highly recalcitrant and extremely invasive in various animal and human tissues where they accumulate *via* the trophic chain.

The toxicity and carcinogenicity of such compounds has been pertinently emphasized elsewhere in this book. What remains to be pointed out however, is the fact that nature, by tinkering genomes for billions of years, has provided the means to a wide variety of living representatives on this planet to accomplish what in our early contemporary times was believed unfeasible: the transformation of hostile xenobiotics to friendly biotics and hence non-toxic compounds. The major role in this beneficial process for humans and animals is played by certain catabolic genes evolved for billions of years under the guidance of natural selection. Only one glance in the pool of these genes is sufficient to reveal their extensive and impressive diversity.

Dispersion and inheritance of such catabolic genes is a result of horizontal transfer literally without limits, within the prokaryotic world in particular. Bacteria have adopted degradation pathways not by generating new enzymes, but mostly by modification, rearrangement and combination of already existing catabolic genes. Taking into consideration that some degradation pathways were already ancient, such as those of alkanes and aromatic compounds (i.e. toluene and naphthalene), a large number and variety of catabolic genes of different pathways and microorganisms have been available for genetic shuffling (Van Der Meer *et al.*, 2001), most of which were evolved by horizontal gene transfer. This combination of genes usually appears as a cluster on a bacterial genome encoding the required degradative enzymes for the whole catabolic pathway of a substrate. This cluster could be subsequently transmitted, as a whole or in part, to other microorganisms that need to catabolize the same or similar compounds. Catabolic gene shuffling in prokaryotic genomes is accomplished by all means of horizontal transfer mechanisms, named transformation, transduction and conjugation. The latter is the most efficient because large parts of DNA carrying clusters of many genes can be co-transferred simultaneously. Horizontal gene transfer occurs through mobile genetic elements, which have the ability to be transferred within the genome of a bacterium and from one bacterium to another, where they are established as self-replicating autonomous plasmids or as integrated elements in the new genome. Mobile genetic elements include transmissible plasmids, transposons, integrons, genomic islands, as well as phages. Recent studies have revealed the existence of new types of xenobiotic catabolic mobile genetic elements, such as catabolic genomic islands, which integrate into the chromosome after transfer. Evidently, such procedures may occur in polluted environments (Springael and Top, 2004).

Therefore, it is not surprising that the last decade a considerable attention has been focused on the study of plasmids and mobile genetic elements involved in the trafficking of genes encoding enzymes for degradation of non-metal pollutants. Early reports on plasmids carrying genes for catabolism of not real xenobiotic compounds present in plants and other food stuffs such as toluene, camphor, salicylic acid and octane appeared in early 1970s (Rheinwald *et al.*, 1973; Chakrabarty, 1972, 1973; Williams and Worsey, 1976). Ever since, a large number of reports on catabolic plasmids and transposons have accumulated and numerous excellent review articles are available, giving detailed descriptions on the biochemistry and genetics of such mobile genetic elements (Nojiri *et al.*, 2009). For instance, Pemberton and Schmidt (2001) review the compatibility types, ecology, transmissibility, stability, interactions and regulation of plasmids encoding catabolic pathways for the degradation of herbicides, PCBs, nitrotoluene, γ-hexachlorocyclohexane, styrene, nylon, trichloroethylene, naphthalene, carbazole and others. The plasmid location of such catabolic genes versus their chromosomal counterparts involved in the degradation of conventional sugars complies with their early evolutionary development (Williams, 2004) as a result of horizontal gene transfer, recombinational genome rearrangements and transposition events (Springael and Top, 2004 and references therein). Therefore, the appearance of transposase genes in the vicinity of plasmid encoded catabolic genes for xenobiotics is not an uncommon occurrence in nature (Williams, 2004).

Catabolic gene clusters have been also localized on linear plasmids harboured in actinobacteria (*Rhodococcus, Terrabacter* and *Arthrobacter nitroguajacolicus*), Gram-positive (*Mycobacterium, Gordonia rubripertincta* and *Nocardioides*), and Gram-negative (*Xanthobacter, Pseudomonas putida* and *Ochrobactrum*), by which they degrade a wide range of compounds, like alkanes and aromatic compounds such as fluorene, dibenzofuran, naphthalene, biphenyls, or alkylbenzenes. The nucleotide sequence of some linear plasmids sequenced so far indicates a high similarity and organization of genes coding for degradative enzymes, implying that horizontal transfer of mobile genetic elements contributed to the spread of such catabolic gene clusters to phylogenetically different bacteria (Fetzner *et al.*, 2007, and references therein).

The aim of this chapter is to highlight plasmids and mobile genetic elements found in a variety of bacterial genera and species, which carry genes or clusters of genes coding for catabolic enzymes involved in the degradation of a number of particular industrial pollutants. Plasmid born catabolism of polycyclic aromatic hydrocarbons has been extensively reviewed recently by others (Van

Der Meer, 2006; Fetzner *et al.*, 2007; Nojiri *et al.*, 2009). Therefore, emphasis is given here to plasmids and mobile genetic elements carrying genes for the catabolism of chlorinated and polychlorinated aliphatic hydrocarbons used as solvents in several industrial applications, agricultural pesticides (such as atrazine, organophosphates, carbamates, phenyl, phenoxy and halocarboxylic pesticides), phthalates used in plastic industry, sulphurinated hydrocarbons of the petrol industry, as well as chlorinated and polychlorinated biphenyls and their intermediates. The origin, evolution and spread of such catabolic mobile genetic elements among different genera are also discussed.

Plasmid encoded catabolism of agro-industrial pollutants

Among agro-industrial pollutants, pesticides are considered the most dangerous and persistent organic chemicals as reported by the Stockholm Convention on Persistent Organic Pollutants (Gilden *et al.*, 2010). Their long lasting use has raised a number of environmental and health concerns, triggering intensive social demands for remediation of all polluted environmental niches, including soil and water. Several indigenous microorganisms have been isolated from polluted sediments and their ability to metabolize pesticides is under study by numerous researchers. The catabolic properties of such microorganisms were obtained from the development of several evolutionary pathways, in which catabolic plasmids play a central role. In this chapter, we highlight the catabolic plasmids involved in the biodegradation of six major groups of pesticides including atrazine, organophosphates, carbamates, phenyl, phenoxy and halocarboxylic pesticides, and we discuss the catabolic genes that they carry, their origin and evolution.

Atrazine

Atrazine (2-chloro-4-ethylamino-6-isopropylamino-1,3,5-*s*-triazine) belongs to *s*-triazine herbicides used for over 50 years against weeds that harm major crops, such as maize. Although atrazine was banned in the European Union in 2004 because of its wide underwater contamination, it still remains the most commonly used herbicide in the world. These facts in addition to its long half-life ranging from 4 to 57 weeks (Mandelbaum *et al.*, 1995, and references therein) pose atrazine as a high risk pollutant.

One of the first reports for complete atrazine degradation appeared in 1995 by Mandelbaum *et al.*, who isolated a *Pseudomonas* sp. strain ADP from a herbicide spill site, that could utilize atrazine as a sole nitrogen source. The catabolic genes responsible for the initial steps of atrazine catabolism named *atzA*, *atzB* and *atzC* were located on a 108 kb self-transmissible plasmid, pADP-1 (De Souza *et al.*, 1998). These genes encode three hydrolases, AtzA, AtzB and AtzC that hydrolyse the C–Cl bond (De Souza *et al.*, 1996), the *N*-ethylamino side chain (Boundy-Mills *et al.*, 1997) and the *N*-isopropyl group (Sadowsky *et al.*, 1998), respectively, liberating cyanuric acid and isopropylamine. In addition, Martinez *et al.* (2001) localized genes *atzD*, *atzE* and *atzF* on pADP-1 encoding enzymes that further mineralize cyanuric acid to carbon dioxide and ammonia. Their highly conserved homologues *atzA*, *atzB* and *atzC* were also found to be plasmid borne in strains of *Pseudaminobacter* sp. (Topp *et al.*, 2000), *Arthobacter* sp. strain TC1 (Sajjaphan *et al.*, 2004) and various strains of *Chelatobacter heintzii* and *Arthrobacter crystallopoietes* (Rousseaux *et al.*, 2002). Nevertheless, the observed correlation of *atz* genes with known transposable DNA elements (IS*1071*) suggests that the wide spread distribution of *atz* genes could be due to their transfer by a combination of transposition and plasmid transfer (Topp *et al.*, 2000; Martinez *et al.*, 2001; Rousseaux *et al.*, 2002; Wackett *et al.*, 2002).

Organophosphates

Organophosphates (OP) are phosphoric or thiophosphoric acid esters. Organophosphates form a wide category of insecticides inhibiting irreversibly acetylcholinesterase, an enzyme essential to nerve function. Parathion, methyl-parathion, malathion, paraoxon, fenitrothion and diazinon are included in this group. They are highly toxic and can be absorbed via inhalation, ingestion and dermal adsorption causing profound health damages, because exposure to organophosphates has been associated with an increased risk of serious neurological diseases (Costa, 2006; Hayden *et al.*, 2010) as well as cancer (Webster *et al.*, 2002).

Various parathion (*O,O* diethyl *O-p*-nitrophenol phosphorothioate) hydrolysing bacteria have been isolated. They were all found to convert parathion to diethylthiophosphoric acid and *p*-nitrophenol (Munnecke *et al.*, 1976) using a hydrolase encoded by the *opd* (for organophosphate degradation) gene carried on large plasmids. In specific, in 1974 Munnecke *et al.* isolated a parathion hydrolysing microorganism in Texas, which was later characterized as *Pseudomonas diminuta* strain MG by Serdar *et al.* (1982). In this strain the parathion hydrolase gene was carried on a 70 kb plasmid, named pCS1 (Serdar *et al.*, 1982). Additionally, Sethunathan *et al.* (1973) isolated a *Flavobacterium* sp. strain ATCC 27551 from a Philippine rice field that degraded diazinon and parathion. Here again the *opd* gene was localized on a plasmid named pPDL2 (39 kb) (Mulbry *et al.*, 1986). Furthermore, Somara *et al.* (1995) isolated a *Flavobacterium balustinum* strain from the agricultural soils of Andhra Pradesh in Anantapur district, India, in which the *opd* gene was carried on an 86 kb plasmid. Hybridization experiments revealed that although the isolated plasmids from the three completely separated geographical sites (U.S.A., Philippines and India) seemed unrelated, the *opd* genes were located on highly conserved regions exhibiting sequence similarity over 98%, a fact strongly implicating that the *opd* genes may be carried on a transposable element (Mulbry *et al.*, 1987). This hypothesis was further supported by the linkage of *opd* to IS elements and transposase genes (Siddavattam *et al.*, 2003). Concerning methyl-parathion (*O,O*-dimethyl *O-p*-nitrophenol phosphorothioate), the *Pseudomonas* sp. strain WBC-3 isolated in China was found to possess a methyl-parathion hydrolase encoded by a *mph* gene carried on the large plasmid pZWL0 (70 kb) (Liu *et al.*, 2005). The observation that *mph* gene is flanked by two copies of the insertion sequence IS*6100*, which is typical for class I transposons, suggested that *mph* gene could be located on a putative catabolic transposon. Nevertheless, the sequence comparison of the MPH with the OPDs of *P. diminuta* strain MG and *Flavobacterium* sp. strain ATCC 27551 presented only 12% similarity at the amino acid level suggesting that they are not evolutionary related (Liu *et al.*, 2005). In regard to fenitrothion (*O,O*-dimethyl *O*-p-nitro-*m*-tolyl phosphorothioate) catabolism, a *Burkholderia* sp. strain NF100 was isolated from soil that had been exposed to fenitrothion for at least two years (Hayatsu *et al.*, 2000). Strain NF100 has a fenitrothion degradation pathway encoded on two plasmids, pNF1 and pNF2 (105 and 33 kb, respectively). Strain NF100 was also found to hydrolyse methyl parathion, parathion, and ENP (*o*-ethyl *O-p*-nitrophenyl phenylphosphonothioate) to 4-nitrophenol, which is further metabolized as a carbon source. However, in this strain information regarding sequence analysis of *opd* genes is not yet available. Thus, much work still needs to be done in order to elucidate the possible divergent evolution of organophosphate hydrolases.

Carbamates

Carbamates are esters of *N*-methyl carbamic acid. The carbamate insecticides form a group of pesticides, which also kill insects by inactivating the enzyme acetylcholinesterase, but, unlike organophosphates, in a reversible manner. The application of carbamates as insecticides began in the 1950s, and approximately 25 kinds of them are in use today. Other representatives of this category include carbofuran (2,3-dihydro-2,2-dimethyl-7-benzofuranyl methylcarbamate), carbaryl (1-naphthyl methylcarbamate) and methomyl [(*E,Z*)-methyl *N*-({methylamino} carbonyl)oxy ethanimidothioate]. Carbofuran is one of the most toxic carbamate pesticides (Feng *et al.*, 1997 and references there in) and therefore it was banned in Canada and European Union. Carbaryl is also toxic (100 mg/kg to 650 mg/kg for mice) and therefore it was also banned in several countries. On the other hand, methomyl was classified by the Environmental Protection Agency as highly toxic to humans and it was characterized as a restricted-use pesticide.

In 1986, Karns *et al.* reported the isolation and characterization of a carbofuran degrading soil isolate *Achromobacter* sp. strain WM111 that could hydrolyse the N-methylcarbamate linkage of carbofuran liberating carbofuran-7-phenol and methylamine, which was further used as a sole nitrogen source. The *mcd* gene (for methylcarbamate degradation) encoding this N-methylcarbamate hydrolase (carbofuran hydrolase) was located on the large plasmid pPDL11 (>100 kb) of *Achromobacter* sp. strain WM111 (Tomasek *et al.*, 1989). The degradation ability of carbofuran, but not of the liberated intermediate carbofuran-7-phenol, was also reported for an unusual type II methylotroph bacterial strain ER2, containing plasmids pER2a and pER2b (120 kb and 130 kb, respectively) (Topp *et al.*, 1993). Hybridization experiments using as probe the *mcd* sequence of *Achromobacter* sp. strain WM111 revealed that its carbofuran hydrolase gene was located on plasmid pER2a. In 1992, a preliminary study on the isolation of a bacterial strain MS2d that could completely degrade carbofuran was reported (Head *et al.*, 1992). This strain, which was not classified in any known genus, harboured seven plasmids (pIH1 to pIH7). A carbofuran hydrolase gene was located on pIH4 (79 kb), while pIH3 (>199 kb) carried genes encoding enzymes responsible for further degradation of carbofuran-7-phenol to carbon dioxide.

Three bacterial strains (named CF06, TA and CD) of *Sphingomonas* sp. were isolated from carbofuran contaminated soils in Washington and Florida (Feng *et al.*, 1997; Ogram *et al.*, 2000). Strains TA and CD were phylogenetically close, whereas CF06 was found to belong to a different species of *Sphingomonas*. The ability of these strains to degrade the carbonyl group, as well as the aromatic ring of carbofuran, was proven to be plasmid borne. Strain CF06 harboured five plasmids (pCF01- pCF05) ranging in size from 5.5 kb to over 200 kb, while strain TA harbours one large plasmid pCTOO1 (100 kb) and strain CD harbours four plasmids (pCD1- pCD4). Hybridization experiments demonstrated that these three systems contain large regions of high similarity, as well as a putative IS element (designated IS*1412*), supporting that in regard to carbofuran catabolism these strains probably share the same ancestor (Ogram *et al.*, 2000).

Regarding plasmid mediated carbaryl catabolism the first report appeared in 1999 by Hayatsu *et al.* (1999), who isolated an *Arthrobacter* sp. RC100 from a carbaryl-treated soil. Strain RC100 contained three plasmids named pRC1, pRC2 and pRC3 of 110, 120 and 130 kb, respectively. As it is proposed, the intermediates of carbaryl metabolism in strain RC100 are 1-naphthol, salicylaldehyde, salicylate, and gentisate. The initial hydrolysis of the N-methylcarbamate linkage was catalysed by an N-methylcarbamate hydrolase encoded by a gene located on pRC1, whereas all the other steps were catalysed by enzymes encoded by pRC2. Hashimoto *et al.* (2002) isolated a *Rhizobium* sp. AC100 capable of converting carbaryl to 1-naphthol. The carbaryl hydrolase was observed to be encoded by the *cehA* gene located on the 25 kb plasmid pAC200. *cehA* was not homologous to *mcd* suggesting that these genes were evolved independently. In regard to methomyl, plasmid mediated degradation was observed for a *Stenotrophomonas maltophilia* strain M1, in which the catabolic gene/s were localized on plasmid PMb, but no further information concerning gene sequences is available (Mohamed *et al.*, 2009).

Therefore, various plasmids harboured in several bacterial strains, geographically and phylogenetically distinct, have been detected to be involved in carbamate degradation. Regarding their carbamate hydrolase gene sequences, different levels of homology to *mcd* gene from *Achromobacter* sp. strain WM111 have been observed, posing challenging questions about this world diversity of carbamate hydrolases (Topp *et al.*, 1993; Parekh *et al.*, 1995, 1997; Ogram *et al.*, 2000).

Phenyl urea (diuron)

Diuron (3-(3,4-dichlorophenyl)–1,1-dimethylurea; DCMU) is a widely used phenyl urea herbicide that inhibits photosynthesis. Degradation of diuron by *Arthrobacter globiformis* strain D47 was proposed to take place through hydrolysis of the urea- carbonyl bond, as accumulation of 3,4-dichloroaniline (Cullington and Walker, 1999) was observed. Strain D47 contained two plasmids, pHRIM620 and pHRIM621, of 47 kb and 34 kb, respectively (Turnbull *et al.*, 2001). Degradation of diuron was found to be plasmid mediated in D47 (Turnbull *et al.*, 2001) and a *puh* (for phenyl urea degradation) gene was located on plasmid pHRIM620.

Phenoxy herbicides

In spite a number of reports link it with human cancer (Zahm *et al.*, 1990; Ibrahim *et al.*, 1991), 2,4 dichlorophenoxyacetic acid (2,4-D) is the most widely used herbicide in the world since the 1940s. Fortunately, it has a short half life in the environment because of the presence of a high number of 2,4-D degrading bacteria belonging to different phylogenetic groups (Fisher *et al.*, 1978). 4-chloro-2-methylphenoxyacetic acid (MCPA), on the other hand, is a non-toxic phenoxy herbicide for which it has recently been demonstrated that it can form complexes with metal ions increasing their bioavailability (Kobylecka *et al.*, 2003).

The first report on plasmid-mediated 2,4-D and MCPA catabolism was by Fisher *et al.* (1978) who observed that the ability of *Alcaligenes paradoxus* strain JMP116 to degrade 2,4-D to 2,4-dichlophenol is due to plasmid pJP1. Additional information on 2,4-D and MCPA plasmid encoded biodegradation was provided by Don and Pemberton (1981) who isolated six catabolic plasmids (pJP2, pJP3, pJP4, pJP5, pJP7, pJP9) harboured in strains characterized as *Alcaligenes paradoxus* and *Alcaligenes eutrophus*. The 2,4-D degradation genes were located on the IncP-1 88-kb plasmid pJP4 of *Wautersia eutropha* (formely *Ralstonia eutropha* or *A. eutropus*) strain JMP134 isolated in Australia, which was found to carry six genes involved in 2,4-D and MCPA degradation (*tfdA-tfdF*). The initial degradation step was catalysed by an α-ketoglutarate dioxygenase encoded by gene *tfdA* liberating 2,4-dichlophenol (Streber *et al.*, 1987; Fukumori and Hausinger, 1993). These intermediates can be degraded by two chlorophenol hydroxylases encoded by the $tfdB_{I}$ and $tfdB_{II}$ genes to produce the respective chlorocatechols. Both TfdB enzymes are needed for growth on 2,4-D, but $TfdB_{I}$ contributes to a significantly higher extent than $TfdB_{II}$ (Ledger *et al.*, 2006). *tfdA* and *tfdB* genes are a characteristic example of recruitment for the evolution of a degradation route. Chlorocatechol is subsequently catabolized to β-ketoadipate by the modified *ortho*-pathway encoded by the isofunctional *tfdCDEF* and $tfdD_{II}C_{II}E_{II}F_{II}$ operons, both essential for the efficient degradation of these compounds by JMP134 (Perkins *et al.*, 1990; Top *et al.*, 1996; Laemmli *et al.*, 2000; Plumeier *et al.*, 2002).

Two other plasmids containing *tfd* genes are the 77 kb pEST4011 of *Achromobacter xylosoxidans* EST4002 (Vedler *et al.*, 2004) and the 99.5 kb pIJB1 of *Burkholderia cepacia* 2a (Xia *et al.*, 1998; Poh *et al.*, 2002), which belong to incompatibility group IncP-1. These plasmids were sequenced and their *tfd* genes were found homologous to those of pJP4, but differently organized (Vedler, 2009). Other plasmid borne 2,4-D catabolic genes exhibiting high similarity with *tfd* located on pJP4 were detected on plasmids pRC10 (45 kb) of *Flavobacterium* sp. strain 50001 (Chaudhry and Huang, 1988) and pMABl (90 kb) of *Pseudomonas cepacia* strain CSV90 (Bhat *et al.*, 1994). Nevertheless, numerous studies are available reporting no or very low homology between *tfd* genes of pJP4 and other plasmid-borne 2,4-D-degrading genes (Fulthorpe *et al.*, 1995; Vallaeys *et al.*, 1998, 1999). These observations, in addition to the detection of IS elements flanking catabolic genes in pJP4 and pEST4011 (Leveau and van der Meer, 1997; Vedler *et al.*, 2004), suggest that these genes or gene cassettes of divergent origin could have been disseminated by horizontal gene transfer.

Alcaligenes paradoxus 2811P, isolated from agricultural soil, was also found to possess a 42.9-kb self-transmissible plasmid, pKA2, which contained the genetic information for the degradation of 2,4-D. In the derivative strain 2811C, isolated from a stock culture, the entire pKA2 plasmid was integrated into the host chromosome as observed by experimental procedures. Another strain in the same field plot but isolated at different time, *Pseudomonas pickettii* 712, found to contain the 40.9-kb plasmid pKA4, almost identical to pKA2. The high genetic homology between these two plasmids indicates that they exist in two different genera as a result of lateral gene transfer (Ka and Tiedje, 1994).

Halocarboxylic herbicides

The soil bacterium *Pseudomonas putida* PP3 was isolated after growth in continuous culture in presence of the herbicide Dalapon (2,2-dichloropropionic acid). Responsible for the degradation of Dalapon was the transposon *DEH* (9.74 kb), which contained a dehalogenase gene and was bordered by two copies of the IS element IS*Ppu12* that has also been found flanking other catabolic genes (Wyndham *et al.*, 1994; Weightman *et al.*, 2001). The existence of such genes on transposons able to perform inter- and intracellular transfer and establishment in the new hosts indicate the flexibility of these organisms to adopt new gene clusters for their survival in the polluted areas.

On the other hand, halocarboxylic acid degradation in general has been observed in *Moraxella* sp. strain B isolated from industrial wastewater in Japan. Strain B contains the 65-kb size plasmid pUO1, which harbours two genes encoding the haloacetate dehalogenases H-1 and H-2. These enzymes possess different activities by means of halogen specificity; H-1 acts better on monofluoroacetate, while H-2 on monochloro-, monobromo-, and monoiodoacetate. The genes for H-1 and H-2, designated *dehHl* and *dehH2*, respectively, showed no homology to each other. The deduced amino acid sequence of the *dehHl* gene showed significant similarity to three hydrolases of *Pseudomonas putida* and a haloalkane dehalogenase of *Xanthobacter autotrophicus* (Kawasaki *et al.*, 1992). The *dehH2* coding region is flanked by two copies of an IS*1071*-like element, which by a specific mechanism of transposition have been integrated into plasmid pUO1 (Wyndham *et al.*, 1994).

Plasmid-born catabolism of phthalates used in plastic industry

Phthalates or phthalate esters are the dialkyl or alkyl aryl esters of phthalic acid (1,2-benzenedicarboxylic acid). They are mainly used as plasticizers, increasing flexibility and durability of plastics, and therefore are involved in the production of a large variety of products used for pharmaceutical, medical, electronic and household applications. They contribute 10–60% of plastic products by weight (Rudel and Perovich, 2008), a quantity corresponding to over 400,000 tones every year. Phthalates are not chemically bound to plastics and thus they can easily accelerate into the environment, a fact that causes serious health concerns owing to their hepatotoxic, teratogenic, and carcinogenic characteristics (Liang *et al.*, 2008, and references therein). Regarding phthalates biodegradation, it has been found that a wide range of microorganisms possess the ability to transform phthalate and its derived esters (Staples *et al.*, 1997, and references therein; Liang *et al.*, 2008, and references therein). The biodegradation of phthalates begins with their de-esterification to phthalate, which is further metabolized to protocatechuate and is ultimately mineralized via an *ortho-* or *meta-* cleavage pathway (Liang *et al.*, 2008, and references there in).

In regard to plasmid born phthalate catabolism, Nomura *et al.* (1989) isolated a phthalate degrading *Pseudomonas putida* strain NMH102–2 from activated sludge of a plastic factory. This strain was found to harbour a large catabolic plasmid named PHT (200 kb) and to degrade phthalate to protocatechuate (PCA) via 4,5-dihydroxyphthalate (DHP) (Nomura *et al.*, 1990). Additionally, in *P. putida* strain NMH102–2 the enzymes involved in the initial steps of phthalate degradation were encoded by five genes (*pht1-pht5*) carried on plasmid PHT (Nomura *et al.*, 1992). Another catabolic plasmid reported to be involved in phthalate degradation was pMOP (226–232 kb) of *Burkholderia cepacia* (formerly *Pseudomonas cepacia*) strain Pc701, isolated from Florida soil, capable of degrading 4-methylphthalate, 4-hydroxyisophthalate and phthalate (Saint and Ribbons, 1990). A 4-methylphthalate-specific permease gene involved in the transportation of both 4-methylphthalate and phthalate in the cell, as well as a 4-methylphthalate reductase gene, were located on pMOP (Saint and Romas, 1996). Another Pseudomonad in which phthalate degradation involves a large catabolic plasmid (140 kb) was *P. fluorescens* strain SKP3 which was isolated from a petroleum-polluted soil sample in Mandras, India (Karpagam and Lalithakumar, 1999). In addition, it was demonstrated that in *Comamonas acidovorans* strain UCC61, which was isolated from sewage sludge, the catabolic pathway of phthalates is mediated by plasmid pOPH1 (270 kb) (Dutton *et al.*, 1995). Strain UCC61 was able to catabolize *o*-phthalate via the 4,5-dihydroxylation pathway and pOPH1 was suggested to carry the genes for oxygenation, dehydrogenation and decarboxylation. Furthermore, the catabolic region of pOPH1 was found to be flanked by repeated sequences (Dutton *et al.*, 1995), but until today there is no evidence that pOPH1 possesses any transposition activity. Another report on plasmid mediated phthalate catabolism was for a 60kb plasmid in *Moraxella* sp. strain VG45 isolated from an oil-field soil sample that also uses the 4,5-dihydroxylation pathway (Rani *et al.*, 1996). The plasmid cured strain did not express any protocatechuate 4,5-dioxygenase, phthalate 4,5-dioxygenase, or 4,5-dihydro-4,5-dihydroxyphthalate dihydrogenase activities, a fact clearly demonstrating the involvement of the 60kb plasmid in phthalate degradation in strain VG45 (Rani *et al.*, 1996). In contrast to the Gram-negative bacteria mentioned, the Gram-positive *Arthrobacter keyseri* (formely *Micrococcus* sp.) strain 12B catabolizes phthalate through cis-3,4-dihydro-3,4 dihydroxyphthalate and 3,4-dihydroxyphthalate (Eaton and Ribbons, 1982). Strain 12B harbours a 130kb plasmid, named pRE1, on which two putative genes *pehA* and *tnpR* were localized encoding a possible phthalate ester hydrolase and a transposon resolvase, respectively, as well as three operons. The *pht* and *pcm* operons encode the conversion of phthalate to protocatechuate and further to pyruvate and oxaloacetate, while the putative *ptr* operon encodes a possible phthalate, protocatechuate, or phthalate ester transporter (Eaton, 2001). However, the presence of a resolvase on pRE1 suggests that its catabolic region is associated with a transposable element (Eaton, 2001). Another report regarding dimethyl-phthalate (DMP) degradation concerns a *Bacillus* sp. which harbours a 60kb plasmid (Niazi *et al.*, 2001). This plasmid was found to encode four iso-esterases mediating the initial de-esterification step of DMP (Niazi *et al.*, 2001).

Plasmid-born degradation of sulphurinated hydrocarbons

Sulphur is the third most abundant element in petroleum since it is contained in quantities ranging from 0.05 to 5% w/w in crude oil or even 14% w/w regarding heavier oil (Van Hamme *et al.*, 2003). The physical/chemical methods for crude oil desulphurization are of high cost and thus biodesulphurization has been of major interest for several research teams worldwide. Here we present an overview of plasmid encoded microbial biodesulphurization pathways focused on the catabolic plasmids and bacterial host strains involved.

In 1985, Monticello *et al.* described two strains of *Pseudomonas* sp., a *Pseudomonas alcaligenes* (isolate DBT2) and *Pseudomonas putida* (isolate DBT4) that could accomplish plasmid mediated sulphur-specific metabolism of dibenzothiophene (DBT) to 3-hydroxy-2-formyl-benzothiophene (HFBT) via the Kodama pathway (Kodama *et al.*, 1973), a mechanism that involves a cleavage of a carbon–carbon bond (Ohshiro and Izumi, 1999; Mohebali and Ball, 2008). The genes for this pathway were proposed by Denome *et al.* (1993) in a soil isolate, *Pseudomonas* sp. strain C18. In this strain, the genes involved in the DBT degradation named *dox* (for DBT oxidation) were found to be carried on a plasmid (75kb) and exhibited sequence homology with other genes involved in the degradation of other aromatic compounds in *P. putida*. In strain C18, *dox* genes were also found responsible for naphthalene and phenanthrene biotransformation (Denome *et al.*, 1993). In *Rhodococcus erythropolis* strain IGTS8, DBT degradation does not occur by destruction of the carbon ring structure; it involves the cleavage of a carbon–sulphur bond, producing 2-hydroxybiphenyl (HBP) (4S pathway) (Ohshiro and Izumi, 1999; Mohebali and Ball, 2008). The genes involved, named *dszA, B* and *C*, were grouped on a 4kb locus of a large linear plasmid (120kb) (Denome *et al.*, 1994) comprising the *dsz* operon. This operon located near insertion sequence elements (IS*1166* and IS*1295*) (Denome and Young, 1995) was found to be plasmid-borne and highly conserved in other *Rhodococci* strains (Denis-Larose *et al.*, 1997; Prince and Grossman, 2003), as well as in *Arthrobacter* sp. strain DS7 (Serbolisca *et al.*, 1999) and other soil aerobic bacterial consortia (Duarte *et al.*, 2001), strongly supporting the idea that the distribution of the *dsz* operon in nature is due to horizontal gene transfer (Mohebali and Ball, 2008).

Plasmid-mediated catabolism of chlorinated and polychlorinated pollutants

Chlorinated compounds are used extensively in human activities such as agriculture and industry and therefore are considered as serious environmental pollutants and dangerous for health. On the other hand, during the last decades it was discovered that many of the chloroorganic pollutants were produced naturally for millions of years, especially by marine organisms, such as seaweeds, sponges, algae etc. (Hay and Fenical, 1996; Gribble, 1998). These organisms produce a variety of halogenated compounds as secondary metabolites, the role of which for the physiology of their producers is largely not understood. Tetrachloroethene and trichloroethene, halogenated propanes, chloromethane, chloroform, chlorophenols, polychlorinated dibenzodioxins, are only a few examples of the variety of compounds that are biosynthesized (Löffler *et al.*, 2003, and references therein).

The chlorinated compounds produced biologically by representatives of all kingdoms and living domains serve to other organisms, mainly bacteria, as nutrient and energy sources. Therefore, although the biological production of chlorinated compounds is rather high, their concentration in the environment is balanced by the action of microorganisms that degrade and maintain them to very low levels. On the other hand, many bacteria living in niches polluted by chlorinated compounds, have gained over the years the ability to utilize them as nutrition and energy sources for their growth and survival under the new environmental conditions. These bacteria have adopted the degradation properties mainly by horizontal gene transfer. It is therefore reasonable that many of the gene clusters responsible for the degradation of these compounds are located on plasmids or other mobile genetic elements, well conserved and spread over various genera.

Different mechanisms have been developed in bacteria for the removal of chlorine from chlorinated pollutants. Their metabolism involves oxidation or reduction reactions of chlorinated hydrocarbons by using them as electron donors or acceptors. The oxygenolytic removal of chloride(s) takes place only under aerobic conditions and is catalysed by dioxygenases or monooxygenases. In hydrolytic dechlorination, which is observed in aliphatic hydrocarbons and aromatic compounds, chlorines are replaced by hydroxyl groups from water. In reductive dechlorination, a chlorine substituent is replaced by hydrogen. Many highly chlorinated environmental pollutants that are resistant to oxygenolytic and hydrolytic dechlorination can be reductively dechlorinated. Hydrolytic and reductive dechlorination can occur under both aerobic and anaerobic conditions. The products of all dechlorination reactions are chloride and protons (Löffler *et al.*, 2003).

Chlorinated and polychlorinated aliphatic hydrocarbons

Chloro-/polychloroalkanes and alkenes (especially chlorinated ethenes), including several known or putative carcinogens, are solvents widely used in industries, where they are disposed in ground water after their use. As water pollutants they are distributed and located at many sites. A large number of microorganisms have been isolated in these polluted areas.

A plasmid involved in the biodegradation of sort-chain chlorinated aliphatic compounds is the 200-kb pXAU1 of the nitrogen-fixing hydrogen bacterium *Xanthobacter autotrophicus* GJ10, which can use 1,2-dichloroethane (DCE) as carbon source. This is a case of hydrolytic dechlorination, in which a haloalkane dehalogenase transforms DCE to 2-chloroethanol, which in turn is oxidized to 2-chloroacetaldehyde by a pyrroloquinoline quinone-dependent alcohol dehydrogenase. The final product is glycolic acid, which enters the central metabolic pathways (Janssen *et al.*, 1985). Plasmid pXAU1 contains at least two genes, *dhlA* and *ald*, coding for haloalkane dehalogenase and chloroethanol dehydrogenase, respectively (Tardif *et al.*, 1991).

Another plasmid encoded hydrolytical dechlorination mechanism is present in *Rhodococcus rhodochrous* NCIMB 13064, isolated in UK from a soil sample of an industrial site long exposed to chlorinated alkanes, which uses 1-chloroalkanes with short chains (C_3–C_9) as carbon sources. The cleavage of the carbon–halogen bond in 1-chlorobutane is catalysed by an inducible hydrolytic haloalkane dehalogenase (DhaA) and results in the formation of n-butanol. The *dhaA* gene was part of a 1-chlorobutane catabolic gene cluster on the auto-transmissible 100-kb plasmid pRTL1. This plasmid carries, among other essential genes, an alcohol dehydrogenase gene (*adhA*) and an aldehyde dehydrogenase gene (*aldA*), whose products catalyse the oxidative conversion of *n*-butanol, the hydrolytic product of 1-chlorobutane, to n-butyric acid, a growth substrate for many bacteria (Kulakova *et al.*, 1995; Poelarends *et al.*, 2000).

Similar examples of degradation pathways with a dehalogenase starting step were found in the 1,3-dichloropropene-utilizing bacterium *Pseudomonas pavonaceae* 170 and the 1,2-dibromoethane-degrading *Mycobacterium* sp. strain GP1, which share a highly conserved *dhaA* gene homologue with that of *R. rhodochrous* NCIMB 13064, although they are phylogenetically distinct bacteria. Analysis of sequences flanking the conserved *dhaA* segment revealed the existence of a putative integrase gene, suggesting a possible integration event in order to obtain catabolic pathways as a response to selection pressure opposed by these man-made chemicals (Poelarends *et al.*, 2000).

In the aerobic degradation of ethane and chloroethene (vinyl chloride, VC) by *Mycobacterium* strain JS60 isolated from contaminated groundwater by enrichment on VC as the sole carbon source, a different process was followed involving the epoxyalkane:coenzyme M (CoM) transferase (EaCoMT). The EaCoMT and alkene monooxygenase genes appear to be carried on a large linear plasmid (Coleman and Spain, 2003). The EaCoMT gene was closely related to homologues of *Xanthobacter* strain Py2, a Gram-negative facultative methylotroph, and *Rhodococcus rhodochrous* (*Rhodococcus corallinus; Nocardia corallina*) B-276, a Gram-positive actinomycete (Saeki *et al.*, 1999; Krum and Ensign, 2001). These strains degrade propylene by epoxidation followed by a sequence of three reactions resulting in epoxide ring opening and carboxylation to form acetoacetate. EaCoMT is part of an unusual epoxide carboxylase enzyme complex consisting of EaCoMT, two stereoselective dehydrogenases and an oxidoreductase/carboxylase (Ensign, 2001). In *Xanthobacter* strain Py2 the propene monooxygenase genes, the epoxide carboxylase system and CoM biosynthesis genes are carried on linear megaplasmids. In *Rhodococcus corallinus* B-276, the gene for an alkene monooxygenase, which allows this organism to grow on propene and to co-oxidize trichloroethene (TCE) lies on the 185-kb linear plasmid pNC30 (Saeki *et al.*, 1999).

Nocardioides sp. JS614, which also uses VC and ethene as sole carbon sources in aerobic growth, degrades them through AkMO (alkene monooxygenase) and EaCoMT (epoxyalkane:coenzyme M transferase). The corresponding genes were found in a putative operon that included CoA transferase, acyl-CoA synthetase, dehydrogenase, and reductase genes. A divergently transcribed gene cluster was located closely to the above potential operon that encoded possible coenzyme M biosynthesis enzymes. These clusters lie most probably on the linear 290-kb plasmid pNoc614, as shown by curing experiments (Mattes *et al.*, 2005).

Chlorinated/polychlorinated biphenyls

Chlorinated/polychlorinated biphenyls (CBs/PCBs) are used extensively in various industrial and commercial applications, owing to their stable structure. Aroclor, Delor and Clophen are a few characteristic PCB mixtures. PCBs have been shown to cause cancer and other serious health effects on animals and humans (Pieper, 2005). The ability for degradation of PCBs lies on the number and the position of chlorines in the molecule. Highly chlorinated biphenyls are usually not degraded aerobically. However, dechlorination can take place reductively under anaerobic conditions. It is shown experimentally that PCBs are used as terminal electron acceptors by chloridogenic bacteria (Löffler *et al.*, 2003). Bacteria that degrade biphenyls can also catabolize PCBs to chlorobenzoic acids by using the BP degrading enzymes through an oxidative route by the biphenyl *meta* ring fission pathway, where a 4-chlorobenzoate dehalogenase, a *p*-hydroxybenzoate hydroxylase and a protocatechuate-4,5-dioxygenase are participating, resulting to the complete mineralization of BP/PCB (Furukawa, 2006). Genes encoding enzymes that convert BPs and PCBs to benzoates and chlorobenzoates, designated as *bph*, have been found on both bacterial chromosomes and plasmids. Their sequences are very well conserved, indicating that they have been extensively spread to many different bacterial populations by horizontal gene transfer.

Ralstonia eutropha (formerly *Alcaligenes eutrophus*) A5 was isolated from PCB-contaminated river sediments in USA for its ability to catabolize 4CBP to 4-chlorobenzoate (4CBA) and biphenyl (BP) to CO_2 via benzoate (BA) by the biphenyl *meta* ring fission pathway (Furukawa and Chakrabarty, 1982; Hooper *et al.*, 1989). Strain A5 originally carried a 75-kb plasmid with all the necessary genes for the complete mineralization of BP and CBP, which during cultivation in lab conditions was modified to a 51-kb plasmid pSS50 by loosing its original degradation genes (Layton *et al.*, 1992). Plasmid pSS50 was able to mediate the transfer of 4-chlorobiphenyl catabolic genes to *Alcaligenes eutrophus* recipients after mating experiments (Springael *et al.*, 1993). In the new hosts, pSS50 derivatives of larger size were isolated containing insertions of A5 chromosomal DNA with the genes involved in the degradation of BP and 4CBP to BA and 4CBA, but not carrying the chlorobenzoate dehalogenase gene. This transmissible element with a size of 59 kb, named Tn*4371* (Furukawa, 2006), was able to integrate as a single contiguous DNA fragment in different locations.

A *Chromobacterium* sp. strain LBS1C1, isolated from the same PCB contaminated reservoir as *Alcaligenes* sp. strain A5, was found to contain a plasmid with similarities to pSS50, named pSS60 (Layton *et al.*, 1992), which carried the 4-chlorobenzoate dehalogenase gene (Burlage *et al.*, 1990). Similarly to pSS50, plasmid pSS60 was responsible for the transmission of an element related to Tn*4371* of *Alcaligenes eutrophus* strains. The new hosts converted 4-chlorobiphenyl to *p*-hydroxybenzoate, which was subsequently degraded through the chromosomal encoded biphenyl *meta* ring fission pathway. The *bph* genes carried on an RP4::Tn*4371* plasmid were transferred in a variety of recipients. Among them, growth on biphenyl was observed in some *Alcaligenes* strains, an *Acinetobacter* species, as well as fluorescent *Pseudomonas* species (Springael *et al.*, 1993, 1994). The *bphC* genes sequenced from these new hosts were very similar to each other and to a *bphC* gene recently isolated from a Gram-positive *Arthrobacter* sp. M5 (Peloquin and Greer, 1993). It is concluded therefore that the *bph* genes located on Tn*4371* are highly conserved between different (chloro) biphenyl-degrading hosts, isolated globally but belonging mainly to the β-proteobacteria (Springael *et al.*, 2001). Tn*4371* seemed initially to behave and transferred as a transposon, being first inserted into a broad-host-range transmissible plasmid. However, because Tn*4371* integrated easily into the recipient's chromosome at specific locations and its organization exhibited a mosaic of several gene clusters, it can be considered as a genomic island (Merlin *et al.*, 1999).

Bacteria of the genus *Rhodococcus* are aerobic Gram-positive nocardioform actinomycetes able to catabolize a wide number of xenobiotic compounds including polychlorinated biphenyls (PCBs). *R. erythropolis* TA421, isolated from an ecosystem of wood-feeding termites, is able to catabolize BP as a sole carbon source and can co-metabolize a wide spectrum of PCBs. TA421 contains seven *bphC* genes expressing 2,3-dihydroxybiphenyl dioxygenase activity, three of which are harboured on the 500-bp linear plasmid pTA421 (*bphC2*, *bphC3*, and *bphC4*). All of the seven *bphC* gene products exhibited much higher substrate specificities for 2,3-dihydroxybiphenyl (2,3-DHBP) than for catechol, 3-methylcatechol, or 4-methylcatechol (Maeda *et al.*, 1995), and thus these gene products are primarily involved in the ring cleavage of 2,3-DHBP. According to experimental observations, *Rhodococcus globerulus* P6 harbours the 650-kb pLP6 and 360-kb pSP6 plasmids, which contain *bphC* homologues of TA421. The *bphC2* homologue is located on both plasmids, while the *bphC3* and *bphC4* homologues are on pSP6. The results indicate that these gene homologues were most probably plasmid-borne in *Rhodococcus* (Kosono *et al.*, 1997). *Rhodococcus* sp. RHA1 is a potent polychlorinated-biphenyl (PCB)-degrading strain and contains three linear plasmids of approximate size 1100 to 300 kb, named pRHL1, pRHL2 and pRHL3. It is believed that the gene clusters *bphA1A2A3A4CB* and *bphDEF* responsible for the initial oxidation of the aromatic ring of biphenyl/polychlorinated biphenyls (PCB) are shared between the two largest plasmids (Masai *et al.*, 1997).

Chlorobenzoic acid

Chlorobenzoates are intermediates of the microbial degradation of CBs/PCBs. Many bacteria use chlorobenzoates as the sole carbon source under aerobic conditions. The chlorine substituent is removed in the initial degradation step of the halogenated compound prior to the cleavage of the aromatic ring system, producing 4-hydroxybenzoate, which is then further degraded by standard degradation pathways. The chlorine substituent can also be removed from the aromatic ring system in an oxygenolytic way by the attack of dioxygenase enzyme systems on the chlorine substituted carbon. The resulting chlorohydrodiols are unstable and spontaneously eliminate HCL to form the corresponding catechols. On the other hand, degradation of CBAs involves chlorocatechols as intermediates. CBAs are first converted into chlorodihydrodihydroxybenzoates by benzoate 1,2-dioxygenase (BZDO) and subsequently converted into the corresponding (chloro)catechols by benzoate dihydrodiol dehydrogenase (DHB dehydrogenase) (Wyndham *et al.*, 1994; Löffler *et al.*, 2003).

One of the bacterial genera that possess degradation activities on CBAs is *Alcaligenes*. *Alcaligenes* sp. CPE3, isolated from a PCB-contaminated soil in Italy, was able to degrade 3-chloro-, 4-chloro-, and 3,4-dichlorobenzoates via the protocatechuate pathway, similarly to *Alcaligenes* sp. strain BR60. The *cbaABC* gene cluster responsible for the degradation properties was located on the 85-kb plasmid pCPE3 (Fava *et al.*, 1993; Gioia *et al.*, 1998). Additionally, *Alcaligenes* sp. strain ALP83 was isolated with degradation properties on 4-chlorobenzoate to 4-hydroxybenzoate, which were associated with a 10-kb fragment carried on plasmid pSS70 (Layton *et al.*, 1992). On the other hand, *Alcaligenes* sp. strain BR60 was isolated from the sediments of Bloody Run Creek, which drained leachate from a chemical landfill site in the Niagara region. This strain can utilize 3CBA as its sole carbon source. The genes encoding the catabolic enzymes of this procedure are carried on a conjugative plasmid, pBRC60, and in specific on the 17-kb transposon Tn*5271* (Fulthorpe and Wyndham, 1991). More recently, *Ralstonia eutropha* (former *Alcaligenes eutrophus*) NH9 was isolated in Japan from soil with the ability to utilize 3-chlorobenzoate as a sole carbon source. It contains the 78-kb plasmid pENH91, which carries the *cbnR-ABCD* genes of the modified *ortho*-cleavage pathway. These genes showed highest homology to the *tcbR-CDEF* genes on plasmid pP51 of the 1,2,4-trichlorobenzene-degrading bacterium *Pseudomonas* sp. strain P51. Two homologous sequences, present as direct repeats in both flanking regions of the *cbnR-ABCD* genes on pENH91, were found to be identical insertion sequences (ISs), designated IS*1600*, which formed a composite transposon designated Tn*5707* (Ogawa and Miyashita, 1995, 1999).

Another genus catabolizing CBA among many xenobiotic pollutants is *Arthrobacter*. For instance, *Arthrobacter* sp. SU degrades 4-chlorobenzoate (4-CBA) to *p*-hydroxybenzoate through a hydrolytic dehalogenation, where the dehalogenase genes are carried on the 120-kb plasmid pASU1 (Schmitz *et al.*, 1992). On the other hand, plasmid pBS1501 seems to be involved in the 4CBA degradation by *Arthrobacter globiformis* KZT1 (Rodrigues *et al.*, 2001).

Pseudomonas strains are well known to possess, among other catabolic pathways, also chlorobenzoate degradation. *Pseudomonas putida* PIII degrades a number of mono-, di-, and trichlorinated benzoates. Two individual dioxygenases participate in this pathway; the benzoate dioxygenase, which converts 3-chlorobenzoate (3-CB), 4-CB, and benzoate to the corresponding catechols via reduction of a dihydrodiol, and the chlorobenzoate-1,2-dioxygenase that converts *ortho*-chlorobenzoates to the corresponding catechols without the need of a functional dioldehydrogenase. The latter is encoded on the 75-kb plasmid pPBIII (Brenner *et al.*, 1993). Experimental evidence suggests the involvement of the *clc* operon in these genetic rearrangements.

Pseudomonas sp. A7, A8, A18 and A19, isolated in PBCs-contaminated areas in the Czech

Republic with the ability to completely catabolize chlorobenzoates, were all shown to contain plasmids carrying part of the genes encoding for this metabolic pathway (Pavlů *et al.*, 1999). Two other distinctly located *Pseudomonas* strains, namely *P. putida* AC858 isolated in the United States in the late 1970s and *Pseudomonas* sp. B13 isolated in Germany in the early 1970s, both capable of catabolizing 3-chlorobenzoic acid, harbour plasmids pAC25 and pB13, respectively, which carry genetic information for the CBA degradation and share close similarities, despite of the different origin of these strains (Chatterjee and Chakrabarty, 1983).

Chlorobenzene/*n*-chlorobenzene

Chlorobenzene was once used for production of DDT and industrial manufacture of phenol. Today, it participates in the production of substances such as pesticides, dyes and rubbers. Pseudomonads contributed to the degradation of chlorobenzenes, as described below. *Pseudomonas* sp. P51 is a Rhine river sediment isolate which utilizes 1,2-dichloro-, 1,4-dichloro-, and 1,2,4-trichlorobenzene as sole carbon and energy sources. P51 carries the 110-kb plasmid pP51, which harbours many catabolic gene clusters. Among them, two regions contained the *tcbCDEF* gene cluster encoding chlorocatechol ring-fission enzymes, and the *tcbAB* genes for chlorobenzene dioxygenase and dehydrogenase catalysing the conversion of chlorobenzenes to chlorocatechols, respectively. The latter was located on a transposable element designated Tn*5280* and exhibited size, organization and function characteristics similar to those of other known bacterial dioxygenases, such as *todC1C2BA* from *P. putida* Fl (Van Der Meer *et al.*, 1991). The 5.1 kb *tcbAB* catabolic gene region is flanked by two almost identical IS. The close location of Tn*5280* to the lower pathway regulator *tcbR* and *tcbCDEF* operon implies that these catabolic genes have been closely clustered on the plasmid (Wyndham *et al.*, 1994).

The strain *P. putida* GJ31 can degrade chlorobenzene via a *meta*-cleavage pathway in which chlorocatechol 2,3-dioxygenase (CbzE) is involved. The gene encoding this enzyme is carried on the 180-kb plasmid pKW1 within the *cbzTEXGS* cluster, which corresponds to a partial (chloro)catechol *meta*-cleavage pathway. Other enzymes encoded by this cluster, which is flanked by transposases, include ferredoxin reductase, chlorocatechol 2,3-dioxygenase, an unknown protein, 2-hydroxymuconic semialdehyde dehydrogenase and glutathione S-transferase (Kunze *et al.*, 2009). pKW1 also contains more transfer elements, transposase genes and other catabolic gene clusters, probably contributing to the complete degradation of chlorobenzene. The organization of pKW1 implies the evolutionary rearrangements that took place in order to establish the necessary catabolic activities (Kunze *et al.*, 2009).

Burkholderia sp. PS12, able to grow on 1,2,4,5- tetrachlorobenzene, encodes a composite transposon, found on the 85-kb plasmid pPS12-1, carrying *tecAB* genes for the degradation of CB into chlorocatechol similar to the *tcbAB* genes of strain P51 (Beil *et al.*, 1999).

Another interesting case is *Comamonas* sp. strain CNB-1 isolated among other genera in a microbial community for its ability to degrade 4-chloronitrobenzene (4CNB). 4CNB is a newer xenobiotic compound used for the production of various chemicals such as drugs and pesticides. The co-existence of chlorine and nitro substitutes makes 4CNB particularly resistant to degradation. The *cnb* catabolic genes coding for aminophenol 1,6-dioxygenase, nitroreductase, mutase and deaminase were located, among other gene models, on a 44.5-kb transposon designated Tn*CNB1*. This new transposon, surrounded by two IS*1071* elements, exhibits high gene rearrangements, implying that various alterations have taken place for the establishment of 4CNB catabolism. Tn*CNB1* was located on the 91-kb plasmid pCNB1 which belongs to the IncP1 group and contains many other properties including transfer (Ma *et al.*, 2007).

Chlorocatechols

As already described, chlorocatechols are intermediate products in the degradation pathways of CBs/PCBs and CBAs. Consequently, bacteria using these compounds as a sole carbon source are able to grow on chlorocatechols as well. A mobile genetic element that should be highlighted here is the transferable *clc* element, unique among the so far known bacterial degraders, hosted by *Pseudomonas* sp. B13, a 3-chlorobenzoate metabolizing strain. The genes for chlorocatechol degradation, similar to those carried on pP51, pPS12–1 and pENH91, were located on *clc*, which additionally harbours a phage P4-like integrase. *clc* is one of the first conjugative integrative elements carrying genes for the biotransformation of xenobiotics. Owing to this ability, it can be transmitted on a chromosome or a plasmid in its host's genome, or excised, converting to a self-transmissible plasmid-like form and transfer automonously to a recipient bacterium by conjugation, where it is integrated again in the genome of the new host (Van Der Meer *et al.*, 2001).

Similar *clc* elements have been located in a variety of bacteria. A *clc* gene cluster for 4-chloro-/3,5-dichlorocatechol degradation and a *clc2* gene cluster for 3-chlorocatechol degradation is located in the 740-kb linear mega-plasmid p1CP of *Rhodococcus opacus* 1CP, which utilizes several aromatic or chloroaromatic compounds as sole carbon sources (König *et al.*, 2004). *Burkholderia xenovorans* LB400 contains an element almost 100% identical to *clc* which additionally carries some other catabolic genes (Chain *et al.*, 2006; Gaillard *et al.*, 2006). Moreover, *Burkholderia* sp. PS12 harbours a chlorocatechol degradation gene module on the 85-kb plasmid pPS12-1 similar to that found on pP51, *clc* and pENH91 (Beil *et al.*, 1999). *Ralstonia* sp. strain JS705, isolated from contaminated groundwater in the USA, is able to grow on chlorobenzene and 3-chlorobenzoate. In this strain, a *clc*-like element was found to contain the *clc* genes for chlorocatechol degradation (Van Der Meer *et al.*, 1998).

As already analysed for *Wautersia eutropha* (*Ralstonia eutropha*) JMP134, growth on the herbicide 2,4-dichlorophenoxyacetate mediates the conversion of the intermediate chlorocatechol into 3-oxoadipate through the $tfdC_ID_IE_IF_I$ and $tfdD_{II}C_{II}E_{II}F_{II}$ gene modules of plasmid pJP4 encoding complete sets of functional enzymes which are all expressed for the transformation of 2,4-D.

Concluding remarks

Bacteria capable of degrading organic compounds such as phenols, toluene, octane, biphenyls, naphthalene etc., naturally released as wastes or byproducts of the metabolism of higher organisms, were always present in the environment. The genes responsible for the degradation of these compounds developed very early in the evolutionary time and provided the original pool upon which natural selection worked to create the present outstanding diversity of biochemical pathways for the degradation of virtually any chemical pollutant. Over the years, new bacteria came into light, as a natural response to anthropogenic selective pressures, with abilities to degrade more complicated compounds released in the environment. The dynamic nature of DNA to mutate, to move and to recombine was the motive force for the players of the game to adjust to the new rules opposed by the dominating *Homo sapiens* species. The interplay of diverse genetic mechanisms armed the catalytic capacity of enzymes to tackle the environmental crisis. Mutagenesis always feeds diversity. But living organisms, especially bacteria, exploit also other intrinsic properties of DNA. For instance, clustering of genes essential for catabolic process is often required for better results. These rearrangements would never be so efficient without the contribution of transposons that bring them together. Moreover, these clusters would never be able to distribute to other bacteria without the help of transmissible mobile genetic elements. The catabolic operons studied so far have been developed and established over the years with the contribution of conjugative plasmids and other mobile genetic elements. By the means of horizontal gene transfer, new combinations of degrading operons entered new hosts, creating novel arrangements of gene clusters able to completely mineralize a variety of xenobiotics.

Bacterial communities of different taxa have acquired a wide range of catabolic genes, operons, clusters and genomic islands by the wide and rapid spread of mobile genetic elements, contributing to the adaptation of the whole neighbourhood to feed upon and hence remove the pollutant. This explains the high conservation of gene clusters among different prokaryotes. What we know today is only the tip of the iceberg. It is almost impossible to mention every mobile genetic element that has been discovered and playing a critical role on the metabolism of pollutants.

However, the combination of plasmids and transposons should be recognized accordingly, and a great deal should be awarded to horizontal gene transfer. Indeed, our knowledge concerning the factors that stimulate the genetic transmission by horizontal transfer, especially in polluted areas, is still very limited and needs to be further studied.

As a final remark, given the extensive diversity of catabolic pathways, especially those associated with plasmids and other mobile genetic elements that have been developed through evolutionary processes within an impressively short time, we may conclude that nature has the solution to remediate polluted sites with only one prerequisite: Humans should realize that they must abandon their damaging activities.

References

Bhat, M.A., Tsuda, M., Horike, K., Nozaki, M., Vaidyanathan, C.S., and Nakazawa, T. (1994). Identification and characterization of a new plasmid carrying genes for degradation of 2,4-dichlorophenoxyacetate from *Pseudomonas cepacia* CSV90. Appl. Environ. Microbiol. *60*, 307–312.

Beil, S., Timmis, K., and Pieper, D. (1999). Genetic and biochemical analysis of the *tec* operon suggests a route for evolution of chlorobenzene degradation genes. J. Bacteriol. *181*, 341–346.

Boundy-Mills, K.L., de Souza, M.L., Mandelbaum, R.T., Wackett, L.P., and Sadowsky, M.J. (1997). The *atzB* gene of *Pseudomonas* sp. strain ADP encodes the second enzyme of a novel atrazine degradation pathway. Appl. Environ. Microbiol. *63*, 916–923.

Brenner, V., Hernandez, B.S., and Focht, D.D. (1993). Variation in chlorobenzoate catabolism by *Pseudomonas putida* PIII as a consequence of genetic alterations. Appl. Environ. Microbiol. *59*, 2790–2794.

Burlage, R.S., Bemis, L.A., Layton, A.C., Sayler, G.S., and Latimer, F. (1990). Comparative genetic organization of incompatibility group P degradative plasmids. J. Bacteriol. *172*, 6818–6825.

Chain, P.S., Denef, V.J., Konstantinidis, K.T., Vergez, L.M., Agulló, L., Reyes, V.L., Hauser, L., Córdova, M., Gómez, L., González, M., *et al.* (2006). *Burkholderia xenovorans* LB400 harbors a multireplicon, 9.73-Mbp genome shaped for versatility. Proc. Natl. Acad. Sci. U.S.A. *103*, 15280–15287.

Chakrabarty, A.M. (1972). Genetic basis of the biodegradation of salicylate in *Pseudomonas*. J. Bacteriol. *112*, 815–823.

Chakrabarty, A.M. (1973). Genetic fusion of incompatible plasmids in *Pseudomonas*. Proc. Natl. Acad. Sci. U.S.A. *70*, 1641–1644.

Chatterjee, D.K., and Chakrabarty, A.M. (1983). Genetic homology between independently isolated chlorobenzene-degradative plasmids. J. Bacteriol. *153*, 532–534.

Chaudhry, G.R., and Huang, G.H. (1988). Isolation and characterization of a new plasmid from a *Flavobacterium* sp. which carries the genes for degradation of 2,4-dichlorophenoxyacetate. J. Bacteriol. *170*, 3897–3902.

Coleman, N.V., and Spain, J.C. (2003). Distribution of the coenzyme M pathway of epoxide metabolism among ethene- and vinyl chloride-degrading *Mycobacterium* strains. Appl. Environ. Microbiol. *69*, 6041–6046.

Costa, L.G. (2006). Current issues in organophosphate toxicology. Clin. Chim. Acta *366*, 1–13.

Cullington, J.E., and Walker, A. (1999). Rapid biodegradation of diuron and other phenylurea herbicides by a soil bacterium. Soil Biol. Biochem. *31*, 677–686.

De Souza, M.L., Sadowsky, M.J., and Wackett, L.P. (1996). Atrazine chlorohydrolase from *Pseudomonas* sp. strain ADP: gene sequence, enzyme purification, and protein characterization. J. Bacteriol. *178*, 4894–4900.

De Souza, M.L., Wackett, L.P., and Sadowsky, M.J. (1998). The *atzABC* genes encoding atrazine catabolism are located on a self-transmissible plasmid in *Pseudomonas* sp. strain ADP. Appl. Environ. Microbiol. *64*, 2323–2326

Denis-Larose, C., Labbe, D., Bergeron, H., Jones, A.M., Greer, C.W., al-Hawari, J., Grossman, M.J., Sankey, B.M., and Lau, P.C. (1997). Conservation of plasmid-encoded dibenzothiophene desulphurization genes in several Rhodococci. Appl. Environ. Microbiol. *63*, 2915–2919.

Denome, S.A., and Young, K.D. (1995). Identification and activity of two insertion sequence elements in *Rhodococcus* sp. strain IGTS8. Gene *161*, 33–38.

Denome, S.A., Stanley, D.C., Olson, E.S., and Young, K.D. (1993). Metabolism of dibenzothiophene and naphthalene in *Pseudomonas* strains: complete DNA sequence of an upper naphthalene catabolic pathway. J Bacteriol. *175*, 6890–6901.

Denome, S.A., Oldfield, C., Nash, L.J., and Young, K.D. (1994). Characterization of the desulfurization genes from *Rhodococcus* sp. strain IGTS8. J. Bacteriol. *176*, 6707–6716.

Don, R.H, and Pemberton, J.M. (1981). Properties of six pesticide degradation plasmids isolated from *Alcaligenes eutrophus* and *Alcaligenes paradoxus*. J. Bacteriol. *145*, 681–686.

Duarte, G.F., Rosado, A.S., Seldin, L., De Araujo, W., and Van Elsas, J. D. (2001). Analysis of bacterial community structure in sulfurous oil- containing soils and detection of species carrying dibenzothiophene desulfurization (dsz) genes. Appl. Environ. Microbiol. *67*, 1052–1062.

Dutton, J.R., Venables, W.A., and Saint, C.P. (1995). *Comamonas acidovorans* UCC61 catabolizes *o*-phthalate via a 4,5 oxygenation pathway that is encoded on a 70 kbp section of plasmid pOPHl bounded by directly repeated sequences. Microbiology *141*, 1673–1682.

Eaton, R.W., and Ribbons, D.W. (1982). Metabolism of dibutylphthalate and phthalate by *Micrococcus* sp. strain 12B. J. Bacteriol. *151*, 48–57.

Eaton, R.W. (2001). Plasmid-encoded phthalate catabolic pathway in *Arthrobacter keyseri* 12B. J. Bacteriol. *183*, 3689–3703.

Ensign, S.A. (2001). Microbial metabolism of aliphatic alkenes. Biochemistry *40*, 5845–5853.

Fava, F., Di Gioia, D., Marchetti, L., Quattroni, G., and Marraffa, V. (1993). Aerobic mineralization of chlorobenzoates by a natural polychlorinated biphenyl-degrading mixed bacterial culture. Appl. Microbiol. Biotechnol. *40*, 541–548.

Feng, X., Ou, L.T., and Ogram, A. (1997). Plasmid-mediated mineralization of carbofuran by *Sphingomonas* sp. strain CF06. Appl. Environ. Microbiol. *63*, 1332–1337.

Fetzner, S., Kolkenbrock, S., and Parschat, K. (2007). Catabolic linear plasmids. In Microbial Linear Plasmids, Meinhardt, F., and Klassen, R., eds. (Springer-Verlag: Berlin Heidelberg), pp. 63–98.

Fisher, P.R., Appleton, J., and Pemberton, J.M. (1978). Isolation and characterization of the pesticide-degrading plasmid pJP1 from *Alcaligenes paradoxus*. J. Bacteriol. *135*, 798–804.

Fukumori, F., and Hausinger, R.P. (1993). Purification and characterization of 2,4-dichlorophenoxyacetate/a-ketoglutarate dioxygenase. J. Biol. Chem. *268*, 24311–24317.

Fulthorpe, R.R., and Wyndham, R.C. (1991). Transfer and expression of the catabolic plasmid pBRC60 in wild bacterial recipients in a freshwater ecosystem. Appl. Environ. Microbiol. *57*, 1546–1553.

Fulthorpe, R.R., Mc Gowan, C., Maltseva, O.V., Holben, W.E., and Tiedje, J.M. (1995). 2,4-Dichlorophenoxyacetic acid-degrading bacteria contain mosaics of catabolic genes. Appl. Environ. Microbiol. *61*, 3274–3281.

Furukawa, K. (2006). Oxygenases and dehalogenases: Molecular approaches to efficient degradation of chlorinated environmental pollutants. Biosci. Biotechnol. Biochem. *70*, 2335–2348.

Furukawa, K., and Chakrabarty, A.M. (1982). Involvement of plasmids in total degradation of chlorinated biphenyls. Appl. Environ. Microbiol. *44*, 619–626.

Gaillard, M., Vallaeys. T., Vorhölter, F.J., Minoia, M., Werlen, C., Sentchilo, V., Pühler, A., and van der Meer, J.R (2006). The *clc* element of *Pseudomonas* sp. strain B13, a genomic island with various catabolic properties. J. Bacteriol. *188*, 1999–2013.

Gilden, R.C., Huffling, K., and Sattler, B. (2010). Pesticides and health risks. J. Obstet. Gynecol. Neonatal. Nurs. *39*, 103–110.

Gioia, D.D., Peel, M., Fava, F., and Wyndham, R.C. (1998). Structures of homologous composite transposons carrying *cbaABC* genes from Europe and North America. Appl. Environ. Microbiol. *64*, 1940–1946.

Gribble, G.W. (1998). Chlorinated compounds in the biosphere, natural production. In Encyclopedia of Environmental Analysis and Remediation, Meyers, R.A., ed. (John Wiley & Sons, Inc.: New York), pp. 972–1035.

Hashimoto, M., Fukui, M., Hayano, K., and Hayatsu, M. (2002). Nucleotide sequence and genetic structure of a novel carbaryl hydrolase gene (*cehA*) from *Rhizobium* sp. strain AC100. Appl. Environ. Microbiol. *68*, 1220–1227.

Hay, M.E., and Fenical, W. (1996). Chemical ecology and marine biodiversity: Insights and products from the sea. Oceanography *9*, 10–20.

Hayatsu, M., Hirano, M., and Nagata, T. (1999). Involvement of two plasmids in the degradation of carbaryl by *Arthrobacter* sp. strain RC100. Appl. Environ. Microbiol. *65*, 1015–1019.

Hayatsu, M., Hirano, M., and Tokuda, S. (2000). Involvement of two plasmids in fenitrothion degradation by *Burkholderia* sp. strain NF100. Appl. Environ. Microbiol. *66*, 1737–1740.

Hayden, K., Norton, M., Darcey, D., Ostbye, T., Zandi, P., Breitner, J., Welsh-Bohmer, K., and For the Cache County Study Investigators (2010). Occupational exposure to pesticides increases the risk of incident AD: The Cache County study. Neurology *74*, 1524–1530.

Head, I.M., Cain, R.B., and Suett, D.L. (1992). Characterization of a carbofuran-degrading bacterium and investigation of the role of plasmids in catabolism of the insecticide carbofuran. Arch. Microbiol. *158*, 302–308.

Hooper, S.W., Dockendorff, T.C., and Sayler, G.S. (1989). Characteristics and restriction analysis of the 4-chlorobiphenyl catabolic plasmid, pSS5O. Appl. Environ. Microbiol. *55*,1286–1288.

Ibrahim, M.A., Bond, G.G., Burke, T.A., Cole, P., Dost, F.N., Enterline, P.E., Gough, M., Greenberg, R.S., Halperin, W.E., McConnell, E., *et al.* (1991). Weight of the evidence on the human carcinogenicity of 2,4-D. Environ. Health Perspect. *96*, 213–222.

Janssen, D.B., Scheper, A., Dijkhuizen, L., and Witholt, B. (1985). Degradation of halogenated aliphatic compounds by *Xanthobacter autotrophicus* GJ10. Appl. Environ. Microbiol. *49*, 673–677.

Ka, J.O., and Tiedje, J.M. (1994). Integration and excision of a 2,4-dichlorophenoxyacetic acid-degradative plasmid in *Alcaligenes paradoxus* and evidence of its natural intergeneric transfer. J. Bacteriol. *176*, 5284–5289.

Karns, J.S., Mulbry, W.W., Nelson, J.O., and Kearney, P.C. (1986). Metabolism of carbofuran by a pure bacterial culture. Pestic. Biochem. Physiol. *25*, 211–217.

Karpagam, S., and Lalithakumar, D. (1999). Plasmid-mediated degradation of *o*- and *p*-phthalate by *Pseudomonas* fluorescens. W. J. Microbiol. Biotechnol. *15*, 565–569.

Kawasaki, H., Tsuda, K., Matsushita, I., and Tonomura, K. (1992). Lack of homology between two haloacetate dehalogenase genes encoded on a plasmid from *Moraxella* sp. strain B. J. Gen. Microbiol. *138*, 1317–1323.

Kobylecka, J., Ptaszynski, B., Rogaczewski, R., and Turek, A. (2003). Phenoxyalkanoic acid complexes. Part I. Complexes of lead (II), cadmium (II) and copper (II) with 4-chloro-2-methylphenoxyacetic acid (MCPA). Thermochim. Acta *407*, 25–31.

Kodama, K., Umehara, K., Shimizu, K., Makatani, S., Minoda, Y., and Yamada K. (1973). Identification of microbial products for dibenzothiophene and its proposed oxidation pathway. Agric. Biol. Chem. *37*, 45–50.

König, C., Eulberg, D., Gröning, J., Lakner, S., Seibert, V., Kaschabek, S.R., and Schlömann, M. (2004). A linear megaplasmid, p1CP, carrying the genes for chlorocatechol catabolism of *Rhodococcus opacus* 1CP. Microbiol. *150*, 3075–3087.

Kosono, S., Maeda, M., Fuji, F., Arai, H., and Kudo, T. (1997). Three of the seven *bphC* genes of *Rhodococcus erythropolis* TA421, isolated from a termite ecosystem, are located on an indigenous plasmid associated with biphenyl degradation. Appl. Environ. Microbiol. *63*, 3282–3285.

Krum, J.G., and Ensign, S.A. (2001). Evidence that a linear megaplasmid encodes enzymes of aliphatic alkene and epoxide metabolism and coenzyme M (2-mercaptoethanesulfonate) biosynthesis in *Xanthobacter* strain Py2. J. Bacteriol. *183*, 2172–2177.

Kulakova, A.N., Stafford, T.M., Larkin, M.J., and Kulakov, L.A. (1995). Plasmid pRTL1 controlling 1 chloroalkane degradation by *Rhodococcus rhodochrous* NCIMB13064. Plasmid 33, 208–217.

Kunze, M., Zerlin, K.F., Retzlaff, A., Pohl, J.O., Schmidt, E., Janssen, D.B., Vilchez-Vargas, R., Pieper, D.H., and Reineke, W. (2009). Degradation of chloroaromatics by *Pseudomonas putida* GJ31: assembled route for chlorobenzene degradation encoded by clusters on plasmid pKW1 and the chromosome. Microbiology *155*, 4069–4083.

Laemmli, C.M., Leveau, J.H.J., Zehnder, A.J.B., and Van Der Meer, J.R. (2000). Characterization of a second *tfd* gene cluster for chlorophenol and chlorocatechol metabolism on plasmid pJP4 in *Ralstonia eutropha* JMP134(pJP4). J. Bacteriol. *182*, 4165–4172.

Layton, A.C., Sanseverino, J., Wallace, W., Corcoran, C., and Sayler, G.S. (1992). Evidence for 4- chlorobenzoic acid dehalogenation mediated by plasmids related to pSS50. Appl. Environ. Microbiol. *58*, 399–402.

Ledger, T., Pieper, D.H., and González, B. (2006). Chlorophenol hydroxylases encoded by plasmid pJP4 differentially contribute to chlorophenoxyacetic acid degradation. Appl. Environ. Microbiol. 72, 2783–2792.

Leveau, J.H.J., and van der Meer, J.R. (1997). Genetic characterization of insertion sequence ISJP4 on plasmid pJP4 from *Ralstonia eutropha* JMP134. Gene *202*, 103–114.

Liang, D-W., Zhang, T., Fang, H.H.P., and He, J. (2008). Phthalates biodegradation in the environment. Appl. Microbiol. Biotechnol, *80*, 183–198.

Liu, H., Zhang, J.J., Wang, S.J., Zhang, X.E., and Zhou, N.Y. (2005). Plasmid-borne catabolism of methyl parathion and *p*-nitrophenol in *Pseudomonas* sp. strain WBC-3. Biochem. Biophys. Res. Com. *334*, 1107–1114.

Löffler, F.E., Cole, J.R., Ritalahti, K.M., and Tiedje, J.M. (2003). Diversity of dechlorinating bacteria. In Dehalogenation: Microbial Processes and Environmental Applications, Häggblom, M.M., and Bossert, I.D., eds. (Dordrecht: Kluwer Academic Publishers), pp. 53–87.

Ma, Y.-F, Wu, J.-F., Wang, S.-Y., Jiang, C.-Y., Zhang, Y., Qi, S.-W., Liu, L., Zhao, G.-P., and Liu, S.-J. (2007). Nucleotide sequence of plasmid pCNB1 from Comamonas strain CNB-1 reveals novel genetic organization and evolution for 4-chloronitrobenzene degradation. Appl. Environ. Microbiol. 73, 4477–4483.

Maeda, M., Chung, S.-Y., Song, E., and Kudo, T. (1995). Multiple genes encoding 2,3-dihydroxybiphenyl 1,2-dioxygenase in the Gram-positive polychlorinated biphenyl-degrading bacterium *Rhodococcus erythropolis* TA421, isolated from a termite ecosystem. Appl. Environ. Microbiol. *61*, 549–555.

Mandelbaum, R.T., Allan, D.L., and Wackett, L.P. (1995). Isolation and characterization of a *Pseudomonas* sp. that mineralizes the *s*-triazine herbicide atrazine. Appl. Environ. Microbiol. *61*, 1451–1457.

Martinez, B., Tomkins, J., Wackett, L.P., Wing, R., and Sadowsky, M.J., (2001). Complete nucleotide sequence and organization of the atrazine catabolic plasmid pADP-1 from *Pseudomonas* sp. strain ADP. J. Bacteriol. *183*, 5684–5697.

Masai, E., Sugiyama, K., Iwashita, N., Shimizu, S., Hauschild, J.E., Hatta, T., Kimbara, K., Yano, K., and Fukuda, M. (1997). The *bphDEF meta*-cleavage pathway genes involved in biphenyl/polychlorinated biphenyl degradation are located on a linear plasmid and separated from the initial *bphACB* genes in *Rhodococcus* sp. strain RHA1. Gene *187*, 141–149.

Mattes, T.E., Coleman, N.V., Spain, J.C., and Gossett, J.M. (2005). Physiological and molecular genetic analyses of vinyl chloride and ethene biodegradation in *Nocardioides* sp. strain JS614. Arch. Microbiol. *183*, 95–106.

Merlin, C., Springael, D., and Toussaint, A. (1999). Tn*4371*: a modular structure encoding a phage-like integrase, a *Pseudomonas*-like catabolic pathway and RP4/Ti-like transfer functions. Plasmid *41*, 40–54.

Mohamed, M.S. (2009). Degradation of methomyl by the novel bacterial strain *Stenotrophomonas maltophilia* M1. Electr. J. Biotech. *12*, 1–6.

Mohebali, G., and Ball, A.S. (2008). Biocatalytic desulfurization (BDS) of petrodiesel fuels. Microbiol. *154*, 2169–2183

Monticello, D.J., Bakker, D., and Finnerty, W.R. (1985). Plasmid-mediated degradation of dibenzothiophene by *Pseudomonas* species. Appl. Environ. Microbiol. *49*, 756–760.

Mulbry, W.W., Karns, J.S., Kearney, P.C., Nelson, J.O., McDaniel, C.S., and Wild, J.R. (1986). Identification of a plasmid borne parathion hydrolase gene from *Flavobacterium* sp. by Southern hybridization with opd from *Pseudomonas diminuta*. Appl. Environ. Microbiol. *51*, 926–930.

Mulbry, W.W., Kearney P.C., Nelson, J.O., and Karns J.S. (1987). Physical comparison of parathion hydrolase plasmids from *Pseudomonas diminuta* and *Flavobacterium* sp. Plasmid *18*, 173–177.

Munnecke, D.M., and Hsieh, D.P.H. (1974). Microbial decontamination of parathion and *p*-nitrophenol in aqueous media. Appl. Microbiol. *28*, 212–217.

Munnecke, D.M., and Hsieh, D.P. (1976). Pathways of microbial metabolism of parathion. Appl. Microbiol. *31*, 63–69.

Niazi, J.H., Prasad, D.T., and Karegoudar, T.B. (2001). Initial degradation of dimethylphthalate by esterases from *Bacillus* species. FEMS Microbiol. Lett. *196*, 201–205.

Nojiri, H., Sota, M., and Shintani, M. (2009). Catabolic plasmids involved in the degradation of polycyclic aromatic hydrocarbons and heteroaromatic compounds. In Microbial Megaplasmids, Schwartz, E., ed. (Springer-Verlag: Berlin), pp. 55–87.

Nomura, Y., Takada, N., and Oshima, Y. (1989). Isolation and identification of phthalate-utilizing bacteria. J. Ferment. Bioeng. *67*, 297–299.

Nomura, Y., Harashima, S., and Oshima, Y. (1990). PHT, a transmissible plasmid responsible for phthalate utilization in *Pseudomonas putida*. J. Ferment. Bioeng. *70*, 295–300.

Nomura Y., Nakagawa M., Ogawa N., Harashima S., and Oshima Y. (1992). Genes in PHT plasmid encoding the initial degradation pathway of phthalate in *Pseudomonas putida*. J. Ferment. Bioeng. *74*, 333–344.

Ogawa, N., and Miyashita, K. (1995). Recombination of a 3-chlorobenzoate catabolic plasmid from *Alcaligenes eutrophus* NH9 mediated by direct repeat elements. Appl. Environ. Microbiol. *61*, 3788–3795.

Ogawa, N., and Miyashita, K. (1999). The chlorocatechol-catabolic transposon Tn*5707* of *Alcaligenes eutrophus* NH9, carrying a gene cluster highly homologous to that in the 1,2,4-trichlorobenzene-degrading bacterium *Pseudomonas* sp. strain P51, confers the ability to grow on 3-chlorobenzoate. Appl. Environ. Microbiol. *65*, 724–731.

Ogram, A.V., Duan, Y.-P., Trabue, S.L., Feng, X., Castro, H., and Ou, L.-T. (2000). Carbofuran degradation mediated by three related plasmid systems. FEMS Microbiol. Ecol. 32, 197–203.

Ohshiro, T., and Izumi, Y. (1999). Microbial desulfurization of organic sulfur compounds in petroleum. Biosc. Biotechnol. Biochem. 63, 1–9.

Parekh, N.R., Hartmann, A., Charnay, M.P., and Fourier, J.C. (1995) Diversity of carbofuran-degrading soil bacteria and detection of plasmid-encoded sequences homologous to the *mcd* gene. FEMS Microbiol. Ecol. *17*, 149–160.

Parekh, N.R., Hartmann, A., and Fournier, J.-C. (1997). PCR detection of the *mcd* gene and evidence of sequence homology between the degradative genes and plasmids from diverse carbofuran-degrading bacteria. Soil Biol. Biochem. *28*, 1797–1804.

Pavlů, L., Vosáhlová, J., Klierová, H., Prouza, M., Demnerová K., and Brenner, V. (1999). Characterization of chlorobenzoate degraders isolated from polychlorinated biphenyl-contaminated soil and sediment in the Czech Republic. J. Appl. Microbiol. *87*, 381–386.

Peloquin, L., and Greer, C.W. (1993). Cloning and expression of the polychlorinated biphenyl-degradation gene cluster from *Arthrobacter* M5 and comparison to analogous genes from Gram-negative bacteria. Gene *125*, 35–40.

Pemberton, J.M., and Schmidt, R., editors (2001). Catabolic plasmids. Encyclopedia of Life Sciences. (John Wiley & Sons, Ltd). Available from www.els.net.

Perkins, E.J., Gordon, M.P., Caceres, O., and Lurquin, P.F. (1990). Organization and sequence analysis of the 2,4-dichlorophenol hydroxylase and dichlorocatechol oxidative operons of plasmid pJP4. J. Bacteriol. *172*, 2351–2359.

Pieper, D.H. (2005). Aerobic degradation of polychlorinated biphenyls. Appl. Microbiol. Biotechnol. *67*, 170–191

Plumeier, I., Pérez-Pantoja, D., Heim, S., González, B., and Pieper, D.H. (2002). Importance of different tfd genes for degradation of chloroaromatics by *Ralstonia eutropha* JMP134. J. Bacteriol. *184*, 4054–4064.

Poelarends, G.J., Kulakov, L.A., Larkin, M.J., Van Hylckama Vlieg, J.E.T., and Janssen, D.B. (2000). Roles of horizontal gene transfer and gene integration in evolution of 1,3-dichloropropene- and 1,2-dibromoethane- degradative pathways. J. Bacteriol. *182*, 2191–2199.

Poh, R.P., Smith, A.R., and Bruce, I.J. (2002). Complete characterisation of Tn*5530* from *Burkholderia cepacia* strain 2a (pIJB1) and studies of 2,4-dichlorophenoxyacetate uptake by the organism. Plasmid *48*, 1–12.

Prince, R.C., and Grossman, M.J. (2003). Substrate preferences in biodesulfurization of diesel range fuels by *Rhodococcus* sp. strain ECRD-1. Appl. Environ. Microbiol. *69*, 5833–5838.

Rani, M., Prakash, D., Sobti, R.C., and Jain, R.K. (1996). Plasmid-mediated degradation of *o*-phthalate and salicylate by a *Moraxella* sp. Biochem. Biophys. Res. Com. *220*, 377–381.

Rheinwald, J.G., Chakarabarty, A.M., and Gunsalus, I.C. (1973). A transmissable plasmid controlling camphor oxidation in *Pseudomonas putida*. Proc. Natl. Acad. Sci. USA *70*, 885–889.

Rodrigues, J.L.M., Maltseva, O.V., Tsoi, T.V., Helton, R.R., Quensen, J.F., Fukuda, M., and Tiedje, J.M. (2001). Development of a *Rhodococcus* recombinant strain for degradation of products from anaerobic dechlorination of PCBs. Environ. Sci. Technol. 35, 663–668.

Rousseaux, S., Soulas, G., and Hartmann, A. (2002). Plasmid localisation of atrazine-degrading genes in newly described *Chelatobacter* and *Arthrobacter* strains. FEMS Microbiol. Eco. *41*, 69–75.

Rudel, R., and Perovich, L. (2008). Endocrine disrupting chemicals in indoor and outdoor air. Atmos. Environ. *43*, 170–181.

Sadowsky, M.J., Tong Z., de Souza M., and Wackett, L.P. (1998). AtzC is a new member of the amidohydrolase protein superfamily and is homologous to other atrazine-metabolizing enzymes. J. Bacteriol. *180*, 152–158.

Saeki, H., Akira, M., Furuhashi, K., Averhoff, B., and Gottschalk, G. (1999). Degradation of trichloroethene by a linear-plasmid-encoded alkene monooxygenase in *Rhodococcus corallinus* (*Nocardia corallina*) B-276. Microbiology *145*, 1721–1730.

Saint, C.P., and Ribbons, D.W. (1990). A catabolic plasmid involved in 4-methyl-*o*-phthalate and 4-hydroxy-*iso*-phthalate degradation in *Pseudomonas cepacia*. FEMS Microbiol. Lett. *69*, 323–328.

Saint, C.P., and Romas, P. (1996). 4-Methyl phthalate catabolism in *Burkholderia* (*Pseudomonas*) *cepacia* Pc701: a gene encoding a phthalate-specific permease forms part of a novel gene cluster. Microbiol. *142*, 2407–2418.

Sajjaphan, K., Shapir, N., Wackett, L.P., Palmer, M., Blackmon, B., Tomkins, J., and Sadowsky, M.J. (2004). *Arthrobacter aurescens* TC1 atrazine catabolism genes *trzN, atzB*, and *atzC* are linked on a 160-kilobase region and are functional in *Escherichia coli*. Appl. Environ. Microbiol. *70*, 4402–4407.

Schmitz, A.K., Gartemann, H., Fiedler, J., Grund, E., and Eichenlaub, R. (1992). Cloning and sequence analysis of genes for dehalogenation of 4-chlorobenzoate from *Arthrobacter* sp. strain SU. Appl. Environ. Microbiol. *58*, 4068–4071.

Serbolisca, L., de Ferra, F., and Margarit, I. (1999). Manipulation of the DNA coding for the desulphurizing activity in a new isolate of *Arthrobacter* sp. Appl. Microbiol. Biotechnol. *52*, 122–126.

Serdar, C.M.S, Gibson, D.T., Munnecke, D.M., and Lancaster, J.H. (1982). Plasmid involvement in parathion hydrolysis by *Pseudomonas diminuta*. Appl. Environ. Microbiol. *44*, 246–249.

Sethunathan, N., and Yoshida, T., (1973). A *Flavobacterium* that degrades diazinon and parathion. Canad. J. Microbiol. *19*, 873–875.

Siddavattam, D., Khajamohiddin, S., Manavathi, B., Pakala, S.B., and Merrick, M., (2003). Transposon-like organization of the plasmid-borne organophosphate degradation (*opd*) gene cluster found in *Flavobacterium* sp. Appl. Environ. Microbiol. *69*, 2533–2539.

Somara, S., and Siddavattam, D. (1995). Plasmid mediated organophosphate pesticide degradation by *Flavobacterium balustinum*. Biochem. Mol. Biol. Int. *36*, 627–631.

Springael, D., and Top., E.M. (2004). Horizontal gene transfer and microbial adaptation to xenobiotics: new types of mobile genetic elements and lessons from ecological studies. TRENDS Microbiol. *12*, 53–58.

Springael, D., Kreps, S., and Mergeay, M. (1993). Identification of a catabolic transposon, Tn*4371*, carrying biphenyl and 4-chlorobiphenyl degradation genes in *Alcaligenes eutrophus* A5. J. Bacteriol. *175*, 1674–1681.

Springael, D., Diels, L., and Mergeay, M. (1994). Transfer and expression of PCB-degradative genes into heavy metal resistant *Alcaligenes eutrophus* strains. Biodegradation *5*, 343–357.

Springael, D., Ryngaert, A., Merlin, C., Toussaint, A., and Mergeay, M. (2001). Occurrence of Tn*4371*-related mobile elements and sequences in (chloro) biphenyl-degrading bacteria. Appl. Environ. Microbiol. *67*, 42–50.

Staples, C.A., Peterson, D.R., Parkerton, T.F., and Adams, W.J. (1997). The environmental fate of phthalate esters: a literature review. Chemosphere *35*, 667–749.

Streber, W.R., Timmis, K.M., and Zenk, M.H. (1987). Analysis, cloning, and high-level expression of 2,4-dichlorophenoxyacetate monooxygenase gene *tfdA* of *Alcaligenes eutrophus* JMP134. J. Bacteriol. *169*, 2950–2955.

Tardif, G., Greer, C.W., Labbe, D., and Lau, P.C.K. (1991). Involvement of a large plasmid in the degradation of 1,2-dichloroethane by *Xanthobacter autotrophicus*. Appl. Environ. Microbiol. *57*, 1853–1857.

Tomasek, P.H., and Karns, J.S. (1989). Cloning of a carbofuran hydrolase gene from *Achromobacter* sp. strain WM111 and its expression in Gram-negative bacteria. J. Bacteriol. *171*, 4038–4044.

Top, E.M., Maltseva, O.V., and Forney, L.J. (1996). Capture of a catabolic plasmid that encodes only 2,4-dichlorophenoxyacetic acid: a-ketoglutaric acid dioxygenase (TfdA) by genetic complementation. Appl. Environ. Microbiol. *62*, 2470–2476.

Topp, E., Hanson, R.S., Ringelberg, D.B., White, D.C., and Wheatcroft, R. (1993). Isolation and characterization of an N-methylcarbamate insecticide-degrading methylotrophic bacterium. Appl. Environ. Microbiol. *61*, 1691–1698.

Topp, E., Zhu, H., Nour, S.M., Houot, S., Lewis, M., and Cuppels, D. (2000). Characterization of an atrazine-degrading *Pseudaminobacter* sp. isolated from canadian and french agricultural soils. Appl. Environ. Microbiol. *66*, 2773–2782.

Turnbull, G.A., Ousley, M., Walker, A., Shaw, E., and Morgan, J.A.W. (2001). Degradation of substituted phenylurea herbicides by *Arthrobacter globiformis* strain D47 and characterization of a plasmid-associated hydrolase gene, *puhA*. Appl. Environ. Microbiol. *67*, 2270–2275.

Vallaeys, T., Albino, L., Soulas, G., Wright, A.D., and Weightman, A.J. (1998). Isolation and characterization of a stable 2,4-dichlorophenoxyacetic acid degrading bacterium, *Variovorax paradoxus* using chemostat culture. Biotechnol. Lett. *20*, 1073–1076.

Vallaeys, T., Courde, L., Mc Gowan, C., Wright, A.D., and Fulthorpe, R.R. (1999). Phylogenetic analyses indicate independent recruitment of diverse gene cassettes during assemblage of the 2,4-D catabolic pathway. FEMS Microbiol. Ecol. *28*, 373–382.

Van Der Meer, J.R., Zehnder, A.J.B., and De Vos, W.M. (1991). Identification of a novel composite transposable element, Tn*5280*, carrying chlorobenzene dioxygenase genes of *Pseudomonas* sp. strain P51. J. Bacteriol. *173*, 7077–7083.

Van Der Meer, J.R., Werlen, C., Nishino, S.F., and Spain, J.C. (1998). Evolution of a pathway for chlorobenzene metabolism leads to natural attenuation in contaminated groundwater. Appl. Environ. Microbiol. *64*, 4185–4193.

Van Der Meer, J.R., Ravatn, R., and Sentchilo, V. (2001). The *clc* element of *Pseudomonas* sp. strain B13 and other mobile degradative elements employing phage-like integrases. Arch. Microbiol. *175*,79–85.

Van Der Meer, J.R. (2006). Evolution of catabolic pathways in *Pseudomonas* through gene transfer. In *Pseudomonas*, Volume 4: Molecular Biology of Emerging Issues, Ramos, J.-L., and Levesque, R.C., eds. (Springer: the Netherlands), pp. 139–236.

Van Hamme, J.D., Singh, A., and Ward, O.P. (2003). Recent advances in petroleum microbiology. Microbiol. Mol. Biol. Rev. *67*, 503–549.

Vedler, E., Vahter, M., and Heinaru, A. (2004). The completely sequenced plasmid pEST4011 contains a novel IncP1 backbone and a catabolic transposon harboring *tfd* genes for 2,4-dichlorophenoxyacetic acid degradation. J. Bacteriol. *186*, 7161–7174.

Vedler, E. (2009) Megaplasmids and the degradation of aromatic compounds by soil bacteria. In Microbial Megaplasmids, Schwarz, E., ed. (Springer-Verlag: Berlin Heidelberg), pp. 33–53.

Wackett, L.P., Sadowsky, M.J., Martinez B., and Shapir, N. (2002). Biodegradation of atrazine and related *s*-triazine compounds: from enzymes to field studies. Appl. Microbiol. Biotechnol. *58*, 39–45.

Webster, L.R., McKenzie, G.H., and Moriarty, H.T. (2002). Organophosphate-based pesticides and genetic damage implicated in bladder cancer. Cancer Genet. Cytogenet. *133*, 112–117.

Weightman, A.J., Topping, A.W., Hill, K.E., Lee, L.L., Sakai, K., Slater, J.H., and Thomas, A.W. (2001). Transposition of *DEH*, a broad-host-range transposon flanked by ISP*pu12*, in *Pseudomonas putida* is associated with genomic rearrangements and dehalogenase gene silencing. J. Bacteriol. *184*, 6581–6591.

Williams, P.A., and Worsey, M.J, (1976). Ubiquity of plasmids in coding for toluene and xylene metabolism in soil bacteria: evidence for the existence of new TOL plasmids. J. Bacteriol. *125*, 818–828.

Williams, P.A. (2004). Catabolic plasmids: fast-track bacterial evolution to combat pollution. Microbiol. Today *31*, 168–170.

Wyndham, R.C., Cashore, A.E., Nakatsu, C.H., and Peel, M.C. (1994). Catabolic transposons. Biodegradation *5*, 323–342.

Xia, X.S,, Aathithan, S., Oswiecimska, K., Smith, A.R., and Bruce, I.J. (1998) A novel plasmid pIJB1 possessing a putative 2,4-dichlorophenoxyacetate degradative transposon Tn*5530* in *Burkholderia cepacia* strain 2a. Plasmid *39*, 154–159.

Zahm, S.H., Weisenburger, D.D., Babbit, P.A., Saal, R.C., Vaught, J.B., Cantor, K.P., and Blair, A. (1990). A case–control study of non-Hodgkin's Lymphoma and the herbicide 2,4-dichlorophenoxyacetic acid (2, 4-D) in Eastern Nebraska. Epidemiology *1*, 349–356.

Microbial Biodegradative Genes and Enzymes in Mineralization of Non-metal Pollutants

10

Nazia Mojib, Jack T. Trevors and Asim K. Bej

Abstract

Bioremediation is an attractive, generally low cost, innovative technology that is a fundamental and sustainable approach to clean up non-metal or organic compounds from contaminated environments. These pollutants include hydrocarbons – principal components of petroleum and fossil fuels; polychlorinated biphenyls (PCBs) – a broad family of man-made organic chemicals also known as chlorinated hydrocarbons; polyaromatic hydrocarbons (PAHs) – industrial pollutants that are significant byproducts of coal, chemical, petroleum processing and refining and toxic pesticides used in agricultural lands. The principle of bioremediation lies in the diverse metabolism of microorganisms to degrade or transform organic compounds to assimilate energy with the help of enzymes encoded by diverse biodegradative genes. This can lead to the efficient removal of a wide range of pollutants and wastes from the environment. This chapter elucidates the structure, diversity and function of the biodegradative genes and enzymes involved in the biodegradation pathways of different contaminants. Also, the use of modern genetic methodologies and genome-based global techniques to better understand the function of these biodegradative genes are briefly discussed. Since, many degradation pathways, along with the enzymes and their respective genes are known and reactions are well understood, a bioinformatics approach in predicting enzymes and reactions involved in biodegradation of new compounds is also examined.

Introduction

Since the second industrial revolution around 1850, a significant increase in the use of an extended list of fossil fuels including petroleum hydrocarbons has impacted our biosphere, environmental health, global climate change, public health, agriculture, transportation and the socioeconomic conditions of the entire world, to name a few examples. The constituents of the petroleum fuels, primarily the hydrocarbons and anthropogenic halogenated compounds, are considered to be 'non-metal organic pollutants' in the environment and harmful to human health. While the use of petroleum and its products now almost inseparable from our everyday life and economic systems, there has been increasing global concern over the environment pollution impacts by these compounds. A balance between the safe use of the petroleum hydrocarbons and effective clean-up strategies or containment of any unintended spill or its byproducts is necessary to reduce the adverse effect of these compounds to the environment and human health. Non-metal bioremediation forms the fundamental and sustainable approach to clean up the contaminated environments. This includes hydrocarbons- principal components of petroleum and fossil fuel, polychlorinated biphenyls (PCBs) – broad group of man-made organic chemicals also known as chlorinated hydrocarbons, polyaromatic hydrocarbons (PAHs) – industrial pollutants that are significant byproduct of coal, chemical, petroleum processing and refining, and toxic pesticides used in agricultural lands. The fundamental principle of bioremediation lies in the diverse metabolism of microorganisms whose genome consists of diverse biodegradative genes to degrade, transform and accumulate organic pollutants in diverse environments (Errampalli *et al.*, 1997; Kanavllil *et al.*, 2007). This leads to effective elimination of a wide range of pollutants and wastes from the contaminated environment. Taking into account the type of pollutants (compounds) that probably contaminate the specific environment and the knowledge of the microorganisms harbouring the biodegradative genes that code for the enzymes capable of degradation of those compounds will help implement bioremediation process to their full potential.

Bioremediation may be achieved by attenuation also known as 'intrinsic bioremediation', which relies on the innate capabilities of naturally occurring microbial communities in the environment to degrade and reduce contaminant concentrations in the environment (Beaudette *et al.*, 2001). Natural attenuation is a passive remedial approach and a major process among the natural mechanisms that account for the reduction of contaminant concentrations (Alexander, 1999). Alternatively, genetically engineered microorganisms (GEMs) originally isolated as wild types from the natural environment and then engineered to degrade targeted contaminants, have experimentally been proven to be effective in degrading some targeted contaminants. For a better understanding of both natural and engineered bioremediation, it is essential to study of the structure, dynamics and function of biodegradative genes in indigenous microbial communities and the biological and geochemical factors that affect the degradation process in the contaminated environments (Atlas, 1981; Gibson and Sayler, 1992). Efforts have been directed to describe different enzymes responsible for catalysis of reactions involved in the degradation of organic compounds in different microorganisms. This has led to elucidation of different pathways involved in biodegradation of variety of toxic compounds. The University of Minnesota Biocatalysis/Biodegradation Database (UM-BBD) stores updated information on microbial biocatalytic reactions and biodegradation pathways for xenobiotic and chemical compounds (Gao *et al.*, 2010; http://umbbd.msi.umn.edu/). This database is very helpful for studies related to understanding the nature of reactions leading to biodegradation of environmental pollutants. The information on the starting and intermediate chemical compounds; the organisms that transform these compounds; the enzymes and the genes are presented with individual reactions and metabolic pathways in this database. UM-BBD (last updated on June 2, 2010) consists of lists of 1233 compounds transformed by 508 microorganisms utilizing 857 enzymes encoded by corresponding genes with 1329 different reactions in 195 different pathways (Gao *et al.*, 2010). The activities of the microbial communities in the contaminated samples in specific ecosystems have been studied by using PCR amplification, gene cloning, terminal-restriction fragment length polymorphism (TRFLP), denaturing gradient gel electrophoresis (DGGE), and *in situ* hybridization. Recently, microbial ecologists have been interested in using modern, rapid, quantitative and cost-effective molecular methods and genomic technology such as real-time PCR and microarray to study the dynamics of microbial communities. A relatively new metagenomic method can potentially help to investigate the microbial populations and their biodegradative genes irrespective of their culturability and taxonomic identities (Diana Marco, 2010). These modern molecular methods hold promise in revealing a comprehensive picture of the compositions of the microbial population in contaminated environments along with the dynamics of biodegradative genes leading to important and much needed contributions in bioremediation with increasing amounts of anthropogenic pollution.

Biodegradation of hydrocarbons

Alkane-degrading enzymes/genes

Several microorganisms are able to use alkanes, the saturated hydrocarbons of different sizes and structures, as a source of carbon and energy (Labinger and Bercaw, 2002). Depending on the chain-length of alkanes, different enzymes are used by microorganisms for their degradation.

Methanotrophs use methane monoxygenases (MMO) for degradation of methane as carbon and energy source (Sazinsky and Lippard, 2006; van Beilen and Funhoff, 2007; Hakemian and Rosenzweig, 2007). Enzymes related to MMOs are used for the degradation of short chain-length alkanes (C_2–C_4) by some bacteria (Dubbels *et al.*, 2007). Propane, butane, medium-chain-length alkanes (C_5–C_{11}) and long-chain length alkanes (>C_{12}) are degraded by AlkB family of alkane hydroxylases or alkane monooxygenases (van Beilen *et al.*, 2003; van Beilen and Funhoff, 2007). In some bacteria, soluble cytochrome P450 alkane hydroxylases enzymes hydroxylates C_5–C_{11} alkanes and for long-chain alkanes (>C_{20}) degradation, different alkane hydroxylases are utilized. The list of different enzymes encoded by the respective genes in different microorganisms is presented in Table 10.1.

Methane monooxygenases (MMOs) and other alkane monooxygenases

Methanotrophs produce two distinct forms of MMOs: more prevalent but less understood membrane-bound particulate methane monooxygenases (pMMO) and less prevalent but more studied, soluble methane monooxygenases (sMMO) (Nakamura *et al.*, 2007). Soluble methane monooxygenases have three subunits: a hydroxylase consisting of three polypeptides arranged as $\alpha2\beta2\gamma2$, a reductase that shuttles electrons from NADH to the hydroxylase, and a regulatory protein that is required for activity along with a diiron-active site. The seven genes involved in the expression of sMMO described in *Methylomicrobium* sp. strain NI are: *mmoX*, *mmoY*, *mmoB*, *mmoZ*, *mmoD*, *mmoC* and *mmoG*. The three genes *mmoXYZ* codes for hydroxylase component, *mmoC* codes for reductase subunit and the regulatory component is encoded by *mmoB* gene. The *mmoG* gene codes for protein involved in regulating the expression and assembly of sMMOs. The function of the protein encoded by *mmoD* gene is still unknown, although its possible role in the assembly of the hydroxylase subunit has been predicted (McDonald *et al.*, 1997; Nakamura *et al.*, 2007). Particulate methane monooxygenases are also composed of three subunits, α (PmoA), β (PmoB) and γ (PmoC) with copper embedded in the active site of the enzyme (Sazinsky and Lippard, 2006; Hakemian and Rosenzweig, 2007). These three subunits are encoded by the three genes *pmoA*, *pmoB* and *pmoC*, respectively described in *Methylomicrobium* sp. strain NI and *Methylococcus capsulatus* (Bath) (Zahn and DiSpirito, 1996; Nakamura *et al.*, 2007).

Pseudomonas butanovora is capable of growing on C_2–C_4 gaseous alkanes as a source of carbon and energy with the help of non-heme iron monooxygenase similar to the sMMO named butane monooxygenases (BMOs) by hydroxylating alkanes (Sluis *et al.*, 2002; Dubbels *et al.*, 2007). The sBMO is a multimeric hydroxylase enzyme in which 61-, 45- and 19-kDa subunits are encoded by *bmoXYZ* genes; a 40 kDa oxidoreductase is encoded by *bmoC* gene; and a 15 kDa regulatory protein is encoded by *bmoB* gene. A 9.6 kDa protein with similarity exclusively to *mmoD* (*orfY*) is encoded by *bmoD* gene. High sequence identity (up to 64%) has been observed between BMOs and MMOs (Sluis *et al.*, 2002).

The AlkB family of alkane hydroxylases

The alkane degradation pathway located on the OCT-plasmid, was first genetically characterized in *P. putida* GPo1 (formerly *Pseudomonas oleovorans* GPo1) (Kok *et al.*, 1989; van Beilen *et al.*, 1994). Since then more than 60 homologues of alkane hydroxylase gene (*alkB*) have been cloned and sequenced in both Gram-positive and Gram-negative bacteria such as *Acinetobacter* sp., *Rhodococcus* sp., *Mycobacterium* sp., *Pseudomonas putida* P1, *P. aureofaciens* and *P. fluorescens*. The *alkB* gene and encoded protein AlkB were found to be diverse in nucleotide as well as amino acid sequences (Sotsky *et al.*, 1994; Ratajczak *et al.*, 1998; Whyte *et al.*, 1998, 2002a,b; Bej *et al.*, 2000; Vomberg and Klinner, 2000; van Beilen *et al.*, 2002; Panicker *et al.*, 2002; Kuhn *et al.*, 2009; Wasmund *et al.*, 2009). Genes encoding alkane hydroxylases (*alks*) were also identified and their active transcriptions were assessed using the polymerase chain reaction (PCR) and Reverse transcriptase-PCR (RT-PCR) in both Gram-positive and Gram-negative bacteria isolated from oil-contaminated Antarctic soil samples

Table 10.1 List of enzymes and their genes, and microorganisms involved in the biodegradation of non-metal or organic compounds

Compounds	Genes (GenBank accession no.)	Enzymes	Pathways/reactions	Bacteria	References
Alkanes					
Methane	*mmoX, mmoY, mmoZ, mmoB, orfY, mmoC* (U81594)	Soluble methane monooxygenases (sMMO)	Methane → methanol	*Methylocystis* sp. strain M	McDonald *et al.* (1997)
	pmoA, pmoB and *pmoC* (AB253367)	Particulate methane monooxygenase (pMMO)	Methane → methanol	*Methylomicrobium* sp. strain NI	Nakamura *et al.* (2007)
Butane	*bmoX, bmoY, bmoB, bmoZ, bmoC, bmoD* (AY093933)	Butane monooxygenase (BMO)	Butane → butanol	*Pseudomonas butanovora*	Sluis *et al.* (2002)
n-alkanes (C_2–C_{20})	*alkB* (AJ245436)	Alkane hydroxylase	Alkane → alcohol	*P. putida*	van Beilen *et al.* (1994), Panicker *et al.* (2010)
C_5–C_{11}	*cytP450 alk* (AJ311718)	Cytochrome P450 alkane hydroxylase	Alkane → alcohol	*Acinetobacter* sp. strain EB104	Maier *et al.* (2001)
Mono- and polyaromatic hydrocarbons					
Xylenes	*xylA* and *xylM* (M37480)	Xylene monooxygenase	m-Xylene → 3-methylbenzyl alcohol	*P. putida*	Suzuki *et al.* (1991)
			o-Xylene → 2-methylbenzyl alcohol		
			p-Xylene → *p*-methylbenzyl alcohol		
Toluenes	*xylA* and *xylM* (M37480)	Xylene monooxygenase	Toluene → benzyl alcohol	*P. putida*	Shaw *et al.* (1992)
			Toluene → 2-hydroxytoluene	*Burkholderia(Pseudomonas) cepacia G4*	
			Toluene → 3-hydroxytoluene	*Pseudomonas pickettii PK01*	
	tom (AF319657)	Toluene 2-monooxygenase	Toluene → 4-hydroxytoluene	*Pseudomonas mendocina KR1*	Shields *et al.* (1995)
			Toluene → toluene-*cis*-1,2-dihydrodiol		

	tbuABC (U04052)	Toluene 3-monooxygenase	Benzo(a)pyrene → Benzo(a)pyrene-*cis*-9,10-dihydrodiol	*Pseudomonas putida*	Olsen *et al*. (1994)
	tmoABCDE (M65106)	Toluene 4-monooxygenase	Anthracene → 1 hydroxy-2-naphthoic acids	*Sphingomonas CHY-1*	Yen *et al*. (1991)
			PCBs → chlorobenzoates		Furukawa and Mirazaki (1986)
Pesticides and herbicides					
Hexachlorocyclohexane (HCH)	*linA, linB, linC, lin D and lin E* (AY903217)	HCH dehydrochlorinase, halidohydrolase, dehydrogenase, reductive dechlorinase, deoxygenase	HCH → dichlorohydroquinone	*S. paucimobilis* strain B90,UT26	Kumari *et al*. (2002)
Atrazine	*thc BCD* (U17130)	Cytochrome P450 system	Atrazine → deisopropylatrazine/ deethylatrazine/hydroxyisopropylatrazine	*Rhodococcus* sp. strain NI86/21	Nagy *et al*. (1995)
	atz ABC (U66917)	Atrazine chlorohydrolase, hydroxyatrazine ethylaminohydrolase, and *N*-isopropylammelide isopropylaminohydrolase	Atrazine → cyanuric acid	*Pseudomonas* sp.	de Souza *et al*. (1998); Aislabie *et al*. (2005)
Endosulfan	*esd* (AF537302)	Esd monooxygenase	β-Endosulfan → endosulfan monoaldehyde and endosulfan hydroxyethe	*Mycobacterium* sp. strain ESD	Sutherland *et al*. (2002)
	ese (DQ124296)	Ese monooxygenase	α or β endosulfan → endosulfan monoalcohol	*Arthrobacter* sp.	Weir *et al*. (2006)
Carbaryl	*cehA* (AB069723)	Carbaryl Hydrolase	Carbaryl → 1-naphthol and methylamine	*Rhizobium* sp. Ac100, *Pseudomonas* sp. strains C4, C5, and C6	Hashimoto *et al*. (2002), Swetha and Phale (2005)

(Panicker *et al.*, 2010). AlkB that hydroxylates alkanes at the terminal position is composed of two soluble electron transfer proteins: rubredoxin (AlkG) and rubredoxin reductase (AlkT). AlkB is believed to have six transmembrane regions with an active site containing four Histidine-containing sequence motifs, which chelates two iron atoms (van Beilen *et al.*, 1992a). Although the organization of the *alkB* genes differs among alkane-degrading bacteria, the two operons consisting of clusters of *alkB* genes of *P. putida* GPo1 seems to be transferred via horizontal gene transfer (Rojo, 2009). Owing to the location of alkane *hydroxylase* at different sites on the chromosome in a single strain, there is variability in the degree of clustering among different bacterial strains. The AlkS, a transcriptional regulator in *Alcanivorax borkumensis*, activate expression of *alkB* and downstream genes when exposed to alkanes (van Beilen *et al.*, 2004; Schneiker *et al.*, 2006). Some bacterial strains contain more than one alkane hydroxylases, which are differentially regulated under specific physiological conditions. Depending on the presence of different types of alkanes, two AlkB related alkane hydroxylases, named AlkMa and AlkMb are differentially regulated in *Acinetobacter* sp. strain M-1(Tani *et al.*, 2001). Similarly, *P. aeruginosa* strains RR1 and PAO1 contain two alkane hydroxylases, AlkB1 and AlkB2, though their substrate range overlaps i.e. AlkB1 oxidizes C_{16}–C_{24} *n*-alkanes while AlkB2 is active on C_{12}–C_{20} *n*-alkanes (Rojo, 2009). Some strains, for example, *A. borkumensis,* have three alkane hydroxylase systems: two AlkB-like alkane hydroxylases and three genes coding for cytochromes P450 alkane hydroxylase (van Beilen *et al.*, 2004; Hara *et al.*, 2004; Schneiker *et al.*, 2006).

Cytochrome P450 family of alkane hydroxylases

One of the ubiquitous proteins found distributed among all kingdoms of life is cytochrome P450 (Cyt P450). Several bacterial strains degrade C_5–C_{11} alkanes using cytochrome P450 alkane hydroxylases. The CYP153A1 in *Acinetobacter* sp. EB104 was the first characterized alkane hydroxylase (Maier *et al.*, 2001) and also found in different bacteria belonging to *Mycobaterium* and *Rhodococcus* genera (van Beilen *et al.*, 2005a, 2006; Sekine *et al.*, 2006). For the transfer of electrons from NADPH to cytochrome, this enzyme requires ferredoxin and a ferredoxin reductase.

Long chain *n*-alkane degrading alkane hydroxylases

Bacteria containing AlkB family and Cytochrome P450 alkane hydroxylases can assimilate long chain alkanes up to C_{20}. But there is different enzyme system involved in the degradation of alkanes larger than C_{20}. A Cu^{2+}-dependent alkane hydroxylase in *Acinetobacter* sp. M1, capable of growth on C_{13}–C_{44} alkanes, has been proposed to function like dioxygenase (Maeng *et al.*, 1996; Tani *et al.*, 2001). A flavin-binding monooxygenase (AlmA), has been found in *Acinetobacter* strain, DSM 17874, which oxidizes C_{20} to $>C_{32}$ alkanes (Throne-Holst *et al.*, 2007). *Acinetobacter* sp. M1 and *A. borkumensis* SK2 contain genes homologous to *almA*. LadA, a long-chain alkane hydroxylase, has been identified in *GeoBacillus thermodenitrificans* NG80-2 (Feng *et al.*, 2007). Some bacterial strains such as *Alcanivorax* spp. can also degrade pristane and phytane branched alkanes in sea water contaminated with oil (Hara *et al.*, 2003).

Mono and polycyclic aromatic hydrocarbons degrading enzymes/genes

One of the components of fuel and industrial pollutant are polyaromatic hydrocarbons (PAHs), consisting of substituted benzenes and naphthalenes along with low levels of highly volatile monocyclic hydrocarbons like benzene, toluene and xylene (Zeiger and Smith, 1998). Some PAHs have been reported to be carcinogens, mutagens and teratogens and therefore, the US Environmental Protection Agency (EPA) has designated 32 PAH compounds as priority pollutants, out of which16 different PAHs are used in the assessment of contamination in environmental samples (EPA, 1998).

Xylene monooxygenases, toluene monooxygenases and dioxygenases

The first enzyme in the xylene degradation pathway that oxidizes xylene to methyl (benzyl) alcohols is xylene monooxygenase that has broad substrate specificity. This enzyme has two different subunits, encoded by *xylM* and *xylA,* and has been characterized in the TOL plasmid pWWO of *P. putida* (Suzuki *et al.,* 1991). This enzyme falls in the category of second class of monooxygenases, which consists of two different subunits, one of which is a hydroxylase. The XylM shows 25% sequence homology with alkane hydroxylases, which catalyses the oxidation of substrates. The first 90 amino acid sequence of XylA shows strong similarity with chloroplast-type ferredoxins and the remaining sequence resemble ferredoxin-NADP+ reductases. Thus, XylM is the catalytic component whereas XylA contains two domains i.e. N-terminal chloroplast-type ferredoxins and a C-terminal ferredoxin-NADP+ reductases (Suzuki *et al.,* 1991; Shaw *et al.,* 1992). There are five different toluene degradation pathways described in bacterial system (Shields *et al.,* 1991). One of the pathways consists of the above described enzyme xylene monooxygenase, which is encoded by *xylM* and *xylA* genes of catabolic TOL plasmid of *P. putida.* This oxidizes toluene to catechol by first converting it to benzyl alcohol, benzaldehyde and benzoate. The next three pathways involve oxidation of toluene by toluene monooxygenases (*Tom*) that hydroxylate toluene at all three, ortho, meta and para positions (Table 10.1). The genes encoding toluene 2-monooxygenase are *tomA* and *tomB* which are described and studied in *Burkholderia* (*Pseudomonas*) *cepacia* G4 (Shields *et al.,* 1995). The toluene 3-monooxygenase encoding genes of *P. pickettii* PK01 are *tbuABC* (Olsen *et al.,* 1994). The toluene 4-monooxygenase gene cluster is slightly more complex consisting of five genes, *tmoABCDE,* and have been studied in *P. mendocina* KR1 (Yen *et al.,* 1991). The fifth pathway involves oxidation of toluene by an enzyme complex of three proteins named toluene 2,3-dioxygenase. The first component is a reductase encoded by *todA, todB* genes, the second component ferredoxin reductase is encoded by *todC1,* and *todC2* gene codes for the third protein i.e. iron sulfur oxidase as described in *P. putida* F1 (Zylstra *et al.,* 1989).

Naphthalene dioxygenases, salicylate hydroxylases and catechol 2,3-dioxygenases

A wide range of polyaromatic hydrocarbons (PAHs) are catabolized by a multicomponent enzyme system naphthalene dioxygenase (NDO). Naphthalene dioxygenase enzyme is encoded by the *ndoB* gene in *P. putida* strain NCIB9816 (Kurkela *et al.,* 1988). In *P. aeruginosa* PAO1, it catalyses naphthalene to cis-1, 2-Dihydroxy-1,2-dihydronaphthalene (Eaton and Chapman, 1992). Naphthalene, a bicyclic aromatic hydrocarbon and its substituted derivatives are commonly found in crude oil and oil products and therefore, microbial metabolism of this compound has been extensively studied. This enzyme has broad substrate specificity. In *Sphingomonas* CHY-1, it catalyses benzo[a]pyrene to benzo[a]pyrene-cis-9,10-dihydrodiol (Jouanneau *et al.,* 2006). Benzo(a)pyrene (BaP) is one of the PAHs that are found in crude oil, coal tar and creosote. At 47% of waste sites containing these materials, BaP has been detected as a soil contaminant at concentrations up to 1300 ppb and is a possible human carcinogen (Aspelin and Grube, 2000). In *Pseudomonas* sp. NCIB9816-4, the naphthalene 1,2-dioxygenase is composed of three components, first is NdoA, a flavin- and iron-containing oxidoreductase is encoded by *ndoA* gene, the second is a ferredoxin encoded by *ndoR* gene and the third component is a heterohexamer $\alpha_3\beta_3$ encoded by *ndoB* (α subunit) and *ndoC* (β subunit) (Ensley and Haigler, 1990; Kauppi *et al.,* 1998). And the α subunit contains a Rieske (2Fe–2S) centre and a catalytic domain with non-heme ferrous ion and a water molecule. Tricyclic aromatic hydrocarbons such as anthracene and phenanthrene found in high concentrations in PAH-contaminated sites and are toxic to fish and algae are also degraded by naphthalene degrading enzymes (Cerniglia, 1992). Naphthalene 1,2-dioxygenase encoded by *nah* genes also catalyses conversion of anthracene to 1-hydroxy-2-naphthoic acids (Menn *et al.,* 1993). Phenanthrene is stepwise converted to salicylic and 1-hydroxy-2-naphthoic acids by phenanthrene 3,4-dioxygenase, dehydrogenase, hydratase-aldolase and isomerase enzymes. All of these enzymes are encoded by *pahABCDEF* (Kiyohara *et al.,* 1994). Naphthalene dioxygenase is the first enzyme in the naphthalene degradation pathway. After it has been degraded to salicylate in *P. stutzeri* AN10, another enzyme named salicylate hydroxylase catalyses salicylate to catechol, which is encoded by *nahW* gene (Bosch *et al.*, 1999). In *P. putida* KT2440, catechol is converted to 2-hydroxy-cis, cis-muconate semialdehyde by catechol 2,3-dioxygenase encoded by *xylE* or *C23DO* gene. Catechol 2,3-dioxygenase gene is encoded by TOL plasmid pWWO of *P. putida* consists of four identical subunits, each containing one ferrous ion (Cerdan *et al.*, 1994).

Classification of multicomponent dioxygenase enzymes

All above discussed dioxygenases fall in the category of aromatic ring hydroxylating dioxygenases (ARHDs). These enzymes convert closed-ring structures of aromatic hydrocarbons into non-aromatic cis-diols by adding dioxygen to the aromatic nucleus (Neidle *et al.*, 1991). According to Batie *et al.* (1992) classification, dioxygenases are classified according to the number of constituent components and type of redox centres (Batie *et al.*, 1991). A more recent classification has been proposed by Nam *et al.* (2001) in which dioxygenases are classified according to the amino acid sequences of the α-subunit of the terminal oxygenase component that allows their phylogenetic affiliation. They classified 54 oxygenase components into four groups: group I consisting of homo-multimer oxygenases with only α-subunit, and group II, III and IV consists of heterodimer oxygenases with both α- and β-subunits. Benzoate/toluate dioxygenases falls under group II, naphthalene/PAH dioxygenases comes under group III and group IV consists of benzene/toluene/biphenyl dioxygenases. Group I consist of broad range oxygenases and therefore exhibit low homology with each other and also contain some components of monooxygenases. We analysed and validated few new dioxygenases using Nam *et al.* classification. At first, the amino acid sequences of α-subunit of different dioxygenases were extracted from National Centre for Biotechnology Information (NCBI) database. Then, pairwise sequence comparison of the extracted sequences with toluene dioxygenase (TodC1) from *Pseudomonas putida* F1 (which worked as a standard) was done with BLOSUM matrix using ClustalW2 software (Larkin *et al.*, 2007) (Fig. 10.1). As a rule, pairwise sequence alignment scores between 0–14 include enzymes in group I, scores between 15–24 include group II enzymes, 25–34 scores include group III and group IV enzymes score lies between 35–100. It was found that the analysed enzymes fall into one of these groups according to the pairwise alignment score with TodC1 enzyme (Fig. 10.1). This validated the identification of these enzymes and robustness of this classification was also established. Therefore, it can be suggested that Nam *et al.* classification system is helpful in preliminary classifying and phylogenetically affiliating the newly identified genes.

Biodegradation of polychlorinated biphenyls

PCBs are produced by direct chlorination of biphenyl and gets concentrated on soils surfaces and sediments causing environmental contamination. It is also known to bioaccumulate leading to decreased wildlife populations. PCBs can be degraded by many bacteria that are capable of degrading PAHs. PCBs are degraded by well-known 2,3-dioxygenase pathway by acting on the 2,3 carbons and a metacleavage, producing benzoate. Thus, four specific enzymes, biphenyl dioxygenase (BphA), dihydrodiol dehydrogenase (BphB), 2,3-dihydroxybiphenyl dioxygenase (BphC) and 2-hydroxyl-6-oxo-6-phenylhexa-2,4-dienoic acid hydrolase (BphD) encoded by respective genes i.e. *bphA, bphB, bphC* and *bphD* are sequentially involved in the oxidative degradation of PCBs into chlorobenzoates and 2-hydroxypenta-2,4-dienoate (Furukawa and Miyazaki, 1986).

Biodegradation of pesticides and herbicides

The chemical substances that are lethal to pests are pesticides while herbicides are the chemicals that kill weeds that hamper the growth of important crops. In broad sense pesticides are insecticides, fungicides, bactericides and herbicides that are

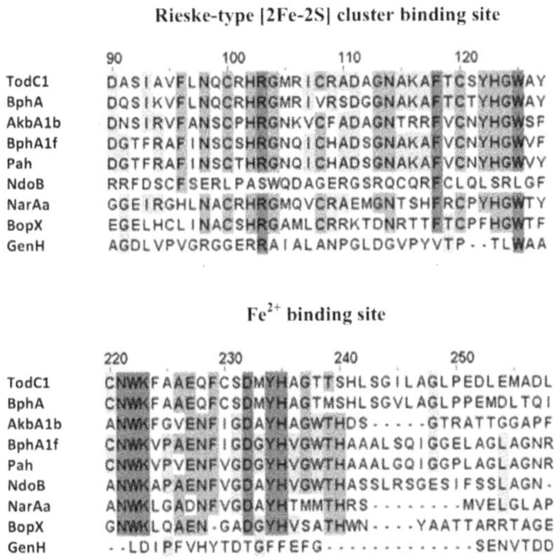

Enzymes	Alignment Score	Group
GenH	6	I
BopX	20	II
NdoB	28	III
NarAa	29	III
Pah	35	IV
BphA1f	35	IV
AkbA1b	35	IV
BphA	66	IV

Figure 10.1 Multiple sequence alignment of the amino acid sequences of the Rieske-type [2Fe–2S] cluster and Fe^{2+} binding site of α-subunit of the terminal oxygenase component of dioxygenases with toluene dioxygenase, TodC1 from *Pseudomonas putida* F1 and their pairwise alignment score and classification in various groups according to Nam *et al.* (2001) are recorded in the adjacent table. The accession numbers of the sequences along with the name of microorganisms in parentheses are as follows: toluene dioxygenase,TodC1 (P13450; *P. putida* F1); biphenyl dioxygenase, BphA (AAC44526; *Pandoraea pnomenusa*); biphenyl/naphthalene dioxygenase, BphA1f (ABM91740; *Sphingobium yanoikuyae*); naphthalene dioxygenase, Pah (CAG17576; *Sphingomonas* CHY-1); alkylbenzene dioxygenase, AkbA1b (AAR90131; *Rhodococcus* sp. strain DK17); naphthalene dioxygenase, NarAa (ABH01029; *Rhodococcus opacus* R7); benzoate dioxygenase, BopX (AF279141; *Rhodococcus* sp. strain 19070); gentisate dioxyenase, GenH (ABH01038; *Rhodococcus opacus* R7); naphthalene dioxygenase, NdoB (AY034588; *Pseudomonas* sp. 30–2).

not only used to hinder and inhibit plant diseases but also to control insect populations in gardens and homes. During the past 3 decades wide-scale application of pesticides and herbicides has augmented crop yields; however, concern has arisen as the presence of these agents in the environment may pose concern about ecosystems and human health. There are various types of pesticides with different chemical structures, including organochlorines, organophosphates, organophosphonates, *s*-triazines, triazinones, carbamates, acetanilides, and sulfonylureas. Among these different types, some are degraded by microorganisms to less harmful compounds, or completely mineralized. Moreover some can be degraded by a single microorganism, but often several microorganisms are necessary to completely degrade them (Schrijver and Mot, 1999).

Lin family of biodegradative genes

Hexachlorocyclohexane (HCH) also called benzene hexachloride (BHC) is a chlorinated cyclic saturated hydrocarbon. BHC actually a mixture of four HCH isomers α, β, γ (lindane), and δ has been widely used for the control of agricultural pests and mosquitoes (Windholz *et al.*, 1976). Although only γ-HCH has insecticidal properties, BHC has been widely used in developing countries because isolating γ-HCH from the mixture will make it costly. In spite of severe restrictions and bans on the use of lindane and BHC in many countries HCH residues can be detected all over the world (Simonich and Hites, 1995; Noren and Meironyte, 2000; Nigam and Siddqui, 2001). The problem arises from α and β isomers of the HCH because of their low solubility in water and high potential of bioaccumulation. Although the γ-HCH is degradable but there is evidence that α and γ isomers of HCH are converted into the β isomer in living organisms (Jensen and Slorach, 1991). *Sphingomonas paucimobilis* strain SS86 (formerly known as *P. paucimobilis*) was reported to degrade γ-HCH (Senoo and Wada, 1989; Thomas *et al.*, 1996). These strains were degrading α-, δ-, and γ-HCH, but were unable to degrade β-HCH because of its unique chemical structure and stereochemistry. However, some reports described *S. paucimobilis* strain B90 that was able to mineralize β-HCH along with α-, δ-, and γ-HCH (Sahu *et al.*, 1990; Johri *et al.*, 1998). In the strain UT26, the key enzyme for γ-HCH degradation is HCH dehydrochlorinase, encoded by the gene *linA*. The other genes important in the γ-HCH degradative pathway are *linB* which encodes halidohydrolase, *linC*, that encodes dehydrogenase, *linD*, and *linE*, that encode a reductive dechlorinase, and a dioxygenase, respectively. All these genes have been cloned and are well characterized from *S. paucimobilis* UT26. The strain B90 of *S. paucimobilis* has been reported to degrade α-, β-, γ-, and δ- isomers of HCH, but this organism lacks the dioxygenase gene *linE* (Kumari *et al.*, 2002).

Atrazine degradation genes

Atrazine (2-chloro-4-ethylamino-6-isopropylamino-1, 3, 5-triazine), one of the major herbicides of the *s*-triazine family has been heavily used all over the world, particularly where maize is grown in abundance (Phillips and McDougall, 1993.). A wide variety of bacteria that are capable of mineralization of the atrazine ring have been isolated (Yanze-Kontchou and Gschwind, 1994; Mandelbaum *et al.*, 1995; Radosevich *et al.*, 1995). The *thcBCD* genes, in *Rhodococcus* sp. strain NI86/21 encode a novel type of cytochrome P450, an electron-supplying system, consisting of a 2Fe–2S ferredoxin (ThcC) and ferredoxin reductase (ThcD) have been reported to degrade atrazine to deisopropylatrazine/deethylatrazine/hydroxyisopropylatrazine (Nagy *et al.*, 1995). The mutant strain of strain NI86/21 named FAJ2027 was unable to degrade atrazine, however, when these genes where transformed into the mutant strain it started degrading atrazine. In *Pseudomonas* sp. the first three enzymatic steps in atrazine degradation have been widely studied (de Souza *et al.*, 1995, 1996; Boundy-Mills *et al.*, 1997; Sadowsky *et al.*, 1998). The genes *atz ABC*, encode enzymes Atrazine chlorohydrolase (AtzA), hydroxyatrazine ethylaminohydrolase (AtzB), and *N*-isopropylammelide isopropylaminohydrolase (AtzC) that convert atrazine to cyanuric acid in a step wise process (de Souza *et al.*, 1998). The atrazine-degradation genes *atzABC* have also been identified in *Arthrobacter nicotinovorans* HIM from soil samples by using PCR (Aislabie *et al.*, 2005). Cloning and DNA sequence analysis of the *atzABC* genes of *Pseudomonas ADP*, *Alcaligenes* SG1, *Ralstonia* M91–3 and *Agrobacterium* J14a revealed significant homology (>99%) in the atrazine degradation genes. This strong similarity suggests a common ancestry among *atz* genes in different genera and these genes have not diverged to a large extent evolutionarily (de Souza *et al.*, 1998).

esd and *ese* Biodegradative genes

The commercial endosulfan an organochlorine insecticide exists as a mixture of two diastereoisomers, α-endosulfan and β-endosulfan, at a ratio of 7:3. Endosulfan isomers are metabolized by *Mycobacterium* sp. strain ESD to endosulfan sulfate and endosulfan monoaldehyde (Sutherland *et al.*, 2002). The α- isomer of endosulfan is primarily oxidized to endosulfan sulfate, and the β isomer is converted to endosulfan monoaldehyde and endosulfan hydroxyether. The *esd* gene encodes a monooxygenase enzyme that uses reduced flavins as cosubstrate, which is provided by a separate flavin reductase enzyme. The enzyme encoded by *esd* is not capable of degrading α-endosulfan and neither the metabolite of endosulfan, endosulfan sulfate. However, both the isomers of endosulfan as well as the toxic metabolite endosulfan sulfate are degraded by the enzyme encoded by gene *ese* from the *Arthrobacter* species. This enzyme is also related to the two-component flavin-dependent monooxygenases whose members are dependent on reduced flavin for their activity. The end product of this degradation is endosulfan monoalcohol and endosulfan hemisulfate. The *ese* gene is located in a group of 10 open reading frames encoding enzymes that have very few sulfur-containing amino acids. These open reading frames were organized into two apparent divergently orientated operons and a gene encoding a putative LysR-type transcriptional regulator (Wier *et al.*, 2006).

'Omics'-based tools in bioremediation

Bioremediation can be studied by different 'omics' -based tools – genomics, transcriptomics, proteomics and metabolomics. All these technologies will help correlate DNA sequences with mRNA, protein and metabolite abundance thereby giving a clear and more true picture of the *in situ* bioremediation. One of the powerful genomic technologies to study the microbial community structure in contaminated soils is the microarray. It has been successfully used for detecting biodegradative genes in soil microcosms and enrichments (Rhee *et al.*, 2004). Owing to the limitations of the culturability of microorganisms in the environmental samples, microarray technology is attracting interest among microbial ecologists. Microarrays used in environmental studies can be divided into three major classes: functional gene arrays (FGAs), community genome arrays (CGAs), and phylogenetic oligonucleotide arrays (Zhou *et al.*, 2002). FGAs can be designed with probes, both PCR-amplified DNA fragments and oligonucleotides corresponding to known diverse biodegradative genes involved in biodegradation of wide variety of compounds by different microorganisms. Whereas, CGAs has probes containing whole genomic DNA and phylogenetic microarrays has short synthetic oligonucleotides from rRNA genes as probes and can be used for the analysis of phylotypes in a microbial community in the contaminated environmental samples. Rhee *et al.* (2004) monitored biodegradation potential and activity by developing comprehensive 50-mer oligonucleotide microarrays containing 1657 probes from all 2402 known genes involved in biodegradation and metal resistance. The functional gene microarrays are based entirely on known genes and pathways reported in individual strain of bacteria, therefore, there is information bias in this technique. Moreover, the specificity and sensitivity of the microarray and correlation of the mRNA and protein expression has always been questioned. Post-genomics techniques such as transcriptomics and proteomics along with metagenomics, which are less restrictive in comparison to gene microarray, can be applied to study biodegradation process with the increasing availability of complete individual genome sequences and metagenomic sequences of environmental samples bypassing the problem of missing out the uncultured microorganisms. These techniques can also reveal how different microorganisms respond to changes in the environment due to the addition of pollutants. Another upcoming and popular post genomic technique is metabolomics which is still in its infant stage. Villas-bôas and Bruheim (2007) in their review article have discussed the potential of various experimental and conceptual approaches developed for metabolomics to be applied in bioremediation research, such as using isotope distribution analysis for elucidation of biodegradation pathways and molecular connectivity analysis, using metabolic footprinting analysis to assess of mineralization process, and the improvement of the biodegradation process via metabolic engineering.

Predicting biodegradation pathway: bioinformatics approach

Numerous studies have already been reported or being done in various laboratories related to bioremediation. And till date, many degradation pathways, along with the enzymes and their respective genes are already known and reactions are well understood. This is possible due to the usage of OMICS- technology in bioremediation studies, which helps in giving researchers a comprehensive and dynamic molecular perspective to view the overall physiology of microorganisms in the presence of contaminants. Therefore, researchers are able to elucidate the mechanism of biodegradation of different compounds by new microorganisms with enzymes and genes involved with the help genomic, transcriptomics and proteomics tools recently. This has led to tremendous increase in the knowledge of microbial degradation pathways and reactions along with the mechanism. As mentioned earlier in the chapter, all these data are compiled under one database i.e. The University of Minnesota Biocatalysis/Biodegradation Database (UM-BBD) (Gao *et al.*, 2010). Using the bioinformatics approach, with the help of these available and previously worked out data, algorithms are fed to machine to predict the biodegradation products and pathways (Wicker *et al.*, 2010). Thus, the hybrid knowledge-based and machine learned data may provide researchers to predict the initial reactions or at least lead them to directed experiments which will save time, money and effort at the same time. Thus, the UM-BBD Pathway Prediction System (PPS) predicts plausible microbial catabolic reactions using substructure searching, a rule-base, and atom-to-atom mapping. The system is able to recognize organic functional groups found in a compound and predict transformations based on biotransformation rules (http://umbbd.msi.umn.edu/predict/) (Wicker *et al.*, 2010). The biotransformation rules are based on reactions found in the UM-BBD database or in the scientific literature. Accuracy of the prediction is more if a compound is similar to compounds whose biodegradation pathways are earlier reported, and the compound forms the sole source of carbon, nitrogen or other essential elements and energy for the microorganisms under the presence of specific environment, which is mainly aerobic with stable and moderate temperature, pH and moisture without any environmental stress such as chemicals or toxins.

Conclusions

Human activities have led to the contamination of both aquatic and terrestrial ecosystems with petroleum hydrocarbons such as alkanes, which are major component of petroleum and fuel (81% are long-chain *n*-alkanes in C_8–C_{17} range), PAHs and PCBs (Zeiger and Smith, 1998). There is continuing usage of pesticides and herbicides in the agricultural sector because of an increasing human population of about 75–80 million annually. Unfortunately, some of these compounds are toxic and can persist in soils and sediments as well as enter the food chain or bioaccumulate within biological tissues or enter ground water leading to possible health problems. Bioremediation can be an effective, generally low cost, innovative environmental technology that has the potential to alleviate numerous contamination problems, and therefore is frequently used for clean up of contaminated sites on the National Priority List (NPL) (USA). Molecular biological and genomic technology has potential to be used in the study of microbial community structure associated with bioremediation of different kinds of environmental contaminants and may also help in monitoring the microbial population dynamics throughout the biodegradation process. For bioremediation to clean up contaminated sites to established guidelines, it is essential to have information on various biodegradative genes and the encoded enzymes involved in the degradation pathway of the targeted compound. With the availability of vast genomic data, and use of genomic tools, efficient bioremediation strategies can be planned and executed. Still there is a paucity of research to elucidate the mechanisms of action of biodegradative enzymes and the regulation of their expression especially in *in situ* conditions. This requires more efforts in the elucidation of molecular mechanisms of degradation of different compounds, which can be similar or different among microorganisms. Thus, it will help to understand how they are mineralizing in nature and also to design efficient bioremediation strategies for the degradation of compounds in diversely contaminated sites. But above all, instead of relying entirely on bioremediation and other restoring technology to clean up the contamination sites, there should be increased environmental awareness and strict regulatory measures should be followed so that past mistakes are not repeated and protect the environment from future contamination and non-sustainable activities. This may require toxicity bioassays/testing (Marwood *et al.*, 1998; Knoke *et al.*, 1999) and risk assessments to be conducted before, during and after the bioremediation program.

Acknowledgements

A. Bej's work was supported by the UAB Faculty Development Award; Tawani Foundation (Col. James Pritzker), Chicago; NASA/NSSTC/VCSI Antarctic expedition (Marty Kress); NCAOR, India (Rasik Ravindra) and AARI/RAE, Russia. J.T.T. is supported by the NSERC (Canada) Discovery Program.

References

Aislabie, J., Bej, A.K., Ryburn, J., Lloyd, N., and Wilkins, A. (2005). Characterization of *Arthrobacter* nicotinovorans HIM, an atrazine-degrading bacterium, from agricultural soil New Zealand. FEMS Microbiol. Ecol. 52, 279–286.

Alexander, M. (1999). Biodegradation and Bioremediation, 2nd ed. (Academic Press, Inc.: San Diego, CA).

Aspelin, A.L., and Grube, A.H. (2000). In Report 733-R-99 C001. U.S. Environmental Protection Agency, Office of Pesticide Programs, Washington, DC.

Atlas, R.M. (1981). Microbial degradation of petroleum hydrocarbons: an environmental perspective. Microbiol Rev. 7, 285–292.

Batie, C.J., Ballou, D.P., and Correll, C.C. (1992). Phthalate dioxygenase reductase and related flavin-iron–sulfur containing electron transferases. Chem. Biochem. Flavoenzy. 3, 543–556.

Beaudette, L., Cassidy, M.B., Habash, M., Kirk, J., Lee, H., and Trevors. J.T. (2001). Bioremediation. Encyclopedia of Environmental Microbiology (John Wiley and Sons: New Jersey, USA).

van Beilen, J.B., and Funhoff, E.G. (2007). Alkane hydroxylases involved in microbial alkane degradation. Appl. Microbiol. Biotechnol. *74*, 13–21.

van Beilen, J.B., Penninga, D., and Witholt, B. (1992). Topology of the membrane-bound alkane hydroxylase of *Pseudomonas* oleovorans. J. Biol. Chem. *267*, 9194–9201.

van Beilen, J.B., Wubbolts, M.G., and Witholt, B. (1994). Genetics of alkane oxidation by *Pseudomonas* oleovorans. Biodegradation 5, 161–174.

van Beilen, J.B., Smits, T.H., Whyte, L.G., Schorcht, S., Rothlisberger, M., Plaggemeier, T., Engesser, K.H., and Witholt, B. (2002). Alkane hydroxylase homologues in Gram-positive strains. Environ. Microbiol. *4*, 676–682.

van Beilen, J.B., Marin, M.M., Smits, T.H., Rothlisberger, M., Franchini, A.G., Witholt, B., and Rojo, F. (2004). Characterization of two alkane hydroxylase genes from the marine hydrocarbonoclastic bacterium *Alcanivorax borkumensis*. Environ. Microbiol. *6*, 264–273.

van Beilen, J.B., Holtackers, R., Luscher, D., Bauer, U., Witholt, B., and Duetz, W.A. (2005) Biocatalytic production of perillyl alcohol from limonene by using a novel *Mycobacterium* sp. Cytochrome P450 alkane hydroxylase expressed in *Pseudomonas putida*. Appl. Environ. Microbiol. *71*, 1737–1744.

van Beilen, J.B., Funhoff, E.G., van Loon, A., Just, A., Kaysser, L., Bouza, M., Holtackers, R., Röthlisberger, M., Li, Z., and Withholt, B. (2006). Cytochrome P450 alkane hydroxylases of the CYP153 family are common in alkane-degrading eubacteria lacking integral membrane alkane hydroxylases. Appl. Environ. Microbiol. 72, 59–65.

Bej, A., Saul, D., and Aislabie, J. (2000). Cold-tolerant alkane-degrading *Rhodococcus* species from Antarctica. Polar Biol. 23, 100–105.

Bosch, R., Moore, E.R., García-Valdés, E., and Pieper, D.H. (1999). NahW, a novel, inducible salicylate hydroxylase involved in mineralization of naphthalene by *Pseudomonas stutzeri* AN10. J. Bacteriol. *181*, 2315–2322.

Boundy-Mills, K., de Souza, M.L., Mandelbaum, R.M., Wackett, L.P., and Sadowsky, M.J. (1997). The atzB gene of *Pseudomonas* sp. strain ADP encodes the second enzyme of a novel atrazine degradation pathway. Appl.Environ. Microbiol. *63*, 916–923.

Cerdan, P., Wasserfallen, A., Rekik, M., Timmis, K.N., Harayama, S. (1994). Substrate specificity of catechol 2,3-dioxygenase encoded by TOL plasmid pWW0 of *Pseudomonas putida* and its relationship to cell growth. J. Bacteriol. *176*, 6074–6081.

Cerniglia, C.E. (1992). Biodegradation of polycyclic aromatic hydrocarbons. Biodegradation *3*, 351–368.

Dubbels, B.L., Sayavedra-Soto, L.A., and Arp, D.J. (2007). Butane monooxygenase of '*Pseudomonas* butanovora': purification and biochemical characterization of a terminal alkane hydroxylating diiron monooxygenase. Microbiology *153*, 1808–1816.

Eaton, R.W., and Chapman, P.J. (1992). Bacterial metabolism of naphthalene: construction and use of recombinant bacteria to study ring cleavage of 1,2-dihydroxynaphthalene and subsequent reactions. J. Bacteriol. *174*, 7542–54.

Ensley, B.D., and Haigler, B.E. (1990). Naphthalene dioxygenase from *Pseudomonas* NCIB 9816. Methods Enzymol. *188*, 46–52.

EPA (1998). Handbook for Air Toxics Emission Inventory Development, Volume I: Stationary Sources. (National Service Center for Environmental Publications, Research Triangle Park, NC.).

Errampalli, D., Trevors, J.T., Lee, H., Leung, K., Cassidy, M., Knoke, K., Marwood, T., Shaw, K., Blair M., and Chung. E. (1997). Bioremediation: a perspective. J. Soil Contaminat. *6*, 207–218.

Feng, L., Wang, W., Cheng, J., Ren, Y., Zhao, G., Gao, C., *et al.* (2007). Genome and proteome of long-chain alkane degrading *GeoBacillus thermodenitrificans* NG80–2 isolated from a deep subsurface oil reservoir. Proc. Natl. Acad. Sci. U.S.A. *104*, 5602–5607.

Furukawa, K.,and Miyazaki, T. (1986). Cloning of a gene cluster encoding biphenyl and chlorobiphenyl degradation in *Pseudomonas* pseudoAlcaligenes. J. Bacteriol. *166*, 392–398.

Gao, J., Ellis, L.B.M., and Wackett, L.P. (2010). The University of Minnesota Biocatalysis/Biodegradation Database: improving public access. Nucleic Acids Res. *38*, D488–D491.

Gibson, D.T., and Sayler, G.S., eds. (1992). Scientific Foundations of Bioremediation: Current Status and Future Needs (American Academy of Microbiology, Washington, D.C).

Hakemian, A.S., and Rosenzweig, A.C. (2007). The biochemistry of methane oxidation. Annu. Rev. Biochem. 76, 223– 241.

Hara, A., Syutsubo, K., and Harayama, S. (2003). Alcanivorax which prevails in oil-contaminated seawater exhibits broad substrate specificity for alkane degradation. Environ. Microbiol. *5*, 746- 753.

Hara, A., Baik, S.H., Syutsubo, K., Misawa, N., Smits, T.H., van Beilen, J.B., and Harayama, S. (2004). Cloning and functional analysis of alkB genes in *Alcanivorax borkumensis* SK2. Environ. Microbiol. *6*, 191–197.

Jensen, A.A., and Slorach, S.A. (1991). Chemical Contaminants in Human Milk (CRC Press, Inc.: Boca Raton, FL).

Johri, A.K., Dua, M., Tuteja, D., Saxena, R., Saxena, D.M., and Lal, R. (1998). Degradation of alpha, beta, gamma, and delta-hexachlorocyclohexane by *Sphingomonas paucimobilis*. Biotechnol. Lett. *20*, 885–887.

Jouanneau, Y., Meyer, C., Jakoncic, J., Stojanoff, V., and Gaillard, J. (2006). Characterization of a naphthalene dioxygenase endowed with an exceptionally broad substrate specificity toward polycyclic aromatic hydrocarbons. Biochemistry *45*, 12380–12391.

Kanavllil, N., Kurissery, S., Leung, K.T., Lee, H., and Trevors, J.T. (2006). Biodegradation and bioremediation of organic pollutants. In Modern Soil Microbiology II, Van Elsas, J.D., Trevors, J.T., and Jansson, J., eds. (Taylor and Francis CRC Press: Boca Raton, FL).

Kauppi, B., Lee, K., Carredano, E., Parales, R.E., Gibson, D.T., Eklund, H., and Ramaswamy, S. (1998). Structure of an aromatic-ring-hydroxylating dioxygenase-naphthalene 1,2-dioxygenase. Structure *6*, 571–586.

Kiyohara, H., Torigoe, S., Kaida, N., Asaki, T., Iida, T., Hayashi, H., and Takizawa, N. (1994). Cloning and characterization of a chromosomal gene cluster, PAH, that encodes the upper pathway for phenanthrene and naphthalene utilization by *Pseudomonas putida* OUS82. J. Bacteriol. *176*, 2439–2443.

Knoke, K.L., Marwood, T.M., Cassidy, M.B., Lee, H., Trevors, J.T., Liu, D., and Seech, A.G. (1999). A comparison of five bioassays to monitor toxicity during bioremediation of pentachlorophenol-contaminated soil. Water Air Soil Pollut. *110*, 157–169.

Kok, M., Oldenhuis, R., van der Linden, M.P., Raatjes, P., Kingma, J., van Lelyveld, P.H., and Witholt, B. (1989). The *Pseudomonas* oleovorans alkane hydroxylase gene. Sequence and expression. J. Biol. Chem. *264*, 5435–5441.

Kuhn, E., Bellicanta, G.V., and Pellizari, V.H. (2009). New alk genes detected in Antarctic marine sediments. Environ. Microbiol. *11*, 669–673.

Kumari, R., Subudhi, S., Suar, M., Dhingra, G., Raina, V., Dogra, C., Lal, S., van der Meer, J.R., Holliger, C., and Lal, R. (2002). Cloning and characterization of lin genes responsible for the degradation of Hexachlorocyclohexane isomers by *Sphingomonas paucimobilis* strain B90. Appl. Environ. Microbiol. *68*, 6021–6028.

Kurkela, S., Lehvaslaiho, H., Palva, E.T., and Teeri, T.H. (1988). Cloning, nucleotide sequence and characterization of genes encoding naphthalene dioxygenase of *Pseudomonas putida* strain NCIB9816. Gene 73, 355–362.

Labinger, J.A., and Bercaw, J.E. (2002). Understanding and exploiting C–H bond activation. Nature *417*, 507–514.

Larkin, M.A., Blackshields, G., Brown, N.P., Chenna, R., McGettigan, P.A., McWilliam, H., Valentin, F., Wallace, I.M., Wilm, A., Lopez, R., *et al.* (2007). ClustalW and ClustalX version 2. Bioinformatics 23, 2947–2948.

Maeng, J.H., Sakai, Y., Ishige, T., Tani, Y., and Kato, N. (1996). Diversity of dioxygenases that catalyze the first step of oxidation of long-chain n-alkanes in *Acinetobacter* sp. strain M-1. FEMS Microbiol. Lett. *141*, 177–182.

Maier, T., Forster, H.H., Asperger, O., and Hahn, U. (2001). Molecular characterization of the 56 kDa CYP153 from *Acinetobacter* sp. EB104. Biochem. Biophys. Res. Commun. *286*, 652–658.

Mandelbaum, R.T., Allan, D.L., and Wackett, L.P. (1995). Isolation and characterization of a *Pseudomonas* sp. that mineralizes the s-triazine herbicide atrazine. Appl. Environ. Microbiol. *61*, 1451–1457.

Marco, D., ed. (2010). Metagenomics: Theory, Methods and Applications. (Caister Academic Press: Universidad Nacional de Cordoba, Argentina).

Marwood, T.M., Suchorski-Tremblay, A., Flemming, C.A., Hodge, V., Knoke, K., Yau, K., Liu, D.L., Seech, A.G., Lee, H., and Trevors, J.T. (1998). Soil toxicity during bioremediation of diesel fuel-spiked soil. Environ. Toxicol. Water Qual. *13*, 117–126.

McDonald, I.R., Uchiyama, H., Kambe, S., Yagi, O., and Murrell, J.C. (1997). The soluble methane monooxygenase gene cluster of the trichloroethylene-degrading methanotroph Methylocystis sp. strain M. Appl. Environ. Microbiol. *63*, 1898–1904.

Menn, F., Applegate, B.M., and Sayler, G.S. (1993). NAH plasmid-mediated catabolism of anthracene and phenanthrene by naphthoic acids. Appl. Environ. Microbiol. *59*, 1938–1942.

Nagy, I., Compernolle, F., Ghys, K., Vanderleyden, J., and De Mot R. (1995). a single cytochrome P450 system is involved in degradation of the herbicides EPTC (S-ethyl dipropylthiocarbamate) and atrazine by *Rhodococcus* sp. strain NI86/21. Appl. Environ. Microbiol. *61*, 2056–2060.

Nakamura, T., Hoaki T., Hanada S., Maruyama, A., Kamagata, Y., and Fuse H. (2007). Soluble and particulate methane monooxygenase gene clusters in the marine methanotroph *Methylomicrobium* sp. strainNI. FEMS Microbiol. Lett. 277, 157–164.

Nam, J.W., Nojiri, H., Yoshida, T., Habe, H., Yamane, H., and Omori, T. (2001). New classification system for oxygenase components involved in ring-hydroxylating oxygenations. Biosci. Biotechnol. Biochem. 65, 254–263.

Neidle, E.L., Hartnett, C., Ornston, L.N., Bairoch, A., Rekik, M., and Harayama, S. (1991) Nucleotide sequences of the *Acinetobacter* calcoaceticus benABC genes for benzoate 1,2-dioxygenase reveal evolutionary relationships among multicomponent oxygenases. J. Bacteriol. *173*, 5385–5395.

Nigam, U., and Siddqui. K. J. (2001). Organochlorine insecticide residues in dairy milk samples collected in Lucknow. India. Bull. Environ. Contam. Toxicol. 66, 678–682.

Noren, K., and Meironyte, D. (2000). Organochlorine and organobromine contaminants in Swedish human milk in perspective of past 20–30 years. Chemosphere 44, 1111–1123.

Olsen, R.H., Kukor, J.J., Kaphammer, B. (1994). A novel toluene-3-monooxygenase pathway cloned from *Pseudomonas* pickettii PKO1 J. Bacteriol. *176*, 3749–3756.

Panicker, G., Aislabie, J., Saul, D., Bej, A.K. (2002). Cold tolerance of *Pseudomonas* 30-3 isolated from oil-contaminated soil, Antarctica. Polar Biol. 25, 5–11.

Panicker, G., Mojib, N., Nakatsuji, T, Aislabie, J., and Bej, A.K. (2010). Occurrence and distribution of capB in Antarctic microorganisms and study of its structure and regulation in the Antarctic biodegradative *Pseudomonas* sp. 30/3. Extremophiles *14*, 171–83.

Phillips, M., and McDougall, J. (1993). Triazine herbicides. In Wood-McKenzie report, Wood McKenzie Ltd, London, UK, Part 1, pp. 10–11.

Phillips, T.M., Seech, A.G., Lee, H., and Trevors, J.T. (2006). Full-scale *in situ* bioremediation of hexachlorocyclohexane contaminated soil. J. Chem. Technol. Biotechnol. *81*, 289–298.

Radosevich, M., Traina, S.J., Hao, Y.-L., and Tuovinen, O.H. (1995). Degradation and mineralization of atrazine by a soil bacterial isolate. Appl. Environ. Microbiol. *61*, 297–302.

Ratajczak, A., Geidorfer, W., Hillen, W. (1998). Alkane hydroxylase from *Acinetobacter* sp. strain ADP-1 is encoded by alkM and belongs to a new family of bacterial integral membrane hydrocarbon hydroxylases. Appl. Environ. Microbiol. *64*, 1175–1179.

Rhee, S.K., Liu, X., Wu, L., Chong, S.C., Wan, X., and Zhou, J. (2004). Detection of genes involved in biodegradation and biotransformation in microbial communities by using 50-mer oligonucleotide microarrays. Appl. Environ. Microbiol. *70*, 4303–4317.

Rojo, F. (2009). Degradation of alkanes by bacteria. Environ. Microb. *11*, 2477- 2490.

Sadowsky, M.J., Tong, Z., de Souza, M.L., and Wackett, L.P. (1998). AtzC is a new member of the amidohydrolase protein superfamily and is homologous to other atrazine-metabolizing enzymes. J. Bacteriol. *180*, 152–158.

Sahu, S.K., Patnaik, K.K., Sharmila, M., and Sethunathan, N. (1990). Degradation of alpha-, beta- and gamma-hexachlorocyclohexane by a soil bacterium under aerobic conditions. Appl. Environ. Microbiol. *56*, 3620–3622.

Sazinsky, M.H., and Lippard, S.J. (2006). Correlating structure with function in bacterial multicomponent monooxygenases and related diiron proteins. Acc. Chem. Res. *39*, 558–566.

Schneiker, S., Martins dos Santos, V.A., Bartels, D., Bekel, T., Brecht, M., Buhrmester, J., Chernikova, T.N., Denaro, R., Ferrer, M., *et al.* (2006). Genome sequence of the ubiquitous hydrocarbon-degrading marine bacterium *Alcanivorax borkumensis*. Nat. Biotechnol. 24, 997–1004.

Schrijver, A.D., and Mot, R.D. (1999).Degradation of pesticides by actinomycetes. Crit. Rev. Microbiol. 25, 85–119.

Sekine, M., Tanikawa, S., Omata, S., Saito, M., Fujisawa, T., Tsukatani, N., Omata, S., Saito, M., Fujisawa, T., Tsukatani, N., Tajima, T., Sekigawa, T., Kosugi, H., Matsuo, Y., *et al.* (2006). Sequence analysis of three plasmids harboured in *Rhodococcus erythropolis* strain PR4. Environ. Microbiol. 8, 334–346.

Senoo, K., and Wada., H. (1989). Isolation and identification of an aerobic γ-HCH-decomposing bacterium from soil. Soil Sci. Plant Nutr. 35, 79–87.

Shaw, J.P., and Harayama, S. (1992). Purification and characterisation of the NADH:acceptor reductase component of xylene monooxygenase encoded by the TOL plasmid pWW0 of *Pseudomonas putida* mt-2. Eur. J. Biochem. *209*, 51–61.

Shields, M.S., Montgomery, S.O., Cuskey, S.M., Chapman, P.J., and Pritchard, P.H. (1991). Mutants of *Pseudomonas cepacia* G4 defective in catabolism of aromatic compounds and trichloroethylene. Appl. Environ. Microbiol. *57*, 1935–1941.

Shields, M.S., Reagin, M.J., Gerger, R.R., Campbell, R., and Somerville, C. (1995). TOM, a new aromatic degradative plasmid from *Burkholderia* (*Pseudomonas*) *cepacia* G4. Appl. Environ. Microbiol. *61*, 1352–1356.

Simonich, S. L., and Hites, R. A. 1995. Global distribution of persistent organochlorine compounds. Science 269, 1851–1854.

Sluis, M.K., Sayavedra-Soto, L.A., and Arp, D.J. (2002). Molecular analysis of the soluble butane monooxygenase from '*Pseudomonas* butanovora'. Microbiology 148, 3617–3629.

Sotsky, J.B., Greer, C.W., and Atlas, R.M. (1994). Frequency of genes in aromatic and aliphatic hydrocarbon biodegradation pathways within bacterial populations from Alaskan sediments. Can. J. Microbiol. *40*, 981–985.

de Souza, M.L., Wackett, L.P., Boundy-Mills, K.L., Mandelbaum, R.T., and Sadowsky, M.J. (1995). Cloning, characterization, and expression of a gene region from *Pseudomonas* sp. strain ADP involved in the dechlorination of atrazine. Appl. Environ. Microbiol. *61*, 3373–3378.

de Souza, M.L., Sadowsky, M.J., and Wackett, L.P. (1996). Atrazine chlorohydrolase from *Pseudomonas* sp. strain ADP: gene sequence, enzyme purification, and protein characterization. J. Bacteriol. *178*, 4894–4900.

de Souza, M.L., Seffernick, J., Martinez, B., Sadowsky, M.J., and Wackett, L.P. (1998). The atrazine catabolism genes atzABC are widespread and highly conserved. J. Bacteriol. *180*, 1951- 954.

Sutherland, T.D., Horne, I., Russell, R.J., and Oakeshott, J.G. (2002). Gene cloning and molecular characterization of a two-enzyme system catalyzing the oxidative detoxification of beta-endosulfan. Appl. Environ. Microbiol. *68*, 6237–6245.

Suzuki, M., Hayakawa, T., Shaw, J.P., Rekik, M., Harayama, S. (1991). Primary Structure of xylene monooxygenase: similarities to and differences from the alkane hydroxylation system. J. Bacteriol. *173*, 1690–1695.

Swetha, V.P., and Phale, P.S. (2005). Metabolism of carbaryl via 1, 2-dihydroxynaphthalene by soil isolates *Pseudomonas* sp. strains C4, C5, and C6. Appl. Environ. Microbiol. *71*, 5951–5956.

Tani, A., Ishige, T., Sakai, Y., and Kato, N. (2001). Gene structures and regulation of the alkane hydroxylase complex in *Acinetobacter* sp. strain M-1. J. Bacteriol. *183*, 1819–1823.

Thomas, J.C., Berger, F., Jacquier, M., Bernikkon, D., Baud-Grasset, F., Truffaul, N., Normand, P., Vogel, T.M., and Simonet, P. (1996). Isolation and characterization of a novel γ-hexachlorocyclohexane-degrading bacterium. J. Bacteriol. *178*, 6049–6055.

Throne-Holst, M., Wentzel, A., Ellingsen, T.E., Kotlar, H.K., and Zotchev, S.B. (2007). Identification of novel genes involved in long-chain n-alkane degradation by *Acinetobacter* sp.strain DSM 17874. Appl. Environ. Microbiol. 73, 3327–3332.

Villas-bôas, S.G., and Bruheim, P. (2007). The potential of metabolomics tools in bioremediation studies. OMICS J. Int. Biol. *11*, 305–313.

Vomberg, A., and Klinner, U. (2000). Distribution of alkB genes within n-alkane-degrading bacteria. J. Appl. Microbiol. *89*, 339–348.

Wasmund, K., Burns, K.A., Kurtböke D.I., and Bourne, D.G. (2009). Novel alkane hydroxylase gene (alkB) diversity in sediments associated with hydrocarbon seeps in the Timor Sea, Australia. Appl. Environ. Microbiol. 75, 7391–7398.

Weir, K.M., Sutherland, T.D., Horne I., Russell, R.J., and Oakeshott J.G. (2006). A single monooxygenase, ese, is involved in the metabolism of the organochlorides endosulfan and endosulfate in an *Arthrobacter* sp. Appl. Environ. Microbiol. *72*, 3524–3530.

Whyte, L.G., Hawari, J., Zhou, E., Bourbonniere, L., Inniss, W.E., Greer, C.W. (1998). Biodegradation of variable-chain-length alkanes at low temperatures by a psychrotrophic *Rhodococcus* sp. Appl. Environ. Microbiol. *64*, 2578–2584.

Whyte, L.G., Smits, T.H., Labbe, D., Witholt, B., Greer, C.W., van Beilen, J.B. (2002a). Gene cloning and characterization of multiple alkane hydroxylase systems in *Rhodococcus* strains Q15 and NRRL B-16531. Appl. Environ. Microbiol. *68*, 5933–5942.

Whyte, L.G., Schultz, A., van Bielen, J.B., Luz, A.P., Pellizari,V., Labbe, D., Greer, C.W. (2002b). Prevalence of alkane monooxygenase genes in Arctic and Antarctic hydrocarbon contaminated and pristine soils. FEMS Microbiol. Ecol. *41*, 141–150.

Wicker, J., Fenner, K., Ellis, L., Wackett, L., and Kramer, S. (2010). Predicting biodegradation products and pathways: a hybrid knowledge-based and machine learning based approach. Bioinformatics *26*, 814–821.

Windholz, M., Budavari, S., Stroumtsos, L.Y., and Fertig, M.N., eds. (1976). The Merck Index, 9th ed. (Merck & Co., Inc.: Rahway, NJ) p. 719–720.

Yanze-Kontchou, C., and Gschwind, N. (1994). Mineralization of the herbicide atrazine as a carbon source by a *Pseudomonas* strain. Appl. Environ. Microbiol. *60*, 4297–4302.

Yen, K.M., Karl, M.R., Blatt, L.M., Simon, M.J., Winter, R.B., Fausset, P.R., Lu, H.S., Harcourt, A.A., and Chen, K.K. (1991). Cloning and characterization of a *Pseudomonas* mendocina KR1 gene cluster encoding toluene-4-monooxygenase. J. Bacteriol. *173*, 5315–5327.

Zahn, J.A., and DiSpirito, A.A. (1996). Membrane-associated methane monooxygenase from Methylococcus capsulatus (Bath). J. Bacteriol. *178*, 1018–1029.

Zeiger, E., and Smith, L. (1998). The first international conference on the environmental health and safety of Jet fuel. Environ. Health Perspect. *106*, 763–764.

Zhou, J.Z., and Thompson, D.K. (2002). Challenges in applying microarrays to environmental studies. Curr. Opin. Biotechnol. *13*, 202–204.

Zylstra, G. J., and Gibson, D.T. (1989). Toluene degradation by *Pseudomonas putida* F1: nucleotide sequence of the tod C1C2BADE genes and their expression in *Escherichia coli*. J. Biol. Chem. *264*, 14940–14946.

Molecular Technologies for Analysis of Petroleum Bioremediation

11

Yuki Kasai

Abstract

Bioremediation is a cost-effective technique for treatment of polluted environments and it involves usage of microorganisms for pollutant degradation. It can be defined as natural attenuation (intrinsic bioremediation), biostimulation (introduction of nutrients and chemicals to stimulate indigenous microorganisms), and bioaugmentation (inoculation with exogenous microorganisms). When carrying out bioremediation, special attention should be paid to its effects on the indigenous microbiota and dispersal and outbreaks of the inoculated organisms. Recent advances in microbial ecology have provided molecular technologies, such as community fingerprinting, molecular detection of degradative genes, and metagenomics, which facilitate the analysis and monitoring of indigenous and inoculated microorganisms in contaminated sites. This chapter outlines these technologies and discusses how they can contribute to bioremediation.

Introduction

Petroleum contamination results from leaking aboveground and underground storage tanks, accidental spills during transport of petroleum products, releases from abandoned manufactured gas plants, other unplanned releases, and industrial processes. Petroleum hydrocarbons, such as gasoline and diesel fuel, consist of complex mixtures of alkanes and aromatics as well as nitrogen-, oxygen-, and sulfur-containing compounds. Because of the negative impact of the released petroleum on health and the environment, it is considered a major contaminant of soil and aquatic environments. Bioremediation, which involves exploitation of the ability of microorganisms to degrade and/or detoxify organic contaminants, has been established as an efficient, economic, versatile, and environmentally sound treatment for treating petroleum-contaminated environments (Norris, 1994). Bioremediation can be defined as monitored natural attenuation (intrinsic bioremediation), biostimulation, and bioaugmentation.

Natural attenuation (NA) is a passive remediation for degradation of pollutants and occurs at most polluted sites. It relies on natural processes to attenuate contaminant concentrations to an acceptable level. The roles of native microorganisms in eliminating contaminants in NA are documented through tests performed at field sites or on-site-derived samples. This is called monitored NA (MNA). MNA has been employed for remediation of soils and groundwaters contaminated with pollutants such as petroleum hydrocarbons, chlorinated solvents, and methyl-t-butyl ether (Roling and van Verseveld, 2002; Rittmann, 2004; Scow and Hicks, 2005). The speed of pollutant removal in MNA is generally slow, and in many cases, biostimulation has been employed to accelerate it. Biostimulation involves modification of the environment to stimulate the existing bacteria for pollutant removal. This can be done by adding various forms of limiting nutrients, such as phosphorus, nitrogen, or carbon (e.g. in the form of molasses), and chemicals that induce the expression of enzymes involved in pollutant removal. Bioaugmentation is defined as introduction of cultured microorganisms into a contaminated environment for the purpose of enhancing bioremediation of pollutants. This process is primarily used to remove pollutants for which degradative microorganisms are not commonly present in the environment (El Fantroussi and Agathos, 2005; Thompson *et al.*, 2005). Although rapid removal of pollutants can be expected through bioaugmentation, it is essential to consider risks associated with the introduction of exogenous organisms, such as ecological impacts. Specific monitoring of the introduced microorganisms is therefore necessary.

Since most bioremediation processes involve the use of bacteria, analyses of the indigenous and introduced organisms can provide valuable information for evaluating the progress of bioremediation. In the 1990s, a variety of molecular ecological methods were developed for the analysis of natural microbial populations, and these methods have greatly advanced our understanding of microbial ecology (Von Wintzingerode *et al.*, 1997; Theron and Cloete, 2000; Dahllo, 2002; Watanabe and Hamamura, 2003; Spiegelman *et al.*, 2005). In such molecular ecological methods, one type of gene, e.g. a 16S rRNA gene or a gene encoding a specific enzyme, is detected simultaneously in different microbial populations by polymerase chain reaction (PCR) and/or hybridization, which obviates the need for isolation and cultivation of microorganisms. Compared with the culture-dependent methods, molecular methods are expected to provide rapid detection a wider range of natural microbial populations. When using molecular methods, however, the users need to know their inherent advantages and disadvantages and select appropriate methods according to the purpose. Furthermore, scientists have recently attempted to introduce metagenomic approaches (a metagenome is defined as a mixture of genomes extracted from different species) in this field (Eyers *et al.*, 2004; van Beilen and Funhoff, 2005); here, pollutant degradation in the environment is considered in the genomic context (e.g. in terms of pollutant degradation enzyme networks).

In this chapter, I summarize some recently developed molecular ecological methods that can be used in bioremediation practice. Some of these methods have already been applied in areas of bioremediation such as community fingerprinting, while others, such as metagenomic approaches, are still in their developmental stages.

Community fingerprinting

Community fingerprinting (also known as molecular fingerprinting and phylogenetic fingerprinting) is a technique for sketching an overall pattern of a microbial community that can be used for rapid comparison of different microbial communities (Theron and Cloete, 2000; Dahllo, 2002; Spiegelman *et al.*, 2005). The community fingerprinting methods used in microbial ecology are summarized in Table 11.1.

In principle, any gene can be used for community fingerprinting. While the 16S rRNA gene is used most frequently for generation of total bacterial community fingerprints in terms of universal distribution, a huge number of sequences is stored in the database (over 150,000 sequences) and the unique primary structure of this gene is suitable for PCR amplification (alternating conserved and variable regions). Alternatively, other phylogenetic markers (e.g. genes encoding the protein elongation factor, heat-shock proteins, glutamine synthetase, ATPases, and topoiosomerases) and genes coding for catabolic enzymes in pollutant degradation pathways (degradative genes) can also be used for community fingerprinting. In particular, the use of degradative genes is considered useful in bioremediation for analysing microbial populations of interest (Watanabe *et al.*, 1998; Oved *et al.*, 2001; Futamata *et al.*, 2003).

Community fingerprinting methods are generally considered to be semiquantitative owing to inherent biases (Von Wintzingerode *et al.*, 1997; Wilson, 1997). Biases can originate from differences in cell rigidness affecting DNA extraction efficiency, variation in the number of gene copies per genome, and sequence heterogeneity in primer (or probe) regions. Nucleic acid isolation from an environmental sample is the most important step in examining the microbial community. Therefore, a primary consideration should be optimization of the DNA/RNA extraction method for each site (Frostegard *et al.*, 1999; Hurt *et al.*, 2001; Martin-Laurent *et al.*, 2001; Stach *et al.*, 2001). In

Table 11.1 Community fingerprinting methods

Method	Procedure	Advantage	Disadvantage	Reference
DGGE/TGGE	PCR and electrophoresis in a polyacrylamide gel with an increasing concentration gradient of denaturing agents or temperature gradient	High resolution (1 bp difference); inexpensive equipment; sequence determination possible	Poor reproducibility; semi-quantitative; several bands from one species; possible containment of several different sequences in one band	Watanabe *et al*. (1998), Roling *et al*. (2001), Kasai *et al*. (2005)
T-RFLP	PCR, restriction enzyme digestion, and separation of terminal fragments using a DNA sequencer	High reproducibility; good gel-to-gel comparison; simple procedure	Expensive equipments; semi-quantitative; several peaks from one species; representation of different sequences by one peak; sequence information unavailable	Liu *et al*. (1997), Cavalca *et al*. (2004);, Macbeth *et al*. (2004), Katsivela *et al*. (2005)
SSCP	PCR, heat denaturation, and polyacrylamide gel electrophoresis	Good gel-to-gel comparison; cheap equipments	Semiquantitative; overestimation of the diversity owing to the presence of multiple conformations of one sequence; sequence information unavailable	Pirie *et al*. (1999), Zumstein *et al*. (2000), Wenderoth *et al*. (2003), Macedo *et al*. (2005)
PCR-ALH	PCR and size separation by polyacrylamide gel electrophoresis	Good gel-to-gel comparison; sequence determination possible; cheap equipments; simple procedure	Semi-quantitative; limited resolution; several bands from one species	Borneman and Triplett (1997), Fisher and Triplett, (1999), Ritchie *et al*. (2000), Zhang *et al*. (2005)
ARDRA	PCR, digestion using restriction endonucleases, and separation of resulting fragments by high-density agarose or acrylamide gel electrophoresis	High reproducibility; simple procedure	Semiquantitative; limited resolution; sequence information unavailable	Watts *et al*. (2001)
DHPLC	PCR and separation of resulting fragments using denaturing HPLC	Rapid, sensitive, and accurate; high reproducibility; conventional equipments	Few applications	
Microarray	Labelling of DNA or RNA fragments and parallel hybridization with fabricated DNA probes	High throughput; good array-to-array comparison; elimination of PCR bias	Expensive; difficulties in hybridization-condition settings; uncertainty in signal specificity; only applicable to known sequences	Koizumi *et al*. (2002), Loy *et al*. (2005), Neufeld *et al*. (2006)

DGGE/TGGE, denaturing or temperature gradient gel electrophoresis; T-FFLP, terminal restriction fragment length polymorphism; SSCP, single-strand conformation polymorphism; PCR-ALH, polymerase chain reaction amplicon length heterogeneity; ARDRA, amplified ribosomal DNA restriction analysis; DHPLC, denaturing high-performance liquid chromatography.

addition, PCR users should design or select PCR primers carefully, since it was reported that use of different primer sets resulted in the detection of different 16S rRNA gene sequence types from the same environmental DNA samples (Watanabe *et al.*, 2002; Baker and Cowan, 2004). It is therefore recommended, at least in the initial stage of a study, that multiple PCR primer sets should be used in order to obtain reliable community fingerprints.

Below I briefly explain several community fingerprinting methods (summarized in Table 11.1) and present examples of their application to bioremediation studies. These examples illustrate how community fingerprinting methods have provided bioremediation studies with information about *in situ* mechanisms of pollutant degradation and the effects of treatments on indigenous microbial biota.

Denaturing gradient gel electrophoresis (DGGE) and its cousin thermal GGE (TGGE) are electrophoretic techniques used to separate PCR-amplified DNA fragments of the same length according to the differences in sequences (melting temperature) (Myers *et al.*, 1987; Muyzer *et al.*, 1993; Muyzer, 1999). Because the melting temperature of a DNA fragment is determined by its G+C content, DGGE and TGGE are acceptable methods for separating DNA fragments based on sequence differences. DGGE uses concentration gradients of denaturant (i.e. urea and formamide) to separate DNA fragments, while TGGE uses temperature gradient. These electrophoreses have also been used for single-nucleotide polymorphism analysis (Myers *et al.*, 1987) as they can separate DNA fragments differing by even one nucleotide. In microbial ecology, PCR-amplified 16S rRNA gene fragments from community DNA are commonly used. When using these methods the user should be aware of their disadvantages (see Table 11.1). For example, (1) many organisms have several 16S rRNA gene copies with slightly different sequences in their genomes, and (2) DNA fragments with different sequences can have the same melting temperature.

Thus far, several studies have used DGGE for analysing bioremediation samples. Roling *et al.* (2001) reported the use of DGGE for analysing NA. They analysed pollutant and redox chemistry in parallel with DGGE profiling of the microbial communities present in a well-defined aquifer contaminated with landfill leachate. Numerical comparison of DGGE banding patterns of bacteria and Archaea in 29 groundwater samples from the aquifer revealed a clear difference between the microbial community structures inside and outside the contaminant plume, indicating that specific microbial populations were established in the contaminated zone. Moreover, the DGGE analysis was combined with sequence analyses of cloned 16S rRNA gene fragments, which showed that specific groups of bacteria contributed to the NA process. A relationship was observed between the dominant redox processes and the bacteria identified, suggesting the presence of zonal variation in the contributing microbial populations. Such information will be useful for predicting the progress of NA in underground aquifers.

In another example, Kasai *et al.* (2005) used slurry microcosm systems to analyse the contribution of biodegradation to NA of petroleum-contaminated soil. Biodegradation potentials were evaluated based on the decay of spiked indicator hydrocarbons, and shifts in the microbial community structure were analysed by DGGE. Comparison of results of the biodegradation test and DGGE results led the authors to suggest that petroleum contamination of the unsaturated zone resulted in the establishment of a fermentative/methanogenic community with substantial hydrocarbon-degrading potential.

Terminal restriction fragment length polymorphism (T-RFLP) is a type of RFLP specifically developed for community fingerprinting. The general procedure includes PCR amplification of target sequences (commonly a region of the 16S rRNA gene) followed by restriction enzyme digestion and gel fractionation of the digested fragments (Liu *et al.*, 1997). In PCR, one primer is fluorescently labelled at the 5′ end, resulting in PCR products being tagged with a fluorescent dye at one terminus. Following restriction enzyme digestion (usually with a tetrameric restriction enzyme), the restricted products are separated by high-resolution (±1 base) gel fractionation in which only the fluorescently tagged terminal fragments are detected. In this procedure, specifically tuned automated DNA sequencers equipped with laser-induced fluorescence detectors have been used in many studies, since this equipment enables detection of as little as 100 attmol of a fluorescent tag (e.g. 6-carboxyfluorescein). Microbial 16S rRNA gene fragments with different terminal restriction sites can be distinguished from each other, and in some cases, the microorganisms present can be phylogenetically identified according to the sizes of the terminal restriction fragments. However, sizing accuracy is reported to be affected by the size calling method employed (Osborn *et al.*, 2000), fragment length, and type of electrophoresis systems used (Nachamkin *et al.*, 2001). On the whole, T-RFLP is best considered as a semiquantitative technique that can be used to gain a rough fingerprint of a whole microbial community; it is unable to identify the phylogenetic positions of microorganisms represented by fragments with unique sizes.

Macbeth *et al.* (2004) analysed a trichloroethene (TCE) residual source area in deep fractured basalt at a US Department of Energy site. They showed that sodium lactate additions resulted in the enrichment of an indigenous microbial community, completely dechlorinating nearly all aqueous-phase TCE to ethene. T-RFLP was used to compare the indigenous community with communities in laboratory enrichment cultures to identify the microbial populations responsible for TCE dechlorination and methane production of the added lactate.

In addition, *in situ* bioremediation of petroleum waste sludge at landfarming sites in a petroleum refinery was studied by monitoring the changes in the petroleum composition of the waste sludge and by T-RFLP, which was used to show temporal changes in the structure of the microbial communities (Katsivela *et al.*, 2005). The authors showed that marked changes in the microbial communities occurred concomitantly with the degradation of petroleum hydrocarbons. In addition, the diversity of the microbial community decreased with increased degradation of the petroleum hydrocarbons contained in the landfarms, suggesting T-RFLP as an efficient indicator of the progress of petroleum biodegradation.

Single-strand conformation polymorphism (SSCP) is a technique that separates single-stranded DNA fragments differing in length and conformation (conformers) by gel electrophoresis. SSCP is generally used to analyse PCR-amplified DNA fragments and was initially used to detect minor sequence variants caused by point mutations in human alleles (Orita *et al.*, 1989). For community fingerprinting of microbial communities, PCR-amplified 16S rRNA gene fragments are separated by SSCP after denaturation at a high temperature (i.e. 95°C for 2–3 minutes) in the presence of denaturants such as formamide (Lee *et al.*, 1996; Schwieger and Tebbe, 1998). Since a single-stranded DNA fragment forms stable conformations according to its sequence, sequence differences contribute to different structures that can be resolved by electrophoresis on a nondenaturing polyacrylamide gel. A DNA fragment can, however, form three major types of structures, i.e. self-folded single-stranded DNA, heteroduplex double-stranded DNA, and homoduplex double-stranded DNA (formed as a result of the high rate of DNA reassociation), which may complicate an SSCP pattern (Lee *et al.*, 1996). Differentiation between single-stranded and double-stranded DNA is possible based on the migration distance and variation in band colours when employing the silver staining method. Similar to T-RFLP, the use of fluorescent dye-labelled PCR primers in combination with analysis using an automated DNA sequencer provides the greatest reproducibility and sensitivity (Pirie *et al.*, 1999; Zumstein *et al.*, 2000).

We can find several example of its application to bioremediation studies, including that described by Wenderoth *et al.* (2003). In their study, microcosms were established using natural groundwater contaminated with 1,2-dichlorobenzene, 1,4-dichlorobenzene, and chlorobenzene, and the effects of different bioremediation strategies, i.e. biostimulation and bioaugmentation, on pollutant removal and indigenous microbiota were evaluated by SSCP. Detailed statistical analyses of SSCP patterns revealed that the community dynamics of the groundwater microbiota were highly reproducible among replications of the same treatment and were specific to the introduced strain, its inoculum size, and the imposed physicochemical conditions. The authors claimed that their findings could contribute to designing better *in situ* bioremediation strategies for chlorobenzene-contaminated groundwater.

Macedo *et al.* (2005) used SSCP to analyse biofilm microbial communities developed at the liquid–liquid interface between polychlorinated biphenyls and water. They monitored the biofilm for 31 days using confocal laser scanning microscopy, SSCP, amplicons of the 16S rRNA genes, and chemical analyses of PCB congeners, revealing that there were several sequential stages of biofilm formation prior to PCB oil colonization. Since the mature biofilm was able to degrade higher-chlorinated PCB congeners, the information should be useful for the biotreatment and bioremediation of PCBs.

PCR amplicon length heterogeneity (PCR-ALH) is a method that distinguishes different microorganisms based on natural variation in the lengths of 16S rRNA gene sequences. As described above, microbial 16S rRNA gene fragments are amplified from different members of a microbial community using PCR primers conserved among microorganisms of interest, and the resulting amplicons are then separated by polyacrylamide gel electrophoresis and visualized using a sensitive staining method (Ritchie *et al.*, 2000). However, since length heterogeneity in the inner regions of the 16S rRNA gene is not sufficient to discriminate diverse microbial populations, the 16S-23S intergenic spacer region in the rRNA operon has also been used in the analysis, i.e. ribosomal intergenic spacer analysis (RISA) (Borneman and Triplett, 1997; Fisher and Triplett, 1999). Like T-RFLP and SSCP, fluorescent-tagged primers and automated DNA sequencer analysis have been used to increase sensitivity and reproducibility in the detection of small amounts of DNA and to separate fragments differing by only 1 or 2 bp, respectively (Fisher and Triplett, 1999; Ritchie *et al.*, 2000). A few examples of its application to bioremediation studies have been reported.

Zhang *et al.* (2005) used RISA to assess the efficacy of using perchlorate (PCE)-reducing bacteria as inoculants in a bioreactor treating PCE-contaminated groundwater. The bioreactor was inoculated with a field-grown suspension of the PCE-respiring bacterium *Dechlorosoma* sp. strain KJ and fed groundwater containing indigenous bacteria and a carbon source amendment. The SSCP analysis, however, could not detect the band representing the strain KJ, although an intense band corresponding to the known PCE degraders of *Dechloromonas* was observed. This result has raised a question about the utility of the strain KJ for PCE bioremediation.

Amplified ribosomal DNA restriction analysis (ARDRA) is commonly used to study microbial diversity that relies on DNA polymorphism (Deng *et al.*, 2008). Clones containing 16S rDNA gene fragments, obtained by applying either universal or genus-specific primer sets, are amplified and digested by restriction endonucleases, and this is followed by separation of the resulting fragments on high-density agarose or acrylamide gels. The emerging profiles are then used either to cluster the community into genotypic groups or for strain typing (Tiedje *et al.*, 1999).

Watts *et al.* (2001) reported a comparative analysis of PCB-dechlorinating communities in enrichment cultures using three different molecular screening techniques: ARDRA, DGGE, and T-RFLP. They found that these methods had different biases, which were apparent from discrepancies in the relative clone frequencies (ARDRA), band intensities (DGGE), or peak heights (T-RFLP) for the same enrichment culture. However, all three methods were effectively quantitative and identified the same organisms (Watts *et al.*, 2001). Overall, in community fingerprinting and preliminary identification, DGGE proved to be the most rapid and effective tool for monitoring microorganisms within a highly enriched culture. The T-RFLP results corroborated the DGGE fingerprint analysis results; however, identification required the additional step of creating a clone library. ARDRA provided an in-depth analysis of the community and this technique detected slight intraspecies sequence variation in 16S rDNA (Watts *et al.*, 2001).

Denaturing high-performance liquid chromatography (DHPLC) is a rapid, sensitive, and accurate method of detecting sequence variation. DHPLC can detect single base-pair mutations within a specific sequence (Taliani *et al.*, 2001). It has been used for the separation and identification of PCR-amplified fragments of 16S rRNA genes for bacterial community analysis, monitoring, and identification (Domann *et al.*, 2003; Belda *et al.*, 2004; Barlaan *et al.*, 2005; Ting *et al.*, 2009).

However, this method has not yet been applied to bioremediation studies.

DNA microarrays (alternatively called DNA microchips or DNA chips) represent a high-throughput format for the highly parallel application of multiple nucleic acid probes by reverse hybridization, and therefore, constitute the most powerful tools in molecular biology. DNA microarrays are routinely applied in the pharmaceutical industry, clinical diagnostics, and various fields of research, such as functional genomics and genetic analysis (Lander, 1999). In recent years, DNA microarrays have also entered the field of microbial ecology (Call *et al.*, 2003; Li and Liu, 2003; Zhou, 2003; Eyers *et al.*, 2004; Lettieri, 2006; Palmer *et al.*, 2006).

In microbial ecology studies, microarrays have most commonly been used to analyse 16S rRNA fragments for phylogenetic fingerprinting of a microbial community (Guschin *et al.*, 1997; Small *et al.*, 2001; Peplies *et al.*, 2004; Palmer *et al.*, 2006). Microbial 16S rRNA fragments can be amplified by PCR from community DNA or RNA prior to microarray hybridization (Wilson *et al.*, 2002; Palmer *et al.*, 2006); however, direct labelling and detection of rRNA fragments has a profound advantage over PCR-mediated methods, i.e. elimination of PCR-generated biases (Call *et al.*, 2003; El Fantroussi *et al.*, 2003; Peplies *et al.*, 2004). In another application of microarrays to microbial ecology, microarray community fingerprinting was coupled with *in situ* labelling of rRNA with a radioisotope-incorporated substrate, which facilitated linkage of the phylogeny of uncultured microorganisms to their functions (Adamczyk *et al.*, 2003).

A preliminary examination of the application of small-scale microarrays to biodegradation experiments was reported by Koizumi *et al.* (2002), who analysed 16S rRNA fragments in mesophilic toluene- and ethylbenzene-degrading consortia and oil-contaminated sediments from the coast of Kuwait. The authors compared the results obtained by microarray detection with those obtained by DGGE fingerprinting and membrane hybridization, and found both consistency and inconsistency in their results. They claimed that although DNA microarray technology has limitations, it will ultimately provide a rapid and high-throughput platform for microbial population analysis.

Loy *et al.* (2005) developed a microarray for the detection of 79 known 16S rRNA sequence types from the order Rhodocyclales. After evaluation using pure cultures of bacteria affiliated with the Rhodocyclales, the authors applied the microarray to the analysis of activated sludge samples and showed the presence of as yet uncultured bacteria related to *Zoogloea*, *Ferribacterium*/*Dechloromonas*, and *Sterolibacterium*. We are interested in this microarray, since the order Rhodocyclales is known to include bacteria that are capable of degrading aromatic hydrocarbons under denitrifying conditions (Fries *et al.*, 1994; Coates *et al.*, 2001; Kasai *et al.*, 2006) and are known to contribute to the NA of gasoline-contaminated aquifers (Hess *et al.*, 1997; Kasai *et al.*, 2006).

Microarray technology has also been used to characterize and compare hexachlorocyclohexane (HCH)-contaminated soils (Neufeld *et al.*, 2006). A microarray carrying probes specific to the 100 most abundant ribosomal sequence tags was designed after the analysis of a clone library of 16S rRNA gene sequences amplified by PCR from contaminated and uncontaminated soils. Application of this habitat-specific microarray allowed detection of probe signals corresponding to the genus *Sphingomonas* to be correlated to HCH contamination. Since the genus *Sphingomonas* contains known HCH degraders, the results suggest that the population detected was enriched *in situ* by HCH contamination and may play a role in HCH degradation. The last two applications used specifically designed microarrays of a relatively small scale compared with a microarray carrying a full set of rRNA gene probes (Wilson *et al.*, 2002; Palmer *et al.*, 2006), raising the possibility of cost-saving options for the application of microarrays to bioremediation processes.

Genetic monitoring of *in situ* bioremediation

Resolution of 16S rRNA gene-targeted methods is generally insufficient for the analysis of functional diversity because of the slow evolution of the rRNA gene. Microbial populations harbouring different functions are therefore detected frequently as an identical phylogenetic type. Lateral transfer of degradative genes occurs frequently, resulting in the frequent mismatch of degradative capabilities and phylogenetic characteristics (Watanabe and Hamamura, 2003). The detection of degradative genes or gene messages is considered to be a useful method of overcoming the limitations associated with 16S rRNA approaches.

Two different approaches are available for investigating the diversity of degradative genes in environmental samples: culture-dependent and culture-independent methods. In culture-dependent methods, bacteria are isolated from environmental samples using a culture medium. Nucleic acid is then extracted from the bacterial culture. In contrast, culture-independent methods employ direct extraction of nucleic acids from environmental samples (Okuta *et al.*, 1998; Watanabe *et al.*, 1998; Lloyd-Jones *et al.*, 1999). To date, a variety of bacteria capable of degrading different types of petroleum hydrocarbons have been isolated (Cook *et al.*, 1983; Watanabe, 2001; Harayama *et al.*, 2004) and degradative genes have been identified. Results of these studies have prompted scientists to design PCR primers that can amplify these degradative genes directly from environmental nucleic acid samples. Culture-independent methods using PCR now appear to offer an alternative technique for identifying novel degradative genes in environmental samples (Widada *et al.*, 2002; Watanabe and Hamamura, 2003). Once a degradative gene has been isolated, the application of the gene can be expanded/incleased to include its use in quantitative PCR or real-time PCR and functional gene probe arrays. Several molecular tools have now been developed to detect and quantify degradative genes in environmental samples (Table 11.2).

Quantitative competitive PCR (cPCR) is based on the incorporation of an internal standard in each PCR reaction. The internal standard (or competitor DNA) should be as similar to the target DNA as possible and should be amplified with the same primer set, yet still be distinguishable from the target, for example, by size. A standard curve constructed using a constant series of competitor DNAs can be used to calculate unknown target DNA concentrations in environmental samples. Mesarch *et al.* (2000) developed a quantitative cPCR system that uses the conserved region of the catechol 2,3-dioxygenase (C23DO) genes as PCR targets, making it possible to enumerate their gene copy number and to correlate their relative abundances to pollutant levels. C23DO genes were enumerated by this technique with a 163-bp competitor in a petroleum-amended soil using the same primer.

Real-time PCR allows continuous monitoring of the sample during PCR using hybridization probes (Harms *et al.*, 2003; Dionisi *et al.*, 2004; Layton *et al.*, 2005) or DNA-binding dyes such as SYBR Green I (Baldwin *et al.*, 2003; Lopez-Gutierrez *et al.*, 2004; Park and Crowley, 2006). Using a real-time PCR method, anaerobic toluene-degrading bacteria were monitored by quantifying the *bssA* gene with a primer–probe set in microcosms inoculated with aquifer sediments and incubated anaerobically with benzene, toluene, ethylbenzene, and xylene (BTEX) as well as nitrate in the presence and absence of ethanol (Beller *et al.*, 2002). Population trends were consistent with observed toluene degrading activity: the microcosms with the most rapid toluene degrading activity had the largest numbers of *bssA* copies; the number increased from 100- to 1000-fold over the first 4 days of incubation, during which period most of the toluene was consumed.

Rhee *et al.* (2004) developed a comprehensive 50 mer-based oligonucleotide microarray containing 1657 probes from all 2402 known genes involved in aromatic hydrocarbon biodegradation and metal reduction and used it to monitor the biodegradation potential and activity of enrichment cultures and soil microcosms supplemented with naphthalene. With this 50-mer functional gene array, they successfully detected changes in microbial community structure in enrichment and soil microcosms depending on the incubation conditions. This study indicates that the DNA microarray will be a powerful technology for monitoring bioremediation processes *in situ*.

It is desirable to examine whether environmental clones of degradative genes are actually expressed *in situ* by detection of their transcription directly as mRNA or via a reporter gene

Table 11.2 Molecular tools for the detection and quantification of degradative genes

Enzyme/protein	Gene	Molecular approach	Comment	Reference
Alkane hydroxylase	*alk*B	PCR, probing	Primers and probes that allow specific amplification of *alk*B genes in Gram-negative and Gram-positive bacteria were designed. Their application to amplifying genes from environmental samples was not tested	Smits *et al*. (1999)
		PCR, probing	Primer sets were designed to separately detect three different AlkB phylotypes. The detection of alkane-degrading bacteria in environments was not tested	Kohno *et al*. (2002)
		cPCR[1)]	Primer sets specific for distinct clusters of *alk*B gene were designed and quantified in hydrocarbon-contaminated and pristine freshwater and soil samples	Heiss-Blanquet *et al*. (2005)
		Real-time PCR	An increase in copy number of alkane hydroxylase genes was monitored during the bioremediation of crude oil-contaminated soil	Kyung-Hwa *et al*. (2009)
Ammonia monooxygenase	*amo*A	PCR, cloning	The *amo*A diversity was examined in rice rhizosphere, activated sludge lake water, and enrichment culture samples, demonstrating the utility of catabolic gene sequences for fine-scale community analysis	Rotthauwe *et al*. (1997)
		In situ PCR	Cells in a nitrifying biofilm possessing amoA genes were visualized by *In situ* PCR, showing the occurrence of nitrifiers at the surface of the biofilm as demonstrated by FISH	Hoshino *et al*. (2001)
Anaerobic aromatic compound degradation	*bzd*Q/ *bcr*A/ *bad*F	PCR, cPCR	A quantitative cPCR system with primers specific for BCR α subunit genes was developed for assessing natural attenuation	Hosoda *et al*. (2005)
Anaerobic toluene degradation	*bss*A	RT PCR	A quantitative real-time PCR system with primers specific for *bss*A, the gene for benzylsuccinate synthase, was developed for assessing natural attenuation	Beller *et al*. (2002)
Arsenate reductase	*ars*C	Real-time PCR	A quantitative PCR system with primers specific for *ars*C gene was developed and applied to some environmental samples	Sun *et al*. (2004)
	*arr*A	PCR, RT-PCR	*arr*A gene specific primers were designed and used for geochemical studies of Hiwee sediments	Malasarn *et al*. (2004)
Biphenyl dioxygenase	*bph*A	Real-time PCR	Primer sets were designed to detect biphenyl dioxygenase genes. The detection of biphenyl-degrading bacteria in environments was not tested	Baldwin *et al*. (2003)
Catechol 2,3-dioxygenase	*xyl*E	PCR, probing, cPCR	An increase in copy number of C23DO genes in response to a toluene spike to soil was detected by cPCR	Mesarch *et al*. (2000)
	C23O subfamily 1.2.A genes	PCR-SSCP, cloning	Primer sets specific for C23O subfamily 1.2.A genes were designed and used to investigate the catabolic gene diversity in several BTEX-contaminated soils by PCR-SSCP	Junca and Pieper (2004)

Table 11.2 continued

Enzyme/protein	Gene	Molecular approach	Comment	Reference
Chlorite dismutase	*cld*	PCR, cloning	Primer sets were designed to detect chlorite dismutase genes. The diversity of cld genes were analysed in chlorite-contaminated and uncontaminated sediments	Bender *et al*. (2004)
Chlorocatechol dioxygenase	*tcb*C, *clc*A	RT-PCR, cloning, RFLP	RFLP analysis of RT-PCR-amplified chlorocatechol 1,2-dioxygenase and catechol 2,3-dioxygenase genes in chlorobenzene-contaminated aquifer.	Alfreider *et al*. (2003)
Denitrification pathway	*nir*S and others	RT-PCR, cloning, probing	The diversity of denitrification genes expressed in river sediment samples was analysed	Nogales *et al*. (2002)
Dibenzothiophene desulfurization pathway	*dsz*ABC	PCR, cloning	*dsz*A genes retrieved by PCR from petroleum-contaminated soil samples were highly homologous to those of *Rhodococcus erythropholis* IGTS8	Duarte *et al*. (2001)
Methane monooxygenase	*mmo*X (sMMO)	PCR, probing	The diversity and abundance of sMMO genes in lake sediment were analysed, suggesting the importance of *Methylomonas*-like sMMO-containing type I methanotrophs in co-metabolic transformation of halogenated solvents in the environment	Auman and Lidstrom (2002)
	Mmo, pmo	PCR, cloning	Molecular diversity of pMMO and sMMO in a TCE-contaminated aquifer was analysed during field tests of biostimulation	Baker *et al*. (2001)
	pmo (pMMO)	PCR, cloning	Microbial populations in TCE-contaminated aquifer after methane biostimulation were analysed by polyphasic approaches, showing that type I methanotrophs became abundant	Pacheco-Oliver *et al*. (2002)
Naphthalene dioxygenase	*pah*Ac, *phn*Ac	PCR, probing	*pah*Ac- and *phn*Ac (*nah*-type PAH dioxygenase)-specific probes and primers were used to characterize PAH dioxygenase genes in the environment, suggesting the presence of unidentified PAH-degrading enzymes in the environment	Widada *et al*. (2002)
	*nah*Ac	PCR, RFLP	RFLP analyses of PCR-amplified 16S rRNA and naphthalene dioxygenase genes in coal-tar-contaminated aquifer were conducted.	Bakermans and Madsen (2002)
	*nah*Ac	Real-time PCR	Primer sets were designed to detect *nah*-type dioxygenase gene. The detection of naphthalene degrading bacteria in environments was not tested	Baldwin *et al*. (2003)
	*nah*Ac	RT-PCR	Expression of naphthalene dioxygenase gene was determined by RT-PCR when phenanthrene was degraded	Marlowe *et al*. (2002)
	*nag*Ac	Real-time PCR, probing, cloning	A quantitative PCR system with primers specific for *nag*Ac gene was developed and correlation with naphthalene concentrations and *nag*Ac-like gene copies in a coal tar-contaminated sediment analysed	Dionisi *et al*. (2004)

Nitrate reductase	*nar*G	Real-time PCR	Quantitative PCR systems with primers specific for *nar*G genes and 16S rRNA genes were developed and correlation with density of nitrate-reducing community and activity in a environmental sample analysed	Lopez-Gutierrez *et al.* (2004)
PAH-induced dioxygenase	*nid*A, *nid*B	PCR, cloning	*nid*A- and *nid*B (PAH dioxygenase of *Mycobacteria*)-specific probes and primers were designed and used for monitoring of on-site soil bioremediation	Hall *et al.* (2005)
PHB depolymerase	*phaZ*	PCR, probing	Molecular detection of PHB-degrading bacteria containing the *phaZ* gene was attempted	Sei *et al.* (2001)
Phenol hydroxylase	*dmp*	PCR, cPCR	A correlation was found between PH genotypes and TCE-degradation activity of soil communities based on changes in dominant PH genotypes	Futamata *et al.* (2001)
Phenol monooxygenase	*Dmp* and others	Real-time PCR	Primer sets were designed to detect phenol monooxygenase genes. The detection of phenol-degrading bacteria in environments was not tested	Baldwin *et al.* (2003)
Toluene dioxygenase	*tod*C	Real-time PCR	Primer sets were designed to detect toluene dioxygenase genes. The detection of toluene-degrading bacteria in environments was not tested	Baldwin *et al.* (2003)
Toluene monooxygenase	*tmo*	PCR, RFLP	The diversity of toluene monooxygenase genes in BTEX-polluted groundwater was analysed, suggesting that horizontal gene transfer of existing genes could have occurred between indigenous microorganisms	Cavalca *et al.* (2004)
Transcriptional activator	*nah*R	PCR, cloning	Homologues of *nah*R encoding a transcriptional activator for naphthalene catabolism operon in *Pseudomonas putida* G7 were detected from soil	Park *et al.* (2002)

BTEX, benzene, toluene, ethylbenzene and xylenes; cPCR, competitive polymerase chain reaction; PAH, polycyclic aromatic hydrocarbons; PHB, poly(3-hydroxybutyrate); RFLP, restriction fragment length polymorphism; RT-PCR, reverse transcription PCR; TCE, trichloroethylene fused to the target gene. Numerous environmental factors can activate or repress gene expression and thereby modulate microbial activity. Marlowe *et al.* (2002) used reverse-transcription PCR (RT-PCR) to evaluate the regulation of the *nah*Ac gene during phenanthrene degradation. Gene expression was successfully determined by extraction of bacterial mRNA followed by RT-PCR amplification of the *nah*Ac gene and an internal housekeeping gene (*rpo*D). The *rpo*D gene, which encodes a housekeeping sigma factor for transcription initiation, served as a baseline control in this gene expression assay. Temporal changes in *nah*Ac expression were shown when phenanthrene was degraded (Marlowe *et al.*, 2002). Thus, monitoring gene expression may allow the detection of subtle changes in the expression of degradative genes because of temporal changes in cell physiology or changes in toxicity associated with the accumulation of intermediates.

An alternative molecular method is to use biomarkers as specific tags for the identification of specific bacteria or gene activity. A biomarker is defined as a DNA sequence introduced into an organism, which confers a distinct genotype or phenotype to enable monitoring in a given environment. Bioluminescent, fluorescent, and chromogenic marker genes, such as *luxAB* (bacterial luciferase), *gfp* (green fluorescent protein, GFP), and *xylE* (catechol 2,3-oxygenase) genes, as well as reporter genes are examples of promising biomarkers (Jansson, 2003).

In many cases, it is difficult to link results obtained using the degradative gene approach to those obtained using the rRNA approach. To solve this problem, combinations of fluorescence *in situ* hybridization (FISH) with *in situ* PCR (IS-PCR) or tyramide signal amplification (TSA) have been used to detect intracellular mRNA in microbial mats (Schonhuber *et al.*, 1999; Hoshino *et al.*, 2001). A TSA-FISH method was developed and applied to *Pseudomonas putida* NCB9816–4 cells actively transcribing naphthalene dioxygenase (*nah*Ac) and related genes in environmental samples from a site contaminated with coal tar; thus, documenting the *in situ* expression of specific genes in individual cells (Bakermans and Madsen, 2002).

Metagenomics

Metagenomics is a fast-growing and diverse field within the biological sciences, directed at obtaining knowledge on the genomes of environmental microbes as well as of entire microbial communities, omitting the cultivation step. The purposes of metagenomics include (1) improving understanding of the ecological processes (e.g. catabolic reactions) proceeding in ecosystems of interest and (2) the mining of novel genes from uncultured microorganisms that may be biotechnologically useful (Handelsman, 2004). For biotechnological applications, two major types of method are traditionally recognized: sequence-based and function-based metagenomics. The sequence-based metagenomics approach relies on prior knowledge about proteins possessing the activity of interest, and screening is performed to find genes that are predicted to encode proteins with specific sequences/folds/domains indicative of their function. Therefore, only new members of known gene families can be identified by this approach. The advantage of this approach is its independence of gene expression and the production of foreign genes in the library host (Lorenz *et al.*, 2002). Several novel functional enzymes, such as alcohol oxidoreductases, diol dehydratases, and enzymes conferring antibiotic resistance, were recovered by employing sequence-based approaches (Knietsch *et al.*, 2003; Banik and Brady, 2008; Morimoto and Fujii, 2009). The function-based metagenomics approach relies on cloning environmental DNA into expression vectors and propagating them in appropriate hosts, followed by appropriate activity screens (Streit *et al.*, 2004; Schmeisser *et al.*, 2007; Craig *et al.*, 2010). After an active clone is identified, the sequence of the clone is determined, the gene of interest and its respective product are further analysed, and their biotechnological potential is explored. The major advantage of the function-based approach is that only full-length genes and functional gene products are detected. A significant limitation of this technique is its dependence on the expression of the target genes and the production of functional gene products in a foreign host, which in most studies is *Escherichia coli*. Recently, the following three types of function-based approach were employed for the screening of metagenomics libraries: (1) direct detection of specific phenotypes of individual clones; (2) heterologous complementation of host strains or mutants; and (3) induced gene expression (Simon and Daniel, 2009). Each approach has strengths and limitations, and together they have enriched our understanding of the uncultured world, providing insight into groups of prokaryotes that would otherwise be entirely unknown. Below I introduce some published metagenomics studies that employ methods that may be used in future to improve our understanding of the processes of bioremediation.

Kube *et al.* (2005) constructed a metagenome library for a microbial mat from the Black Sea shelf and determined nucleotide sequences for several clones. In one clone they identified all genes required for the complete pathway of anaerobic benzoate degradation to take place, including the four subunits of the key enzyme benzoyl-CoA reductase (*bcrCBAD*), which catalyses the ATP-driven 2-electron reduction of the aromatic ring. Genes for a keto acid:acceptor oxidoreductase (*korABC*) and a ferredoxin (*fdx*), which are required for the generation of a suitable electron donor, were also detected. Most of the identified catabolic gene products were similar to their respective orthologues from the denitrifying freshwater bacterium *Azoarcus evansii*, and the operon was similarly organized. On the basis of these findings, the authors suggested that the metabolic diversity of the Black Sea mat is wider than is currently believed and that aromatic hydrocarbons can be anaerobically degraded in the mat.

The molecular detection of genes for 16S rRNA and particulate methane monooxygenase (*pmo*) has revealed the abundant presence of putative methanotrophic bacteria (USC alpha) in upland soils, although their physiology is unknown in terms of their resistance to cultivation (Ricke *et al.*, 2005). Ricke *et al.* (2005) screened a metagenomic library from forest soil using *pmo* PCR to obtain genomic fragments derived from USC alpha bacteria. Analyses of a positive clone containing a 42-kb genomic contig revealed that the *pmo* operon structure and the predicted secondary structure of deduced pMMO in the USC alpha bacteria were highly similar to those of *Methylocapsa acidiphila* B2, suggesting that the USC alpha bacteria are also methanotrophs adapted to the trace level of methane in the atmosphere.

Uchiyama *et al.* (2005) introduced a substrate-induced gene expression screening system (SIGEX) for the identification of novel catabolic genes. In the SIGEX scheme, an operon-trap *gfp* expression vector is used to construct a library, which enables subsequent high-throughput screening for GFP-expressed positive clones with the aid of fluorescence-activated cell sorting. This method has been applied to the identification of genes involved in aromatic hydrocarbon degradation from a metagenome library constructed from an oil-contaminated groundwater (Uchiyama *et al.*, 2005). It was shown that in metagenome fragments isolated by means of SIGEX with benzoate as the induction substrate, open reading frames homologous to known catabolic genes were highly enriched and some of them were substantially similar (over 70%) to genes in the known benzoate degradation pathway. A novel cytochrome P450 converting 4-hydroxybenzoate to protocatechuate was also identified in a SIGEX-selected metagenome fragment. As suggested by Uchiyama *et al.* (2005), using SIGEX screening includes several merits as follows: (1) it facilitates the isolation of a catabolic gene whose enzymatic activity is not easily expressed in a cloning host, and (2) substrates for an unknown enzyme can be deduced from the induction substrate used in the SIGEX screening.

Poretsky *et al.* (2005) reported an environmental transcriptome approach to the analysis of genes expressed in marine and freshwater bacterioplankton communities. In this approach, environmental mRNA, obtained from total RNA by subtractive hybridization of rRNA, was reverse-transcribed, amplified with random primers, and cloned. Sequence analyses indicated that mRNAs belonged to diverse taxonomic groups, including both bacteria (mainly α- and γ-Proteobacteria) and Archaea (mainly Euryarchaeota). Many transcripts could be linked to environmentally important processes, such as sulfur oxidation, the assimilation of C1 compounds, and the acquisition of nitrogen via polyamine degradation. The authors suggested that environmental transcriptomics is a means of exploring functional gene expression within natural microbial communities without bias towards known sequences.

As described above, scientists have developed different types of metagenomic approaches that are potentially applicable to the analysis of bioremediation processes. Unlike genetic methods, such as PCR amplification of one type of degradative gene, these metagenomic approaches allow simultaneous retrieval of different types of genes (i.e. genes coding for different functions) from a polluted site. Therefore, it would be of great interest to reconstruct *in situ* degradative pathways based on metagenomic information. If this can be done in a reasonably quantitative manner, metagenomics will become a vital technique in bioremediation studies.

Conclusions

Methods of analysing microorganisms in the natural environment have shifted from cultivation and isolation towards molecular detection and identification, enabling microbial ecologists to believe that they are now better capable of understanding the structure and function of microbial communities. This paradigm shift has affected bioremediation practices, providing ways for assessing the potential and progress of these technologies and forecasting their endpoints. Community fingerprinting is the most useful of the molecular ecological methods, and I believe that its application will be more useful if it is combined with the use of key degradative genes rather than phylogenetic markers. As summarized in Table 11.2, PCR systems for many degradative genes are now available, although only a few of them have been coupled with community fingerprinting methods. In order to design better primers for the PCR amplification of degradative genes, we need to continue our efforts to clone a wide range of genes coding for degradative enzymes. In such studies, it is also necessary to analyse the functions of putative degradative genes found in genome and metagenome sequences.

We anticipate that the next paradigm shift will emerge when the use of metagenomic approaches will enable us to understand the degradative pathways (or degradative enzyme networks) that are present and/or operating at polluted sites. Unfortunately, however, the current status of metagenomics facilitates the retrieval of only a small proportion of the entire gene sets necessary for the reconstruction of degradative pathways. We therefore face the challenge of further technical development of metagenomics so that it will at least be possible to identify the central degradative pathway in a given ecosystem. The techniques that will become available will provide information that will helps us assess the effects and risks of bioremediation treatments, such as the emergence of toxic metabolites.

Acknowledgement

This work was in part supported by Grant-in-Aid for Scientific Research (C) 21580118.

References

Adamczyk, J., Hesselsoe, M., Iversen, N., Horn, M., Lehner, A., Nielsen, P.H., Schloter, M., Roslev, P., and Wagner, M. (2003). The isotope array, a new tool that employs substrate-mediated labeling of rRNA for determination of microbial community structure and function. Appl. Environ. Microbiol. *69*, 6875–6887.

Alfreider, A., Vogt, C., and Babel, W. (2003). Expression of chlorocatechol 1, 2-dioxygenase and chlorocatechol 2, 3-dioxygenase genes in chlorobenzene-contaminated subsurface samples. Appl. Environ. Microbiol. *69*, 1372–1376.

Auman, A.J., and Lidstrom, M.E. (2002). Analysis of sMMO-containing type I methanotrophs in Lake Washington sediment. Environ. Microbiol. *4*, 517–524.

Baker, G.C., and Cowan, D.A. (2004). 16 S rDNA primers and the unbiased assessment of thermophile diversity. Biochem. Soc. Trans. *32*, 218–221.

Baker, P.W., Futamata, H., Harayama, S., and Watanabe, K. (2001). Molecular diversity of pMMO and sMMO in a TCE-contaminated aquifer during bioremediation. FEMS Microbiol. Ecol. *38*, 161–167.

Bakermans, C., and Madsen, E.L. (2002). Diversity of 16S rDNA and naphthalene dioxygenase genes from coal-tar-waste-contaminated aquifer waters. Microb. Ecol. *44*, 95–106.

Baldwin, B.R., Nakatsu, C.H., and Nies, L. (2003). Detection and enumeration of aromatic oxygenase genes by multiplex and real-time PCR. Appl. Environ. Microbiol. *69*, 3350–3358.

Banik, J.J., and Brady, S.F. (2008). Cloning and characterization of new glycopeptides gene clusters found in an environmental DNA megalibrary. Proc. Natl. Acad. Sci. U.S.A. *105*, 17273–17277.

Barlaan, E.A., Sugimori, M., Furukawa, S., and Takeuchi, K. (2005). Profiling and monitoring of microbial populations by denaturing high-performance liquid chromatography. J. Microbiol. Methods *61*, 399–412.

Belda, E., Sentandreu, V., and Silva, F.J. (2004). Identification and separation of PCR products based on their GC content by denaturing high-performance liquid chromatography. J. Chromatogr. *811*, 263–268.

Beller, H.R., Kane, S.R., Legler, T.C., and Alvarez, P.J. (2002). A real-time polymerase chain reaction method for monitoring anaerobic, hydrocarbon-degrading bacteria based on a catabolic gene. Environ. Sci. Technol. *36*, 3977–3984.

Bender, K.S., Rice, M.R., Fugate, W.H., Coates, J.D., and Achenbach, L.A. (2004). Metabolic primers for detection of (per)chlorate-reducing bacteria in the environment and phylogenetic analysis of cld gene sequences. Appl. Environ. Microbiol. *70*, 5651–5658.

Borneman, J., and Triplett, E.W. (1997). Molecular microbial diversity in soils from eastern Amazonia: evidence for unusual microorganisms and microbial population shifts associated with deforestation. Appl. Environ. Microbiol. *63*, 2647–2653.

Call, D.R., Borucki, M.K., and Loge, F.J. (2003). Detection of bacterial pathogens in environmental samples using DNA microarrays. J. Microbiol Methods 53, 235–243.

Cavalca, L., Dell Amico, E., and Andreoni, V. (2004). Intrinsic bioremediability of an aromatic hydrocarbon-polluted groundwater: diversity of bacterial population and toluene monooxygenase genes. Appl. Microbiol. Biotechnol. *64*, 576–587.

Coates, J.D., Chakraborty, R., Lack, J.G., O'Connor, S.M., Cole, K.A., Bender, K.S., and Achenbach, L.A. (2001). Anaerobic benzene oxidation coupled to nitrate reduction in pure culture by two strains of *Dechloromonas*. Nature *411*, 1039–1043.

Cook, A.M., Grossenbacher, H., and Hutter, R. (1983). Isolation and cultivation of microbes with biodegradative potential. Experientia 39, 1191–1198.

Craig, J.W., Chang, F.Y., Kim, J.H., Obiajulu, S.C., and Brady, S.F. (2010). Expanding small-molecule functional metagenomics through parallel screening of broad-host-range cosmid environmental DNA libraries in diverse proteobacteria. Appl. Environ. Microbiol. *76*, 1633–1641.

Dahllo, F.I. (2002). Molecular community analysis of microbial diversity. Curr. Opin. Biotechnol. *13*, 213–217.

Deng, W., Xi, D., Mao, H., and Wanapat, M. (2008). The use of molecular techniques based on ribosomal RNA and DNA for rumen microbial ecosystem studies: a review. Mol. Biol. Rep. 35, 265–274.

Dionisi, H.M., Chewning, C.S., Morgan, K.H., Menn, F.-M., Easter, J.P., and Sayler, G.S. (2004). Abundance of dioxygenase genes similar to *Ralstonia* sp. strain U2 nagAc is correlated with naphthalene concentrations in coal tar-contaminated freshwater sediments. Appl. Environ. Microbiol. *70*, 3988–3995.

Domann, E., Hong, G., Imirzalioglu, C., Turschner, S., Kuhle, J., Watzel, C., Hain, T., Hossain, H., and Chakraborty, T. (2003). Culture-independent identification of pathogenic bacteria and polymicrobial infections in the genitourinary tract of renal transplant recipients. J. Clin. Microbiol. *41*, 5500–5510.

Duarte, G.F., Rosado, A.S., Seldin, L., de Araujo, W., and van Elsas, J.D. (2001). Analysis of bacterial community structure in sulfurous-oil-containing soils and detection of species carrying dibenzothiophene desulfurization (dsz) genes. Appl. Environ. Microbiol. *67*, 1052–1062.

El Fantroussi, S., and Agathos, S.N. (2005). Is bioaugmentation a feasible strategy for pollutant removal and site remediation? Curr. Opin. Microbiol. *8*, 268–275.

El Fantroussi, S., Urakawa, H., Bernhard, A.E., Kelly, J.J., Noble, P.A., Smidt, H., Yershov, G.M., and Stahl, D.A. (2003). Direct profiling of environmental microbial populations by thermal dissociation analysis of native rRNAs hybridized to oligonucleotide microarrays. Appl. Environ. Microbiol. *69*, 2377–2382.

Eyers, L., George, I., Schuler, L., Stenuit, B., Agathos, S.N., and El Fantroussi, S. (2004). Environmental genomics: exploring the unmined richness of microbes to degrade xenobiotics. Appl. Microbiol. Biotechnol. *66*, 123–130.

Fisher, M.M., and Triplett, E.W. (1999). Automated approach for ribosomal intergenic spacer analysis of microbial diversity and its application to freshwater bacterial communities. Appl. Environ. Microbiol. *65*, 4630–4636.

Fries, M.R., Chee-Sandford, J., and Tiedje, J.M. (1994). Isolation, characterization, and distribution of denitrifying toluene degraders form a variety of habitats. Appl. Environ. Microbiol. *60*, 2802–2810.

Frostegard, A., Courtois, S., Ramisse, V., Clerc, S., Bernillon, D., Le Gall, F., Jeannin, P., Nesme, X., and Simonet, P. (1999). Quantification of bias related to the extraction of DNA directly from soils. Appl. Environ. Microbiol. *65*, 5409–5420.

Futamata, H., Harayama, S., and Watanabe, K. (2001). Group-specific monitoring of phenol hydroxylase genes for a functional assessment of phenol-stimulated trichloroethylene bioremediation. Appl. Environ. Microbiol. *67*, 4671–4677.

Futamata, H., Harayama, S., Hiraishi, A., and Watanabe, K. (2003). Functional and structural analyses of trichloroethylene-degrading bacterial communities under different phenol-feeding conditions: laboratory experiments. Appl. Microbiol. Biotechnol. *60*, 594–600.

Guschin, D.Y., Mobarry, B.K., Proudnikov, D., Stahl, D.A., Rittmann, B.E., and Mizabekov, A.D. (1997). Oligonucleotide microchips as genosensors for determinative and environmental studies in microbiology. Appl. Environ. Microbiol. *63*, 2397–2402.

Hall, K., Miller, C.D., Sorensen, D.L., Anderson, A.J., and Sims, R.C. (2005). Development of a catabolically significant genetic probe for polycyclic aromatic hydrocarbon-degrading mycobacteria in soil. Biodegradation *16*, 475–484.

Handelsman, J. (2004). Metagenomics: application of genomics to uncultured microorganisms. Microbiol. Mol. Biol. Rev. *68*, 669–685.

Harayama, S., Kasai, Y., and Hara, A. (2004). Microbial communities in oil-contaminated seawater. Curr. Opin. Biotechnol. *15*, 205–214.

Harms, G., Layton, A.C., Dionisi, H.M., Gregory, I.R., Garrett, V.M., Hawkins, S.A., Robinson, K.G., and Sayler, G.S. (2003). Real-time PCR quantification of nitrifying bacteria in a municipal wastewater treatment plant. Environ. Sci. Technol. *37*, 343–351.

Heiss-Blanquet, S., Benoit, Y., Marechaux, C., and Monot, F. (2005). Assessing the role of alkane hydroxylase genotypes in environmental samples by competitive PCR. J. Appl. Microbiol. *99*, 1392–403.

Hess, A., Zarda, B., Hahn, D., Häner, A., Stax, D., Höhener, P., and Zeyer, J. (1997). *In situ* analysis of denitrifying toluene- and m-xylene-degrading bacteria in a diesel fuel-contaminated laboratory aquifer column. Appl. Environ. Microbiol. *63*, 2136–2141.

Hoshino, T., Noda, N., Tsuneda, S., Hirata, A., and Inamori, Y. (2001). Direct detection by in situ PCR of the amoA gene in biofilm resulting from a nitrogen removal process. Appl. Environ. Microbiol. *67*, 5261–5266.

Hosoda, A., Kasai, Y., Hamamura, N., Takahata, Y., and Watanabe, K. (2005). Development of a PCR method for the detection and quantification of benzoyl-CoA reductase genes and its application to monitored natural attenuation. Biodegradation *16*, 591–601.

Hurt, R.A., Qiu, X., Wu, L., Roh, Y., Palumbo, A.V., Tiedje, J.M., and Zhou, J. (2001). Simultaneous recovery of RNA and DNA from soils and sediments. Appl. Environ. Microbiol. *67*, 4495–4503.

Jansson J.K. (2003). Marker and reporter genes: illuminating tools for environmental microbiologists. Curr. Opin. Microbiol. *6*, 310–316.

Junca, H., and Pieper, D.H. (2004). Functional gene diversity analysis in BTEX contaminated soils by means of PCR-SSCP DNA fingerprinting: comparative diversity assessment against bacterial isolates and PCR-DNA clone libraries. Environ. Microbiol. *6*, 95–110.

Kasai, Y., Takahata, Y., Hoaki, T., and Watanabe, K. (2005). Physiological and molecular characterization of a microbial community established in unsaturated, petroleum-contaminated soil. Environ. Microbiol. *7*, 806–818.

Kasai, Y., Takahata, Y., Manefield, M., and Watanabe, K. (2006). RNA-stable isotope probing and isolation of anaerobic benzene-degrading bacteria from gasoline-contaminated groundwater. Appl. Environ. Microbiol. *72*, 3586–3592.

Katsivela, E., Moore, E.R., Maroukli, D., Strompl, C., Pieper, D., and Kalogerakis, N. (2005). Bacterial community dynamics during *in situ* bioremediation of petroleum waste sludge in landfarming sites. Biodegradation *16*, 169–180.

Knietsch, A., Waschkowitz, T., Bowien, S., Henne, A., and Daniel, R. (2003). Metagenomes of complex microbial consortia derived from different soils as sources for novel genes conferring formation of carbonyls form short-chain polyols on *Escherichia coli*. J. Mol. Microbiol. Biotechnol. *5*, 46–56.

Kohno, T., Sugimoto, Y., Sei, K., and Mori, K. (2002). Design of PCR primers and gene probes for general detection of alkane-degrading bacteria. Microbes Environ. *17*, 114–121.

Koizumi, Y, Kelly, J.J., Nakagawa, T., Urakawa, H., El Fantroussi, S., Al Muzaini, S., Fukui, M., Urushigawa, Y., and Stahl, D.A. (2002). Parallel characterization of anaerobic toluene- and ethylbenzene-degrading microbial consortia by PCR-denaturing gradient gel electrophoresis, RNA-DNA membrane hybridization, and DNA microarray technology. Appl. Environ. Microbiol. *68*, 3215–3225.

Kube, M., Beck, A., Meyerdierks, A., Amann, R., Reinhardt, R., and Rabus, R. (2005). A catabolic gene cluster for anaerobic benzoate degradation in methanotrophic microbial Black Sea mats. Syst. Appl. Microbiol. *28*, 287–294.

Kyung-Hwa, B., Yoon, B.D., Cho, D.H., Kim, B.H., Oh, H.M., and Kim, H.S. (2009). Monitoring bacterial population dynamics using real-time PCR during the bioremediation of crude-oil-contaminated soil. J. Micobiol. Biotechnol. *19*, 339–345.

Lander, E.S. (1999). Array of hope. Nat. Genet. *21*, 3–4.

Layton, A.C., Dionisi, C.H., Kuo, H.W., Robinson, K.G., Garrett, V.M., Meyers, A., and Sayler, G.S. (2005). Emergence of competitive dominant ammonia-oxidizing bacterial populations in a full-scale industrial wastewater treatment plant. Appl. Environ. Microbiol. *71*, 1105–1108.

Lee, D.H., Zo, Y.G., and Kim, S.J. (1996). Nonradioactive method to study genetic profiles of natural bacterial communities by PCR-single-strand conformation polymorphism. Appl. Environ. Microbiol. *62*, 3112–3120.

Lettieri, T. (2006). Recent applications of DNA microarray technology to toxicology and ecotoxicology. Environ. Health Perspect. *114*, 4–9.

Li, E.S.Y., and Liu W.T. (2003). DNA microarray technology in microbial ecology studies – principle, applications and current limitations. Microbes Environ. *18*, 175–187.

Liu, W.T., Marsh, T.L., Cheng, H., and Forney, L.J. (1997). Characterization of microbial diversity by determining terminal restriction fragment length polymorphisms of genes encoding 16S rRNA. Appl. Environ. Microbiol. *63*, 4516–4522.

Lloyd-Jones, G., Laurie, A.D., Hunter, D.W.F., and Fraser, R. (1999). Analysis of catabolic genes for naphthalene and phenanthrene degradation in contaminated New Zealand soils. FEMS Microbiol. Ecol. *29*, 69–79.

Lopez-Gutierrez, J.C., Henry, S., Hallet, S., Martin-Laurent, F., Catroux, G., and Philippot, L. (2004). Quantification of a novel group of nitrate-reducing bacteria in the environment by real-time PCR. J. Microbiol. Methods *57*, 399–407.

Lorenz, P., Liebeton, K., Niehaus, F., and Eck, J. (2002). Screening for novel enzymes for biocatalytic processes: accessing the metagenome as a resource of novel functional sequence space. Curr. Opin. Biotechnol. *13*, 572–577.

Loy, A., Schulz, C., Lucker, S., Schopfer-Wendels, A., Stoecker, K., Baranyi, C., Lehner, A., and Wagner, M. (2005). 16S rRNA gene-based oligonucleotide microarray for environmental monitoring of the betaproteobacterial order 'Rhodocyclales'. Appl. Environ. Microbiol. *71*, 1373–1386.

Macbeth, T.W., Cummings, D.E., Spring, S., Petzke, L.M., and Sorenson, K.S.Jr. (2004). Molecular characterization of a dechlorinating community resulting from in situ biostimulation in a trichloroethene-contaminated deep, fractured basalt aquifer and comparison to a derivative laboratory culture. Appl. Environ. Microbiol. *70*, 7329–7341.

Macedo, A.J., Kuhlicke, U., Neu, T.R., Timmis, K.N., and Abraham, W.R. (2005). Three stages of a biofilm community developing at the liquid–liquid interface between polychlorinated biphenyls and water. Appl. Environ. Microbiol. *71*, 7301–7309.

Malasarn, D., Saltikov, C.W., Campbell, K.M., Santini, J.M., Hering, J.G., and Newman, D.K. (2004). arrA is a reliable marker for As(V) respiration. Science *306*, 455.

Marlowe, E.M., Wang, J.M., Pepper, I.L., and Mayer, R.M. (2002). Application of a reverse transcription-PCR assay to monitor regulation of the catabolic *nahAc* gene during phenanthrene degradation. Biodegradation *13*, 251–260.

Martin-Laurent, F., Philippot, L., Hallet, S., Chaussod, R., Germon, J.C., Soulas, G., and Catroux, G. (2001). DNA extraction from soils: old bias for new microbial diversity analysis methods. Appl. Environ. Microbiol. *67*, 2354–2359.

Mesarch, M.B., Nakatsu, C.H., and Nies, L. (2000). Development of catechol 2,3-dioxygenase-specific primers for monitoring bioremediation by competitive quantitative PCR. Appl. Environ. Microbiol. *66*, 678–683.

Morimoto, S., and Fujii, T. (2009). A new approach to retrieve full lengths of functional genes from soil by PCR-DGGE and metagenome walking. Appl. Microbiol. Biotechnol. *83*, 389–396.

Muyzer, G. (1999). DGGE/TGGE a method for identifying genes from natural ecosystems. Curr. Opin. Microbiol. *2*, 317–322.

Muyzer, G.E., DeWall, C., and Uitterlinden, A.G. (1993). Profiling of complex microbial populations by denaturing gradient gel electrophoresis analysis of polymerase chain reaction amplified genes coding for 16S rRNA. Appl. Environ. Microbiol. *59*, 695–700.

Myers, R.M., Maniatis, T., and Lerman, L.S. (1987). Detection and localization of single base changes by denaturing gradient gel electrophoresis. Meth. Enzymol. *155*, 501–527.

Nachamkin, I., Panaro, N.J., Li, M., Ung, H., Yuen, P.K., Kricka, L.J., and Wilding, P. (2001). Agilent 2100 bioanalyzer for restriction fragment length polymorphism analysis of the *Campylobacter jejuni* flagellin gene. J. Clin. Microbiol. *39*, 754–757.

Neufeld, J.D., Mohn, W.W., and de Lorenzo, V. (2006). Composition of microbial communities in hexachlorocyclohexane (HCH) contaminated soils from Spain revealed with a habitat-specific microarray. Environ. Microbiol. *8*, 126–140.

Nogales, B., Timmis, K.N., Nedwell, D.B., and Osborn A.M. (2002). Detection and diversity of expressed denitrification genes in estuarine sediments after reverse transcription-PCR amplification from mRNA. Appl. Environ. Microbiol. *68*, 5017–5025.

Norris, R.D. (1994). Handbook of Bioremediation. (CRC Press: Boca Raton, FL).

Okuta, A., Ohnishi, K., and Harayama, S. (1998). PCR isolation of catechol 2,3-dioxygenase gene fragments from environmental samples and their assembly into functional genes. Gene *212*, 221–228.

Orita, M., Iwahana, H., Kanazawa, H., Hayashi, K., and Sekiya, T. (1989). Detection of polymorphisms of human DNA by gel electrophoresis as single-strand conformation polymorphisms. Proc. Natl. Acad. Sci. U.S.A. *86*, 2766–2770.

Osborn, A.M., Moore, E.R.B., and Timmis, K.N. (2000). An evaluation of terminal restriction fragment length polymorphism (T-RFLP) analysis for the study of microbial community structure and dynamics. Environ. Microbiol. *2*, 39–50.

Oved, T., Shaviv, A., Goldrath, T., Mandelbaum, R.T., and Minz, D. (2001). Influence of effluent irrigation on community composition and function of ammonia-oxidizing bacteria in soil. Appl. Environ. Microbiol. *67*, 3426–3433.

Pacheco-Oliver, M., McDonald, I.R., Groleau, D., Murrell, J.C., and Miguez, C.B. (2002). Detection of methanotrophs with highly divergent *pmoA* genes from Arctic soils. FEMS Microbiol Lett. *209*, 313–319.

Palmer, C., Bik, E.M., Eisen, M.B., Eckburg, P.B., Sana, T.R., Wolber, P.K., Relman, D.A., and Brown, P.O. (2006). Rapid quantitative profiling of complex microbial populations. Nucleic Acids Res. *34*, e5.

Park, J.W., and Crowley, D.E. (2006). Dynamic changes in nahAc gene copy numbers during degradation of naphthalene in PAH-contaminated soils. Appl. Microbiol. Biotechnol. *72*, 1322–1329.

Park, W., Padmanabhan, P., Padmanabhan, S., Zylstra, G.J., and Madsen, E.L. (2002). nahR, encoding a LysR-type transcriptional regulator, is highly conserved among naphthalene-degrading bacteria isolated from a coal tar waste-contaminated site and in extracted community DNA. Microbiology *148*, 2319–29.

Peplies, J., Lau, S.C., Pernthaler, J., Amann, R., and Glockner, F.O. (2004). Application and validation of DNA microarrays for the 16S rRNA-based analysis of marine bacterioplankton. Environ. Microbiol. *6*, 638–645.

Pirie, F.J., York, D.F., Motala, A.A., and Omar, M.A. (1999). Fluorescent automated single-stranded conformation (F-SSCP) analysis is able to detect a point mutation at the extreme 5′ end of a PCR product. Clin. Biochem. *32*, 481–484.

Poretsky, R.S., Bano, N., Buchan, A., LeCleir, G., Kleikemper, J., Pickering, M., Pate, W.M., Moran, M.A., and Hollibaugh, J.T. (2005). Analysis of microbial gene transcripts in environmental samples. Appl. Environ. Microbiol. *71*, 4121–4126.

Rhee, S.K., Liu, X., Wu, L., Chong, S.C., Wan, X., and Zhou, J. (2004). Detection of genes involved in biodegradation and biotransformation in microbial communities by using 50-mer oligonucleotide microarrays. Appl. Environ. Microbiol. *70*, 4303–4317.

Ricke, P., Kube, M., Nakagawa, S., Erkel, C., Reinhardt, R., and Liesack, W. (2005). First genome data from uncultured upland soil cluster alpha methanotrophs provide further evidence for a close phylogenetic relationship to *Methylocapsa acidiphila* B2 and for high-affinity methanotrophy involving particulate methane monooxygenase. Appl. Environ. Microbiol. *71*, 7472–7482.

Ritchie, N.J., Schutter, M.E., Dick, R.P., and Myrold, D.D. (2000). Use of length heterogeneity PCR and fatty acid methyl ester profiles to characterize microbial communities in soil. Appl. Environ. Microbiol. *66*, 1668–1675.

Rittmann, B.E. (2004). Definition, objectives, and evaluation of natural attenuation. Biodegradation *15*, 349–357.

Roling, W.F., and van Verseveld, H.W. (2002). Natural attenuation: what does the subsurface have in store? Biodegradation *13*, 53–64.

Roling, W.F., van Breukelen, B.M., Braster, M., Lin, B., van Verseveld, H.W. (2001). Relationships between microbial community structure and hydrochemistry in a landfill leachate-polluted aquifer. Apple Environ. Microbiol. *67*, 4619–4629.

Rotthauwe, J.H., Witzel, K.P., and Liesack, W. (1997). The ammonia monooxygenase structural gene amoA as a functional marker: molecular fine-scale analysis of natural ammonia-oxidizing populations. Appl. Environ. Microbiol. *63*, 4704–4712.

Schmeisser, C., Steele, H., and Streit, W.R. (2007). Metagenomics, biotechnology with non-culturable microbes. Appl. Microbiol. Biotechnol. *75*, 955–962.

Schonhuber, W., Arda, B., Eix, S., Rippka, R., Herdman, M., Ludwig, W., and Amman, R. (1999). In situ identification of cyanobacteria with horseradish peroxidase-labeled, rRNA-tagged oligonucleotides probes. Appl. Environ. Microbiol. *65*, 1259–1267.

Schwieger, F., and Tebbe, C.C. (1998). A new approach to utilize PCR-single-strand conformation polymorphism for 16S rRNA gene-based microbial community analysis. Appl. Environ. Microbiol. *64*, 4870–4876.

Scow, K.M., and Hicks, K.A. (2005). Natural attenuation and enhanced bioremediation of organic contaminants in groundwater. Curr. Opin. Biotechnol. *16*, 246–253.

Sei, K., Nakao, M., Mori, K., Ike, M., Kohno, T., and Fujita, M. (2001). Design of PCR primers and a gene probe for extensive detection of poly (3-hydroxybutylate) (PHB)-degrading bacteria possessing fibronectin type III linker type-PHB depolymerases. Appl. Microbiol. Biotechnol. *55*, 801–806.

Simon, C., and Daniel, R. (2009). Achievements and new knowledge unraveled by metagenomics approaches. Appl. Microbiol. Biotechnol. *85*, 265–276.

Small, J., Call, D.R., Brockman, F.J., Straub, T.M., and Chandler, D.P. (2001). Direct detection of 16S rRNA in soil extracts by using oligonucleotide microarrays. Appl. Environ. Microbiol. *67*, 4708–4716.

Smits, T.H., Rothlisberger, M., Witholt, B., and van Beilen, J.B. (1999). Molecular screening for alkane hydroxylase genes in Gram-negative and Gram-positive strains. Environ. Microbiol. *1*, 307–317.

Spiegelman, D., Whissell, G., and Greer, C.W. (2005). A survey of the methods for the characterization of microbial consortia and communities. Can. J. Microbiol. *51*, 355–386.

Stach, J.E., Bathe, S., Clapp, J.P., and Burns, R.G. (2001). PCR-SSCP comparison of 16S rDNA sequence diversity in soil DNA obtained using different isolation and purification methods. FEMS Microbiol. Ecol. *36*, 139–151.

Streit, W.R., Daniel, R., and Jaeger, K.E. (2004). Prospecting for biocatalysts and drugs in the genomes of non-cultured microorganisms. Curr. Opin. Biotechnol. *15*, 285–290.

Sun, Y., Polishchuk, E.A., Radoja, U., and Cullen, W.R. (2004). Identification and quantification of *arcs* genes in environmental samples by using real-time PCR. J. Microbiol. Methods *58*, 335–349.

Taliani, M.R., Roberts, S.C., Dukek, B.A., Pruthi, R.K., Bichols, W.L., and Heit, J.A. (2001). Sensitivity and specificity of denaturing high pressure liquid chromatography for unknown protein C gene mutations. Genet. Test *5*, 39–44.

Theron, J., and Cloete, T.E. (2000). Molecular techniques for determining microbial diversity and community structure in natural environments. Crit. Rev. Microbiol. *26*, 37–57.

Thompson, I.P., van der Gast, C.J., Ciric, L., and Singer, A.C. (2005). Bioaugmentation for bioremediation: the challenge of strain selection. Environ. Microbiol. *7*, 909–915.

Tiedje, J.M., Asuming-Brempong, S., Nusslein, K., Marsh, T.L., and Flynn, S.J. (1999). Opening the black box of soil microbial diversity. Appl. Soil Ecol. *13*, 109–122.

Ting, L., Stroot, P.G., and Oerther, D.B. (2009). Reverse transcription of 16S rRNA to monitor ribosome-synthesizing bacterial populations in the environment. Appl. Environ. Microbiol. 75, 4589–4598.

Uchiyama, T., Abe, T., Ikemura, T., and Watanabe, K. (2005). Substrate-induced gene-expression screening of environmental metagenome libraries for isolation of catabolic genes. Nat. Biotechnol. 23, 88–93.

Van Beilen, J.B., and Funhoff, E.G. (2005). Expanding the alkane oxygenase toolbox: new enzymes and applications. Curr. Opin. Biotechnol. *16*, 308–314.

Von Wintzingerode, F., Gobel, U.B., and Stackebrandt, E. (1997). Determination of microbial diversity in environmental samples: pitfalls of PCR-based rRNA analysis. FEMS Microbiol. Rev. *21*, 213–229.

Watanabe, K. (2001). Microorganisms relevant to bioremediation. Curr. Opin. Biotechnol. *12*, 237–241.

Watanabe, K., and Hamamura, N. (2003). Molecular and physiological approaches to understanding the ecology of pollutant degradation. Curr. Opin. Biotechnol. *14*, 289–295.

Watanabe, K., Teramoto, M., Futamata, H., and Harayama, S. (1998). Molecular detection, isolation, and physiological characterization of functionally dominant phenol-degrading bacteria in activated sludge. Appl. Environ. Microbiol. *64*, 4396–4402.

Watanabe, K., Kodama, Y., Hamamura, N., and Kaku, N. (2002). Diversity, abundance, and activity of archaeal populations in oil-contaminated groundwater accumulated at the bottom of an underground crude oil storage cavity. Appl. Environ. Microbiol. *68*, 3899–3907.

Watts, J.E.M., Wu, Q., Schreiner, S.B., May, H.D., and Sowers, K.R. (2001). Comparative analysis of polychlorinated biphenyl-dechlorinating communities in enrichment cultures using three different molecular screening techniques. Environ. Microbiol. 3, 710–719.

Wenderoth D.F., Rosenbrock P., Abraham W.R., Pieper D.H., and Hofle M.G. (2003). Bacterial community dynamics during biostimulation and bioaugmentation experiments aiming at chlorobenzene degradation in groundwater. Microb. Ecol. *46*, 161–176.

Widada, J., Nojiri, H., Kasuga, K., Yoshida, T., Habe, H., and Omori, T. (2002). Molecular detection and diversity of polycyclic aromatic hydrocarbon-degrading bacteria isolated from geographically diverse sites. Appl. Microbiol. Biotechnol. 58, 202–209.

Wilson, I.G. (1997). Inhibition and facilitation of nucleic acid amplification. Appl. Environ. Microbiol. *63*, 3741–3751.

Wilson, K.H., Wilson, W.J., Radosevich, J.L., DeSantis, T.Z., Viswanathan, V.S., Kuczmarski, T.A., and Andersen, G.L. (2002). High-density microarray of small-subunit ribosomal DNA probes. Appl. Environ. Microbiol. *68*, 2535–2541.

Zhang, H., Logan, B.E., Regan, J.M., Achenbach, L.A., and Bruns, M.A. (2005). Molecular assessment of inoculated and indigenous bacteria in biofilms from a pilot-scale perchlorate-reducing bioreactor. Microbial. Ecol. *49*, 388–398.

Zhou, J. (2003). Microarrays for bacterial detection and microbial community analysis. Curr. Opin. Microbiol. *6*, 288–294.

Zumstein, E., Moletta, R., and Godon, J.J. (2000). Examination of two years of community dynamics in an anaerobic bioreactor using fluorescence polymerase chain reaction (PCR) single-strand conformation polymorphism analysis. Environ. Microbiol. 2, 69–78.

Ex situ Bioremediation of Contaminated Soils: From Biopiles to Slurry-phase Bioreactors

12

Nicolas Kalogerakis

Abstract

Although *in situ* bioremediation technologies for the treatment of contaminated soils are economically attractive, *ex situ* approaches are more often used for surface contaminated soils (typical depths less than 5 m) since they allow a much tighter control of the bioremediation process and provide better estimates of the residual contamination at the end of the treatment period. *Ex situ* bioremediation is the method of choice for hot spot treatment if they are reasonably accessible. In this chapter, the classical *ex situ* technologies (landfarming, composting, biopiling and slurry-phase bioremediation) are presented with a few examples of innovative modifications that enhance their productivity and/or effectiveness.

Ex situ bioremediation typically refers to the methods applied for the remediation of excavated contaminated soils. Besides slurry-phase bioremediation where the soil is mixed with water and other nutrients in mechanically agitated bioreactors, *ex situ* bioremediation includes solid phase bioremediation that covers the techniques of landfarming and the various forms of composting, namely windrows, biopiles and in-vessel composting. In the following sections, the above methods are presented focusing on design principles and on experience gathered from their application.

Ex situ solid-phase bioremediation

Landfarming

Landfarming is a simple technology that has been extensively used for treatment of hydrocarbon contaminated soils and petroleum waste sludge from refineries all over the world. The contaminated soil is excavated and items like stones, metal, plastic and other large items are removed. Subsequently, the soil is spread over a rather large area lined with high-density polyethylene (HDPE) or other impermeable material (Fig. 12.1). Usual fertilizers are added to compensate the lack of key nutrients like nitrogen and phosphorous. The aim is to stimulate the indigenous microbial populations and facilitate the aerobic biodegradation of the organic pollutants. Sufficient oxygen is provided by periodic tiling or ploughing to a depth of 10–40 cm depending on the type of soil (e.g. clay soils are more difficult to aerate) using typical agricultural equipment. In addition, a leachate collection and a leachate treatment system may also be necessary to prevent groundwater contamination from the landfarm (Genes and Cosentini 1993; Eweis *et al.*, 1998).

Design parameters

For a successful landfarming operation, the following process parameters should be checked and adjusted accordingly if needed.

Soil pH

The optimal pH is around 7 (neutral) and hence, the pH should be maintained within the range 6–8. Soil pH can be adjusted by adding lime to raise it or elemental sulphur to lower it (Roig *et al.*, 2004; EPA, 2004).

Soil temperature

Microbial activity in the soil decreases significantly as temperature falls below 10°C and practically ceases at temperatures below 5°C. Since soil

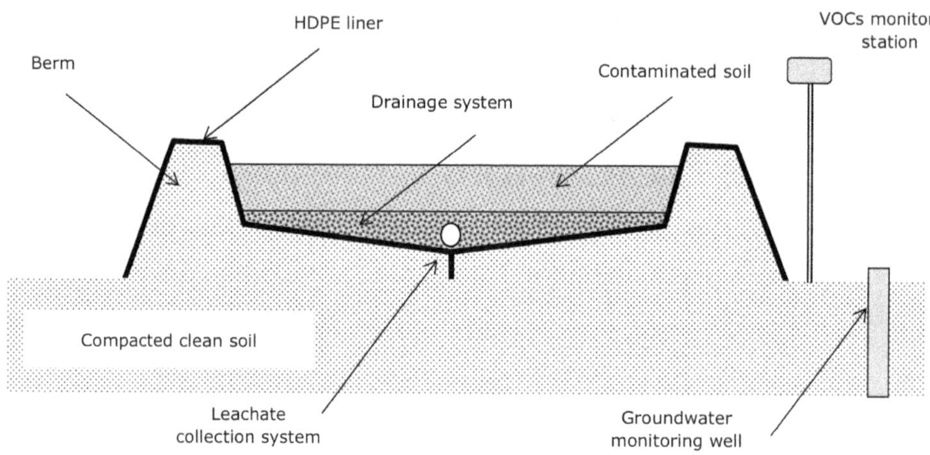

Figure 12.1 Schematic diagram of a landfarming facility (not to scale). HDPE, high-density polyethylene; VOC, volatile organic compound.

temperature follows ambient temperature variations the rate of contaminant degradation will diminish during the winter months. Microbial degradation will commence once the soil temperature returns in the range of 10 to 45°C. The so called 'landfarming season' corresponds to the period of the year when the ambient temperature is in the range of 10–45°C. In cold conditions, soil can be covered to maintain the temperature in the optimum range for bioremediation (15–30°C). Landfarming should be used in regions where the landfarming season is at least four months long otherwise other remediation alternatives should be examined (e.g. biopiles). Paudyn *et al.* (2008) in a field trial in the Canadian arctic region clearly demonstrated enhanced bioremediation when fertilizer was added and also significant hydrocarbon losses due to tilling. They found that despite the low temperatures the rate of bioremediation was similar to the rate of volatilization in the field trial.

Moisture content

Soil microbes require moisture for growth. The optimal range for soil moisture is between 40% and 85% of the water-holding capacity of the soil. Excessive soil moisture should be avoided as it blocks the movement of air through the soil which provides oxygen to the aerobic bacterial populations. Excessive accumulation of moisture occurs often in landfarms with high precipitation or poor drainage. On the other hand, periodic addition of moisture may be required whenever the soils become dry as a result of evaporation, particularly during the periodic tiling/ploughing operations to increase the aeration of the landfarms.

If the location of the landfarming facilities is in an area with an annual rainfall of 80 cm or higher during the landfarming season, a plastic cover or equivalent should be used to avoid an excessive moisture content for prolonged periods of time.

Nutrient concentrations

Besides the carbon source which is available in excess in contaminated soils, microorganisms require sufficient amounts of nitrogen and phosphorus to support cell growth. The C:N:P ratio should be maintained in the range 100:10:1 to 100:1:0.5 through the addition of fertilisers. Rough conservative estimates of the total amount of N and P required can be obtained from the stoichiometry of the overall biodegradation reactions.

Soil texture

Soil amendments like gypsum and bulking materials like sawdust or straw, may be required to improve the soil texture particularly when the soil tends to clump together and becomes difficult to aerate with poor drainage. These problems are typically encountered with clays.

Contaminant volatility

The presence of volatile organic compounds (VOCs) in the contaminated soils or waste sludge to be treated can pose a significant health risk because of evaporation, which is intensified during tilling or ploughing operations. Hejazi *et al.* (2003) examined the health risks posed to personnel during landfarming operations in hot, arid regions. From their exposure study, they showed that landfarming activity poses serious onsite risks for the workers particularly for the initial period of the loading/spreading of the contaminated soils or sludge. If the loading takes place on a continuous basis, significant evaporation is sustained for a longer time. Furthermore, tilling activities enhance volatilization, and this results in an increased risk to field personnel. As a result, a detailed risk assessment analysis should be conducted in the design of the landfarm to ensure that it does not pose a significant risk to onsite and offsite receptors. Safety guidelines developed for onsite landfarming activities must be strictly followed.

Contaminant concentration and toxicity

Very high concentrations of organic contaminants or heavy metals may be toxic or inhibit the growth of the soil bacteria. For example the concentration of total petroleum hydrocarbons (TPH) should be less than 25,000 ppm and heavy metals should not exceed 2500 ppm. If the contamination is higher, then thorough mixing with clean soil or other materials (wood chips, bark, pit moss, etc.) should be considered to reduce inhibitory/toxic effects.

Biodegradation tests

If there is no experience with the particular soil and contaminants, it is highly recommended to perform a series of biodegradation experiments which can show whether the indigenous bacterial populations can degrade the contaminants. In addition, these laboratory experiments enable us to determine the conditions for optimal microbial growth, potential lack of natural nutrients and minerals and the expected biodegradation rates. The latter is very important in determining the period required to obtain the desirable decontamination.

Biodegradation tests can be performed either in shaker flasks or in pans. Flask experiments are more suitable for slurry-phase bioremediation whereas pan experiments can better indicate landfarming effectiveness as contaminated soil is used directly in steel or glass pans as microcosms resembling landfarms more closely.

Landfarm monitoring

The state of the contaminated soils in the landfarm should be monitored at least quarterly to gather data on pH, moisture content, bacterial population, nutrient content and contaminant concentrations. In order to maintain maximum degradation rates, these measurements can be used to adjust tiling/ploughing frequency, fertiliser application frequency, soil moisture and soil pH to optimal values. Finally, periodic monitoring for groundwater contamination should also be performed.

Treated contaminants

Landfarming has been extensively used for many years for the treatment of oily wastes such as various biodegradable sludges, wastes from gas plants, creosote contaminated soils and soils contaminated with all types of hydrocarbons from accidental spills during handling and/or processing. Several classical examples with large amounts of spilled hydrocarbons have been reviewed by Alexander (1999) and Eweis *et al.* (1998). Landfarming has also been successfully used for the treatment of pesticides (Fiorenza *et al.*, 1991). PCP (pentachlorophenol) degradation in excess of 95% and PAHs (polycyclic aromatic hydrocarbons) reduction by about 75% have been reported over a 5-month period by landfarming (Nyer, 1992). PAHs are often encountered in refinery sludges and can be degraded by several types of microorganisms (Katsivela *et al.*, 2003, 2005). Recently, Atagana (2004) reported that combination of biostimulation and bioaugmentation during landfarming could enhance the bioremediation of soils heavily contaminated with creosote.

An eight-month field bioremediation experiment using fresh and composted sewage sludge was carried out by Ros *et al.* (2010) on an aged hydrocarbon-contaminated semi-arid soil. It was demonstrated that organic amendments encourage hydrocarbon degradation in polluted soils owing to the stimulation of microbial population size and activity they produce. Organic amendments with little stabilized organic matter, rich in easily biodegradable C substrates and nutrients (as is the case of fresh sewage sludge), were found particularly beneficial. In addition, it was observed that although bacteria (Actinobacteria) are the main actors in hydrocarbon degradation, the fungal community (particularly Ascomycota members) could play an important role in degrading recalcitrant oil hydrocarbons.

It is worth to note that the very extensive use of landfarming employed in the 1950s and 1960s by the petroleum industry has been reduced owing to more stringent environmental regulations regarding contamination of groundwater and health risks to workers from the volatilization of VOCs.

Advantages and disadvantages

The primary advantage of landfarming is low capital and operating costs. It is also rather simple to design and operate. In addition, under optimal conditions, the treatment times can be relatively short (6 months to 2 years). It is effective on organic pollutants that exhibit slow biodegradation rates.

On the other hand, the main disadvantages include: requirement for a large land area for treatment, VOCs potentially posing a significant health problem, requirement for impermeable liners particularly in areas with increased rainfall and vulnerability to elevated heavy metal concentrations (>2500 ppm) that could inhibit microbial growth. Finally, overall contaminant reductions greater than 95% or final contaminant concentrations less than 0.1 ppm are rather difficult to achieve.

Innovative modifications

There are two innovative modifications that have been employed: (i) enhanced landfarming and (ii) deep landfarming. Enhanced landfarming involves the use of biostimulation and bioaugmentation at the onset of the landfarming operations to decrease the overall degradation time. Deep landfarming is essentially landfarming of contaminated soils using forced aeration (air injection) at the centre of long soil mounds.

Composting

Composting is a very old technology that has been extensively used for the treatment of agricultural wastes (plant and animal waste) as well as municipal solid wastes (garbage) and activated sludge from wastewater treatment plants. Composting is the aerobic oxidation of moist organic material yielding biologically stable compounds known as composts which can be used as natural soil amendment for agricultural purposes. In a typical composting operation, heat is accumulated from the exothermic biodegradation reactions and the temperature rises substantially to about 55°C. At these temperatures, pathogens are also killed.

In the context of treating contaminated soils, composting has been employed only during the past few decades. It should be pointed out that contaminated soils cannot be composted alone, because most of the time the organic load of the contaminants is not sufficiently high to raise the temperature as it occurs in composting of organic wastes. In addition, there is a lack of key nutrients (N and P). Therefore, composting of contaminated soils requires the addition of relatively large amounts of non-hazardous organic matter like animal manure or agricultural wastes. These bulking agents improve soil texture, water-holding capacity and aeration of the compost. Soil amendments are particularly needed for biodegradation of xenobiotics (Miller *et al.*, 2004).

Composting systems

The classical composting system configurations are three: (i) windrows, (ii) biopiles (aerated static piles) and (iii) in-vessel composting.

Windrows

In windrows, the contaminated soil is mixed with the bulking agents and piled in long mounds (windrows) as shown in Fig. 12.2. The cross-sectional area of the windrow should be large enough so that the generated heat by the biodegradation processes is not lost to the surroundings but a good portion of it is retained in the pile and increases the temperature of the compost. Typical widths are between 3 and 4 m whereas the height

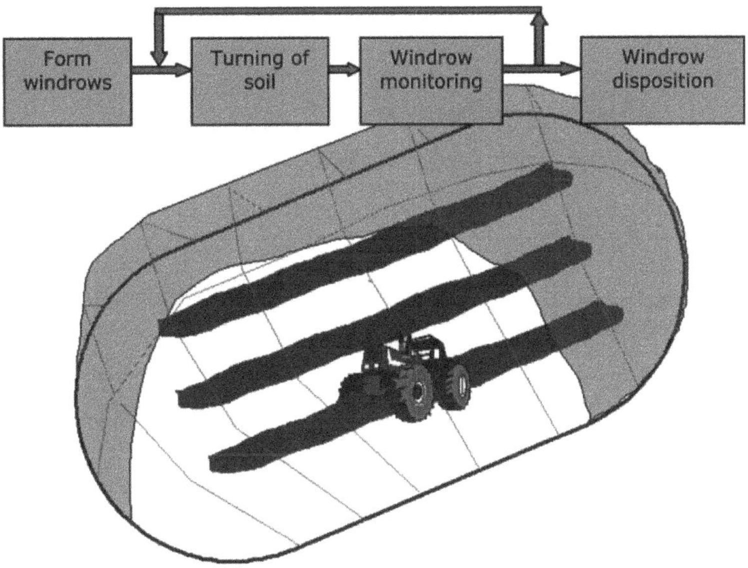

Figure 12.2 Schematic representation and flowchart of activities in windrow composting of contaminated soil (adapted from Van Deuren *et al.*, 2002).

is about 1–1.5 m. Aeration is accomplished by regular mixing of the soil. Turning of the soil is done as often as once a day to once a month (Eweis *et al.*, 1998). It should be noted that regular mixing improves uniformity of aeration, but it can cause odours and loss of moisture.

Biopiles

Biopiles represent a more intensive version of composting which is particularly useful when time and space are limited. The biopile technology maximizes biodegradation of the contaminants by better control of oxygen, moisture and nutrient (N and P) levels within the compost pile. Biopile treatment is a full-scale technology in which excavated soils are mixed with soil amendments and placed on a treatment area that includes leachate collection systems and some form of forced aeration (Fig. 12.3). Leachate from the biopile is either collected for further treatment and disposal or sprayed back onto the pile to maintain the soil moisture at the desired level. The treatment area should generally be covered and have an impermeable liner to minimize the risk of contaminants leaching into soil and groundwater. Soil piles have an air distribution system buried under the soil to pass air through the soil either by vacuum or by injection. The soil piles can be up to 6 m high (generally not recommended; typical height up to 2–3 m).

Soil piles are covered with plastic (HDPE) to control runoff, evaporation and volatilization and to promote solar heating. Aeration equipment usually includes blowers or vacuum pumps which will be attached to the aeration piping manifold. Recently, Li *et al.* (2004) proposed a new aeration system with vertical piping within the biopiles to enhance the transfer of oxygen under passive or forced aeration conditions. If there are VOCs in the soil that will volatilize into the air stream, use of vacuum pumps is preferable as the air sucked from the soil pile is treated to remove or destroy the VOCs before they are discharged to the atmosphere.

In-vessel composting

In-vessel composting refers to closed solid phase bioreactors where contaminated soil is continuously or intermittently mixed. The regular mixing improves the contact between the microbes and the organic contaminants. VOCs are much better controlled and heat losses to the environment are minimized. Oxygenation is provided through forced aeration. The moisture content and the

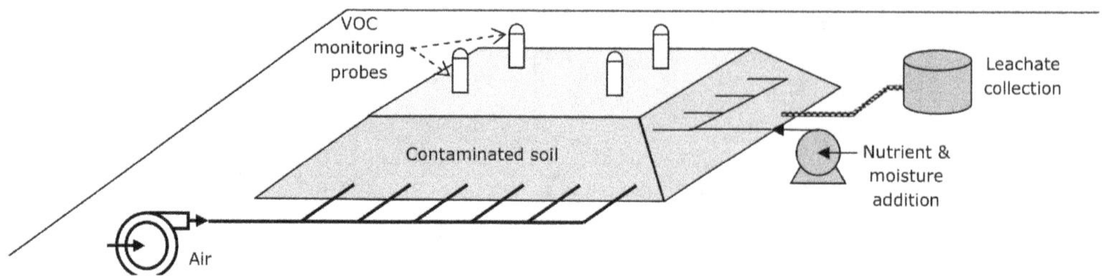

Figure 12.3 Main components in a biopile facility for soil treatment (adapted from US EPA 510-R-04–002). VOC, volatile organic compound.

temperature control can be readily accomplished in such a closed system. However, the obvious advantages of in-vessel composting must overweigh the requirement of increased capital investment.

Design parameters (windrows and biopiles)

For a successful composting operation, all process parameter guidelines described earlier for landfarming operations should be followed here too, keeping in mind the following comments.

In cold and very cold climates composting as opposed to landfarming, is a feasible treatment option when special actions are taken. For example, the windrows or biopiles are kept inside a green house; the injected air in the biopiles is heated; bioaugmentation with bacteria that can function at low temperatures (Mohn *et al.*, 2001; Thomassin-Lacroix *et al.*, 2002) is employed. Instead of a greenhouse structure, Schoefs *et al.* (1998) propose the use of sealed double white polyethylene membrane covers to maintain suitable temperatures year round.

In biodegradation tests estimation of total heterotrophic bacteria should also be done from samples prior to composting. If the total count is greater than 1000 colony-forming units per gram of dry soil (CFU/g dry soil), the operation is expected to be effective. If however, the count is less than 1000 CFU/g of dry soil it might be an indication of toxic conditions. Bioaugmentation is not necessary most of the time (numerous examples have been reported in the literature validating this observation, e.g. Jorgensen *et al.*, 2000; Li *et al.*, 2002). However, the initial biodegradation of recently contaminated soils with xenobiotic compounds may benefit from bioaugmentation by accelerating the initial biodegradation rates.

Because of the elevated temperature and the forced aeration (or mixing of the windrows), the presence of VOCs in the contaminated soils requires special attention. More specifically, the use of a vacuum pump should be preferred (instead of air injection) to aerate the biopiles and the exhaust gases should be treated prior to their discharge in the atmosphere.

For the remediation of contaminated soils from certain organic pollutants (e.g. PAHs, pentachlorophenol, creosote), fungi (particularly white rot fungi) might offer an attractive alternative to indigenous bacteria-based soil treatment. White rot fungi, like *Phanerochaete chrysosporium*, *Phanerochaete sorbida*, *Pleurotus ostreatus*, *Trametes versicolor* or *Bjerkandera* sp. BOS55, have shown the capability to degrade high molecular weight PAHs (five to seven rings) and other compounds that are recalcitrant to bacterial attack (Lamar *et al.*, 1994; Gramss *et al.*, 1999; Reddy and Mathew, 2001; Eggen, 1999). Unfortunately, the rate of degradation is often slow and the cost of preparing the required inocula must be further reduced from current levels to make fungi-based composting competitive. The use of spent mushroom compost (SMC) is a promising medium for inocula development (Eggen, 1999). Recently, Chiu *et al.* (2009) examined the effect of SMC on petroleum residues degradation in industrial soils. The field application showed that SMC treatment did not cause any nuisance to the environment. It was found that SMC treatment not only enhances the degradation of the persistent organo-pollutants but also reduces toxicity of the treated soil. In several occasions, it has been observed that the degradation of organic pollutants by white rot fungi can be complemented with the action of indigenous bacterial populations (Lang *et al.*, 1997; Kotterman *et al.*, 1998).

Biosurfactants and bioemulsifiers

Microbial bioemulsifiers are produced by a wide variety of diverse microorganisms and have very different chemical structures and surface properties (Calvo *et al.*, 2009). The fact that biosurfactants have a biological origin implies a better biocompatibility and good microbial biodegradability. The biological origin is of great interest, especially when there is extensive interference with the environment, for example for tertiary petroleum recovery, for the decontamination of oil-polluted areas, for crop protection and for the cosmetic and pharmaceutical sectors. According to their chemical structure, bioemulsifiers may be classified into the following main groups (Calvo *et al.*, 2009): (a) glycolipids (e.g. rhamnolipids), (b) amino acid-containing bioemulsifiers (e.g. surfactin), (c) polysaccharide–lipid complexes (e.g. emulsan), (d) protein-like substances (e.g. liposan).

Biodegradation of hydrocarbons in soil can be efficiently enhanced by addition or by *in situ* production of biosurfactants. The presence of biosurfactants in soil may produce a positive effect in form of stimulation of dissolution or desorption rates, solubilization or even emulsification of hydrocarbons (Norman *et al.*, 2002). Although, some biodegradable surfactants can compete with the pollutant as carbon source for the microorganisms provoking a preferential degradation of the surfactant and a lower removal of the contaminant (see review by Robles-González *et al.*, 2008). Finally, it should be pointed out that degradation of hydrocarbon can only be enhanced by surfactant when the process is under rate limiting conditions (Norman *et al.*, 2002).

Compost monitoring

The state of the contaminated soils in the compost should be regularly monitored to gather data on pH, moisture content, bacterial population, nutrient content and contaminant concentrations in the soil. However, in biopiles settling over time can result in formation of air channels, leading to 'low oxygen regions' within the pile where aerobic biodegradation rates will be significantly reduced and hence, the residual contaminant concentration in the soil may be quite different if anaerobic degradation does not occur. Therefore, biopiles require more rigorous verification sampling compared mixed piles (windrows or landfarms) as no further mixing takes place after their construction.

Enhanced bioremediation of petroleum hydrocarbons in two biopiles was investigated by Mao *et al.* (2009) and was quantified by high-performance liquid chromatography (HPLC) followed by comprehensive two-dimensional gas chromatography (GCXGC). The attenuation of 34 defined hydrocarbon classes was calculated by HPLC–GCXGC analysis of representative biopile samples at start-up and after 18 weeks of biopile operation. In general, a-cyclic alkanes were most efficiently removed from the biopiles, followed by mono-aromatic hydrocarbons. Cycloalkanes and polycyclic aromatic hydrocarbons (PAHs) were more resistant to degradation. A-cyclic biomarkers farnesane, trimethyl-C13, norpristane, pristane and phytane dropped to only about 10% of their initial concentrations. On the other hand, C_{29}–C_{31} hopane concentrations remained almost unaltered after 18 weeks of biopile operation, confirming their resistance to biodegradation. They are thus reliable indicators to estimate attenuation potential of petroleum hydrocarbons in biopile processed soils (Mao *et al.*, 2009).

Treated contaminants

Composting has been successfully applied to remediate soils contaminated with biodegradable organic pollutants. Windrow composting has been demonstrated as an effective technology for treatment of explosives-contaminated soil. During a field demonstration TNT reduction reached 99.7% in 40 days of operation, with the majority of removal occurring in the first 20 days of operation (Van Deuren *et al.*, 2002). Maximum removal efficiencies were 99.8% and 96.8% for RDX and HMX respectively.

In many studies, soils contaminated with petroleum hydrocarbons or PAHs have been successfully remediated through the use of composting technologies. For instance, Jorgensen *et al.* (2000) have reported a 70% reduction in mineral oil content of diesel or lubricant oil-contaminated soil within a 5-month period using biopile technology. Li *et al.* (2002), in a field scale project for the bioremediation of soil contaminated with crude oil, reported that TPH degradation was mainly controlled by the composition of aromatic hydrocarbons, asphaltene and resin. The degradation of crude oil (with initial contaminations ranging from 25,800 to 77,200 mg/kg dry weight) was found to be between 38% and 57% after 53 days of operation.

Biopile technology has been successfully used for the remediation of hydrocarbon-contaminated soils in the Artic. For example, Mohn *et al.* (2001) reported that extensive hydrocarbon removal occurred after one summer using biopiles where TPH levels were reduced from 196 to less than 10 mg/kg of dry soil at one site and from 2109 to 195 mg/kg dry soil at the other site. Filler *et al.* (2001) designed and installed a thermally enhanced biopile at a diesel-contaminated gravel pad in Prudhoe Bay where it was shown that the important parameters in extending the period of beneficial microbial activity in Arctic biopiles are temperature and substrate availability. More recently, Thomassin-Lacroix *et al.* (2002) used biostimulation and bioaugmentation for the on-site treatment of weathered diesel fuel in Arctic soil. Using biopiles they found that TPH were reduced from 2900 to 500 mg/kg of dry soil within a period of 65 days.

On the other hand, biopile technology has also been successfully used in a dessert environment. In particular, Al-Daher *et al.* (1998) examined the efficacy of windrow technology for the bioremediation of damaged desert environment in Kuwait. The showed that the windrow soil pile system resulted in a significant reduction of oil contamination within 10 months. In a related project, Balba *et al.* (1998) examined the bioremediation of oil-contaminated desert soil in Kuwait employing three different approaches (landfarming, windrow composting piles and static biopiles). Up to 82.5% and 90.5% reduction in TPH and total alkanes was observed respectively after a 12-month period with biopiles proving to be the best system for this particular case.

Several researchers have examined the applicability and efficiency of various amendments in composting of contaminated soils. For example, Wan *et al.* (2003) examined the effect of organic waste amendments on degradation of PAHs in soil. Their study positively supports that pig manure is an effective organic additive to composts for the bioremediation of PAH-contaminated soils. Van Gestel *et al.* (2003) quantified the effect of adding biowaste in composting of diesel oil-contaminated soil by showing that the first-order rate constant of diesel degradation in the biowaste mixture was four times higher than in the soil at room temperature, and 1.2 times higher than in the soil at composting temperature. Recently, Atagana (2004) showed in microcosm studies that co-composting of PAH-contaminated soil with poultry manure was capable of reducing large concentrations of high molecular weight PAHs from to levels below 1 mg/kg of dry soil within 19 months.

Antizar-Ladislao *et al.* (2004) reviewed research to date on the use of composting approaches for bioremediation of PAH-contaminated waste. They examined the relative effectiveness of different composting approaches, determined optimal composting operation conditions, and identified the limitations and advantages of using composting relative to other solutions. Recently, Gandolfi *et al.* (2009) examined the influence of a high quality compost amendment on two soils contaminated with diesel oil and PAHs. The addition of compost was effective in enhancing the degradation of diesel oil and four-ring PAHs and in reducing soil toxicity, while its effect on soil genotoxicity was quite limited. The microbial community of both soils was affected by the amendment. Thus, it was hypothesized that compost was a source of both nutrients and of microorganisms, which were active in the process.

Co-composting has also successfully been used for the remediation of PCP-contaminated soils as reported by Bentham and McClure (2003). It was found that PCP concentrations were reduced by 80% from 68 to 11 mg/kg within a 6-week period. Walter *et al.* (2004) in a series of soil microcosms examined the use of white rot fungi for the bioremediation of PCP-contaminated soil from a timber treatment site. Their results clearly showed that the New Zealand *T. versicolor* isolates mineralized PCP and that degradation of PCP in non-sterile soil was higher in the presence of white-rot fungi than in soil without white-rot fungus.

Cunningham *et al.* (2004) found that bioaugmentation with microorganisms immobilized in polyvinyl alcohol (PVA) can be very effective for the bioremediation of diesel-contaminated soil by composting. In terms of percentage removal of diesel after 32 days, the immobilized systems were found to be the most successful, with greatest removal in a co-immobilisation system containing PVA-entrapped microorganisms and a synthetic oil absorbent. Immobilization technology has been successfully used in many biotechnological applications. In the years to come, composting may prove to be another application area.

Advantages and disadvantages

Biodegradation of contaminants in soils, waste sludge or dredged sediments is generally faster by composting compared with landfarming. Composting requires a smaller area than landfarming and generally it is more suitable for colder climates, as the heat generated during the biodegradation of the contaminants is retained within the compost pile. Other advantages of composting systems include: simplicity in design and implementation; not very costly; short treatment times and vapour emissions much easier to control.

On the other hand, the main disadvantages are similar to those for landfarms, namely: requirement for a large land area for treatment although it is less than that required for landfarms, VOCs are still a problem that needs to be taken care of, requirement for impermeable liners and vulnerability to elevated heavy metal concentrations (>2500 ppm) as microbial growth could be inhibited. Finally, overall contaminant reductions greater than 95% or final contaminant concentrations less than 0.1 ppm are difficult to achieve with composting too.

Innovative modifications

An interesting innovative modification is the biocell technology which has been developed for small scale decontamination projects (volume of contaminated soils typically less than 40 m^3) and essentially it is an in-vessel composting system without the mixing of the soils. A schematic diagram of the system is provided in Fig. 12.4.

The main characteristics of a biocell system are:

- Biocells are not as susceptible to setbacks from rainfall or cold weather as landfarms.
- Biocells normally have an aeration blower to inject or draw air (oxygen) into the contaminated soil and perforated hoses to provide H_2O (moisture) and/or nutrients. Vapour emissions can be readily treated if necessary (GAC, biofilter).
- Biocells are normally equipped with more complex sensors than composting units to promote optimization.
- Typically, biocells are used for sites that have a lower volume of contaminated soil and generally have higher unit costs in comparison to landfarming. They are normally employed when time and space is limited.

Ex situ slurry-phase bioremediation

In slurry-phase bioremediation, excavated contaminated soil or sludge or dredged sediment is placed in a bioreactor where sufficient water or wastewater is added to form a slurry of desired mean density. Aerobic conditions are maintained throughout the bioreactor by air sparging and thorough agitation. The energy requirements for mixing the slurry (i.e. keeping the soil particles in suspension) increases sharply with mean density. Solids loading into the bioreactor is typically between 10% and 30% (w/w) depending on the level of contamination. As the slurry is mixed, the bioavailability of the contaminants to the degrading microbes is increased. In addition, continuous mixing and aeration breaks up soil clumps and enhances contaminant dissolution into the aqueous phase. Typical bioreactor sizes range from 30 to 1400 m^3 and are usually fitted with mixing devices. Aerobic bioreactors are equipped with spargers or diffusers whereas ancillary equipment includes gas emissions treatment (if required), vessels for nutrient and pH conditioning of the slurry, etc. (Robles-González *et al.*, 2008).

Biodegradation rates can be further optimized by adding nutrients (N and P) and maintaining dissolved oxygen, pH and operating temperature

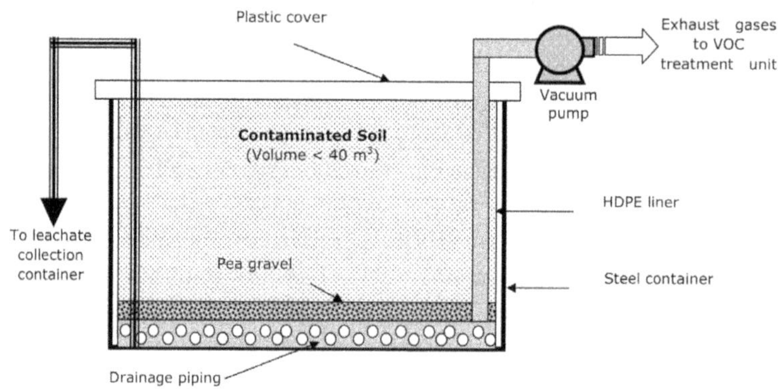

Figure 12.4 Schematic diagram of a biocell treatment unit. HDPE, high-density polyethylene; VOC, volatile organic compound.

to optimal values for microbial growth. By design, slurry-phase bioreactors provide the highest level of contact between pollutants, microbes, oxygen, water and supplemented nutrients.

Pre-treatment of contaminated soils

Pre-treatment of contaminated soil is often a necessary step that results in substantial cost benefits and increased process feasibility (Eweis *et al.*, 1998). Usually rocks, gravel and even large sand particles are separated from the soil to be treated in the bioreactor including large debris (plastic components, construction materials or mechanical parts, wood sticks, etc.).

Soil washing is a physical process often employed as a first step. Large particles can often be disposed as clean after soil washing which in turn improves the overall process economics substantially. As a result of pre-treatment, the volume of the solids to be handled in the bioreactor can be significantly reduced (often more than 70%).

A feasible, effective, and inexpensive alternative for cleaning up such soils might be represented by surfactant aided *ex situ* soil-washing technology (Berselli *et al.*, 2006). Surfactants may enhance the recovery of hydrophobic pollutants sorbed to soils through the soil rollup mechanism (at concentrations below the critical micelle concentration, CMC) and micelle solubilization (at concentrations above CMC). In general, the most effective results are obtained by employing synthetic commercial surfactants at concentrations well above their CMCs. However, under such conditions, the surfactants tend to persist and exert toxic effects both in the washed soils and in the resulting effluents (Berselli *et al.*, 2006). The same authors pointed out for the first time that deoxycholic acid and bovine bile can be advantageously applied in the washing of actual site (chloro)aromatic contaminated soils. They are able to exert marked pollutant mobilization potential without displaying the adverse effects on the soil chemical composition and ecotoxicity exhibited by Triton X-100. In addition, because of their biodegradability, they are able to sustain the soil bacteria in charge for pollutant degradation and, in turn, the aerobic bioremediation of soil washing effluents. Thus, the replacement of chemical surfactants with such relatively cheap, biodegradable, and non-toxic pollutant-mobilizing agents may contribute to improve the overall sustainability of the conventional soil-washing technology (Berselli *et al.*, 2006).

Bioreactor operation (SBR)

A special repeated batch operation, called sequencing batch reactor (SBR), is the usual mode of operation for slurry phase bioremediation. The sequence of events that take place is shown in Fig. 12.5 and are given below:

1. Contaminated solids, sediments or sludge is placed in the bioreactor.
2. Water, nutrients and microbes (first time only) are added.
3. The slurry is mixed and aerated continuously

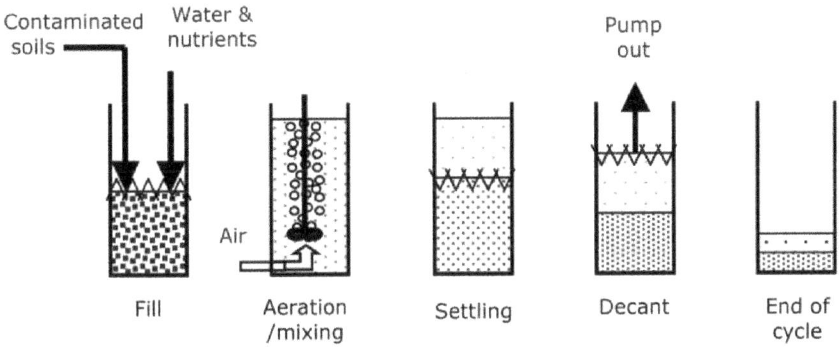

Figure 12.5 Schematic diagram of the slurry-phase bioreactor operation steps in the SBR mode.

until the desired degradation of the contaminants is achieved.

4 Mixing and aeration are stopped and solids are allowed to settle.
5 Solids and liquid are removed.
6 A portion of the slurry is left as seed for subsequent runs.

Preparation of microbial inocula for introduction into the slurry phase bioreactor for the very first SBR cycle has been discussed in detail by Eweis *et al.* (1998).

Treated contaminants

Slurry-phase bioreactors can be used to biodegrade non-halogenated semivolatile and volatile organic compounds (SVOCs and VOCs) in excavated soils or dredged sediments. Slurry-phase bioremediation has also been successfully applied for treating PAH-contaminated soils (Pinelli *et al.*, 1997; Saponaro *et al.*, 2002).

Castaldi (2003) proposed for the treatment of petroleum waste sludges, the use of a continuous flow, multistage slurry-phase bioremediation system that operates at relatively short residence times with minimal loss of volatile constituents. The higher molecular weight PAHs, including the four-ring compounds of pyrene, benzo[a] anthracene, and chrysene, were effectively treated with removal efficiencies greater than 90%. More recently, Fava *et al.* (2004) reported on the beneficial role of humic substances and soya lecithin on the aerobic slurry-phase bioremediation of a soil historically contaminated with PAHs. White rot fungi have also been used successfully in the slurry-phase bioremediation of PAH-contaminated soils (Zheng and Obbard, 2002).

Ex situ treatment of simulated pyrene-contaminated soil was studied by Mohan *et al.* (2008) in slurry phase bioreactors operated in periodic discontinuous batch mode under anoxic–aerobic–anoxic–anoxic conditions and demonstrated the efficacy of their *ex situ* treatment methodology with respect to soil bound pyrene. Use of domestic sewage as augmented inoculum enhanced pyrene degradation which also indicates its compatibility with native soil microflora (Mohan *et al.*, 2008).

With the addition of suitable supplements, halogenated VOCs and SVOCs, pesticides, and PCBs can be treated under cometabolic conditions. In many investigations slurry-phase bioremediation was successful in treating PCP-contaminated soils (e.g. Mueller *et al.*, 1991; Rutherford *et al.*, 1998). Fava *et al.* (2002, 2003) examined the use of randomly methylated beta-cyclodextrins (RAMEB) to enhance the aerobic biodegradation of PCBs (Aroclor 1260) in contaminated soils. RAMEB, which is inexpensive, biodegradable and non-toxic enhanced significantly PCB bioavailability and degradation.

Military explosive compounds can also be treated. Aerobic slurry-phase bioreactor technology was used to reduce TNT, HMX, and RDX concentration in soil at Joliet Army Ammunition Plant, IL (Van Deuren *et al.*, 2002). Slurry-phase bioremediation demonstrated a removal efficiency of greater than 99% and a high degree of mineralization.

The degradation of di-ethylhexyl phthalate (DEHP), one of the more recalcitrant phthalate esters, was studied by Shailaja *et al.*, 2008)in five different slurry bioreactors operated under different conditions with contaminated soil (1 mg-DEHP/g soil). It was found that bioaugmentation with a *Pseudomonas aeruginosa* bacterial strain isolated in the degradation of DnBP was also effective for the degradation of DEHP (Shailaja *et al.*, 2008).

Slurry phase bioremediation can also be used treat soils contaminated with pesticides. Plangklang and Reungsang (2010) investigated the degradation of carbofuran (2,3-dihydro-2,2-dimethylbenzofuran-7-yl methyl carbamate) which is a broad-spectrum insecticide widely used in agriculture to control insects and nematodes. Bioaugmentation with immobilized *B. cepacia* PCL3 was found as the most effective treatment resulting in the highest percentage of carbofuran removal (96.97%).

The dechlorination and mineralization of pentachlorophenol (PCP) was investigated by Yang *et al.* (2009) by combining two different anaerobic microbial populations, a PCP-dechlorinating culture capable of the reductive dechlorination of PCP to phenol and phenol degrading cultures able to mineralize phenol under sulfate- or iron-reducing conditions. This study proposes an anaerobic technology for the mineralization of PCP that uses anaerobic–anaerobic microbial combinations of PCP-dechlorinating and phenol-degrading enrichment cultures, instead of the traditional anaerobic–aerobic combination. Using two different combinations, simultaneous or sequential, the combined microbial activities mineralized PCP to CO_2 and CH_4 under anaerobic conditions (Yang *et al.*, 2009).

Advantages and disadvantages

In general, the treatment times are shorter compared with other methods and less readily degradable pollutants can be treated satisfactorily as the environmental conditions are much more tightly controlled to near optimal levels.

The main disadvantage of slurry-phase bioremediation is cost. Pre-treatment costs can also be significant when dealing with non-homogeneous contaminated soils that contain different size solids. Furthermore, soils that contain large amounts of clays can be difficult to handle during pre-treatment and in the bioreactor. Finally, the portion of the wastewater that cannot be recycled at the end of the slurry-phase treatment must be disposed in an environmentally acceptable manner.

Process improvements

One of the rate limiting steps that results in overall low biodegradation rates is due to the limited bioavailability of the absorbed contaminants in the soil. In order to increase the bioavailability either of the following options may be adopted:

1. Surfactants can be added together with nutrients at the beginning of the SBR cycle. Concentrations below the critical micelle concentration (CMC) do not have any noticeable effect on bioavailability. On the other hand, much higher concentrations than CMC may have a negative effect on the biodegradation rate owing to toxicity.
2. Alternatively solvents can also be used for increasing the availability and bioavailability of low solubility-, hydrophobic pollutants in soil remediation. In particular, immiscible (in water) and non-degradable solvents with affinity for hydrophobic contaminants can help in attracting the molecules of contaminants adsorbed onto soil, transfer the contaminant into the solvent phase, and afterwards to facilitate the exchange of contaminant between the solvent to the aqueous phase where microorganisms can finally degrade the pollutant (Janikowski *et al.*, 2002; Robles-Gonzalez *et al.*, 2008).
3. Use of biosurfactants is considered a better choice. They act as emulsifiers that facilitate desorption of hydrophobic contaminants. High biosurfactant levels can be accomplished either by maintaining a high microbial concentration resulting in high production of biosurfactants or by adding commercially available biosurfactants directly into the bioreactor. In general, biosurfactants are more expensive compared with usual chemical agents.

Another reason for slow overall biodegradation rates is the recalcitrance of specific organic contaminants or complex mixtures of them. The strategy to improve the overall process is based primarily on selecting a combination of different processes. Typical examples are:

1 A sequence of biological processes, for example, anaerobic followed by aerobic treatment. The optimum sequence and the switching point is determined by performing a series of experiments during the biodegradation test phase.
2 Oxidizing agents can be added to the slurry to partially break down recalcitrant contaminants. Fenton's reagent has been successfully used to enhance high molecular weight PAHs degradation (Brown *et al.*, 1995).
3 A combination of chemical/biological processes is the most promising alternative. In particular, the use of advanced oxidation processes (AOPs) as a pre-treatment step where the powerful hydroxyl radicals ($OH^{\cdot}$) break down the contaminants into more easily biodegradable species is highly recommended.
4 Finally, bioaugmentation may prove to be a very useful strategy for xenobiotics particularly at the start up of the operation (Wang *et al.*, 2004).

Nano *et al.* (2003) have recently proposed the sequential use of slurry-phase bioreactors followed by solid-phase bioremediation to improve overall economics and effectiveness of treatment of diesel-decontaminated soils. They showed that a low hydraulic retention time and a low slurry recycle ratio allows for a better utilization of the bioreactor volume. A 7-day slurry-phase bioreactor treatment was shown to provide enough contaminant removal allowing the soil drawn from the slurry-phase bioreactors to be fed effectively to the solid-phase bioreactors for completing the soil clean up. Similarly, Kuyukina *et al.* (2003) reported on the use of slurry phase bioremediation followed by landfarming for the remediation of crude oil-contaminated soil. Slurry-phase biotreatment resulted in an 88% reduction in oil concentration after 2 months. The resulting reactor product, containing approximately 25 g/kg of total recoverable petroleum hydrocarbons (TRPH), was then loaded into landfarming cells where further decontamination to the level of 1.0–1.5 g/kg of TRPH was achieved after 5–7 weeks.

Bioaugmentation and biostimulation of *ex situ* bioremediation

Bioaugmentation is defined as the inoculation of contaminated soil, sediment or sludge with isolated strains or consortia with specific organic compound degrading capabilities to enhance *in situ* or *ex situ* bioremediation applications. Bioaugmentation is primarily used to overcome the problem of lack of microbes expressing the appropriate catabolic potential to mineralize target contaminants. Bioaugmentation is equivalent to increasing the metabolic capabilities of the microbiota present in the soil (Dejonghe *et al.*, 2001).

A bioaugmentation attempt is considered successful if the rate and per cent of pollutant removal is increased and if the augmented species are not outgrown by the indigenous populations. The routine use of bioaugmentation is rather controversial as many occasions of positive and negative attempts can be found in the literature. In many situations unfavourable environmental conditions may prevail for microbial growth and hence, very limited pollutant degradation activity occurs. However, several of the indigenous populations may have the capability to degrade the target contaminants although they could be in very small numbers as the pollutant is a substrate not commonly found in the natural ecosystem and unfavourable environmental conditions prevent the growth of the indigenous population. In such cases, biostimulation of the indigenous micro-organisms with the addition of appropriate nutrients usually results in satisfactory removal of the target organic compounds within a reasonable time frame (Vinas *et al.*, 2002; Rahman *et al.*, 2003; Atagana 2003; Straube *et al.*, 2003).

Furthermore, comparing the effectiveness of bioaugmentation alone, in such cases, the results will probably be poor whereas if we were to combine biostimulation and bioaugmentation the improvement over biostimulation alone will most probably be marginal if the indigenous populations are fast growing. The usual conclusion from such experiments is 'bioaugmentation is not needed'. These type of experimental findings have been reported by Cunningham *et al.* (2004), Simon *et al.* (2004), Saponaro *et al.* (2002), Margesin and Schinner (1999, 1998), Demque *et al.* (1997) among others. Aldrett *et al.* (1997) in a multiparameter test showed that only three out of 13 commercially available bioaugmentation products were successful in enhancing microbial degradation of crude oil under their specific environmental conditions. A discussion of the conditions that dictate the use of bioaugmentation or/and biostimulation is given by Kalogerakis (2005).

As a final note, one should be very careful whenever evaluating the performance of *ex situ* solid-phase bioremediation projects. The main reason is that any measured decrease in the contaminant concentration in the soil is not necessarily due to bioremediation. There are many other processes that can result in a decrease of measured contaminant levels giving a false indication of an effective bioremediation. This 'apparent' bioremediation could be due to loss of volatiles (e.g. BTEX, TCE), dilution of the contaminants or their sorption to organic bulking agents!

Conclusions

Ex situ bioremediation methods are and will continue to be more commonly used than *in situ* ones if the contaminated soil is reasonably easily accessible. Windrows and biopiles are the most frequently used methods for soil remediation, particularly for hydrocarbon contaminated soils. Both are reasonably well understood processes (especially for petroleum hydrocarbons), but it is still difficult to accurately predict end-points. Often it is not clear whether contaminant reduction is due to biodegradation, volatilization (e.g. BTEX), leaching or dilution.

Proprietary additives are somewhat unnecessary for a vast majority of contaminants! Standard fertilisers and bulking agents can do the job of stimulating the indigenous populations. Bioaugmentation is quite often not necessary. However, the initial biodegradation of recently contaminated soils with xenobiotic compounds may benefit from bioaugmentation by accelerating the initial biodegradation rates. In situations where the time period for bioremediation is of essence, a suitable bioaugmentation strategy may prove to be quite beneficial for several contaminants.

If we extrapolate current industrial practice, one should expect the following trends to continue in the future: (i) the use of windrow composting and biopiles will continue to grow; and (ii) slurry phase bioreactors will find only niche applications (e.g. hot spot treatment) because of higher costs; however, they have proved in many instances more cost effective than incineration, solvent extraction and thermal desorption (Robles-González *et al.*, 2008).

References

Al-Daher, R., Al-Awadhi, N., and El-Nawawy, A. (1998). Bioremediation of damaged desert environment using the windrow soil pile system in Kuwait. Environ. Int. *24*, 175–180.

Aldrett, S., Bonner, J.S., Mills, M.A., Autenrieth, R.L., and Stephens, F.L. (1997). Microbial degradation of crude oil in marine environments tested in a flask experiment. Water Res. *31*, 2840–2848.

Alexander, M. (1999). Biodegradation and Bioremediation, 2nd edn (Academic Press, San Diego, CA).

Antizar-Ladislao, B., Lopez-Real, J.M., and Beck, A.J. (2004). Bioremediation of polycyclic aromatic hydrocarbon (PAH)-contaminated waste using composting approaches. Crit. Rev. Env. Sci. Tech. *34*, 249–289.

Atagana, H.I. (2003). Bioremediation of creosote-contaminated soil: a pilot-scale landfarming evaluation. World J. Microb. Biot. *19*, 571–581.

Atagana, H.I. (2004). Bioremediation of creosote-contaminated soil in South Africa by landfarming. J. Appl. Microbiol. *96*, 510–520.

Balba, M.T., Al-Daher, R., Al-Awadhi, N., Chino, H., and Tsuji, H. (1998). Bioremediation of oil-contaminated desert soil: The Kuwaiti experience. Environ. Int. *24*, 163–173.

Bentham, R.H., and McClure, N.C. (2003). A novel laboratory microcosm for cocomposting of pentachlorophenol contaminated soil. Compost Sci. Util. *11*, 311–320.

Berselli, S., Benitez, E., Fedi, S., Zannoni, D., Medici, A., Marchetti, L., and Fava, F. (2006). Development and assessment of an innovative soil-washing process based on the use of cholic acid-derivatives as pollutant-mobilizing agents. Biotechnol. Bioeng. *93*, 761–770.

Brown, K.L., Davilla, B., and Sanseverino, J. (1995). Combined chemical and biological oxidation of slurry-phase PAHs. Proceedings of the 88th Annual

Meeting and Exhibition of Air and Waste Management Association (San Antonio, TX).

Calvo, C., Manzanera, M., Silva-Castro, G.A., Uad, I., and González-Lopez, J. (2009). Application of bioemulsifiers in soil oil bioremediation processes. Future prospects. Sci. Total Environ. *407*, 3634 – 3640.

Castaldi, F.J. (2003). Tank-based bioremediation of petroleum waste sludges. Environ. Progr. *22*, 25–36.

Chiu, S.-W., Gao, T., Chan, C.S.-S., and Ho, C.K.-M. (2009). Removal of spilled petroleum in industrial soils by spent compost of mushroom *Pleurotus pulmonarius*. Chemosphere *75*, 837–842.

Cunningham, C.J., Ivshina, I.B., Lozinsky, V.I., Kuyukina, M.S., and Philp, J.C. (2004). Bioremediation of diesel-contaminated soil by microorganisms immobilised in polyvinyl alcohol. Int. Biodeter. Biodegr. *54*, 167–174.

Dejonghe, W., Boon, N., Seghers, D., Top, E.M., and Verstraete, W. (2001). Bioaugmentation of soils by increasing microbial richness: missing links. Environ. Microbiol. *3*, 649–657.

Demque, D.E., Biggar, K.W., and Heroux, J.A. (1997). Land treatment of diesel contaminated sand. Can. Geotech. J. *34*, 421–431.

Eggen, T. (1999) Application of fungal substrate from commercial mushroom production – *Pleurotus ostreatus* – for bioremediation of creosote-contaminated soil. Int. Biodeter. Biodegr. *44*, 117–126.

Environmental Protection Agency (2004). How to Evaluate Alternative Cleanup Technologies for Underground Storage Tank Sites: A Guide for Corrective Action Plan Reviewers. (Document: EPA 510-R-04–002).

Eweis, J.B., Ergas, S.J., Chang, D.P.Y., and Schroeder, E.D., eds. (1998). Bioremediation Principles (Boston, MA: McGraw Hill)

Fava, F., Di Gioia, D., Marchetti, L., Fenyvesi, E., and Szejtli, J. (2002). Randomly methylated beta-cyclodextrins (RAMEB) enhance the aerobic biodegradation of polychlorinated biphenyl in aged-contaminated soils. J. Incl. Phenom. Macro. *44*, 417–421.

Fava, F., Bertin, L., Fedi, S., and Zannoni, D. (2003). Methyl-beta-cyclodextrin-enhanced solubilization and aerobic biodegradation of polychlorinated biphenyls in two aged-contaminated soils. Biotechnol. Bioeng. *81*, 381–390.

Fava, F., Berselli, S., Conte, P., Piccolo, A., and Marchetti, L. (2004). Effects of humic substances and soya lecithin on the aerobic Bioremediation of a soil historically contaminated by polycyclic aromatic hydrocarbons (PAHs). Biotechnol. Bioeng. *88*, 214–223.

Filler, D.A., Lindstrom, J.E., Braddock, J.F., Johnson, R.A., and Nickalaski, R. (2001) Integral biopile components for successful bioremediation in the Arctic. Cold Reg. Sci. Technol. *32*, 143–156.

Fiorenza, S., Duston, K.L., and Ward, C.H. (1991) Decision-making – is bioremediation a viable option? J. Hazard. Mater. *28*, 171–183.

Gandolfi, I., Sicolo, M., Franzetti, A., Fontanarosa, E., Santagostino, A., and Bestetti, G. (2010). Influence of compost amendment on microbial community and ecotoxicity of hydrocarbon-contaminated soils. Bioresource Technol. *101*, 568–575.

Genes, B.R., and Cosentini, C.C. (1993). Bioremediation of polynuclear aromatic hydrocarbon contaminated soils at three sites. In Hydrocarbon Contaminated Soils, Calabrese, E.J., and Kostecki, P.T., eds. (Lewis Publishers: Chealsea, MI), pp. 161–175.

Gramss, G., Voigt, K.D., and Kirsche, B. (1999). Degradation of polycyclic aromatic hydrocarbons with three to seven aromatic rings by higher fungi in sterile and nonsterile soils. Biodegradation *10*, 51–62.

Hejazi, R.F., Husain, T., Khan, F.I. (2003). Landfarming operation of oily sludge in arid region – human health risk assessment. J. Hazard. Mater. B99, 287–302.

Janikowski, T., Velicogna, D., Punt, M., and Daugulis, A. (2002). Use of a two-phase partitioning bioreactor for degrading polycyclic aromatic hydrocarbons by a *Sphingomonas* sp. Appl. Microbiol. Biotechnol. *59*, 368–376.

Jorgensen, K.S., Puustinen J., and Suortti, A.M. (2000). Bioremediation of petroleum hydrocarbon-contaminated soil by composting in biopiles. Environ. Pollut. *107*, 245–254.

Kalogerakis, N. (2005). *Ex situ* bioremediation of contaminated soils. In Soil and Sediment Remediation, Liens, P., Grotenhuis, T., and Tabak, H., eds. (London, UK: IWA Publishing) pp. 151–175.

Katsivela, E., Moore, E.R.B., and Kalogerakis, N. (2003). Biodegradation of aliphatic and aromatic hydrocarbons: specificity among bacterial isolated from refinery waste sludge. Water Air Soil Poll. Focus *3*, 103–115.

Katsivela, E., Moore, E.R.B., Maroukli, D., Strömpl, C., Pieper, D., and Kalogerakis, N. (2005). *In situ* bioremediation of petroleum waste sludge in landfarming sites. Biodegradation *16*, 169–180.

Kotterman, M.J.J., Vis, E.H., and Field, J.A. (1998). Successive mineralization and detoxification of benzo[a]pyrene by white rot fungus *Bjerkandera* sp. strain BOS55 and indigenous microflora. Appl. Environ. Microbiol. *64*, 2853–2858.

Kuyukina, M.S., Ivshina, I.B., Ritchkova, M.I., Philp, J.C., Cunningham, C.J., and Christofi, N. (2003). Bioremediation of crude oil-contaminated soil using slurry-phase biological treatment and landfarming techniques. Soil Sediment Contam. *12*, 85–99.

Lamar, R.T., Davis, M.W., Dietrich, D.M., and Glaser, J.A. (1994). Treatment of a pentachlorophenol- and creosote-contaminated soil using the lignin degrading fungus *P. sorbida*: a field demonstration. Soil Biol. Biochem. *26*, 1603–1611.

Lang, E., Eller, G., and Zadrazil, F. (1997). Lignocellulose decomposition and production of lignolytic enzymes during interaction of white rot fungi with soil microorganisms. Microb. Ecol. *34*, 1–10.

Li, L., Cunningham, C.J., Valerie Pas, Philp, J.C., Barry, D.A., and Anderson, P. (2004). Field trial of a new aeration system for enhancing biodegradation in a biopile. Waste Manage. *24*, 127–137

Li, P.J., Sun, T.H., Stagnitti, F., Zhang, C.G., Zhang, H.R., Xiong, X.Z., Allinson, G., Ma, X.J., and Allinson,

M. (2002). Field-scale bioremediation of soil contaminated with crude oil. Environ. Eng. Sci. *19*, 277–289.

Mao, D., Lookman, R., Van DeWeghe, H., Van Look, D., Vanermen, G., De Brucker, N., and Diels, L. (2009). Detailed analysis of petroleum hydrocarbon attenuation in biopiles by high-performance liquid chromatography followed by comprehensive two-dimensional gas chromatography. J. Chromatogr. A. *1216*, 1524–1527

Margesin, R., and Schinner, F. (1999). Biological decontamination of oil spills in cold environments. J. Chem. Technol. Biot. *74*, 381–389.

Margesin, R., and Schinner, F. (1998). Low-temperature bioremediation of a waste water contaminated with anionic surfactants and fuel oil. Appl. Microbiol. Biotechnol. *49*, 482–486.

Mohn, W.W., Radziminski, C.Z., Fortin, M.C., Reimer, K.J. (2001). On site bioremediation of hydrocarbon-contaminated Arctic tundra soils in inoculated biopiles. Appl. Microbiol. Biotechnol. *57*, 242–247.

Mueller, J.G., Lantz, S.E., Blattmann, B.O., and Chapman, P.J. (1991). Bench-scale evaluation of alternative biological treatment processes for the remediation of pentachlorophenol-contaminated and creosote-contaminated materials – slurry-phase bioremediation. Env. Sci. Tech. *25*, 1055–1061.

Nano, G., Borroni, A., and Rota, R. (2003). Combined slurry and solid-phase bioremediation of diesel contaminated soils. J. Hazard. Mater. B*100*, 79–94.

Norman, R.S., Frontera-Suau, R., and Morris, P.J. (2002). Variability in *Pseudomonas aeruginosa* lipopolysaccharide expression during crude oil degradation. Appl. Environ. Microbiol. *68*, 5096–103.

Nyer, E.K. (1992). Treatment of organic contaminants, physical/chemical methods. In Bioremediation: The State of Practice in Hazardous Waste Remediation Operations, seminar sponsored by Air and Waste Management Association and HWAC, Pittsburgh, PA, January 1992.

Paudyn, K., Rutter, A., Rowe, R.K., and Poland, J.S. (2008). Remediation of hydrocarbon contaminated soils in the Canadian Arctic by landfarming. Cold Reg. Sci. Technol. *53*, 102–114.

Pinelli, D., Fava, F., Nocentini, M., and Pasquali, G. (1997). Bioremediation of a polycyclic aromatic hydrocarbon-contaminated soil by using different aerobic batch bioreactor systems. J. Soil Contam. *6*, 243–256.

Plangklang, P., and Reungsang, A. (2010). Bioaugmentation of carbofuran by *Burkholderia cepacia* PCL3 in a bioslurry phase sequencing batch reactor. Process Biochem. *45*, 230–238.

Rahman, K.S.M., Rahman, T.J., Kourkoutas, Y., Petsas, I., Marchant, R., and Banat, I.M. (2003). Enhanced bioremediation of n-alkane in petroleum sludge using bacterial consortium amended with rhamnolipid and micronutrients. Bioresource Technol. *90*, 159–168.

Reddy, C.A., and Mathew, Z. (2001). Bioremediation potential of white rot fungi. In Fungi in Bioremediation, Gadd, G.M., ed. (Cambridge University Press: Cambridge, U.K.) pp. 52–78.

Robles-González, I.V., Fava, F., and Poggi-Varaldo, H.M. (2008). A review on slurry bioreactors for bioremediation of soils and sediments. Microb. Cell Fact. 7: doi:10.1186/1475-2859-7-5

Ros, M., Rodriguez, I., Garcia, C., and Hernandez, T. (2010). Microbial communities involved in the bioremediation of an aged recalcitrant hydrocarbon polluted soil by using organic amendments. Bioresource Technol. *101*, 6916–6923.

Roig, A., Cayuela, M.L., and Sanchez-Monedero, M.A. (2004). The use of elemental sulphur as organic alternative to control pH during composting of olive mill wastes. Chemosphere 57, 1099–1105.

Rutherford, P.M., Banerjee, D.K., Luther, S.M., Gray, M.R., Dudas, M.J., McGill, W.B., Pickard, M.A., and Salloum, M.J. (1998). Slurry-phase bioremediation of creosote and petroleum-contaminated soils. Environ. Technol. *19*, 683–696.

Saponaro, S., Bonomo, L., Petruzzelli, G., Romele, L., and Barbafieri, M. (2002). Polycyclic aromatic hydrocarbons (PAHs) slurry phase bioremediation of a manufacturing gas plant (MGP) site aged soil. Water Air Soil Poll. *135*, 219–236.

Schoefs, O., Deschenes, L., and Samson, R. (1998). Efficiency of a new covering system for the environmental control of biopiles used for the treatment of contaminated soils. J. Soil Contam. *7*, 753–771.

Shailaja, S., Venkata Mohan, S., Rama Krishna, M., and Sarma, P.N. (2008). Degradation of di-ethylhexyl phthalate (DEHP) in bioslurry phase reactor and identification of metabolites by HPLC and MS. Int. Biodeter. Biodegr. *62*, 143–152.

Simon, M.A., Bonner, J.S., Page, C.A., Townsend, R.T., Mueller, D.C., Fuller, C.B., and Autenrieth, R.L. (2004). Evaluation of two commercial bioaugmentation products for enhanced removal of petroleum from a wetland. Ecol. Eng. *22*, 263–277.

Straube, W.L., Nestler, C.C., Hansen, L.D., Ringleberg, D., Pritchard, P.H., and Jones-Meehan, J. (2003). Remediation of polyaromatic hydrocarbons (PAHs) through landfarming with biostimulation and bioaugmentation. Acta Biotechnol. *23*, 179–196.

Thomassin-Lacroix, E.J.M., Eriksson, M., Reimer, K.J., and Mohn, W.W. (2002). Biostimulation and bioaugmentation for on-site treatment of weathered diesel fuel in Arctic soil. Appl. Microbiol. Biotechnol. *59*, 551–556.

Van Deuren, J., Lloyd, T., Chhetry, S., Liou, R., and Peck, J. (2002). Remediation Technologies Screening Matrix and Reference Guide, 4th ed. (U.S. Army Environmental Center, Aberdeen Proving Ground, MD).

Van Gestel, K., Mergaert, J., Swings, J., Coosemans, J., and Ryckeboer, J. (2003). Bioremediation of diesel oil-contaminated soil by composting with biowaste. Environ. Pollut. *125*, 361–368.

Venkata Mohan, S., Prasanna, D., Purushotham Reddy, B., and Sarma, P.N. (2008). Ex situ bioremediation of pyrene contaminated soil in bio-slurry phase reactor operated in periodic discontinuous batch mode:

Influence of bioaugmentation. Int. Biodeter. Biodegr. *62*, 162–169.

Vinas, M, Grifoll, M., Sabate, J., and Solanas, A.M. (2002). Biodegradation of a crude oil by three microbial consortia of different origins and metabolic capabilities. J. Ind. Microbiol. Biot. *28*, 252–260.

Walter, M., Boul, L., Chong, R., and Ford, C. (2004). Growth substrate selection and biodegradation of PCP by New Zealand white-rot fungi. J. Environ. Manage. *71*, 361–369.

Wan, C.K., Wong, J.W.C., Fang, M., and Ye, D.Y. (2003). Effect of organic waste amendments on degradation of PAHS in soil using thermophillic composting. Environ. Technol. *24*, 23–30.

Wang, J.L., Mao, Z.Y., Han, L.P., and Qian, Y. (2004). Bioremediation of quinoline-contaminated soil using bioaugmentation in slurry-phase reactor. Biomed. Environ. Sci. *17*, 187–195.

Yang, S., Shibata, A., Yoshida, N., and Katayama, A. (2009). Anaerobic mineralization of pentachlorophenol (PCP) by combining PCP-dechlorinating and phenol-degrading cultures. Biotechnol. Bioeng. *102*, 81–90.

Zheng, Z., and Obbard, J.P. (2002). Polycyclic aromatic hydrocarbon removal from soil by surfactant solubilization and *Phanerochate chrysosporium* oxidation. J. Environ. Qual. *31*, 1842–1847.

Index

A

Other Books of Interest

www.ingramcontent.com/pod-product-compliance
Lightning Source LLC
LaVergne TN
LVHW062119200926
848484LV00007B/1720